Next Generation Mobile Broadcasting

OTHER TELECOMMUNICATIONS BOOKS FROM AUERBACH

Ad Hoc Mobile Wireless Networks: Principles, Protocols, and Applications
Subir Kumar Sarkar, T.G. Basavaraju, and C. Puttamadappa
ISBN 978-1-4665-1446-1

Communication and Networking in Smart Grids
Yang Xiao (Editor)
ISBN 978-1-4398-7873-6

Delay Tolerant Networks: Protocols and Applications
Athanasios V. Vasilakos, Yan Zhang, and Thrasyvoulos Spyropoulos
ISBN 978-1-4398-1108-5

Emerging Wireless Networks: Concepts, Techniques and Applications
Christian Makaya and Samuel Pierre (Editors)
ISBN 978-1-4398-2135-0

Game Theory in Communication Networks: Cooperative Resolution of Interactive Networking Scenarios
Josephina Antoniou and Andreas Pitsillides
ISBN 978-1-4398-4808-1

Green Communications: Theoretical Fundamentals, Algorithms and Applications
Jinsong Wu, Sundeep Rangan, and Honggang Zhang
ISBN 978-1-4665-0107-2

Green Communications and Networking
F. Richard Yu, Xi Zhang, and Victor C.M. Leung (Editors)
ISBN 978-1-4398-9913-7

Green Mobile Devices and Networks: Energy Optimization and Scavenging Techniques
Hrishikesh Venkataraman and Gabriel-Miro Muntean (Editors)
ISBN 978-1-4398-5989-6

Handbook on Mobile Ad Hoc and Pervasive Communications
Laurence T. Yang, Xingang Liu, and Mieso K. Denko (Editors)
ISBN 978-1-4398-4616-2

Intelligent Sensor Networks: The Integration of Sensor Networks, Signal Processing and Machine Learning
Fei Hu and Qi Hao (Editors)
ISBN 978-1-4398-9281-7

IP Telephony Interconnection Reference: Challenges, Models, and Engineering
Mohamed Boucadair, Isabel Borges, Pedro Miguel Neves, and Olafur Pall Einarsson
ISBN 978-1-4398-5178-4

LTE-Advanced Air Interface Technology
Xincheng Zhang and Xiaojin Zhou
ISBN 978-1-4665-0152-2

Media Networks: Architectures, Applications, and Standards
Hassnaa Moustafa and Sherali Zeadally (Editors)
ISBN 978-1-4398-7728-9

Multihomed Communication with SCTP (Stream Control Transmission Protocol)
Victor C.M. Leung, Eduardo Parente Ribeiro, Alan Wagner, and Janardhan Iyengar
ISBN 978-1-4665-6698-9

Multimedia Communications and Networking
Mario Marques da Silva
ISBN 978-1-4398-7484-4

Near Field Communications Handbook
Syed A. Ahson and Mohammad Ilyas (Editors)
ISBN 978-1-4200-8814-4

Next-Generation Batteries and Fuel Cells for Commercial, Military, and Space Applications
A. R. Jha, ISBN 978-1-4398-5066-4

Physical Principles of Wireless Communications, Second Edition
Victor L. Granatstein, ISBN 978-1-4398-7897-2

Security of Mobile Communications
Noureddine Boudriga, ISBN 978-0-8493-7941-3

Smart Grid Security: An End-to-End View of Security in the New Electrical Grid
Gilbert N. Sorebo and Michael C. Echols
ISBN 978-1-4398-5587-4

Transmission Techniques for 4G Systems
Mário Marques da Silva
ISBN 978-1-4665-1233-7

Transmission Techniques for Emergent Multicast and Broadcast Systems
Mário Marques da Silva, Americo Correia, Rui Dinis, Nuno Souto, and Joao Carlos Silva
ISBN 978-1-4398-1593-9

TV White Space Spectrum Technologies: Regulations, Standards, and Applications
Rashid Abdelhaleem Saeed and Stephen J. Shellhammer
ISBN 978-1-4398-4879-1

Wireless Sensor Networks: Current Status and Future Trends
Shafiullah Khan, Al-Sakib Khan Pathan, and Nabil Ali Alrajeh
ISBN 978-1-4665-0606-0

Wireless Sensor Networks: Principles and Practice
Fei Hu and Xiaojun Cao
ISBN 978-1-4200-9215-8

AUERBACH PUBLICATIONS
www.auerbach-publications.com
To Order Call: 1-800-272-7737 • Fax: 1-800-374-3401
E-mail: orders@crcpress.com

Next Generation Mobile Broadcasting

Edited by David Gómez-Barquero

CRC Press
Taylor & Francis Group
Boca Raton   London   New York

CRC Press is an imprint of the
Taylor & Francis Group, an **informa** business

CRC Press
Taylor & Francis Group
6000 Broken Sound Parkway NW, Suite 300
Boca Raton, FL 33487-2742

First issued in paperback 2016

Version Date: 20121214

ISBN 13: 978-1-138-19978-1 (pbk)
ISBN 13: 978-1-4398-9866-6 (hbk)

Library of Congress Cataloging-in-Publication Data

Next generation mobile broadcasting / edited by David Gómez-Barquero.
   pages cm
   Includes bibliographical references and index.
   ISBN 978-1-4398-9866-6 (hardback)
   1. Mobile television. 2. Mobile computing. 3. Multimedia communications. I. Gómez-Barquero, David.

   TK6678.3.N49 2013
   621.388'5--dc23   2012044520

Visit the Taylor & Francis Web site at
http://www.taylorandfrancis.com

and the CRC Press Web site at
http://www.crcpress.com

To my family

Contents

SECTION I   MOBILE BROADCASTING WORLDWIDE

SECTION III   NEXT GENERATION HANDHELD DVB TECHNOLOGY: MULTIPLE-INPUT MULTIPLE-OUTPUT (MIMO) PROFILE

SECTION IV   NEXT GENERATION HANDHELD DVB TECHNOLOGY: HYBRID TERRESTRIAL– SATELLITE PROFILE

SECTION V NEXT GENERATION HANDHELD DVB TECHNOLOGY: HYBRID TERRESTRIAL–SATELLITE MIMO PROFILE

List of Institutions

1. Aalto University, Espoo, Finland
2. Alcatel Lucent Bell Labs, Paris, France
3. British Broadcasting Corporation - Research & Development (BBC R&D), London, United Kingdom
4. Centre for Wireless Communications, University of Oulu, Oulu, Finland
5. Communications Research Center (CRC), Ottawa, Canada
6. Effnet, Luleå, Sweden
7. Electronics and Telecommunications Research Institute (ETRI), Daejeon, South Korea
8. Ericsson Eurolab, Herzogenrath, Germany
9. European Space Agency, European Space Research and Technology Centre (ESTEC), Noordwijk, the Netherlands
10. Fraunhofer Heinrich Hertz Institute (HHI), Berlin, Germany
11. Fraunhofer Institut für Integrierte Schaltungen (IIS), Erlangen, Germany
12. Institut National des Sciences Appliquées, Institut d'Electronique et de Télécommunications de Rennes, Rennes, France
13. Japan Broadcasting Corporation (NHK), Tokyo, Japan
14. LG Electronics, Seoul, South Korea
15. Mitsubishi Electric Corporation, Information Technology R&D Center, Kanagawa, Japan
16. Mitsubishi Electric R&D Centre Europe, Rennes, France
17. mmbi, Inc., Tokyo, Japan
18. Nokia, Espoo, Finland
19. NTT DOCOMO, Tokyo, Japan
20. Panasonic R&D, Langen, Germany
21. Orange Labs Networks, Rennes, France
22. Rai Research and Technical Innovation Centre, Torino, Italy
23. Samsung, Seoul, South Korea
24. Samsung Electronics Research Institute, Surrey, United Kingdom
25. Shanghai Jiao Tong University, Shanghai, China
26. Sinclair Broadcast Group, Hunt Valley, Maryland

27. Sony Semiconductor Design Centre, Basingstoke, United Kingdom
28. Tampere University of Technology, Tampere, Finland
29. TeamCast, Rennes, France
30. Technische Universität Braunschweig, Braunschweig, Germany
31. Telecom Bretagne, Brest, France
32. Teracom, Stockholm, Sweden
33. University of Luxembourg, Luxembourg City, Luxembourg
34. University of the Basque Country UPV/EHU, Bilbao, Spain
35. University of Turku, Turku, Finland
36. Universitat Politècnica de València, Valencia, Spain

List of Figures

Editor

David Gómez-Barquero earned his MSc in telecommunications engineering from the Universitat Politècnica de València (UPV), Spain, and the University of Gävle, Sweden, in 2004, and his PhD in telecommunications from UPV in 2009. During his doctoral studies, he was a guest researcher at the Royal Institute of Technology, Sweden, the University of Turku, Finland, and the Technical University of Braunschweig, Germany. He also did an internship at Ericsson Eurolab, Aachen, Germany. During 2010 and 2011, he was a postdoctoral guest researcher at the Fraunhofer Heinrich Hertz Institute (HHI) in Berlin, Germany. During the second half of 2012, he was a visiting professor at the Sergio Arboleda University of Bogota, Colombia.

Dr. Gómez-Barquero currently serves as a senior researcher at the Institute of Telecommunications and Multimedia Applications (iTEAM) of UPV, where he leads a research group working on multimedia broadcasting, in particular on the optimization of 3GPP MBMS (multimedia broadcast multicast services) and especially DVB (digital video broadcasting) systems. Since 2008 he has been actively participating in the digital television standardization forum, DVB. He participated in the validation of the second generation digital terrestrial TV technology, DVB-T2, and in the standardization processes of its mobile profile, known as T2-Lite, and its handheld evolution, known as DVB-NGH. He also contributed to the DVB-T2 implementation guidelines and coedited the DVB bluebook on upper layer forward error correction as an invited expert. He was also involved in the promotion and adoption of DVB-T2 in Colombia.

Contributors

Mark A. Aitken
Sinclair Broadcast Group
Hunt Valley, Maryland, USA

Pablo Angueira
University of the Basque Country
UPV/EHU
Bilbao, Spain

Amaia Añorga
Fraunhofer Institut für Integrierte
Schaltungen (IIS)
Erlangen, Germany

Pantelis-Daniel Arapoglou
European Space Agency (ESA)
European Space Research and
Technology Centre (ESTEC)
Noordwijk, the Netherlands

Sam Atungsiri
Sony Semiconductor Design Centre
Basingstoke, United Kingdom

Byungjun Bae
Electronics and Telecommunication
and Research Institute (ETRI)
Daejeon, South Korea

Staffan Bergsmark
Teracom
Stockholm, Sweden

Gorka Berjón-Eriz
University of the Basque Country
UPV/EHU
Bilbao, Spain

Arnaud Bouttier
Mitsubishi Electric Research and
Development Centre Europe
Rennes, France

Marco Breiling
Fraunhofer Institut für Integrierte
Schaltungen (IIS)
Erlangen, Germany

Benjamin Bross
Fraunhofer Heinrich Hertz Institute
(HHI)
Berlin, Germany

Damien Castelain
Mitsubishi Electric Research and
Development Centre Europe
Rennes, France

Cristina Ciochină
Mitsubishi Electric Research and
Development Centre Europe
Rennes, France

Matthieu Crussière
Institut National des Sciences
   Appliquées
Institut d'Electronique et de
   Télécommunications de Rennes
Rennes, France

Catherine Douillard
Telecom Bretagne
Brest, France

Iñaki Eizmendi
University of the Basque Country
   UPV/EHU
Bilbao, Spain

Gérard Faria
TeamCast
Saint-Grégoire, France

Christian Gallard
Orange Labs Networks
Rennes, France

Jordi Joan Giménez
Universitat Politècnica de València
Valencia, Spain

Pedro F. Gómez
Universitat Politècnica de València
Valencia, Spain

David Gómez-Barquero
Universitat Politècnica de València
Valencia, Spain

David Gozálvez
Universitat Politècnica de València
Valencia, Spain

Yunfeng Guan
Shanghai Jiao Tong University
Shanghai, China

Fumihiro Hasegawa
Mitsubishi Electric Corporation
Kanagawa, Japan

Dazhi He
Shanghai Jiao Tong University
Shanghai, China

Camilla Hollanti
Aalto University
Helsinki, Finland

Jörg Huschke
Ericsson Eurolab
Herzogenrath, Germany

Hongsil Jeong
Samsung
Seoul, South Korea

Tero Jokela
University of Turku
Turku, Finland

Sylvaine Kerboeuf
Alcatel Lucent Bell Labs
Paris, France

Kwang Yong Kim
Electronics and Telecommunication
   and Research Institute (ETRI)
Daejeon, South Korea

Young Su Kim
Electronics and Telecommunication
   and Research Institute (ETRI)
Daejeon, South Korea

Ryo Kitahara
NTT DOCOMO
Tokyo, Japan

Carl Knutsson
Effnet
Luleå, Sweden

Woo-Suk Ko
LG Electronics
Seoul, South Korea

Lukasz Kondrad
Tampere University of Technology
Tampere, Finland

Hun Hee Lee
Electronics and Telecommunication
  and Research Institute (ETRI)
Daejeon, South Korea

José M. Llorca
Universitat Politècnica de València
Valencia, Spain

Jaime López-Sánchez
Universitat Politècnica de València
Valencia, Spain

Vittoria Mignone
Rai Research and Technical Innovation
  Centre
Torino, Italy

Sangchul Moon
LG Electronics
Seoul, South Korea

Toshiharu Morizumi
mmbi, Inc.
Tokyo, Japan

Peter Moss
British Broadcasting Corporation -
  Research & Development (BBC
  R&D)
London, United Kingdom

Alain Mourad
Samsung Electronics Research Institute
Staines, United Kingdom

Charbel Abdel Nour
Telecom Bretagne
Brest, France

Tomoyuki Ohya
mmbi, Inc.
Tokyo, Japan

Mihail Petrov
Panasonic Research and Development
Langen, Germany

Mai-Anh Phan
Ericsson Eurolab
Herzogenrath, Germany

Tuck Yeen Poon
British Broadcasting Corporation -
  Research & Development (BBC
  R&D)
London, United Kingdom

Jörg Robert
Technische Universität Braunschweig
Braunschweig, Germany

Hitoshi Sanei
Japan Broadcasting Corporation
  (NHK)
Tokyo, Japan

M.R. Bhavani Shankar
University of Luxembourg
Luxembourg

Jean-Luc Sicre
Orange Labs Networks
Rennes, France

Mike Simon
Sinclair Broadcast Group
Hunt Valley, Maryland, USA

Yun Jeong Song
Electronics and Telecommunication
  and Research Institute (ETRI)
Daejeon, South Korea

Erik Stare
Teracom
Stockholm, Sweden

Masayuki Takada
Japan Broadcasting Corporation
  (NHK)
Tokyo, Japan

Visa Tapio
Centre for Wireless Communications
University of Oulu
Oulu, Finland

Shinichiro Tonooka
mmbi, Inc.
Tokyo, Japan

Jani Väre
Nokia
Helsinki, Finland

David Vargas
Universitat Politècnica de València
Valencia, Spain

Manuel Vélez
University of the Basque Country
  UPV/EHU
Bilbao, Spain

Yiyan Wu
Communications Research Center
  (CRC)
Ottawa, Canada

Akira Yamada
NTT DOCOMO
Tokyo, Japan

Kyu Tae Yang
Electronics and Telecommunication
  and Research Institute (ETRI)
Daejeon, South Korea

Joungil Yun
Electronics and Telecommunication
  and Research Institute (ETRI)
Daejeon, South Korea

Wenjun Zhang
Shanghai Jiao Tong University
Shanghai, China

Jan Zöllner
Technische Universität Braunschweig
Braunschweig, Germany

MOBILE BROADCASTING WORLDWIDE

Chapter 1

Next-Generation Mobile Multimedia Broadcasting

David Gómez-Barquero, Pablo Angueira,
and Yiyan Wu

Contents

1.1 Introduction

Mobile wireless communications are in constant evolution due to the continuously increasing requirements and expectations of both users and operators. Mass multimedia* services have been for a long time expected to generate a large amount of data traffic in future wireless networks [1]. Mass multimedia services are, by definition, purposed for many people. In general, it can be distinguished between the distribution of any popular content over a wide area and the distribution of location-dependent information in highly populated areas. Representative examples include the delivery of live video streaming content (like sports competitions, concerts, or news) and file download (multimedia clips, digital newspapers, or software updates). The most representative service is mobile TV, which is commonly believed to bring great opportunities for society, just as the fixed TV and mobile telephony already have. Commercial trials have revealed a strong consumer interest [2].

Traditionally, cellular systems have focused on the transmission of data intended for a single user using a dedicated point-to-point (p-t-p) radio bearer, not addressing the distribution of popular content to a large number of users simultaneously. Hence, cellular systems have been designed for unicast services delivered through p-t-p dedicated connections for each individual user, even if the same content should be delivered to many users. Unicast systems can easily support a wide range of services, as each user may consume a different service. As each connection is intended for a single user, it is possible to optimize the transmission parameters for each user individually, and resources are also employed only when a user is actively consuming a service. The main drawback of unicast is its unfavorable scaling with

---

* The global digitalization process of the communication and information technologies makes possible to combine multiple forms of information content (e.g., text, audio, and video) into a so-called multimedia service. Moreover, due to the intrinsic characteristics of digital data, the information can be easily produced, reproduced without any quality degradation, stored, and manipulated.

the number of users when delivering the same content to many users at the same time. This fact limits the maximum number of users cellular systems can handle, since both radio and network resources are physically limited.

Multicast and broadcast* are more appropriate transport technologies to cope with high numbers of users consuming the same content simultaneously. Multicast/ broadcast wireless transmissions employ a common point-to-multipoint (p-t-m) radio bearer for all users, which allows delivering the same content to an unlimited number of users within the covered area. The main drawback is that user-specific adaptation of the transmission parameters is generally not possible, and the transmission has to be designed to serve the worst-case user.

Within this context, several mobile broadcast technologies were developed to support large-scale consumption of mass multimedia services such as mobile TV [3]. However, the adoption of mobile TV services did not fulfill the initial expectations due to the lack of a successful business model and the high costs associated with the deployment of new mobile broadcasting networks. Hence, in many countries, mobile TV services either did not reach the market or were launched and subsequently closed.

The third-generation (3G) cellular standard was enhanced with Multimedia Broadcast Multicast Services (MBMS)† [4], introducing new p-t-m radio bearers and multicast support in the core network. MBMS can lead to a better utilization of the existing radio resources provided by 3G, enabling the provision of new multimedia multicast services. However, its reduced capacity being shared with voice, messaging, and data services limits significantly its potential to provide mass multimedia services. 3G MBMS has not been commercially deployed, and since its introduction in 3G, Release 6 is under continuous enhancement.

A higher potential than MBMS was shown by new digital terrestrial broadcast technologies specifically designed for mobile services. Existing broadcasting infrastructure characterized by very tall towers and large transmission power, and the significant chunk of allocated spectrum in the ultrahigh frequency (UHF) band represent extremely valuable resources. Moreover, digital terrestrial TV (DTT) networks make use of single-frequency networks (SFNs), where all transmitters operate at the very same frequency and receivers combine signals coming from several transmitters. SFNs achieve a huge spectrum efficiency compared to multi-frequency networks (MFNs), where the same-frequency channel cannot be reused over a long distance because of the co-channel interference.

---

* Multicasting and broadcasting describe different, although closely related, scenarios. Whereas broadcast transmissions are intended for all users in the service area, multicast transmissions are addressed to a specific group of users (usually called the multicast group). Thus, broadcast can be considered as a particular case of multicast.

† MBMS is split into the MBMS bearer service and the MBMS user service, in such a way that it is possible to integrate p-t-p and p-t-m radio bearers in a transparent way to the MBMS service layer. Thus, it is possible to deliver an MBMS service with p-t-p transmissions. For the sake of clarity, we associate MBMS only to p-t-m transmissions and HSDPA to p-t-p transmissions.

In addition to physical and link layer specifications, upper-layer components were also developed for complementing the mobile broadcasting networks with a bidirectional interactivity path offered by the cellular systems. This way, the strengths of mobile broadcasting networks to deliver efficiently very popular content with the personalization, interaction, and billing features of cellular networks could be combined. It should be pointed out that there was no convergence in networks, since both mobile broadcasting and cellular networks were used separately and independently [5]. Convergence was implemented at services, platform, and multimode terminal levels that support both radio access technologies [6].

The most relevant mobile broadcasting technology in Europe at that time was Digital Video Broadcast—Handheld (DVB-H) [7]. After a slow start, and with a clear endorsement of the European Commission, several DVB-H networks were deployed across Europe. Italy was the first country to launch commercial services in 2006. Commercial DVB-H services were also on air in the main urban areas of Finland, Switzerland, Austria, the Netherlands, Hungary, and Albania, but they have been progressively switched off.

A similar process occurred with Media Forward Link Only (FLO) [8], a mobile TV technology developed in the United States, which experienced a similar standard development and rollout failure.

The assumption that investments in a dedicated mobile broadcasting network, such as DVB-H or Media FLO, would be justified by market demand for mobile TV services proved to be wrong. Consumers were not prepared to pay monthly fees for mobile TV services, and the demand was not sufficiently high to offset the costs associated with building and operating the networks. Receiving devices also failed to materialize in mass-market numbers, and the receivers that appeared on the market were limited in choice and did not appeal to consumers. Furthermore, there was no regulatory framework that would have obliged all players in the service chain (broadcasters, broadcast network operators, cellular operators) to work together.

The only relative success stories among the first generation of mobile broadcasting systems are Integrated Services Digital Broadcasting—Terrestrial (ISDB-T) in Japan [9] and Terrestrial—Digital Multimedia Broadcasting (T-DMB) in South Korea [10]. But in these two cases, mobile broadcasting services are a demand from the national regulators. It should be pointed out that the networks are not yet commercially profitable. Only free-to-air mobile television services are available, and the incomes achieved with advertising do not compensate the costs for running and maintaining the networks. An important advantage of ISDB-T is that mobile services are provided in-band, using a narrow portion of the channel bandwidth, while the rest of the bandwidth is dedicated to the transmission of stationary services. This way, there is no need to deploy a dedicated mobile broadcasting network.

Today, the wide availability of smart phones and tablets has renewed the interest in mobile TV. Consumers have now appropriately sized screens to watch video on the move, and they are experiencing mobile multimedia applications over 3G cellular networks with better experience than ever. But 3G networks were not designed to efficiently provide mass multimedia services. As demand grows, the capacity limits of 3G networks will be reached. According to several forecasts [11], it is expected that mobile data traffic demand will explode in the next years, representing mobile video more than 50% of the total traffic. Although the next generation of cellular networks, also known as 3G long-term evolution (LTE) [12], is starting to be deployed around the world providing improved coverage and system capacity, as well as increased data rates and reduced latency, specific mobile broadcast technologies may be required in the near future.

Within this context, a new generation of mobile broadcasting technologies has emerged, incorporating the latest advances in wireless communications that provide significant performance improvements compared to the first-generation mobile broadcast systems. Nevertheless, questions remain regarding the commercial prospects for any mobile broadcasting standard given the limited success of similar standards in the past.

This chapter provides an introduction to the book about the current state of the art in mobile multimedia broadcasting. The book is structured in two parts. The first part provides an overview of the new-generation mobile broadcasting technologies available worldwide. The second part of the book is devoted to the latest technology known as Digital Video Broadcasting—Next-Generation Handheld (DVB-NGH), which is expected to significantly outperform all existing technologies in both capacity and coverage. The rest of the chapter is structured as follows. Section 1.2 describes the challenges of mobile multimedia broadcasting. Section 1.3 provides a brief review of all first-generation mobile broadcasting technologies, highlighting the main characteristics of each technology and the lessons learned. Section 1.4 provides a brief overview of the new-generation mobile broadcasting technologies that are covered in more detail in the first part of the book. Section 1.5 describes the current trends toward a worldwide mobile multimedia broadcasting standard. Finally, the chapter is concluded in Section 1.6.

1.2 Challenges of Mobile Multimedia Broadcasting

1.2.1 Cost of Providing Mobile Multimedia Broadcasting Services

The delivery of mass multimedia content to mobile and portable devices is a very challenging task, as both massive scalability and reliable transfer are required. Moreover, nowadays, it is acknowledged that a thriving mass mobile multimedia market will not develop unless services are provided at low cost, making them affordable to users and profitable to operators.

The most important bottleneck is the cost of the wireless infrastructure, which is proportional to the data rate provided to the user terminal in a system with wide coverage [13]. Wireless communication systems are confined to the rate/range trade-off provided by the "Shannon bound" of communication theory.* This is due to the fact that a minimum amount of energy is required to reliably transfer every bit of information. The higher the data rate, the less energy is available for every bit at the transmitter, and thus the feasible communication distance is smaller.[†]

As a consequence, providing broadband access over a wide area inevitably requires a large number of access points (base stations), which implies larger investments. It has been shown that the number of access points required grows about linearly with the data rate provided [14], meaning that the cost per transmitted bit is virtually constant. Hence, higher user data rates directly translate into higher cost.

Traditional network planning for digital broadcast systems is based on a worst-case analysis, which results in over-dimensioning the network infrastructure. Such worst-case reception conditions must be supported in order to ensure signal reception in very adverse conditions and maximize the potential number of subscribers that can receive the content. Furthermore, mobile channels are significantly more challenging than traditional fixed broadcasting scenarios due to the user mobility, the utilization of more compact antennas, and the reception at ground level.[‡] In order to provide good coverage levels for mobile broadcasting services, it is necessary to deploy dense networks with a large number of sites. This penalty is particularly evident for high coverage targets (e.g., over 90% of service area locations) [15]. The reason is that it is especially costly to guarantee coverage to the last few percent of the worst-served locations.

The adoption of turbo codes and low-density parity check (LDPC) codes in second-generation systems such as DVB-T2 (Terrestrial Second Generation) already achieves a capacity within 1 dB of the Shannon limit in AWGN [16], and thus, major gains cannot be expected by means of improved forward error correction (FEC). Therefore, the use of diversity in the time, frequency, and space domains is necessary in the next generation of mobile broadcasting systems to achieve a significant performance increase compared to the current generation. One of the largest improvements in the next generation of mobile broadcasting systems is expected to come from the incorporation of multiple-input multiple-output (MIMO) techniques [17]. The implementation of multiple antennas in both ends of

---

* According to the Shannon bound, data rates R increase linearly with spectrum bandwidth W, but only logarithmically with the signal-to-interference-plus-noise ratio (SINR). For an additive white Gaussian channel, $R = W \cdot \log_2(1 + SINR)$.

[†] For a given transmit power and bandwidth, carrier frequency, and radio access technology. The carrier frequency also strongly affects the area coverage that can be served with an access point for a given bandwidth and data rate, as the propagation path loss increases as a function of the carrier frequency.

[‡] The link budget difference between fixed rooftop reception and pedestrian outdoor reception due to antenna gain and height loss is over 30 dB in dense urban scenarios [14]. Pedestrian indoor and vehicular (in-car) reception present additional losses due to building and vehicle penetration.

the communication channel is twofold; on the one hand, it improves the reception robustness by means of better diversity, and on the other hand, it provides a higher capacity by exploiting the multiplexing capabilities of the MIMO channel.

1.2.2 Mobile Broadcasting Business Case

The expenses associated with building, maintaining, and servicing a mobile broadcasting network can be high, but that does not mean that a viable business model is not possible [18]. Broadcast is economically advantageous because the infrastructure cost is amortized over the user base. As the user base grows, the cost per subscriber falls. Note that this is unlike unicast cellular architectures, where the cost per supported user is fixed. Hence, everything comes down to consumer demand and population density.*

Reaching the break-even point to recover the costs for infrastructure depends on the user population and the average revenue per user. For a given technology, the business will be thus more likely to succeed in populated areas and where there is a high consumer demand. This fact may prevent achieving nationwide coverage, limiting the deployment to urban areas. The main business aspects when introducing a new mobile broadcast technology are the necessary capital expenditures (CAPEX) and operating expenditures (OPEX) together with the consumer demand. Basically, the high-level questions are the following:

■ How much investment is needed to deploy a mobile broadcast network, and what is the annual operational cost?
■ To what extent are consumers willing to pay for mobile broadcasting services?

A complete answer to these questions requires considering many aspects related to technology, business, and regulations, which is out of the scope of this book. But it is clear that nowadays paying for mobile TV content is not a mass impulse, despite the fact that early DVB-H and Media FLO trials were encouraging to operators and willingness to pay was high [2]. Today, it is commonly accepted that mass-market demand for mobile multimedia entertainment is conditioned to the provision of these services at low cost.

1.2.3 Business Models and Regulatory Barriers

The successful deployment of new mobile broadcasting systems is also conditioned by several business- and regulatory-related aspects. The main problem is that broadcast and cellular networks belong to different industries, each of them developed in a different business environment under very different value chains

---

* From this simple discussion, it should be clear that if more than one network is built, the profit potential would shrink proportionally.

and revenue models. Mobile broadcasting might give rise to several business models with various possible roles [19]. The balance between TV broadcasters, cellular network operators, and broadcast network operators will depend on the local regulatory regime. A good description of the country-specific implementations can be found in Reference 20. Assuming that no player will be able to control the complete value chain in most countries, broadcasters and cellular network operators may need to work together, but need a suitable business model in an extremely complex business environment, each playing to their strengths.

A collaborative model can be achieved with the presence of a wholesaler of the mobile broadcasting service acting as datacast operator. This model is considered to provide the best chances to establish a successful business model for mobile broadcasting and leaves potential retailers negotiate a sustainable business case enabled by cooperation [21]. The key objective is to have one neutral service provider (wholesaler), controlling a single mobile broadcasting network and sharing the scarce spectrum resource with several cellular network operators and content providers. The wholesaler can be a consortium of cellular operators, a broadcast network operator (due to its neutral position in the market and technical competence), or any third party. The service provider is responsible for the technical service provision and acts as a facilitator for the cellular operators in the aggregation of content and the usage of broadcast transmission capacity. This model enables fully interactive services and a common bill for users, since the cellular operators bill their customers for the mobile broadcasting service. Moreover, it also overcomes the inefficient usage of spectrum resources because cellular operators share a common service package, being also possible to differentiate their service offerings to some extent.

1.3 A Look Back to the First-Generation Mobile Broadcasting Systems

1.3.1 Digital Video Broadcasting: Terrestrial

Digital Video Broadcasting—Terrestrial (DVB-T) is the European standard for DTT [22], the most widely deployed DTT system in the world. It was primarily designed for fixed rooftop reception, and it presents an important degradation in mobile channels [23]. The main limiting factor is not the inter-carrier interference caused by the Doppler spread, but rather the lack of time interleaving at the physical layer. DVB-T is also not practical for handheld battery-powered devices because receivers have to decode the whole multiplex. Nevertheless, in DVB-T networks dimensioned for portable indoor reception (e.g., Germany), it is possible to provide commercial services to vehicles. In this case, it is possible to implement antenna diversity at the receivers to compensate the performance degradation under mobility conditions, reducing the required signal level and increasing the maximum supported speed [23].

1.3.2 Digital Video Broadcasting: Handheld

DVB-H was introduced almost exclusively as a link layer on top of DVB-T [7,24]. This way, it is possible to share the same network infrastructure (e.g., transmitters, antennas, and multiplexers). The main features introduced were a discontinuous transmission technique known as time-slicing, which reduces the power consumption of terminals, and an optional FEC mechanism at the link layer called Multi Protocol Encapsulation FEC (MPE-FEC), which ensures more robust transmissions under mobility and impulsive interference conditions. MPE is the adaptation protocol used to transport Internet Protocol (IP) streams over MPEG-2 DVB-T transport streams (TSs).

With time-slicing, data are periodically transmitted in bursts (maximum size is 2 Mb), in such a way that each burst contains information of the time difference to the next burst of the same service. Terminals synchronize to the bursts of the desired service and switch their receivers front-end off when bursts of other services are being transmitted. With MPE-FEC, each burst is encoded with a Reed–Solomon (RS) code at the link layer to generate parity data that are transmitted with the source IP data to compensate for potential transmission errors. MPE-FEC provides interleaving durations in the order of 100–200 ms.

DVB-H is a transmission standard with an IP interface that specifies the physical and link layers. Two sets of specifications were developed to contribute with the higher-layer protocols to build a complete end-to-end system, known as IP datacast (IPDC) and Open Mobile Alliance—Mobile Broadcast Services Enabler Suite (OMA-BCAST), specifying, among others, content delivery protocols for streaming and file download services, the electronic service guide for service discovery, and mechanisms for service purchase and protection. IPDC describes all components required for complementing a DVB-H network with a bidirectional interactivity path offered by the cellular systems. OMA-BCAST is similar to IPDC but with a different emphasis. In particular, it integrates a DVB-H network into a cellular operator infrastructure.

One of the major concerns about the rollout of DVB-H is the network infrastructure cost. Due to the fact that DVB-H terminals suffer from much more severe propagation conditions than DVB-T, a considerably large number of new sites are required for the installation of additional DVB-H transmitters or repeaters (gap-fillers), complementing the existing broadcasting towers and forming very dense SFNs [17], which implies very large investments in infrastructure.

1.3.3 Digital Video Broadcasting: Satellite to Handheld

Digital Video Broadcasting–Satellite to Handheld (DVB-SH) was designed to enable the provision of mobile broadcasting services through hybrid satellite–terrestrial networks in the S-band [25]. This hybrid architecture is probably the most suited for mass multimedia broadcasting, because it allows nationwide coverage in a fast and efficient way thanks to the satellite transmission (e.g., one geostationary satellite is capable of providing coverage to all Western Europe), whereas the terrestrial component ensures coverage in urban areas where direct reception of the satellite signal is not possible.

DVB-SH employs Orthogonal Frequency Division Multiplexing (OFDM) for the complementary terrestrial network, and OFDM or Time Division Multiplexing (TDM) for the satellite component. OFDM allows the deployment of hybrid SFNs, where both network components employ identical transmission settings. The TDM waveform for the satellite component is meant for hybrid MFN networks. Single-carrier modulations such as TDM are better suited for satellite transmissions compared to multicarrier modulations such as OFDM because they are characterized by a high peak power, forcing the onboard high-power amplifiers to work far from the saturation point. In both cases, seamless hand-over is possible.

DVB-SH also incorporates longtime interleaving with fast zapping support. The Land Mobile Satellite channel is characterized by long signal outages due to the blockage of the line of sight with the satellite caused by tunnels, buildings, trees, etc. [26], which can be compensated only with a longtime interleaving duration (e.g., in the order of 10 s). But longtime interleaving with fast zapping is also of interest for terrestrial transmissions, because it allows exploiting the time diversity of the mobile channel in both, high-speed (e.g., vehicles and trains), and low-speed (e.g., pedestrian) reception scenarios [27]. DVB-SH specifies two complementary FEC schemes capable of providing longtime interleaving with fast zapping support: one at the physical layer using a turbo code with a convolutional interleaver (CI) the and other at the link layer known as Multi Protocol Encapsulation inter-burst FEC. The CI can provide fast zapping with longtime interleaving using a uniform-late profile or a uniform-early profile [28]. The uniform-late profile has the advantage that full protection is progressively and smoothly achieved with time. The robustness after zapping is given by the proportion of parity data transmitted in the late part. The larger the size of the late part, the better the performance after zapping. However, the overall performance in mobile channels is reduced because of the use of a nonuniform time interleaving. The size of the late part represents a trade-off between overall performance in mobile channels and performance after zapping.

The turbo-code physical layer FEC scheme is the one employed in 3G cellular systems, and it performs closer to the Shannon limit than the FEC scheme adopted in all first-generation mobile broadcasting systems, based on the concatenation of a convolutional code with an RS code. Hence, the DVB-SH standard is sometimes considered a 1.5-generation technology.

Despite the fact that two DVB-SH satellites were launched in the United States and Europe in 2008 and 2009, respectively, today no commercial services are provided. A major challenge for deploying DVB-SH services is the amount of sites and infrastructure investments required to provide coverage in urban scenarios.

1.3.4 Media Forward Link Only

Media FLO is a proprietary technology, developed by Qualcomm [8]. It was designed to provide mobile multimedia services, including traditional real-time contents as well as non-real-time. Media FLO was designed aiming to be overlaid upon a generic cellular network with independent network infrastructure. The receiving devices would also be furnished with separate modules for mobile broadcast and cellular access.

The system is based on COFDM with a fixed 4K fast Fourier transform (FFT). The coding scheme is a combination of block and convolutional codes. An RS code is used as the outer (block) coder. The inner coder is the turbo code used in cdma2000 but with different puncturing schemes. The modulation schemes possible are QPSK and 16QAM (conventional and hierarchical modulation), and both time and frequency interleaving are exploited at the physical layer. The system capacities range from 2.8 up to 11.2 Mbps, depending upon the selected bandwidth (5, 6, 7, and 8 MHz), and modulation and coding options. The standard was designed for the VHF/UHF and L bands, but the main target band was the 700 MHz band (specifically to be rolled-out in USA). The average service bit rate of Media FLO is around 200–250 kbps using MPEG-4/advanced video coding (AVC) video coding. This yields around 20 real-time services within a 6 MHz bandwidth.

Media FLO arranges services into multicast logical channels (MLCs). An MLC contains one or several components of a service that can be identified at the physical layer and decoded independently. This service encapsulation technique into the physical layer enables time-sliced access to services and their components and thus provides efficient management of receiver battery life. Moreover, the system included an option called "layered services" based on a hierarchical modulation, targeting two different resolutions (each one in a different MLC) according to the received signal level. Delivering local services in SFNs is possible in Media FLO. The system delivers the services (MLCs) encapsulated in OFDM super frames, composed of different OFDM symbol groups (frames) arranged sequentially in time. A super frame consists of a group of 1200 OFDM symbols. The first 18 convey receiver signaling information as well as service metadata. After this block, four frames are transmitted, each one with resources for wide area and local services. The super frame ends with an additional block of symbols for signaling, transmitter identification, and receiver positioning functions.

Media FLO services were commercially launched in 2007 by different partnerships formed by Qualcomm and cellular operators. The network was deployed with high-power UHF transmitters operating in two channels (716–728 MHz) and provided coverage to approximately 70 million inhabitants of the major U.S. cities. The system was not commercially successful and the network was wound up in 2011.

1.3.4.1 Integrated Services Digital Broadcasting: Terrestrial One-Seg

The ISDB system was designed for high-quality audio, video, and associated data [9]. Originally, it was developed by and for Japan. In 2006, Brazil adopted a slightly modified version of the standard, known as ISDB-T$_B$* [29], which has been the terrestrial broadcast choice in most countries of South America.

The ISDB-T standard is based on MPEG-2 for audio/video coding and multiplexing, and it also includes AVC H.264 for mobile services. The system uses Band-Segmented Transmission—OFDM (BST-OFDM) modulation, frequency-domain and time-domain interleaving, and concatenated RS and convolutional code.

The BST-OFDM modulation scheme was designed for hierarchical transmission and partial spectrum reception. In ISDB-T, a 6 MHz bandwidth channel is divided into 14 segments of 6/14 MHz = 428.57 kHz. Thirteen segments are used for data transmission and one segment is used as guard band. In the 13 data segments, it is possible to transmit up to three simultaneous programs in three hierarchical layers with different transmission parameters and hence a specific robustness. The most common configuration employs one segment for mobile services (this is usually referred to as One-Seg), and 12 segments to transmit HDTV programs for fixed reception.

If compared to DVB-T, ISDB-T employs the same error correction scheme. Hence, their performance is very similar in static channels. However, ISDB-T makes use of physical time interleaving (four options are available, being possible to provide time interleaving duration up to 427.5 ms), which makes ISDB-T outperform DVB-T in mobile conditions.

Commercial One-Seg services were launched in Japan in 2006. As of February 2012, approximately 124 million cellular phones had been sold. Despite such good figures, it should be pointed out that the service is not yet commercially profitable. But broadcasters are required to provide One-Seg services by the national regulator.

1.3.4.2 Terrestrial: Digital Multimedia Broadcasting

T-DMB was the first commercialized mobile broadcasting technology in the world in December 2005 [10,30]. It was designed to provide different types of mobile multimedia services including audio and video service (with a target resolution equivalent to CIF (common intermediate format) or QVGA (quarter video graphics array) for small-size displays, 2–7 in.), interactive data service, as well as audio-only services.

T-DMB is based on the Eureka-147 Digital Audio Broadcasting (DAB) system, which was modified in order to suit the more restrictive requirements

---

* The main differences of ISDB-T$_B$ compared to ISDB-T are source code (H.264 for video and MPEG-4 HE AAC for audio); middleware (Ginga); and adoption of VHF/UHF bands for channels 7–13 and 14–69, respectively.

of a system delivering video and associated audio and multimedia information (e.g., T-DMB aims at a BER threshold down to 10^{-8} in comparison to the 10^{-4} threshold of DAB). T-DMB includes an RS code with a byte interleaver to improve the transmission robustness. The rest of modules of the physical layer remain unchanged from the original DAB specification. The RF bandwidth is kept to 1.5 MHz, and the current allocation plan in South Korea is within the VHF band (176–216 MHz) with 12 MHz (six multiplexes) for Seoul area and 6 MHz (three multiplexes) in other regions of the country. The system adopted MPEG-4 H.264 as the source coding for video streams and Bit Sliced Arithmetic Coding (BSAC) for audio material.

Compared to other first-generation mobile broadcasting systems, T-DMB has a lower spectral efficiency because DAB was originally designed for audio and its related data service. The maximum data rate of T-DMB in 1.5 MHz is approximately 1.6 Mbps (the effective throughput only 1 Mbps). T-DMB (along with ISDB-T) is the only success story of first-generation mobile broadcasting technologies. The advantages of T-DMB are a relatively simple service deployment due to the structural inheritance of the Eureka-147 DAB system that has been in commercial service for the past decade in many European countries and that the cost for establishing T-DMB network is relatively low due to simplicity of the system and its optimal frequency band. Since the commercial launch of T-DMB, more than 55 million receivers had been sold in South Korea as of June 2011.

1.3.4.3 3G Multimedia Broadcast Multicast Services

MBMS provides a seamless integration of multicast and broadcast transport technologies into the existing 3G networks and service architectures, reusing much of the existing 3G functionalities [4]. In Release 6, new diversity techniques were introduced to cope against fast fading and to combine transmissions from multiple cells, as well as an additional FEC mechanism at the application layer based on Raptor coding [31]. MBMS supports the use of long transmission time intervals (TTIs), up to 80 ms, to increase the time diversity against fast fading due to mobility of terminals, and the combination of transmissions from multiple cells (sectors) to obtain a macro diversity gain. Longer TTIs have the advantage of an increased time interleaving at the physical layer at the expense of longer delays increasing the network latency. But the unidirectional nature of MBMS hides these delays from the user perception. Two combining strategies are supported: selective combining and soft combining. With selective combining, signals received from different cells are decoded individually, such that terminals select each TTI the correct data (if any). With soft combining, the soft bits received from the different radio links prior to decoding are combined. Soft combining results in higher improvements because it provides also a power gain (the received power from several cells is added coherently), but it is more difficult to implement because it requires synchronizing the transmissions between cells.

Raptor codes are a computationally efficient implementation of fountain codes that achieve close to ideal performance and allows for a software implementation without the need of dedicated hardware even in handheld devices [32]. They have been standardized as application-layer FEC codes for MBMS, for both streaming and file download services, to handle packet losses in the transport network and in the radio access network.

The first release of MBMS offers a limited capacity, and it shares the same 3G cell resources with voice, messaging, and data services. Furthermore, because of the WCDMA (wide-code division multiplex access) technology employed in 3G, MBMS services produce interference to the other services. Hence, mobile operators have prioritized investments to upgrade their networks using p-t-p solutions (e.g., with HSDPA or LTE) rather than investing in MBMS.

Release 8, also known as integrated Mobile Broadcast (iMB), provides improved coverage and capacity performance, including SFN operation mode [33]. iMB can be deployed in the unpaired time-division duplex 3G spectrum, which is currently unused, without impacting current 3G voice and data services (which employ the paired frequency-division duplex, FDD, 3G spectrum). However, it requires the deployment of a dedicated network, and hence it still looks for supporters inside 3GPP ecosystem.

1.4 New-Generation Mobile Multimedia Broadcasting Systems

1.4.1 Japanese ISDB-Tmm and ISDB-T$_{SB}$ Mobile Broadcasting Systems

ISDB-Tmm (Multi Media) [34] and ISDB-T$_{SB}$ (Sound Broadcasting) [35] are evolutions of the ISDB-T One-Seg standard, which allow the transmission of high-quality video and audio coding, and efficient file delivery. Both systems have been designed to provide real-time and non-real-time services, and they are compatible with One-Seg terminals. Hence, the protocol stacks are very similar to that of first-generation systems.

ISDB-Tmm and ISDB-T$_{SB}$ target the VHF band, which in Japan has been traditionally used for analogue television. In particular, ISDB-Tmm targets the upper part of the VHF band (207.5–222 MHz), whereas ISDB-T$_{SB}$ targets the lower part of the VHF band (90–108 MHz). ISDB-T$_{SB}$ aims to provide regional multimedia broadcasting services, where different content is broadcasted for each region. ISDB-Tmm, on the other hand, is a national system where the same content is broadcasted to the whole country. Compared to One-Seg that employs a single-frequency segment, ISDB-Tmm allows using for mobile services any combination of the 13 data segments. ISDB-T$_{SB}$ allows using one or three segments.

ISDB-Tmm was commercially launched in April 2012 in the main cities of Japan by the operator MMBI. MMBI was founded in 2009, and its shareholders include mobile network operators as well as broadcasters, being thus a good example of potential cooperation between the cellular and the broadcast worlds. At the time of writing (July 2012), the starting date of the commercial launch of ISDB-T$_{SB}$ is under discussion.

An overview of the Japanese ISDB-T mobile broadcasting technologies *One-Seg*, *ISDB-T$_{BS}$*, and *ISDB-Tmm* can be found in Chapter 2.

1.4.2 South Korean Advanced T-DMB Mobile Broadcasting System

AT-DMB is a backward-compatible evolution of the T-DMB standard, which enhances the system capacity [36]. The evolution of portable devices with larger screen sizes, as well as the popularity of large screens in trams, buses, trains, and even cars created a need for enhanced resolution services, and, thus, larger bitrates than the ones provided by T-DMB. The target devices of T-DMB were small and mobile devices with screens from 4 to 7 in. For AT-DMB, the target is a 24 in. device with a resolution equivalent to VGA (720×480).

The main technologies adopted in AT-DMB are hierarchical modulation [37], scalable video coding (SVC) [38], and improved channel coding with turbo-coding. AT-DMB introduces an additional transmission layer to increase the system capacity devoted only to AT-DMB receivers [39]. The capacity gain of AT-DMB depends on the hierarchical modulation scheme employed, but it can increase the data rate up to 2 Mbps in 1.536 MHz bandwidth, twice the data rate of T-DMB.

The hierarchical modulation is characterized by two independent streams, known as high-priority stream and low-priority stream, with different robustness levels. The high-priority stream is compatible with legacy T-DMB receivers, because the video streams maintain the legacy MPEG-4 H.264 AVC codec used in T-DMB. The low-priority stream targets only AT-DMB receivers, which can decode both streams. With SVC, the source video is encoded into a base layer, transmitted in the high-priority stream, and an enhancement layer, transmitted in the low-priority stream. The enhancement layer provides higher resolution (up to 900 kbps that is only decoded if the reference base layer content is available). It should be pointed out that the coverage level of the low-priority stream is smaller than the reference coverage of the high-priority stream. Furthermore, the high-priority stream suffers a coverage degradation because the low-priority stream acts as noise in the transmitted signal. Nevertheless, AT-DMB can still provide wide service coverage with standard video quality while improving the service quality in good coverage areas reusing the existing T-DMB infrastructure.

An overview of the *advanced T-DMB* mobile broadcasting technology can be found in Chapter 3.

1.4.3 American ATSC-M/H Mobile Broadcasting System

Advanced Television Systems Committee Mobile/Handheld (ATSC M/H) is a set of add-ons and modifications to the ATSC DTT standard in order to provide audio, video, and datacast services to mobile and portable receivers. ATSC M/H has been designed to keep backward compatibility with the legacy-fixed receivers. The multimedia services (both multimedia streaming and file downloading are supported) are embedded within the existing ATSC signal structure, in a way that fixed receivers can access traditional SDTV/HDTV content while the mobile and portable devices have access to the ATSC M/H content. The balance between mobile/fixed services is a choice of the broadcasters. The standard is not tied to a specific configuration, but possible approaches to allocate resources of the total ATSC capacity (19.4 Mbps) would be [40] as follows:

- 1 HDTV program (13–15 Mbps), 2–3 mobile services (4–6 Mbps), metadata (0.4 Mbps)
- 1 SDTV programs (3–4 Mbps), additional SD multicast (2–4 Mbps), 5–8 mobile services (11–14 Mbps), metadata (0.4 Mbps)

The obvious advantage of the system is the reuse of the existing broadcast infrastructure and the fact of not requiring additional spectrum to provide mobile broadcast services. Considering the fact that ATSC has conveyed free-to-air content, the mobile and portable services are likely to be also free-to-air material. Nevertheless, the standard provides the tools for conditional access and a possible pay-tv business model.

The system architecture has been described by three system layers: presentation layer, management layer, and physical layer. The presentation layer defines the application protocols and the codecs used to pack the multimedia content associated to each service. ATSC M/H video streams are based on the MPEG-4 toolbox (ITU-R H.264) utilizing a layered approach that addresses different types of devices with a variety of different screen resolutions [38].

The system addresses multiple resolutions using two video layers. The first one, called base layer, is equivalent to the AVC baseline profile. The second one is an enhancement layer that uses the SVC features of the video coding standard. The first group of receivers is devices that support 416×240 resolution and 16:9 aspect ratio. The second family of receivers is in-car devices that support two enhanced resolutions: 624×360 and 832×480. The audio coder in ATSC M/H is High-Efficiency Advanced Audio Coding (HE AAC v2). The presentation layer also defines the synchronization of graphical elements and video and audio contents.

The management layer encompasses signaling, file delivery, service description and discovery tools, as well as conditional access and digital rights management functions. This layer also describes the encapsulation, timing, and transport protocols. The transport sub-layer in ATSC M/H is based on IP/User Datagram

Protocol (UDP). The system has abandoned the usual MPEG-2 TS structure, and all services are delivered directly through UDP packets. The timing is managed using the Real-Time Protocol. This protocol provides defined packet ordering and synchronization among the different streams that form a service. The service announcement and signaling are provided redundantly by OMA-BCAST and a specific Fast Information Channel (FIC). The OMA-BCAST service guide provides long-term service description data, and the FIC contains the signaling required for immediate access to current services.

The services are sent using time-slicing techniques replacing some of the MPEG-2 transport packets from the legacy ATSC stream and allocating the payloads in those segments to carry the ATSC mobile DTV data in a way that legacy receivers will ignore. In addition, the system incorporates long training sequences, to improve equalization performance under mobile conditions.

An overview of the *ATSC-M/H* mobile broadcasting technology can be found in Chapter 4.

1.4.4 Chinese DTMB and CMMB Mobile Multimedia Broadcast Systems

China has the largest TV market in the world, and its proportion of television audience using terrestrial method is pretty high (above 70%). China has developed its own DTT broadcasting standard named Digital Terrestrial Multimedia Broadcasting (DTMB) to provide high-quality multimedia services including mobile broadcasting. On the other hand, the technology Chinese Mobile Multimedia Broadcast System (CMMB) is intended to provide TV and radio services to vehicles and handheld terminals with small screens.

DTMB allows both single-carrier and multicarrier options to cope with the needs of the developing stages of radio and television in different areas of China and its vast population. Instead of using a cyclic prefix as in classical OFDM, DTMB uses time-domain pseudo-noise sequences as frame headers to mitigate inter-symbol interference. These sequences are also used as training sequence for the equalizer at the receivers, avoiding the need of introducing known carrier pilots for channel estimation and improving the effective throughput. Two types of channel estimation are defined for each mode (an all-time domain processing approach and a hybrid time- and frequency-domain processing approach), but the only difference between single-carrier and multicarrier modes is an inverse FFT. Furthermore, it achieves a faster channel acquisition compared to, for example, DVB-T, since this can be done directly in the time domain. The FEC code employed in DTMB is based on the concatenation of an LDPC and a Bose–Chaudhuri Hocquenghem (BCH), which provides better performance than the concatenated RS and convolutional code adopted in the first-generation digital terrestrial systems.

The CMMB standard architecture was originally based on a hybrid terrestrial–satellite infrastructure, similar as DVB-SH, to cover the whole extension of China [41]. The system was mainly designed not only to deliver multimedia information but also to support emergency and disaster relief announcement features. The system is based on frequency assignments in two different bands. The direct satellite broadcasting segment utilizes the S band. The coverage provided by the satellites is irregular within cities, and in consequence, there must be a complementary network to serve urban areas. The complementary network can either work rebroadcasting the satellite signals in the S band with terrestrial gap-fillers or broadcasting the same contents using channels of the UHF band. A third frequency band, the Ku band, can be used to distribute the contents to the UHF transmitters, either using the same or using different spatial infrastructure. The physical layer of the CMMB system is based on OFDM, and this standard was the first to include LDPC in combination with RS. CMMB was commercially launched in 2006, but only the terrestrial network component in the UHF band has been deployed.

An overview of the Chinese *DTMB* digital terrestrial broadcasting standard and the mobile broadcasting technology *CMMB* can be found in Chapter 5.

1.4.5 Digital Video Broadcast: Second-Generation Terrestrial

DVB-T2 is the most advanced DTT system in the world, providing at least 50% capacity increase over its predecessor [16]. Although DVB-T2 primarily targets static and portable reception, it shows very good performance in mobile scenarios [42]. DVB-T2 introduces several improvements over DVB-T that provide a higher robustness in mobile scenarios, including better FEC, rotated constellations, time interleaving, and distributed Multiple-Input Single-Output (MISO). Furthermore, DVB-T2 supports reduced power consumption (time-slicing) and per-service configuration of transmission parameters, including modulation, coding, and time interleaving, thanks to the concept of physical layer pipes (PLPs). Multiple PLPs allow for the provision of services targeting different user cases in the same-frequency channel, such that there is no need to build a dedicated network for the delivery of mobile services. In addition, a mobile profile known as T2-Lite has been included in the DVB-T2 specification to reduce the complexity of receivers and to improve the coexistence of fixed and mobile services.

DVB-T2 inherits the FEC coding scheme from DVB-S2 based on the concatenation of LDPC and BCH codes with two different codeword lengths (16,200 and 64,800 bits), which provide a performance close to the channel capacity in AWGN channel. DVB-T2 features rotated constellations as a manner to improve the performance of the system in fading channels [43]. A time interleaver has been also included at the physical layer in order to combat impulsive noise and benefit from time diversity in mobile scenarios [44]. It can provide interleaving durations ranging from few milliseconds up to several seconds. The time interleaving is also very flexible and offers different trade-offs in terms of transmission robustness (time diversity), latency, and power saving. For the first time in terrestrial broadcasting systems, DVB-T2 includes a distributed MISO technique, based on the Alamouti code [45], for improving the reception in SFNs. Distributed MISO reuses the existing network infrastructure by operating across the antennas of different transmitters and, therefore, can be used without any additional cost for network operators.

1.4.5.1 Mobile Profile of DVB-T2: T2-Lite

The simultaneous provision of static and mobile DVB-T2 services in the same-frequency channel is limited by the fact that the FFT size and the pilot pattern are common for the whole T2 multiplex. Static services are generally transmitted with large FFTs and sparse pilot patterns in order to achieve a high spectral efficiency in static channels. However, reception at high velocities requires the utilization of smaller FFTs and more dense pilot patterns.

The T2-Lite specification brings together a subset of DVB-T2 with some extensions specifically aimed at addressing receivers that are incorporated in portable or handheld devices. The main objective of T2-Lite is to allow multiple FFT modes and pilot patterns to be employed in the same-frequency channel by means of future extension frames (FEFs). FEFs were incorporated into the DVB-T2 standard to allow for the inclusion of future improvements in modulation technology. FEFs can be multiplexed along with regular T2 frames without impacting the operation of legacy DVB-T2 receivers (and vice versa). This way, T2-Lite mobile services can be transmitted in FEFs with small FFTs and dense pilot patterns while traditional static DVB-T2 services can still benefit from large FFTs and sparse pilot patterns.

T2-Lite is based on the same core technologies as DVB-T2, but it uses only a limited number of available modes that are best suited for mobile reception. By avoiding modes that require more complexity and larger amounts of storage memory, receivers can be designed that are less power consuming and with a smaller silicon size. The silicon chip size of a T2-Lite demodulator is approximately 50% smaller than a DVB-T2 demodulator. In addition, more robust code rates have been added to offer additional robustness.

The T2-Lite profile will be a default module in DVB-T2 receivers because it has been introduced as an integral part of the DVB-T2 standard in the version 1.3.1 of the specification.

Hence, it is expected that T2-Lite chipsets will be widely available, at a low price, and large numbers of receivers will be in the market following the worldwide rollout of DVB-T2 networks. This may encourage service providers to address this market. Furthermore, T2-Lite requires only a small initial investment because it can be combined into an existing DVB-T2 network without the need of building a dedicated network. This may encourage service operates to launch services gradually. A gradual increase in the availability of T2-Lite services and receivers could lead to the incorporation in the future of the mobile/portable market with T2-Lite-only receivers with significantly reduced chip size.

An overview of the *DVB-T2* and *T2-Lite* mobile broadcasting technologies can be found in Chapter 6.

1.4.6 Digital Video Broadcast: Next-Generation Handheld

DVB-NGH is the handheld evolution of the second-generation DTT standard DVB-T2. DVB-NGH introduces new technological solutions that together with the high performance of DVB-T2 make DVB-NGH a real next-generation mobile multimedia broadcasting technology.

Since DVB-T2 already incorporates capacity-achieving codes such as LDPCs, major further gains cannot be achieved by means of improved FEC codes. The combined utilization of time, frequency, and space diversity is a key instrument for reducing the costs associated with the deployment of mobile broadcasting networks while providing the necessary capacity for the delivery of new services [46]. Mobile broadcasting systems currently encompass time diversity across hundreds of milliseconds, frequency diversity inside a single-frequency channel, and space diversity with multiple antennas at the receiver side. So far, DVB-SH is the only mobile broadcasting system that exploits time diversity across several seconds by means of long TI with fast zapping, and no system fully exploits the diversity in the frequency domain across several RF channels or in the space domain with multiple antennas at the transmitter and the receiver side. DVB-NGH is the first system to incorporate the use of diversity in the three domains by incorporating at the physical layer long TI with fast zapping, time-frequency slicing (TFS) [47], and MIMO. DVB-NGH is the first 3G broadcasting system because it allows for the possibility of using MIMO antenna schemes to overcome the Shannon limit of single-antenna wireless communications.

Other new technical elements of DVB-NGH are the following: SVC with multiple PLPs, header compression mechanisms for both IP and TS packets, new FEC coding rates for the data path (down to one-fifth), nonhomogeneous constellations for 64-QAM and 256-QAM, four-dimensional rotated constellations for QPSK, improved time interleaving in terms of zapping time, end-to-end latency and memory consumption, improved physical layer signaling in terms of robustness, capacity, and overhead, a novel distributed MISO transmit diversity scheme for SFNs, and efficient provisioning of local content in SFNs. DVB-NGH also enables to complement the terrestrial coverage in the UHF band with an optional satellite component in the L- or S-frequency bands. In addition to the MFN configuration, an SFN configuration with both networks operating at the same frequency in the L or S bands is also possible.

Although DVB-NGH will technically outperform both DVB-H and DVB-SH in terms of capacity and coverage, one of its main advantages is that there is no need to deploy a dedicated terrestrial network to start providing DVB-NGH services, because it is possible to reuse existing DVB-T2 infrastructure to start providing DVB-NGH services in-band a DVB-T2 multiplex. Thanks to the FEFs of DVB-T2, it is possible to efficiently share the capacity of one-frequency channel in a time division manner between DVB-T2 and DVB-NGH, with each technology having specific time slots.

An overview of the *DVB-NGH* mobile broadcasting technology can be found in Chapter 7.

The second part of this book (Chapters 11 through 24) is devoted to provide a detailed insight of the DVB-NGH technology. Chapter 11 describes the main features of the Bit-Interleaved Coded Modulation module in the DVB-NGH standard and provides some performance results allowing this new system to be compared with the first-generation DVB-H. Chapter 12 focuses on the time interleaving adopted in the terrestrial profiles of DVB-NGH and discusses the different trade-offs related to the use of time interleaving: latency, time diversity, and power consumption. Chapter 13 provides an overview of the transmission technique time-frequency slicing, describes its implementation in DVB-NGH, and provides illustrative performance evaluation results about the coverage gain with field measurements and physical layer simulations. Chapter 14 provides an overview of the DVB-NGH logical frame structure and details the operation of a DVB-T2 multiplex with FEFs. Chapter 15 provides an overview of the physical layer signaling in DVB-NGH and presents illustrative performance evaluation results comparing the robustness of the signaling of DVB-T2 and DVB-NGH. Chapter 16 provides an overview of the system and upper layers of DVB-NGH, for both TS and IP profiles. Chapter 17 provides an overview of the different overhead reduction methods specified in DVB-NGH. Chapter 18 describes the complementary techniques adopted DVB-NGH for providing global and local contents in SFN topologies: Hierarchical modulation (HM-LSI) and orthogonal local services insertion (O-LSI) techniques. Chapter 19 discusses the potential advantages offered by the use of MIMO and the background to its adoption in DVB-NGH. The characteristics of the MIMO channel to a portable device are described and the channel model resulting from a bespoke measurement campaign. A discussion of use cases is followed by consideration of potential concerns and their solution. Chapter 20 describes the MIMO rate 1 codes adopted in DVB-NGH: distributed Alamouti and enhanced SFN (eSFN), and Chapter 21 describes the MIMO rate 2 code adopted in DVB-NGH, known as enhanced Spatial Multiplexing—Phase Hopping (eSM-PH). Chapter 22 provides an overview of the hybrid terrestrial–satellite profile of DVB-NGH and describes the main technical solutions specifically developed to deal with the characteristic problems of mobile satellite reception and to seamlessly combine the signals coming from the two networks with a single tuner. Chapter 23 describes and evaluates the single-carrier orthogonal frequency multiplexing modulation that was selected along with pure OFDM to implement the transmissions on the DVB-NGH satellite link. Chapter 24 provides an overview of the hybrid terrestrial–satellite MIMO profile of DVB-NGH.

1.4.7 3GPP Enhanced MBMS

3GPP has upgraded the MBMS specification as part of Release 9 of the LTE standard to embed a broadcasting mode in the next-generation cellular system. The flexible and efficient support of multicast/broadcast services was an important design goal of LTE from the start. Furthermore, E-MBMS is part of the continued further evolution of LTE standardization in 3GPP. The short cycles of issuing new standard releases ensure timely implementation of technology opportunities and market demands.

The main goals for LTE are [48] as follows:

■ Increased cell edge and cell average throughput
■ Improved spectrum flexibility
■ Simplified architecture
■ Efficient support of MBMS

The most important innovations in E-MBMS with respect to 3G MBMS are the use of OFDM modulation in the physical layer (which increases the spectrum efficiency), the support of larger bandwidths, and efficient and flexible support for SFN. In E-MBMS, the SFN area size can be tailored geographically from a single cell up to nationwide and areas of different size can overlap, all on the same carrier frequency. Another important advantage is that multicast/broadcast and unicast delivery methods are multiplexed in time, such that there is no need for a dedicated E-MBMS network [49]. E-MBMS uses an evolved network architecture that supports multicast/broadcast transmissions and user individual services on the same carrier and to coordinate the allocation of radio resources across all cells participating in the particular SFN.

An overview of the *E-MBMS* mobile broadcasting technology can be found in Chapter 8.

1.5 Current Trends toward a Worldwide Mobile Multimedia Broadcasting Standard

1.5.1 Common Broadcasting Specification between DVB and 3GPP

In February 2010, as a response to the DVB-NGH Call for Technologies, the mobile network operator Orange proposed 3GPP E-MBMS waveform as a candidate for DVB-NGH. The DVB standardization group did not warmly welcome this proposal. However, understanding the opportunity it offered to reach the world of mobile devices (smart phones and tablets) and to provide a solution to absorb the ever-increasing video traffic on mobile networks, the digital TV standardization forum contacted 3GPP organization in November 2010, asking to consider a potential collaboration in the area of mobile broadcasting. A joint workshop took place in March 2011 in Kansas City (the United States) for mutual presentations of 3GPP and DVB standardization activities. In May 2011, in Xian (China), the creation of a study item was proposed to the 3GPP technical subgroup dealing with services (3GPP SA1). The objective was to "study the feasibility of, and creating common service requirements and use cases for, a common broadcast specification which can be used in a 3GPP mobile communications network and a broadcasting network that is based on DVB or other similar standards." This proposal, introducing, for the first time, the concept of Common Broadcasting Specifications (CBS), was discussed but not accepted by 3GPP SA1, due to lack of support from mobile operators. Nevertheless, on the DVB side and after the finalization of the DVB-NGH specifications, the "CBS" topic should be addressed again.

Chapter 9 highlights the win-win situation for each actor of the multimedia value chain, which should result from such a broadband/broadcast cooperation, and presents preliminary studies about a *common broadcast specification* of state-of-the-art 3GPP and DVB standards to provide a broadcast overlay optimized for mobile and operated in conjunction with a broadband unicast access.

1.5.2 Future of Broadcast Television

Future of Broadcast Television (FoBTV, www.fobtv.org) is an industry consortium to define the future needs, seek unification of various standards, and promote technology sharing to reduce overlap. The 13 founding members of the FoBTV are as follows:

1. Advanced Television Systems Committee (ATSC)
2. Canadian Broadcast Corporation (CBC)
3. Communications Research Centre Canada (CRC)
4. Digital Video Broadcasting Project (DVB)
5. European Broadcasting Union (EBU)
6. Electronics and Telecommunications Research Institute (ETRI)
7. Globo TV-Brazil
8. IEEE Broadcast Technology Society (IEEE-BTS)
9. National Association of Broadcasters (NAB)
10. National Engineering Research Center of Digital TV of China (NERC-DTV)
11. NHK Science and Technology Research Laboratories (NHK)
12. Public Broadcasting Service (PBS)
13. Brazilian Society of Television Engineers (SET)

The first FoBTV Summit was held in Shanghai, China, in November 2011. Over 250 broadcast industry executives, researchers, and engineers signed a joint declaration, which calls for global cooperation to define new requirements, unify various standards, and promote sharing of technologies to benefit developed and underdeveloped countries and conserve resources. All delegates agree that a global approach to the future of terrestrial TV broadcasting would be an ideal method to avoid competing standards, overlap, and inefficient deployment of new services.

A common theme throughout the summit was that broadcasting—the transmission of information to an unlimited number of listeners and viewers is the most spectrum-efficient means for wireless delivery of popular real-time and file-based content. The signatories of the declaration believe that the broadcasting and TV industries will continue to evolve and play a critical role in bringing both information and entertainment to everyone.

While television has prospered, it has not been possible for the world to take full advantage of the convenience and economies of scale of a single broadcast standard. Even in the digital age, splintering of different standards and methods of broadcast TV transmission makes it difficult to share information and entertainment globally. During the Shanghai Summit, world broadcasting leaders established a framework for cooperation to chart the future course of terrestrial television broadcasting.

As a result of the FoBTV conference, signatories to the declaration agree to three major initiatives:

1. Define the requirements of future terrestrial broadcast systems
2. Explore unified terrestrial broadcast standards
3. Promote global technology sharing

The FoBTV initiative was formally launched in April 2012 in Las Vegas, the United States, where technical executives from 13 television broadcast organizations from around the world completed signing a landmark memorandum of understanding (MOU) to officially form the global FoBTV Initiative.

The signatories of the FoBTV MOU believe that terrestrial broadcasting is uniquely important because it is wireless (supports receivers that can move), infinitely scalable (p-t-m and one-to-many architecture), local (capable of delivering geographically local content), timely (provides real-time and non-real-time delivery of content), and flexible (supports free-to-air and subscription services).

The attribute of wireless delivery of media content to a potentially unlimited number of receivers makes terrestrial broadcasting a vital technology all over the world. Broadcasting is, in fact, the most spectrum-efficient wireless delivery means for popular real-time and file-based media content.

The MOU underscores the goals of the FoBTV Initiative, which include the following:

■ Developing future ecosystem models for terrestrial broadcasting taking into account business, regulatory, and technical environments
■ Developing requirements for next-generation terrestrial broadcast systems
■ Fostering collaboration of digital TV development laboratories

- Recommending major technologies to be used as the basis for new standards
- Requesting standardization of selected technologies (layers) by appropriate standards of development organizations

FoBTV is a voluntary, nonprofit association that is open to any organization that signs the MOU. FoBTV has a Management Committee that includes representatives of the founding members. FoBTV also has a Technical Committee that is responsible for solicitation and evaluation of technical proposals and recommending major technologies to be used as the basis for new standards. Participation in the work of FoBTV Technical Committee is open to all MOU-signatory organizations that have a direct and material interest in the work of FoBTV.

1.6 Conclusions

This introductory chapter of the book provides an overview about the past, present state-of-the-art to the day of writing (August 2012), and future in mobile multimedia broadcasting. The book is structured in two parts. The first part provides an overview of the new-generation mobile broadcasting technologies available worldwide. The following technologies are covered: the evolutions of the Japanese mobile broadcasting standard ISDB-T One-Seg *ISDB-Tmm* and *ISDB-T$_{SB}$*, the evolution of the South-Korean T-DMB mobile broadcasting technology *AT-DMB*, the American mobile broadcasting standard *ATSC-M/H*, the Chinese broadcasting technologies *DTMB* and *CMMB*, the second-generation DTT European standard *DVB-T2*, its mobile profile *T2-Lite* and its handheld evolution *DVB-NGH*, and the multicast/broadcast extension of the fourth-generation *LTE* cellular standard *E-MBMS*. This first part of the book ends with an overview of new High-Efficiency Video Coding standard, which is expected to provide significantly improved coding efficiency as compared with the current state-of-the art MPEG-4 AVC video coding.

The second part of the book is devoted to the latest technology DVB-NGH, which is expected to significantly outperform all existing technologies in both capacity and coverage. DVB-NGH introduces new technological solutions that together with the high performance of DVB-T2 make DVB-NGH a real next-generation mobile multimedia broadcasting technology. For example, DVB-NGH is the first system to incorporate the use of diversity in the three domains by incorporating at the physical layer longtime interleaving with fast zapping, TFS, and MIMO.

References

1. B. Karlson, A. Bria, J. Lind, P. Lönnqvist, and C. Norlin, *Wireless Foresight: Scenarios of the Mobile World in 2015*, Wiley, New York, 2003.
2. Mason, Mobile TV—Results from the DVB-H trial in Oxford, *EBU Technical Review*, April 2006.

3. B. Furht and S. Ahson (eds.), *Handbook of Mobile Broadcasting: DVB-H, DMB, ISDB-T and MediaFLO*, CRC Press, Boca Raton, FL, 2008.
4. F. Hartung et al., Delivery of broadcast services in 3G networks, *IEEE Transactions on Broadcasting*, 52(1), 188–199, March 2007.
5. W. Tuttlebee, D. Babb, J. Irvine, G. Martinez, and K. Worrall, Broadcasting and mobile telecommunications: Interworking not convergence, *EBU Technical Review*, January 2003.
6. N. Niebert, A. Schieder, J. Zander, and R. Hancock, *Ambient Networks: Co-operative Mobile Networking for the Wireless World*, Wiley, New York, 2007.
7. G. Faria, J. A. Henriksson, E. Stare, and P. Talmola, DVB-H: Digital broadcast services to handheld devices, *Proceedings of the IEEE*, 94(1), 194–209, January 2006.
8. M. R. Chari et al., FLO physical layer: An overview, *IEEE Transactions on Broadcasting*, 52(1), 145–160, March 2007.
9. G. Bedicks and C. Akamine, Overview of ISDB-T: One-segment reception, in *Handbook of Mobile Broadcasting: DVB-H, DMB, ISDB-T, and MediaFLO*, B. Furht and S. A. Ahson, eds., Chapter 4, CRC Press, Taylor & Francis Group, Auerbach, Germany, June 2008.
10. S. Cho et al., System and services of terrestrial digital multimedia broadcasting (TDMB), *IEEE Transactions on Broadcasting*, 52(1), 171–178, March 2007.
11. Cisco (R) Systems Inc., CISCO Visual Networking Index: Global data forecast update, 2011–2016, Technology report, February 2012.
12. A. Osseiran, J. F. Monserrat, and W. Mohr (eds.), *Mobile and Wireless Communications for IMT-Advanced and Beyond*, Wiley, New York, 2011.
13. J. Zander, On the cost structure of future wideband wireless access, *Proceedings of the IEEE VTC Spring*, Phoenix, AZ, 1997.
14. J. J. Delmas and P. Bretillon, Mobile broadcast technologies—Link budgets, Broadcast Mobile Convergence (BMCO) Forum, Technology report. Available online at http://www.oipf.tv, February 2009.
15. A. Bria and D. Gómez-Barquero, Scalability of DVB-H deployment on existing wireless infrastructure, *Proceedings of the IEEE PIMRC*, Berlin, Germany, 2005.
16. L. Vangelista et al., Key technologies for next-generation terrestrial digital television standard DVB-T2, *IEEE Communications Magazine*, 47(10), 146–153, October 2009.
17. L. Zheng and D. N. C. Tse, Diversity and multiplexing: A fundamental tradeoff in multiple-antenna channels, *IEEE Transactions on Information Theory*, 49(5), 1073–1096, 2003.
18. D. Gómez-Barquero, Cost efficient provisioning of mass mobile multimedia services in hybrid cellular and broadcasting systems, PhD dissertation, Universitat Politècnica de València, Valencia, Spain, 2009.
19. C. Sattler, Mobile broadcast business models—A state of the art study, BMCO forum white paper, November 2006.
20. H. Mittermayr and C. Sattler, Mobile broadcast business models—Country-specific implementations, BMCO forum white paper, Broadcast Mobile Convergence (BMCO) Forum, Technology report. Available online at http://www.oipf.tv, February 2009.
21. C. Sattler, Best practice regulatory framework for mobile TV, BMCO forum white paper, Broadcast Mobile Convergence (BMCO) Forum, Technology report. Available online at http://www.oipf.tv, June 2008.
22. U. Ladebusch and C. Liss, Terrestrial DVB (DVB-T): A broadcast technology for stationary portable and mobile use, *Proceedings of the IEEE*, 94(1), 183–193, January 2006.

23. J. López-Sánchez, D. Gómez-Barquero, D. Gozálvez, and N. Cardona, On the provisioning of mobile digital terrestrial TV services to vehicles with DVB-T, *IEEE Transactions on Broadcasting*, 58, (early access article, available at http://www.ieeexplore.org) December 2012.

24. D. Coquil, G. Hölbling, and H. Kosch, An overview of the emerging digital video broadcasting—Handheld (DVB-H) technology, in *Handbook of Mobile Broadcasting: DVB-H, DMB, ISDB-T, and MediaFLO*, B. Furht and S. A. Ahson, eds., Chapter 1, CRC Press, Taylor & Francis Group, Auerbach, Germany, June 2008.

25. I. Andrikopoulos et al., An overview of digital video broadcasting via satellite services to handhelds (DVB-SH) technology, in *Handbook of Mobile Broadcasting: DVB-H, DMB, ISDB-T, and MediaFLO*, B. Furht and S. A. Ahson, eds., Chapter 2, CRC Press, Taylor & Francis Group, Auerbach, Germany, June 2008.

26. F. Pérez-Fontán et al., Statistical modelling of the LMS channel, *IEEE Transactions on Vehicular Technology*, 50(6), 1549–1567, 2001.

27. D. Gómez-Barquero, P. Unger, T. Kürner, and N. Cardona, Coverage estimation for multiburst FEC mobile TV services in DVB-H systems, *IEEE Transactions on Vehicular Technology*, 59(7), 3491–3500, September 2010.

28. European Telecommunication Standards Institute (ETSI), ETSI TS 102 584 v1.2.1, Digital video broadcasting (DVB): DVB-SH implementation guidelines. Available online at http://www.etsi.org, January 2011.

29. G. Bedicks Jr. et al., Results of the ISDB-T system tests, as part of digital TV study carried out in Brazil, *IEEE Transactions on Broadcasting*, 52(1), 38–44, March 2011.

30. C. Katsigiannis et al., An overview of digital multimedia broadcasting for terrestrial (DMB-T), in *Handbook of Mobile Broadcasting: DVB-H, DMB, ISDB-T, and MediaFLO*, B. Furht and S. A. Ahson, eds., Chapter 3, CRC Press, Taylor & Francis Group, Auerbach, Germany, June 2008.

31. T. Stockhammer et al., Application layer forward error correction for mobile multimedia broadcasting, in *Handbook of Mobile Broadcasting: DVB-H, DMB, ISDB-T, and MediaFLO*, B. Furht and S. A. Ahson, eds., Chapter 10, CRC Press, Taylor & Francis Group, Auerbach, Germany, June 2008.

32. A. Shokrollahi, Raptor codes, *IEEE Transactions on Information Theory*, 52(6), 2251–2567, June 2006.

33. Z. Wang, P. Darwood, and N. Anderson, End-to-end system performance of iMB, *Proceedings of the IEEE BMSB*, Shangai, China, 2010.

34. A. Yamada, H. Matsuoka, T. Ohya, and R. Kitahara, Overview of ISDB-Tmm services and technologies, *Proceedings of the IEEE BMSB*, Erlangen, Germany, 2011.

35. T. Ikeda, Transmission system for ISDB-T$_{SB}$ (digital terrestrial sound broadcasting), *Proceedings of the IEEE*, 94(1), 257–260, January 2006.

36. J. H. Lee, J.-S. Lim, S. W. Lee, and S. Choi, Development of advanced terrestrial DMB system, *IEEE Transactions on Broadcasting*, 56(1), 28–35, March 2010.

37. H. Jiang and P. A. Wilford, A hierarchical modulation for upgrading digital broadcast systems, *IEEE Transactions Broadcasting*, 51(2), 223–229, June 2005.

38. H. Schwarz, D. Marpe, and T. Wiegand, Overview of the scalable extension of the H.264/MPEG-4 AVC video coding standard, *IEEE Transactions Circuits and Systems for Video Technology*, 17(9), 1103–1120, September 2007.

39. H. Choi, I. H. Shin, J.-S. Lim, and J. W. Hong, SVC application in advanced T-DMB, *IEEE Transactions on Broadcasting*, 55(1), 51–61, March 2009.

40. Advanced Television Systems Committee A/154:2011, *ATSC Recommended Practice: Guide to the ATSC Mobile DTV Standard*, April 2011.
41. State Administration of Radio, Film, and Television of China (SARFT), Chinese mobile multimedia broadcasting technical white paper, Technology report, June 2008.
42. I. Eizmendi et al., Next generation of broadcast multimedia services to mobile receivers in urban environments, *Springer Signal Processing: Image Communication*, 27(8), 925–933, 2012.
43. C. Abdel Nour and C. Douillard, Rotated QAM constellations to improve BICM performance for DVB-T2, *Proceedings of the IEEE ISSSTA*, Bologna, Italy, 2008.
44. D. Gozálvez, D. Gómez-Barquero, D. Vargas, and N. Cardona, Time diversity in mobile DVB-T2 systems, *IEEE Transactions on Broadcasting*, 57(3), 617–628, September 2011.
45. S. M. Alamouti, A simple transmit diversity technique for wireless communications, *IEEE Journal on Selected Areas in Communications*, 16(8), 1451–1458, October 1998.
46. D. Gozálvez, Combined time, frequency and space diversity in multimedia mobile broadcasting systems, PhD dissertation, Universitat Politèctica de València, Valencia, Spain, 2012.
47. J. J. Giménez, D. Gozálvez, D. Gómez-Barquero, and N. Cardona, A statistical model of the signal strength imbalance between RF channels in a DTT network, *IET Electronic Letters*, 48(12), 731–732, June 2012.
48. H. Ekström et al., Technical solutions for the 3G long-term evolution, *IEEE Communications Magazine*, 44(3), 38–45, March 2006.
49. J. F. Monserrat, J. Calabuig, A. Fernández-Aguilella, and D. Gómez-Barquero, Joint delivery of unicast and E-MBMS services in LTE networks, *IEEE Transactions on Broadcasting*, 58(2), 157–167, June 2012.

Chapter 2

An Overview of the ISDB-T One-Seg Broadcasting, ISDB-T$_{SB}$ and ISDB-Tmm

Masayuki Takada, Hitoshi Sanei,
Tomoyuki Ohya, Toshiharu Morizumi, Akira Yamada,
Ryo Kitahara, and Shinichiro Tonooka

Contents

2.1 ISDB-T One-Seg Broadcasting

2.1.1 Introduction

In Japan, "One-Seg," digital terrestrial television broadcasting (DTTB) targeted at handheld terminals, was launched on April 1, 2006. Until then, television was primarily watched at home via fixed reception, but with the start of One-Seg, it became possible to watch television anywhere and at any time with a handheld terminal. The most common devices are cellular phones with One-Seg receiving functions, and the unification of telephones and television made it possible to use the functions already offered by cellular phones, such as Internet connection functionality and the acquisition of GPS location data, in conjunction with digital broadcast technology. As of February 2012, approximately 124 million cellular phones with One-Seg receiving functions had been shipped, while their numbers constituted 80% of monthly shipments. In a short period of time, the One-Seg broadcasting service has become very successful. In this section, the transmission scheme, services, video coding technology, and audio coding technology, etc., will be explained.

2.1.2 Transmission Scheme

2.1.2.1 Overview

The Integrated Services Digital Broadcasting-Terrestrial (ISDB-T) system has been adopted in Japan as the DTTB system [1]. ISDB-T uses a segmented multicarrier modulation method called Band Segmented Transmission-Orthogonal Frequency Division Multiplexing (BST-OFDM), which consists of 13 segments within 1 channel (transmission bandwidth of 5.6 MHz) for transmission. The modulation method and error correction coding rate can be changed for each segment, and hierarchical transmission of up to three layers is possible. In addition, the center segment can be assigned for handheld reception as a partial reception segment and can provide services such as the broadcast of low-frame rate and low-resolution pictures, audio, and data to handheld terminals.

A feature of ISDB-T is that it is highly resistant to the multipath interference that occurs as a result of signals reflecting off buildings as well as the phasing interference that occurs during mobile/handheld reception. This is due to the employment of frequency and time interleaving in addition to OFDM.

The transmission scheme system for ISDB-T system is shown in Figure 2.1. The data sources for signals such as video, audio, and data are encoded as MPEG-2 Video, MPEG-4 AVC/H.264, MPEG-2 AAC, etc. After they are multiplexed, they are transmitted as OFDM signals through the channel encoder. In the channel encoder, after the transport stream (TS) is re-multiplexed, error correction, carrier modulation, interleaving, and framing are performed, so as to create the OFDM signal. All of the standardized ISDB-T parameters are listed in Table 2.1, while the parameters that are defined by actual operating standards are listed in Table 2.2 [2]. In an actual ISDB-T system, transmission Mode 3 and a guard interval of 1/8 are used, and in the handheld reception layer, an encoding rate of 2/3 and a time interleave of $I = 4$ (0.43 s) are used.

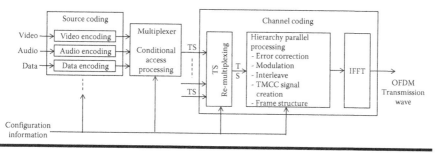

Figure 2.1   Transmission scheme for ISDB-T DTTB.

Table 2.1   ISDB-T Transmission Parameters

Transmission Parameter	Mode 1	Mode 2	Mode 3
Number of OFDM segments	13		
Bandwidth	5.6 MHz		
Carrier spacing (kHz)	3.968	1.984	0.992
Number of carriers	1405	2809	5617
Number of carriers/segment	108	216	432
Control carrier, etc.	SP, CP, TMCC, AC[a]		
Modulation method	QPSK, 16QAM, 64QAM, DQPSK		
Effective symbol length	252 μs	504 μs	1.008 ms
Guard interval length	Effective symbol lengths of 1/4, 1/8, 1/16, 1/32		
Inner code	Convolutional code (coding rates = 1/2, 2/3, 3/4, 5/6, 7/8)		
Outer code	(204, 188) Reed–Solomon code		

[a] SP, scattered pilot; CP, continual pilot; TMCC, transmission and multiplexing configuration control; AC, auxiliary channel.

Table 2.2   Usable Transmission Parameters Defined by the Operating Procedures

Layer	Fixed Reception Layer		Handheld Reception Layer	
Mode/guard interval ratio	Mode 3 1/4, 1/8, (1/16)[a]			
	Mode 2 1/4, (1/8)[a]			
Time interleaving	I = 1 (0.11 s), 2 (0.22 s), 4 (0.43 s) (Mode 3)			
	I = 2 (0.11 s), 4 (0.22 s), 8 (0.43 s) (Mode 2)			
Modulation	64QAM	16QAM[b]	16QAM	QPSK
Coding rate	7/8–1/2	2/3, 1/2	1/2	2/3, 1/2

[a] Usage is difficult with currently established stations but may be possible in the future after frequencies are reassigned.
[b] Can be used only in urgent situations such as in the event of emergencies and/or natural disasters.

Table 2.3   Example of Data Rates for ISDB-T (Mode 3, Guard Interval of 1/8)

	Handheld Reception Layer (1 Segment)			
Modulation	QPSK		16QAM	
Coding rate	1/2	2/3	1/2	
Transmission capacity (kbps)	312.1	416.1	624.1	
	Fixed Reception Layer (12 Segments)			
Modulation	64QAM			
Coding rate	1/2	2/3	3/4	5/6
Transmission capacity (Mbps)	11.2	14.9	16.8	18.7

Table 2.4   Example of Bit Rate for One-Seg Content (QPSK 2/3)

Content	Date Rate (kbps)
Video	244
Audio	55
Data	55
Closed caption	5
EPG	20
Control, etc.	37
Total	416

Example data rates for ISDB-T are listed in Table 2.3. When the transmission parameters for handheld reception (1 segment) and fixed reception (12 segments) are QPSK 2/3 and 64QAM 3/4, respectively, the transmission capacities are 416.1 kbps and 16.8 Mbps. An example breakdown of the data rates in One-Seg is shown in Table 2.4.

2.1.2.2 Hierarchical Transmission

In ISDB-T, it is possible to set a maximum of three hierarchical transmission layers. One-Seg is one layer. When operating One-Seg, two-layer or three-layer transmission is employed. Two-layer transmission consists of 1 segment in the center of

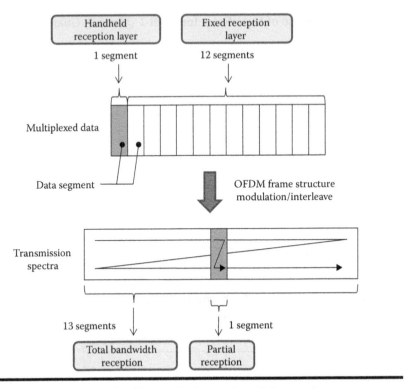

Figure 2.2   ISDB-T hierarchical transmission and partial reception.

the band being allocated to One-Seg while the other 12 segments are allocated to fixed reception. In this case, the One-Seg frequency interleaving is performed only within the segment, and an indication of partial reception is transmitted. Figure 2.2 illustrates hierarchical transmission for ISDB-T and partial reception.

Transmission and Multiplexing Configuration Control (TMCC) signals transmit the transmission parameters for each layer, such as modulation method and error correction coding rate, as well as a partial reception flag that indicates that the center segment is a partial reception segment, etc. TMCC is assigned to special OFDM carriers. The results of TMCC decoding indicate whether the central segment is the One-Seg signal.

2.1.2.3 Multiplexing

ISDB-T uses MPEG-2 system multiplexing [3], but in order to enable partial reception in only one segment, restrictions are imposed on the multiplexing. Specifically, transmission timing is restricted for the TS packets (TSPs) that send out the program clock reference (PCR) signal. The PCR signal is used to correct the system clock in the receiver that is used to fix the display timing of a program.

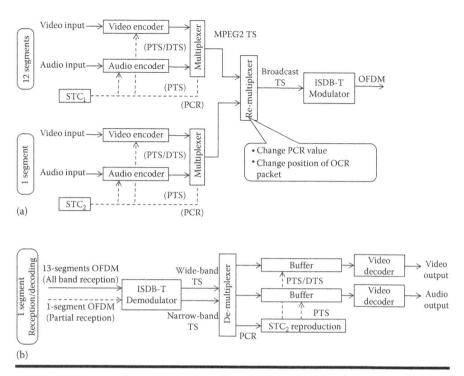

Figure 2.3 Transmission and reception system diagram for ISDB-T. (a) For "transmission" and (b) for "reception."

In order to understand PCR transmission timing, it is necessary to understand the basic concept of the video and audio presentation synchronization method in MPEG-2 systems. Figure 2.3 is a system diagram of ISDB-T transmission and reception focusing on this point. In MPEG-2 systems, there is a system time clock (STC), which is a 27 MHz reference clock in both the transmission station and the receiver. Then, to match the receiver's STC to that of the transmission station, a reference value called PCR is multiplexed in the TS and sent. The receiver synchronizes it with a phase-locked loop (PLL) in the internal clock with the received PCR. Consequently, to stabilize the PLL, the PCR is synchronized with the PLL in the internal clock. Therefore, to stabilize the PLL, it is necessary for the PCR to be transmitted within the prescribed margin of error in a fixed cycle.

MPEG-2 Video, MPEG-4 AVC/H.264, MPEG-2 AAC, etc., which are used for video and audio encoding, are variable-length coding methods, and the amount of encoded information varies greatly depending on the content. Even in situations where the amount of encoded information varies, to present the video and audio with the correct timing, a presentation time stamp (PTS) or a decoding time stamp (DTS), which indicates the display time referenced to STC in the encoding process, are added to the video and audio data and then sent. During the decoding process, the received video and audio data are stored in a buffer, and then either displayed or decoded at the PTS or DTS timing referenced to the reproduced STC. To indicate the correct timing in the receiver, it is necessary to correctly and stably preserve the reproduced STC in the receiver.

2.1.2.4 Wide-Band TS and Narrow-Band TS

Hierarchical transmission and reception are illustrated in Figure 2.2. The TS that is reproduced after decoding all of the 13 segment signals in a band is called the wide-band TS, and the TS that is reproduced after decoding the center segment only is called the narrow-band TS. Programs broadcast using the center segment need to be watched by car-mounted receivers, which receive all 13 segments. Furthermore, these programs need to be watched by handheld receivers, which receive only the center segment to decrease power consumption.

Figure 2.4 shows the wide- and narrow-band TS focusing on the PCR packet, which is sent with the central segment. The wide-band TS is a broadcast TS with an approximately 32.5 MHz transmission clock with a null packet inserted in the receiver, while the narrow-band TS is a broadcast TS with an approximately 2.0 MHz transmission clock with a null packet inserted in the receiver. To stably maintain the receiver's reproduced STC, it is necessary for the PCR packet in the narrow-band TS to appear in the same cycle as that of the wide-band TS. As can be

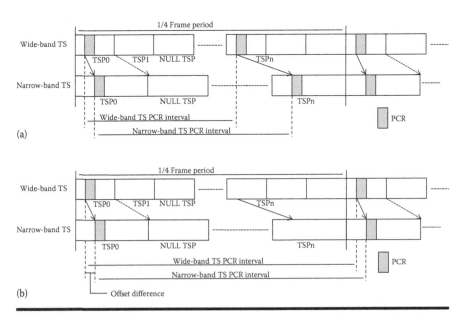

(a)

(b)

Figure 2.4 PCR packet interval of wide-band and narrow-band TS (Mode 3). (a) When PCR packet transmission is not controlled. (b) When PCR packet transmission is controlled.

seen in Figure 2.4a, however, if the PCR transmission timing is not controlled on the transmitter side, it will appear in a different cycle. In this state, the STC cannot be correctly reproduced in a partial reception receiver.

In Mode 3 (refer to Table 2.1, carrier spacing of approximately 1 kHz), the packets placed in each layer and the null packets are repeated in the frame at the same placement pattern four times, so it becomes a regulation where one PCR packet is sent to each pattern. When controlled in this way, the intervals of the PCR packets between the wide-band and narrow-band TS become the same, such that even a partial reception receiver can correctly reproduce the STC.

2.1.3 Media Coding

2.1.3.1 Video Coding

2.1.3.1.1 MPEG-4 AVC/H.264 Method Profile and Level

One-Seg uses the MPEG-4 AVC/H.264 method [4] for video encoding. The MPEG-4 AVC/H.264 method is a broad-range video encoding method that can be applied to a range of applications, such as broadcasting, videophones, and Internet transmissions. MPEG-4 AVC/H.264 is defined by four profiles corresponding to their use (Figure 2.5). The Baseline profile is for handheld terminals and videophones, the Extended profile is for streaming, the Main profile is for transmissions and accumulated media, while the High profile is envisioned for high-quality image use such as in studios. The encoding tools for each possible use are defined. For the Baseline profile, in addition to the MPEG-4 AVC/H.264 method basic encoding tool, an error tolerance tool

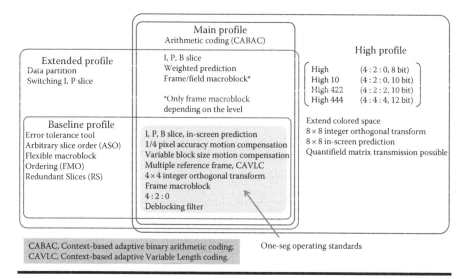

Figure 2.5   MPEG-4 AVC/H.264 profile.

can also be used. In the Extended profile, in addition to the Baseline profile, encoding tools suited to bidirectional prediction and interlaced video, etc., can be used. In addition, the Main profile has the Extended profile added to it and includes arithmetic encoding that makes a higher level of compression possible. However, the encoding tools for error tolerance, etc., are not included in the Main profile. In the High profile, 10 bit video and 4:2:2 video can be encoded. Note that, in the MPEG-4 AVC/H.264 format, there is no profile in which all of the encoding tools can be used.

In the MPEG-2 format, levels are defined by the screen size. With MPEG-4 AVC/H.264 as well, levels are similarly defined, from Level 1, which handles video from the Quarter Common Intermediate Format (176 pixels × 144 lines) level, to Level 5, which handles video of a size exceeding Hi-Vision (1920 pixels × 1080 lines). Furthermore, sublevels are defined and indicated by decimals at each level. For example, 1.2 is the level that handles Quarter Video Graphics Array (QVGA, 320 pixels × 240 lines).

2.1.3.1.2 One-Seg Operating Standards

One-Seg uses the MPEG-4 AVC/H.264 format Baseline profile, Level 1.2, and sets the operating standards that restrict the parameter range, taking the load on the decoder into consideration. Table 2.5 shows the major operating standards for One-Seg. There are two types of screen sizes used for One-Seg: 320 × 180 pixels (16:9) and 320 × 240 pixels (4:3), while square pixels having a pixel aspect ratio of 1:1 are used. The minimum frame interval is 1/15 s (consequently, the maximum frame rate is 15 fps). The instantaneous decoding refresh (IDR) intra-frame is a frame that can commence video decoding when the decoder is turned on or after the channel is changed, and in One-Seg, the maximum IDR intra-frame insertion interval is 5 s, and the recommended interval is 2 s.

Table 2.5 Major Operating Standards for One-Seg Service

Item	Specifications
Video format	320 pixels × 180 lines (when 16:9)
	320 pixels × 240 lines (when 4:3)
Minimum frame interval	1/15 s
IDR intra-frame insertion interval	Normally 2 s (maximum 5 s)
Number of multiple reference frames	Maximum of 3
Motion vector search range	Maximum ± 128 pixels (both horizontal and perpendicular)

2.1.3.2 Audio Coding

One-Seg uses advanced audio coding (AAC) technology, as well as AAC+ spectral band replication (SBR) encoding technology, an extension technology. AAC encoding is one of the highest efficiency encoding method formats that takes hearing characteristics into consideration. Furthermore, by combining the AAC and SBR encoding formats, it is possible to transmit high-quality audio at low bit rates.

2.1.3.2.1 Advanced Audio Coding

The standardized AAC encoding parameters are listed in Table 2.6 [5], and the encoding parameters used in ISDB-T in Japan are listed in Table 2.7. The AAC standard can accommodate 12 types of sampling frequencies, and the channel composition is standard, accommodating everything from mono to seven channels. Also the placement of the speakers can be designated on the encoder: for example, two speakers in the front and two speakers to the side or the rear, etc. Also the coefficient for down mixing the multichannel signal to two channels can be designated in the encoder and then sent to the decoder, and it is possible to control the mixing balance with a playback device having only two channels.

2.1.3.2.2 SBR, an Extension Technology of AAC

SBR is a technology that expands the playback band of low bit rate-encoded audio signals and improves the audible quality. The format in which this SBR is applied

Table 2.6　MPEG-2 AAC (ISO/IEC13818-7) Audio Encoding Standard

Sampling frequency (kHz)	96, 88.2, 64, 48, 44.1, 32, 24, 22.05, 16, 12, 11.025, 8
Number of audio channels	1 (mono), 2, 3 (3/0), 4 (3/1), 5.1 (3/2 + LFE), 7.1 (5/2 + LFE)

Table 2.7　Audio Encoding Standards for ISDB-T

Audio encoding	MPEG-2 AAC (ISO/IEC 13818–7) LC (low complexity) profile
Sampling frequency (kHz)	48, 44.1, 32 (24, 22.05, 16)[a]
Number of quantified bits	16 bits and over
Audio channel number	1 (mono), 2, 3 (3/0), 4 (3/1), 5.1 (3/2 + LFE), as well as these combinations
Bit rate (reference value)	144 kbps/stereo

[a] The values in parentheses do not include BS digital broadcasting or wide-band CS digital broadcasting.

to AAC encoding technology (henceforth called AAC+SBR) has been standardized as MPEG-4 audio [6], and in addition, standardization has also been performed so that MPEG-2 AAC and SBR can be combined [7].

Through these standardizations, the sound quality at low bit rates of 64 kbps and lower has greatly improved. It is expected that, in addition to the field of broadcasting, AAC+SBR will also be widely adopted in the future for music distribution using 3G+ cellular telephones, as well as handheld audio/visual devices.

Furthermore, AAC+SBR is also called aacPlus or High-Efficiency AAC (HE-AAC), but in particular, HE-AAC is the name of the profile when SBR is applied to MPEG-4 AAC.

2.1.3.2.3 AAC Operating Specifications

The operating specifications for AAC and AAC+SBR for One-Seg are listed in Table 2.8. There are two AAC sampling frequencies: 48 kHz and a half sample of 24 kHz, but when the SBR option is in effect, operation is performed only at an AAC sampling frequency of 24 kHz. When the SBR is actually operating and the receiver has an AAC+SBR decoder installed, a wide-band sound decoded at a sampling frequency of 48 kHz will be reproduced. On the other hand, even in receivers that do not support SBR decoding, it is possible to reproduce AAC-decoded audio at a sampling frequency of 24 kHz, although the sound will be narrow-band.

Also, in AAC+SBR, operation at a low bit rate of around 48 kbps was the most effective and is the encoding technology that is best suited to the One-Seg service. However, if backward compatibility is taken into consideration, we can assume that a bit rate of 64 kbps/stereo would be effective.

2.1.4 Broadcasting Equipment and Partial Reception Technology

2.1.4.1 Broadcasting Equipment

An outline of the ISDB-T transmission equipment used by the broadcasting center (main station) and regional stations is shown in Figure 2.6. The transmission equipment in the broadcasting center is a large-range system that handles the provision

Table 2.8 One-Seg Audio Encoding Standards

Sampling frequency	48 kHz, 24 kHz (fixed at 24 kHz while SBR is in use)
Number of audio channels	1 (mono), 2 (stereo/mono)
Bit rate	48 kHz: 24–256 kbps (mono), 32–256 kbps (stereo)
	24 kHz: 24–96 kbps (mono), 32–96 kbps (stereo)

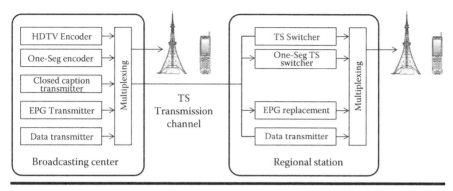

Figure 2.6   Outline of transmission equipment for ISDB-T.

of full-service DTTB and nationwide deployment, but here only the major elements will be listed. Each device performs real-time control based on program composition data.

ISDB-T was first applied to the HDTV service, and facilities were installed for that purpose, and subsequently, additions and improvements were made to the devices for One-Seg. First, H.264 encoders were added for low-frame-rate and low-resolution pictures for handheld terminal use. Besides this, additional improvements were performed for closed captions, EPG, data broadcasting, etc. This content is multiplexed, transmitted to the transmitter site, and simultaneously broadcast to the fixed reception HDTV service and the handheld-reception One-Seg service.

On the other hand, the multiplexed signals sent from the broadcasting center to the regional stations are transmitted using TS transmission lines. The regional stations have a relatively simple structure where they use TS switchers to switch between net and local. Regarding the One-Seg service that provides low-frame-rate and low-resolution pictures for handheld terminals, switching is performed using TS switchers for One-Seg programs. For both HDTV and One-Seg, EPG is replaced in order to accommodate local broadcasts, with each station transmitting its own data.

2.1.4.2 Partial Reception and Reducing Power Consumption

In a One-Seg handheld terminal running on a battery, reducing the power consumption is a major issue. In ISDB-T broadcasting, frequency interleaving for One-Seg is performed in one segment, and all of the One-Seg information can be obtained by receiving a portion of the OFDM signal (the one segment in the center of the 13-segment signal). Here, partial reception and reducing power consumption will be explained using Mode 3 as an example.

The ISDB-T signal is demodulated using a fast Fourier transform (FFT). As a general demodulation method, the reception signal is converted to the intermediate frequency of 8.127 MHz and then quadrature demodulated by sampling it at 32.5 MHz, which is four times the FFT sampling clock, to obtain a complex signal.

By decimating that signal to 1/4, a complex signal with a sampling of 8.127 MHz is obtained, and by performing complex FFT on 8192 points, 5617 carrier signals are demodulated.

On the other hand, when only the central segment is demodulated (partial reception), the reception signal with an intermediate frequency of 1.016 MHz (or 508 kHz) is quadrature demodulated with a sampling clock of 4064 MHz (or 2032 MHz), and through decimation, a complex base band sampling signal of 1016 MHz (or 508 kHz) is obtained, and by performing complex FFT on 1024 (or 512) points, 433 (432 for the one-segment carrier and the one SP carrier directly above) carrier signals are demodulated. Demodulation with low power consumption is possible since it is able to demodulate with a low sampling clock (a clock equal to 1/8 of the 13-segment reception) using a small-sized FFT.

In addition, when the battery has a capacity of approximately 850 mAh, the battery life for One-Seg reception is about four continuous hours, including One-Seg demodulation, display of the video image on the LCD, and audio generation.

2.1.5 Service Aspects

Since in ISDB-T the same programs are simultaneously broadcast to One-Seg and 12 segments, it is possible to watch the same content as that of the fixed reception HDTV service such as news, weather reports, sports, and drama at any time (see Figure 2.7). Also, with features like clear video and audio, One-Seg is expected to function as an information tool in the event of emergencies such as disasters. In addition, with closed captioning, TV programs can be enjoyed in places where the volume must be turned right down, such as on the train.

With the data broadcasting service, it is possible to view news and weather information while watching television. Information is viewed using the up/down and select keys on the receiver. The total transmission bit rate for One-Seg is 416 kbps, and the bit rate that can be used for One-Seg data transmissions is 50–70 kbps. It is significantly lower compared to the data broadcasting bit rates for BS digital broadcasts and ISDB-T fixed reception broadcasts using 12 segments (1.5–2.5 and 1–1.5 Mbps, respectively). Considering the operability of One-Seg data reception, the time for completing data acquisition after selecting the channel should be within 4 s. The amount of data at 50 kbps over 4 s is 25 kB. Therefore, data transmission programs are mainly for text-based content.

When the content consists of too much data due to the use of images, it is not possible to send all of the data over a broadcast channel. To supplement this, there is a system whereby telecommunications are used to connect to the broadcast station's server and the data content is displayed via download. The handheld terminal's Internet connection function is used to connect to the server listed as a link in the data broadcasting content, and the downloaded BML or image files are displayed. In those cases where links are followed from the broadcast data, the downloaded content may be displayed, and then displayed on the same screen as the broadcast video.

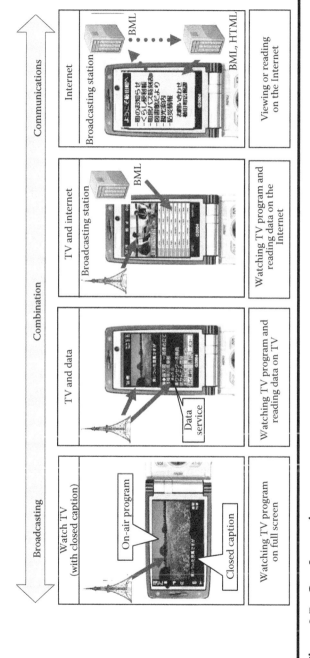

Figure 2.7 One-Seg services.

On the other hand, for the link button in the data broadcasting content provided by the broadcaster, when Internet content provided by other than the broadcaster is downloaded and displayed, in order to distinguish it from the data broadcasting content, it is recommended that the broadcast video and data content be hidden and that only the Internet content be displayed on the screen.

2.1.6 Emergency Warning Broadcast System

The number of cellular phone subscribers is 125.56 million (as of the end of December 2011), more than 98% of the population. Therefore, the provision of One-Seg receiving functions is regarded as being necessary for cellular phones to play a major role particularly during emergencies or in the event of a disaster. The emergency warning broadcasting system features handheld terminals with the One-Seg function that automatically wake up when a disaster such as an earthquake or a tsunami occurs and provide essential information to the users. Unlike communication networks that have the potential for congestion, information can be instantly transmitted to a large number of One-Seg handheld terminals. Therefore, as shown in Figure 2.8, automatically waking up the handheld terminals with an emergency warning broadcasting signal is also an extremely important issue for the disaster preparedness of national and local governments.

To receive emergency warning information, low power consumption while waiting (in standby mode) is necessary. In the ISDB-T format, as shown in Figure 2.9, an emergency warning information signal is placed in the 26th TMCC bit. During an emergency warning broadcast, that bit becomes "1."

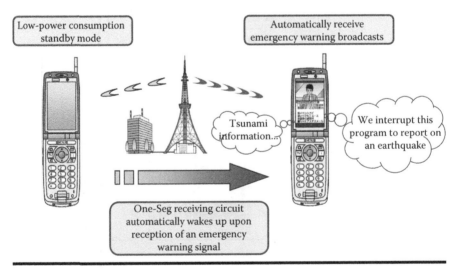

Figure 2.8   Handheld terminal automatically waking up upon reception of emergency warning broadcast.

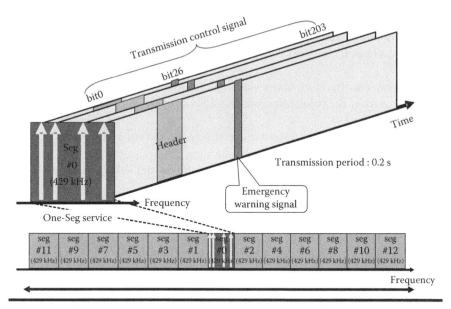

Figure 2.9 **Placement of emergency warning signals in ISDB-T signal.**

To receive an emergency warning broadcast, the terminal must always be watching for this signal, and when the bit becomes "1," it presents the content of the broadcast program while broadcasting an alarm indicating that it is an emergency warning broadcast. Since it is necessary to constantly watch for the emergency warning signal, this must be possible without creating a major power drain on the battery-operated handheld terminal.

To reduce energy consumption in standby mode, the NHK Science & Technology Research Laboratories (STRL) developed a receiver device that selects and receives only the TMCC carrier from among the 432 One-Seg carriers and intermittently monitors the emergency warning broadcasting signal, which is broadcast approximately every 0.2 s. It is estimated that, with this circuit, compared to the energy consumption when the One-Seg receiver circuits are used as is, low energy consumption at a rate of about 1/10 will be possible.

2.2 ISDB-T$_{SB}$ Sound Broadcasting

2.2.1 Introduction

In Japan, the frequency band between 90 and 108 MHz that was previously used for VHF analog TV channels 1–3 is being studied for use for the mobile multimedia broadcasting (MMB) system for mobile or handheld terminals (VHF-low MMB system). This system is intended to provide broadcasting services to regional areas (local block areas, not nationwide) and is expected to contribute to the securing of regional promotion and regional information.

The requirements for VHF-low MMB include those listed as follows:

- Services can freely combine video, audio, and data and provide file-based services over the broadcast waves in addition to ordinary broadcasting (real-time services)
- Diversity and flexibility for advanced services and expandability to accommodate future new services
- Prompt broadcast or transmission of wake-up control signals and messages to the target receivers in the event of an emergency
- Compatibility with One-Seg services
- Functions for protecting intellectual property rights
- Contribution to efficient use of frequencies
- Consistency or compliance with international standards

ISDB-T_{SB} has been selected as a system that satisfies these requirements for the MMB using the VHF-low band. In this section, the transmission scheme, media encoding, file-based broadcasting system, and the assumed services for the ISDB-T_{SB} system are described.

2.2.2 Transmission Scheme (Connected Transmission)

2.2.2.1 Features of ISDB-T_{SB}

From the view point of facilitating the development of receivers for VHF-low MMB, it is important that transmission system for VHF-low MMB has commonality with the transmission system for One-Seg, which is the mobile TV service and a part of the DTTB service using ISDB-T. Therefore, the ISDB-T_{SB} transmission system was selected. ISDB-T_{SB} has the same segment structure as ISDB-T. ISDB-T_{SB} uses "OFDM segments," i.e., units of OFDM carriers corresponding to 6/14 MHz. A segment group is formed from one or three segments, so the format is called the 1-segment or 3-segment format, respectively. The transmission spectrum is formed by combining a number of 1-segment and 3-segment formats without guard bands (see Figure 2.10).

The 3-segment format supports hierarchical transmission, which is a two-layer transmission consisting of one segment in the center of the band and two other segments. Transmission parameters such as the carrier modulation scheme, encoding rate of the inner code, and the time interleaving length can be specified separately for each layer. By performing frequency interleaving only within the center segment, it is possible for 1-segment receivers to receive part of the service transmitted with the center segment. The transmission parameters for the 1-segment and 3-segment formats are listed in Tables 2.9 and 2.10, respectively.

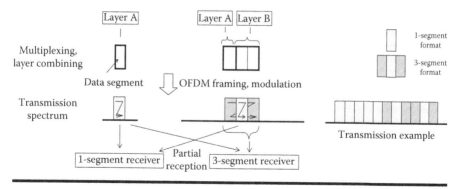

Figure 2.10 Partial reception and transmission spectrum in the ISDB-T$_{SB}$ system.

Table 2.9 Transmission Parameters for the ISDB-T$_{SB}$ 1-Segment Format

Transmission Parameter	Mode 1	Mode 2	Mode 3
Bandwidth	428.57… kHz (6 MHz × 1/14)		
Carrier spacing	3.97 kHz (250/63)	1.98 kHz (125/63)	0.99 kHz (125/126)
Number of carriers			
Total	109	217	433
Data	96	192	384
Pilot	13	25	49
Effective symbol length (µs)	252	504	1008
FFT sampling rate	64/63 = 1.0158… MHz		
Guard interval length	1/4, 1/8, 1/16, 1/32		
Number of symbols/frame	204		
Modulation method	QPSK, 16QAM		
Inner code	Convolutional code (coding rates = 1/2, 2/3)		
Outer code	(204, 188) Reed–Solomon code		
Time interleaving	0, 4, 8, 16, 32	0, 2, 4, 8, 16	0, 1, 2, 4, 8

Table 2.10   Transmission Parameters for the ISDB-T$_{SB}$ 3-Segment Format

Transmission Parameter	Mode1	Mode 2	Mode 3
Bandwidth	1285.71… kHz (6 MHz×3/14)		
Carrier spacing	3.97 kHz (250/63)	1.98 kHz (125/63)	0.99 kHz (125/126)
Number of carriers			
Total	325	649	1297
Data	288	576	1152
Pilot	37	73	145
Effective symbol length (μs)	252	504	1008
FFT sampling rate	128/63 = 2.0317… MHz		
Guard interval length	1/4, 1/8, 1/16, 1/32		
Number of symbols/frame	204		
Modulation method	QPSK, 16QAM		
Inner code	Convolutional code (coding rates = 1/2, 2/3)		
Outer code	(204, 188) Reed–Solomon code		
Time interleaving	0, 4, 8, 16, 32	0, 2, 4, 8, 16	0, 1, 2, 4, 8

2.2.2.2 Multiplexing

The ISDB-T$_{SB}$ multiplexing scheme is based on the MPEG-2 TS. The protocol stack is shown in Figure 2.11. The transmission scheme adopted for real-time broadcasting is the same as that used for ISDB-T. The transmission scheme adopted for file-based broadcasting that does not use Internet Protocol (IP) packets is the Digital Storage Media Command and Control data carousel transmission scheme that is used for current data broadcasting [8]. The transmission scheme adopted for file-based broadcasting using IP packets is the transmission method that efficiently encapsulates MPEG-2 TS by using the header compression method that is used in the Type Length Value (TLV) multiplexing scheme [9] or Robust Header Compression (ROHC) [10], Figure 2.12. As a result, ISDB-T$_{SB}$ provides compatibility with the current DTTB system and transmits not only IPv4 packets but also IPv6 packets. The transparent transmission of IP packets makes it possible to

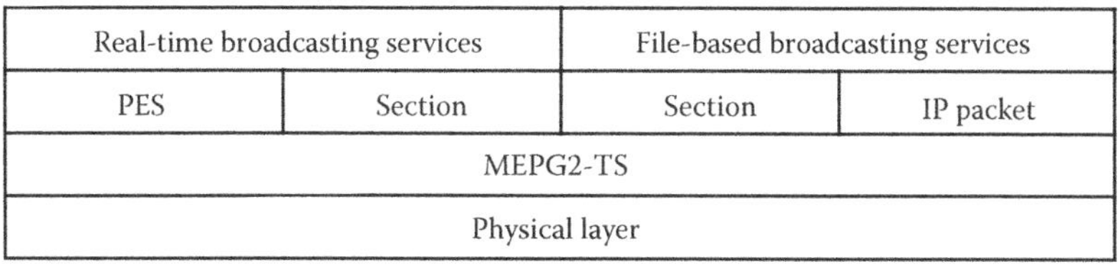

Real-time broadcasting services		File-based broadcasting services	
PES	Section	Section	IP packet
MEPG2-TS			
Physical layer			

Figure 2.11   ISDB-T$_{SB}$ protocol stack for ISDB-T$_{SB}$.

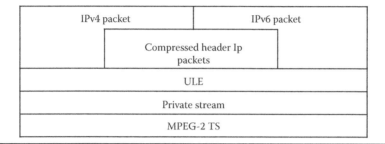

Figure 2.12   ISDB-T$_{SB}$ protocol stack for the IP packet multiplexing scheme.

transmit any type of IP packet, including the expanded header packets. However, the IP header that may not be necessary for the broadcast transmission path may lead to an increase in the transmission overhead. Therefore, for IP packets that have an IP/User Datagram Protocol (UDP) header, which is the packet format usually used for one-way transmission, the header is compressed to suppress the increase in the transmission overhead. Because it is assumed that receivers with a communication function will become the mainstream, as well as receivers without a communication function, so ISDB-T$_{SB}$ can use ROHC, which can be used together with communication functions, and the TLV multiplexing scheme, which eliminates the communication unit and realizes a highly efficient header compression rate.

2.2.3 Media Coding

Video and audio coding schemes should be as compatible as possible with One-Seg to share receivers. So ISDB-T$_{SB}$ uses video coding method complying with ITU-T Recommendation H.264 | ISO/IEC 14496-10 and audio coding method complying with MPEG-2 AAC Audio (ISO/IEC13818-7) and MPEG Surround (ISO/IEC 23003-1). The encoding parameters for the real-time services and file-based services are shown in Tables 2.11 and 2.12, respectively. MPEG Surround is backward compatible with MPEG-2 AAC, so an MPEG Surround audio encoding code stream can be decoded as stereo or monaural audio by an MPEG-2 AAC decoder.

Table 2.11   ISDB-T$_{SB}$ Media Coding for Real-Time Broadcasting

Video coding	
Codec	ITU-T H.264/MPEG-4 AVC
Profile	Baseline
Level (maximum)	1.3
Resolution (maximum)	320×180 (QVGA)
Frame rate (maximum)	30 fps
Audio coding	
Codec	MPEG-2 AAC+SBR+PS, Surround
Number of channels (maximum)	5.1

Table 2.12   ISDB-T$_{SB}$ Media Coding for File-Based Broadcasting

Video coding	
Codec	ITU-T H.264/MPEG-4 AVC
Profile	Main
Level (maximum)	3
Resolution (maximum)	720×480 (525SD)
Frame rate (maximum)	30 fps
Audio coding	
Codec	MPEG-2 AAC+SBR+PS, Surround
Number of channels (maximum)	5.1

2.2.4 *File-Based Broadcasting*

The file-based broadcasting service is based on the premise that programs are not watched until after being received and stored on the receiver. Two types of transmission methods for file-based broadcasting are proposed, one being the data carousel transmission method and the other being the IP transmission method. File-based broadcasting services using the data carousel transmission method are regarded as being an extension of the current data broadcasting services. On the other hand, file-based broadcasting using the IP packet transmission method is assumed to provide diverse new services. The data carousel transmission method is explained in the next section.

2.2.4.1 Data Carousel Transmission Method

For cellular phone terminals, broadcasters may provide a downloading or streaming program via a communication line. But when communication networks are used to provide such services to many recipients, the transmission time may increase and the quality of the services may degrade because of the transmission capacity and facilities provided by the communication networks. On the other hand, file-based broadcasting services using broadcast waves can transmit programs to many users at the same time. It is also assumed that there is a demand for file-based services designed for in-vehicle receivers and other handheld terminals that do not have communication functions. Therefore, the data carousel transmission method for file-based broadcasting services using only broadcast waves is being studied (Figure 2.13).

The broadcasting station divides program files into units called Download Data Blocks (DDBs) and then further divides the DDB into TSPs, which are transmitted sequentially. The receiver recomposes the DDBs from the received TSPs and then decodes the program file from the stored DDB. If one or more of the TSPs that constitute a DDB is not received, the receiver discards that DDB. VHF-low MMB, on the other hand, assumes reception in automobiles, trains, or other moving vehicles or while walking. In those cases, the received power is decreased due to radio attenuation by buildings, etc., and the received signal becomes depleted due to multipath fading. In a mobile environment in which reception is unstable, failure in file downloading due to lost TSPs is assumed, so the broadcasting station broadcasts the same files repeatedly. The receiver stores received DDBs successfully and complements missing DDBs in subsequent broadcasts.

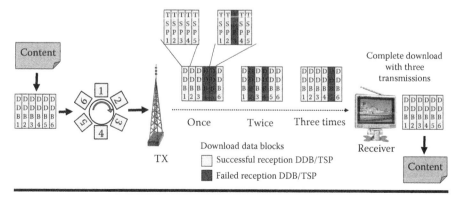

Figure 2.13 **Data carousel transmission method.**

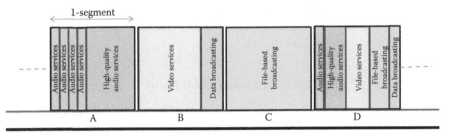

Figure 2.14   Example of segment use in VHF-low MMB.

2.2.5 Service Aspects

VHF-low MMB provides a range of services, including (A) multiple audio services, (B) video services, and (C) file-based services (see Figure 2.14). Multiple audio services are provided within one segment. Video services with data broadcasting are provided in the same way as for One-Seg, but it is possible to have a video frame rate of up to 30 fps. As one example of a file-based service, high-quality and high-data-rate programs are transmitted over more time than the program playback time. This approach enables video services beyond the data rate that can be achieved with real-time broadcasting like (A) and (B). Another example is high-speed downloading at a low data rate, such as audio programs or text-only news programs. Furthermore, it is possible to transmit different services such as audio, video, data broadcasting and file-based services by multiplexing them within one segment like (D).

2.3 ISDB-Tmm Multimedia Broadcasting

The Integrated Services Digital Broadcasting for Terrestrial Multimedia (ISDB-Tmm) is an MMB system based on ISDB-T. The physical layer transmission system of ISDB-Tmm is compatible with ISDB-T, and it has excellent performance even under severe wireless channel conditions of mobile environments. ISDB-Tmm provides high-quality audio visual streaming services, file-based multimedia broadcasting services, and integrated communications and broadcasting services to mobile terminals. This section gives an overview of the services and technologies of ISDB-Tmm.

2.3.1 Transmission Scheme

ISDB-Tmm is based on ISDB-T, and it offers excellent reception performance in mobile environments and is resilient against multipath propagation. It has enhanced multimedia capabilities for, for example, large media file delivery.

2.3.1.1 Radio Transmission

The ISDB-Tmm transmission system can utilize any combination of the 13-segment format, which has a 6/7/8 MHz bandwidth and is compatible with ISDB-T, and 1-segment format, which has 1/13th the bandwidth of the 13-segment format and is compatible with One-Seg and digital radio (Table 2.13). The remaining parameters are shared with ISDB-T and ISDB-T$_{SB}$. Japan has allocated 14.5 MHz from 207.5 to 222 MHz in the VHF band to ISDB-Tmm. This spectrum became available when analog terrestrial TV broadcasting ends in March 2012. Two 13-segment formats and seven 1-segment formats are fit into the bandwidth (Figure 2.15).

2.3.1.2 Multiplex

Figure 2.16 shows the ISDB-Tmm protocol stack. The protocol stack for real-time service is designed to be compatible with ISDB-T, considering the implementation to the One-Seg capable terminals. Moreover, since long-duration interleaving is good for large file transmissions, the file-based service uses the AL-FEC (Application Layer—Forward Error Correction) on File Delivery over Unidirectional Transport (FLUTE), in addition to the radio physical layer error correction.

Real-time services and file-based services are packetized and multiplexed into an MPEG-2 TS. The transmission bit rate can be instantaneously optimized without having to allocate a fixed transmission rate for each service.

2.3.2 File-Based Broadcasting

For the file-based broadcasting, protocols were added on the existing ISDB-T that are required for the file download service over MPEG-2 systems such as AL-FEC, FLUTE [10], UDP/IP [11–13], Robust Header Compression (ROHC) [14], and Unidirectional Light-weight Encapsulation (ULE) [15]. The process from contents file to MPEG-2 TS is shown in Figure 2.17.

2.3.2.1 Technologies for File Download Service

2.3.2.1.1 AL-FEC

AL-FEC is applied to the file-based broadcasting to improve the robustness against transmission channel impairments. The overview of the AL-FEC mechanism for the ISDB-Tmm is shown in Figure 2.18.

The content file is split into source symbols s1–s6 with a predefined byte length. The AL-FEC encoder creates parity symbols p1–p6 from the source symbols. Then the transmitter sends both source and parity symbols to the receiver. Even if the receiver does not receive the entire source symbols, the lost symbols can be recovered from successfully received source and parity symbols. For the efficient

Table 2.13   ISDB-Tmm Transmission System

Spectrum Usage	Any Combination of 13-Seg/1-Seg Format
Bandwidth/segment (Bws)	6000/14 = 428.57 kHz, 7000/14 = 500 kHz, 8000/14 = 571.428 kHz
Number of radiated carriers/ segment	108 (Mode 1), 216 (Mode 2), 432 (Mode 3)
Modulation	OFDM (DQPSK, QPSK, 16QAM, 64QAM)
FEC	Outer Code: Reed–Solomon (204, 188)
	Inner code: Convolutional code (7/8, 5/6, 3/4, 2/3, 1/2)
Hierarchical transmission	Possible
SFN operation	Possible
Active symbol duration	6, 7, 8 MHz
	252, 216, 189 µs (Mode 1)
	504, 432, 378 µs (Mode 2)
	1008, 864, 756 µs (Mode 3)
Carrier spacing (Cs)	6, 7, 8 MHz
	Bws/108 = 3.968, 4.629, 5.271 kHz (Mode 1)
	Bws/216 = 1.984, 2.361, 2.645 kHz (Mode 2)
	Bws/432 = 0.992, 1.157, 1.322 kHz (Mode 3)
Guard interval duration	1/4, 1/8, 1/16, 1/32 of active symbol duration
	63, 31.5, 15.75, 7.875 µs (Mode 1: 6 MHz)
	126, 63, 31.5, 15.75 µs (Mode 2: 6 MHz)
	252, 126, 63, 31.5 µs (Mode 3: 6 MHz)
Transmission frame duration	204 OFDM symbols
Inner interleaving	Intra and inter segments interleaving (frequency interleaving). Symbol-wise convolutional interleaving 0, 380, 760, 1520 symbols (time interleaving).
Outer interleaving	Byte-wise convolutional interleaving, I = 12

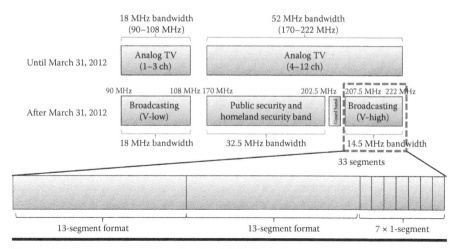

Figure 2.15 Frequency allocation for ISDB-Tmm in Japan.

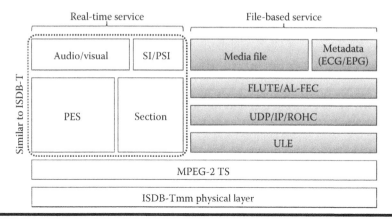

Figure 2.16 ISDB-Tmm protocol stack. AL-FEC, application layer-forward error correction; ECG, electronic contents guide; EPG, electronic program guide; FLUTE, file delivery over unidirectional transport; IP, internet protocol; PES, packetized elementary stream; PSI, program specific information; ROHC, robust header compression; SI, service information; TS, transport stream; UDP, user datagram protocol; ULE, unidirectional lightweight encapsulation.

file-based transmission of large size contents over error-prone broadcast channels, the ISDB-Tmm employs low-density parity check-based AL-FEC and achieves higher reception probability than conventional carousel-type repetitive transmission. Figure 2.19 shows the advantage of the AL-FEC against the conventional carousel-type repetition under the same transmission overhead (coding rate = 1/2).

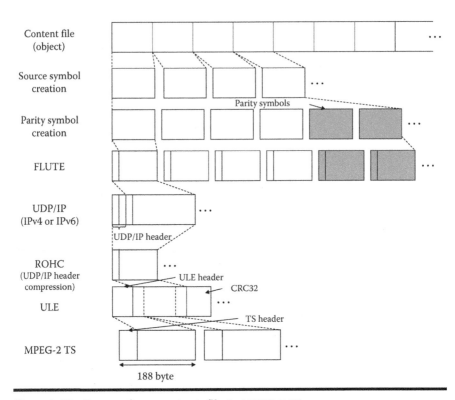

Figure 2.17   Process from contents file to MPEG-2 TS.

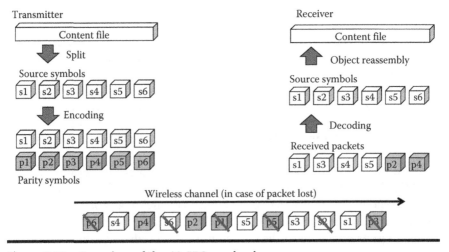

Figure 2.18   Overview of the AL-FEC mechanism.

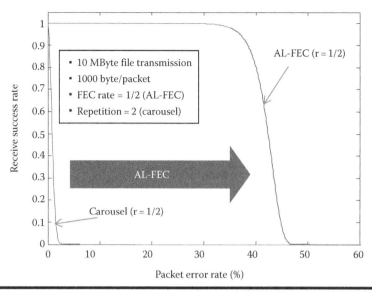

Figure 2.19 High-efficiency file transmission by AL-FEC.

2.3.2.1.2 FLUTE

FLUTE is a session management protocol for unidirectional transmission channel. An example of the mapping from FLUTE session into MPEG-2 TS is shown in Figure 2.20. In ISDB-Tmm, media content consists of a set of file components, such as manifest file, which describes file management information, and the

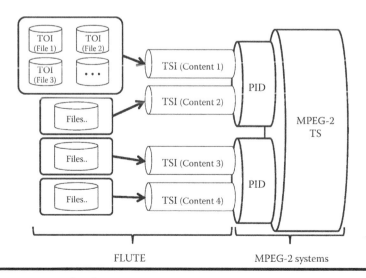

Figure 2.20 Mapping from FLUTE to MPEG-2 systems.

associated movie, audio, and hypertext files. Each file component is managed by Transport Object Identifier (TOI). The media contents are managed by Transport Session Identifier (TSI). The TSI is mapped into Packet Identifier (PID), and transmitted over the MPEG-2 Systems. In addition, File Delivery Table (FDT) instance is periodically transmitted to notify receivers of file attributions of the ongoing sessions.

2.3.2.1.3 UDP/IP/ROHC

The session information and the content data are packetized by adding UDP header and IP header by FLUTE. Taking into account the low possibility of the information being changed in UDP/IP headers, ISDB-Tmm uses the ROHC header compression algorithm. By ROHC, IP packet flows with the same header information (next header, source/destination IP addresses, and port numbers) are managed as one context. ROHC enables effective packet compression by transmitting full information of UDP/IP header only when the contents of UDP/IP headers are updated. By applying ROHC, 28 bytes of UDP/IP headers are compressed down to 1 byte.

2.3.2.1.4 ULE

In order to transmit IP packets over the MPEG-2 Systems, the ULE encapsulation protocol is applied. For this purpose, the EtherType value in the ULE header for ROHC is assigned as "$0 \times 22F1$" by IEEE.

2.3.2.1.5 File Repair over Communication Networks

When a receiver cannot reassemble the whole contents after AL-FEC decoding over the broadcasting channels, the file repair procedures can be executed over the bidirectional communication networks (Figure 2.21). In such a case, the receiver requests the lost parts of the source symbols to the file repair server using HTTP content-range options over communication networks, such as cellular networks (Figure 2.22). By applying HTTP for file repair, widely used World Wide Web server can be used as the file repair server.

An enhancement of this basic file repair procedures are introduced by ISDB-Tmm for minimizing packet traffic over the communication networks by requesting the minimum set of source symbols. This algorithm suppresses approximately 50% of the traffic load for file repair [16].

2.3.3 Media Coding

For real-time broadcasting, audio and visual media can be transmitted in the same way as in conventional digital TV broadcasting. As shown in Table 2.14,

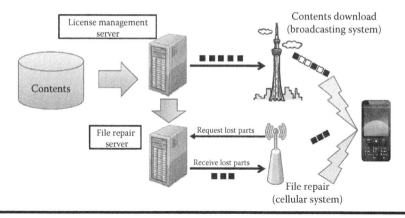

Figure 2.21   ISDB-Tmm file repair overview.

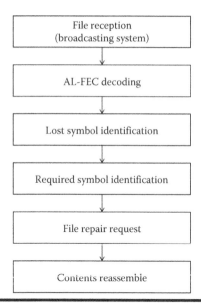

Figure 2.22   Process of file repair in ISDB-Tmm.

ISDB-Tmm has high-quality media coding, including multichannel audio, using ITU-T H.264 and MPEG Audio.

File-based broadcasting is capable of sending large media files encoded at high bit rates, such as H.264 high profile level 4.2 (Table 2.15).

ISDB-Tmm can keep up to the state-of-the-art new applications and new file formats for mobile phone terminals, as it embodies Multipurpose Internet Mail Extensions-type information as a flexible framework for handling terminals' and mobile carriers' specific options.

Table 2.14   ISDB-Tmm Media Coding for Real-Time Broadcasting

Video coding	
Codec	ITU-T H.264/MPEG-4 AVC
Profile	Baseline or Main
Level (maximum)	3.0
Resolution	720 × 480 (525SD)
Frame rate (maximum)	30 fps
Audio coding	
Codec	MPEG-2 AAC + SBR + PS, Surround
Number of channel	Monaural/Stereo/5.1ch
Sampling rate (kHz)	48, 44.1, 32, 24, 22.05, 16

Table 2.15   ISDB-Tmm Media Coding for File-Based Broadcasting

Video coding	
Codec	ITU-T H.264/MPEG-4 AVC
Profile	High
Level (maximum)	4.2
Resolution	1980 × 1080 (1080HD)
Frame rate (maximum)	60 fps
Audio coding	
Codec	MPEG-2 AAC + SBR + PS, Surround
	MPEG-4 ALS/SLS (option)
Number of channel	Monaural/Stereo/5.1ch
Sampling rate (kHz)	48, 44.1, 32, 24, 22.05, 16

2.3.4 Metadata

Two types of metadata are defined for ISDB-Tmm, as follows:

▪ Electronic Program Guide/Electronic Content Guide (EPG/ECG) metadata: This metadata describes information related to EPG and ECG used for content navigation (browsing/searching, purchasing, viewing, etc.)

■ Transmission control metadata: This metadata describes information related to reception, download, and file repair necessary for file-based broadcasting content

These two types of metadata enable integrated handling of real-time broadcast and file-based broadcasting services. Metadata is expressed in the eXtensible Markup Language.

Based on the specifications already defined in ARIB STD-B38 [17], ETSI TV-Anytime Phase 1 [18] and Phase 2 [19], the metadata elements in EPG/ECG have been enhanced for multimedia broadcast operations. The EPG/ECG metadata mainly describes program schedules including detailed content information, such as title, content synopsis, fees, usage conditions, and scene descriptions. The enhancements made for multimedia broadcasting also enable coupon information related to content to be described and distributed to receiving devices.

Based on the specification [20] defined in 3GPP Multimedia Broadcast/ Multicast Service, the transmission control metadata has also been enhanced for multimedia broadcasting. The transmission control metadata describes not only a set of control parameters used in FLUTE sessions for file-based broadcasting services but also another set of control parameters used in repair sessions for erroneously received content. These multimedia broadcasting enhancements also enable flexible program scheduling (e.g., splitting a program into three parts to be transmitted in the morning, at noon, and in the evening) and also allow load balancing (e.g., peak shift and peak shaping) of communication traffic generated by repair sessions for erroneously received content.

As shown in Figure 2.23, receiving devices can get metadata via either the broadcast network or the cellular network. In the broadcast network, metadata encoded in Binary format for MPEG-7 [21] is delivered. On the other hand, in the cellular network, metadata is received as web contents from the metadata server operated by the broadcasters. Especially for metadata transmission via broadcasting, considering the bandwidth required for metadata transmission and how often receiving devices download metadata and so forth, two methods have been defined—continuous small-volume metadata transmission and scheduled large-volume metadata transmission.

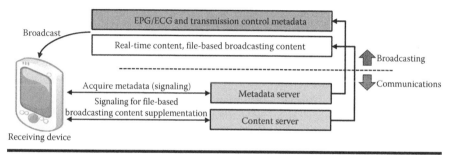

Figure 2.23   Content/metadata delivery system.

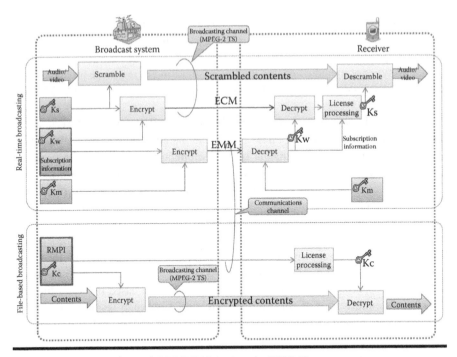

Figure 2.24 Overview of CAS/DRM system in ISDB-Tmm.

2.3.5 CAS/DRM

The Conditional Access System (CAS) is used to restrict access to the real-time service; user subscription information is used to descramble the broadcasted content. The Digital Rights Management system (DRM) is used by the file-based service to enable restricted playback of encrypted content that is stored in the terminal. In both cases, the contents are broadcasted through the broadcasting channel, and the license information is transmitted through the communication channel. Figure 2.24 shows an overview of the access control system.

The relevant content security schemes are listed and compared with ISDB-T in Table 2.16.

2.3.6 Service Aspects

Multimedia broadcasting enables access to contents and services without the users having to care about broadcast schedules or place of use. It expands the capabilities of DTTB with communication technology for mobile receivers. As shown in Figure 2.25, there are two types of multimedia broadcasting: (1) high-quality real-time broadcasting offering video, audio, and data; and (2) file-based broadcasting that can store various combinations of video, audio, images, text, and data.

Table 2.16   Comparison of Content Security Schemes (Access Control)

	ISDB-Tmm	*ISDB-T*
Real-time broadcasting		
Scrambling	Scrambling key change (several seconds to several tens of seconds) MPEG2-TSP unit	
Encryption algorithms	Select from MULTI2 64 bit, AES 128 bit and Camellia 128 bit	MULTI2 64 bit
Standardization	MIC Notice No. 302	
	ARIB (Association of Radio Industries and Businesses) STD-B25	
File-based broadcasting		
Encryption scheme	Mainly in units of content, usually files	Not used
Encryption algorithms	Not specified	Not used
Standardization	ARIB STD-B25: AES 128 bit or equivalent recommended	ARIB STD-B25: at least DES or equivalent recommended
ITU international standardization status	Nothing directly, but it is possible to refer to ARIB STD-B25 from BT.1306 System C	

ISDB-Tmm provides coordinated broadcasting and communication services; its broadcasting channels can transmit a large amount of content, and its communication channels can be used by viewers to purchase licenses, recommend content, and post messages (Figure 2.26). When content cannot be completely transmitted over the broadcasting channel, the remaining portion of the content can be obtained from the communication channel.

2.4   Conclusions

In this chapter, the family of ISDB-T-based DTTB for handheld devices, such as One-Seg, ISDB-T$_{SB}$, and ISDB-Tmm, is described. The systems have common transmission parameters, using 13/3/1 segment OFDM radio modulation.

In Japan, ISDB-T One-Seg broadcasting was launched on April 1, 2006, using a part of ISDB-T HDTV broadcasting signal. As of February 2012, approximately 124 million One-Seg capable cellular phones have been shipped.

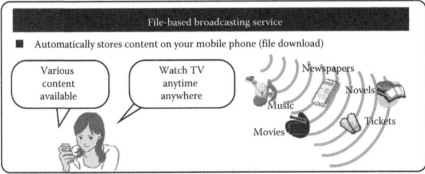

Figure 2.25 ISDB-Tmm services.

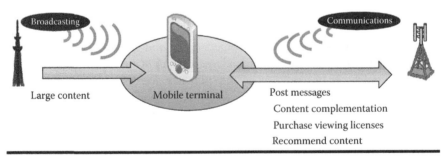

Figure 2.26 ISDB-Tmm coordinated broadcasting and communication service.

ISDB-T$_{SB}$ has been selected as a system providing broadcasting services to regional areas, using VHF-low (90–108 MHz) spectrum band. The launch of the ISDB-T$_{SB}$ broadcasting is currently under discussion. ISDB-Tmm is used for VHF-high (207.5–222 MHz) nationwide MMB services. The commercial service was launched on April 1, 2012.

References

1. Association of Radio Industries and Businesses (ARIB) STD-B31, Transmission system for digital terrestrial television broadcasting, November 2005.
2. Association of Radio Industries and Businesses (ARIB) TR-B14, Operational guidelines for digital terrestrial television broadcasting, May 2006.
3. ISO/IEC 13818-1:2000, Information technology—Generic coding of moving pictures and associated audio information: Systems, 2000.
4. ISO/IEC 14496-10:2003, Information technology, Coding of audio-visual objects—Part 10: Advanced video coding | ITU-T recommendation H.264, Advanced video coding for generic audiovisual services, 2003.
5. ISO/IEC 13818-7/2003; Advanced audio coding (AAC), 2003.
6. ISO/IEC 14496-3/2001 AMD1; Bandwidth extension, 2003.
7. ISO/IEC 13818-7/2003 AMD1; Embedding of bandwidth extension, 2004.
8. Association of Radio Industries and Businesses (ARIB) STD-B24, Data coding and transmission specification for digital broadcasting, May 2006.
9. Rec. ITU-R BT. 1869, Multiplexing scheme for variable-length packets in digital multimedia broadcasting systems, March 2010.
10. T. Paila, M. Luby, R. Lehtonen, V. Roca, and R. Walsh, FLUTE—File delivery over unidirectional transport, IETF RFC3926, October 2004.
11. J. Postel, User datagram protocol, IETF RFC 768, August 1980.
12. J. Postel, Internet protocol, IETF RFC 791, September 1981.
13. S. Deering and R. Hinden, Internet protocol, version 6 (IPv6) specification, IETF RFC 1883, December 1995.
14. C. Bormann et al., RObust header compression (ROHC): Framework and four profiles: RTP, UDP, ESP, and uncompressed, IETF RFC 3095, July 2001.
15. G. Fairhurst and B. Collini-Nocker, Unidirectional lightweight encapsulation (ULE) for transmission of IP datagrams over an MPEG-2 transport stream (TS), IETF RFC 4326, December 2005.
16. H. Matsuoka, A. Yamada, and T. Ohya, Low density parity check code extensions applied for broadcast-communication integrated content delivery, *Proceedings of the ITC Specialist Seminar on Multimedia Applications—Traffic, Performance and QoE*, Miyakazi, Japan, 2010.
17. Association of Radio Industries and Businesses (ARIB) STD-B38, Coding, transmission and storage specification for broadcasting system based on home servers, July 2012.
18. ETSI TS 102 822-5 v1.1.1, Broadcast and on-line services: Search, select, and rightful use of content on personal storage systems ("TV-anytime phase 1"); Part 5: Rights management and protection (RMP) information for broadcast applications, March 2005.
19. ETSI TS 102 822-8 v1.5.1, Broadcast and on-line services: Search, select, and rightful use of content on personal storage systems ("TV-anytime"); Part 8: Phase 2—Interchange data format, November 2011.
20. 3GPP TS 26.346 v.11.0.0, Multimedia broadcast/multicast service (MBMS); Protocols and codecs, March 2012.
21. ISO/IEC 23001-1:2006, MPEG systems technologies—Part 1: Binary MPEG format for XML, 2006.

Chapter 3

Overview of the South Korean Advanced T-DMB Mobile Broadcasting System

Kyu Tae Yang, Young Su Kim,
Byungjun Bae, Kwang Yong Kim, Joungil Yun,
Hun Hee Lee, and Yun Jeong Song

Contents

3.1 Introduction

T-DMB (Terrestrial–Digital Multimedia Broadcasting) is the first commercialized terrestrial mobile broadcasting technology in the world. T-DMB was designed to provide various types of multimedia services in mobile environment including audio and video (A/V) service, interactive data service, and audio-only service, based on the Eureka-147 DAB system [1–3]. Moreover, it is expected that broadcasting and telecommunication convergence service could be realized when T-DMB is linked with mobile communication networks such as cellular or WiMAX. There are several broadcasting systems available in the global mobile broadcasting market. Although each system has its own strengths and weaknesses, T-DMB has distinguishable characteristics compared to the others.

One of the benefits for adopting T-DMB is an easiness of service employment due to the structural inheritance of the Eureka-147 DAB that has been in commercial service for the past decade in many European countries. The other benefit is that the cost for establishing T-DMB network is relatively low due to simplicity of the system and its optimal frequency band.

On the other hand, from the frequency efficiency point of view, T-DMB has a weakness compared with other mobile broadcasting systems because DAB was originally designed for audio and its related data service. Accordingly, it has been argued that the number of available services within the allocated frequency band is not sufficient enough to support a wide range of business opportunities.

The project "Development of Advanced T-DMB (AT-DMB) Technology" was launched to overcome such a weakness and enhance the system performance while guaranteeing backward compatibility with the existing T-DMB. AT-DMB was successfully accomplished after researching and developing it for more than 3 years.

This chapter provides an overview of the AT-DMB technology. The rest of the chapter is structured as follows. Section 3.2 describes the background of AT-DMB. Section 3.3 presents the key technologies of AT-DMB. Section 3.4 presents illustrative performance evaluation results of AT-DMB from simulations and field trials. Finally, the chapter is concluded with Section 3.5.

3.2 Background

3.2.1 Concept of AT-DMB

Since T-DMB system is based on the Eureka-147 DAB system, it has the characteristics of the DAB system such as good mobile reception, power-efficient transmission, low power dissipating receiver, etc. AT-DMB also has these characteristics because it simply adds a newly modulated signal on T-DMB signal. Due to this modulation mechanism, AT-DMB can increase its data rate up to 2.304 Mbps in 1.536 MHz bandwidth, twice the data rate of T-DMB.

Since the existing T-DMB uses π/4-DQPSK (differential quadrature phase-shift keying) modulation, AT-DMB adopted 16QAM (quadrature amplitude modulation) and 8APSK (amplitude phase-shift keying) using the hierarchical modulation scheme.

Figure 3.1 shows the basic concept of AT-DMB. As shown in the figure, there are two layers called the base layer (BL) and the enhancement layer (EL). The BL is the T-DMB channel and the EL is the channel added for AT-DMB. Because of the hierarchical modulation, the two layers can deliver different contents each. Commercial T-DMB receivers can get only the BL signal, but AT-DMB receivers can get both the BL and the EL signals, providing thus more services. The data rate of AT-DMB depends on its applied hierarchical modulation scheme. When 16QAM is used, the EL can carry the same amount of data with the BL. But when 8APSK is used, the EL can carry half the amount of data with the BL.

3.2.2 Standardization Status

Since the specification on T-DMB video service [1] was released in June 2005 as an ETSI standard, DAB voice applications [4] became an ETSI standard

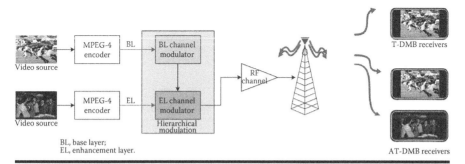

Figure 3.1 Basic concept of AT-DMB.

in November 2008. T-DMB middleware called DMB MATE (Multimedia Application Terminal Environment) was also announced as an ETSI standard in August 2009 in succession [5,6]. These were standards on principal data services led by Korea. The standard on T-DMB transmission mechanism and video service was released by ITU-R Recommendation in December 2007 [7].

AT-DMB transmission mechanism was released in June 2009 as a domestic standard of Korea [8,9]. In the process of the standardization, various opinions from research institutes, broadcasting stations, manufacturing companies, and universities were reflected in order to improve the quality of the standard.

After that, the standard on AT-DMB scalable video service [10,11], which would be expected to be one of the most attractive data services, was released in December 2009 as a domestic standard of Korea.

As a first stage of AT-DMB's international standardization, Korea contributed the technology of AT-DMB to ITU-R Study Group 6 in April 2009. As a result, the existing ITU-R report on mobile broadcasting technologies was revised [12]. As a second stage, Korea's domestic AT-DMB transmission and scalable video service standards have been in ITU-R standardization since September 2009. Finally, the specification on AT-DMB scalable video service was approved as a new service standard at March 2011 through the revision of the existing ITU-R recommendation [13]. The specification on AT-DMB transmission mechanism was released in March 2012 as Recommendation ITU-R BT.2016 [14].

3.3 Key Technologies in AT-DMB

Figure 3.2 shows the block diagram of AT-DMB transmission system. As illustrated, AT-DMB consists of two layers: the BL and the EL.

The upper part of the diagram is the BL section, which is identical to the conventional T-DMB transmission system. The lower part of the diagram is the EL section added by AT-DMB technology. As compared to the BL, the EL adopts new technologies such as hierarchical modulation and turbo coding.

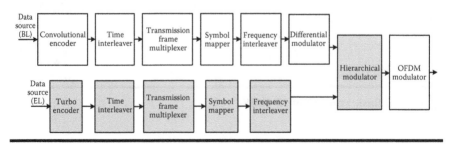

Figure 3.2   Block diagram of the AT-DMB transmission system.

3.3.1 Hierarchical Modulation Scheme

In order to achieve higher data rate while maintaining backward compatibility to conventional T-DMB receivers, a hierarchical modulation method is applied [15]. Hierarchical modulation of AT-DMB is characterized by a constellation ratio and a hierarchical modulation mode. The constellation ratio, α, is defined as follows:

$$\alpha = \frac{a}{b}, \tag{3.1}$$

where

   a is minimum distance between constellation points in two farthest quadrants
   b is maximum distance between constellation points in a quadrant

There are two modulation modes for the EL in AT-DMB: mode B and mode Q. In mode B, binary phase-shift keying (BPSK) modulation is used. In mode Q, quadrature phase-shift keying (QPSK) modulation is used.

   The value of α represents a trade-off between data throughput and service coverage (i.e., reception performance). The defined values of α in AT-DMB are 1.5, 2.0, 2.5, and 3.0. The hierarchical modulation mode is relevant only to the modulation scheme of the EL. Regardless of the hierarchical modulation mode, the BL fully follows π/4 DQPSK modulation method used in T-DMB. Figure 3.3

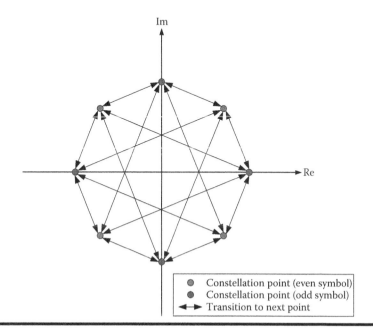

Figure 3.3   π/4-DQPSK constellation diagram of T-DMB.

shows the constellation diagram of π/4-DQPSK modulation. The π/4-DQPSK modulation is widely used in many applications because the signal can be differentially detected and symbol transitions in the constellation diagram do not pass through the origin [16]. Thus, it requires less complex receiver implementation and provides better spectral characteristics than other QPSK techniques. Figure 3.4 shows the constellation diagram of AT-DMB.

3.3.1.1 Hierarchical Modulation Mode B (BPSK)

In mode B, a π/4-DQPSK-modulated BL symbol and a BPSK-modulated EL symbol are combined into a single symbol as follows:

$$S_H^B = S_{BL}\left(1 + \frac{1}{1+\alpha}S_{EL}^B\right), \tag{3.2}$$

where

S_{BL} is the π/4-DQPSK-modulated BL symbol
S_{EL}^B is the BPSK-modulated EL symbol

The magnitude of both S_{BL} and S_{EL}^B is 1. Figure 3.4a shows the constellation diagram of mode B symbols. The variables *a* and *b* denote the spacing between the groups of constellation points and the spacing between individual constellation points in a group, respectively.

3.3.1.2 Hierarchical Modulation Mode Q (QPSK)

Mode Q uses QPSK for EL symbol mapping. QPSK-modulated EL symbol should be rotated by −π/4 before combining with π/4-DQPSK-modulated BL symbol. Therefore, a symbol of hierarchical modulation mode Q shall be mapped according to the following relation:

$$S_H^Q = S_{BL}\left(1 + \frac{1}{1+\alpha}S_{EL}^Q \cdot e^{-j\frac{\pi}{4}}\right), \tag{3.3}$$

where S_{EL}^Q is the QPSK-modulated EL symbol and has a magnitude of 1. Figure 3.4b shows the constellation diagram of mode Q symbols. From the figure, we can expect that the reception performance of mode B is better than mode Q.

3.3.2 Turbo Coding

AT-DMB adopted punctured duo-binary turbo coding [17] for more powerful error correction of EL channel while convolutional coding is still used in BL channel for backward compatibility with T-DMB. Figure 3.5 shows the structure of the turbo

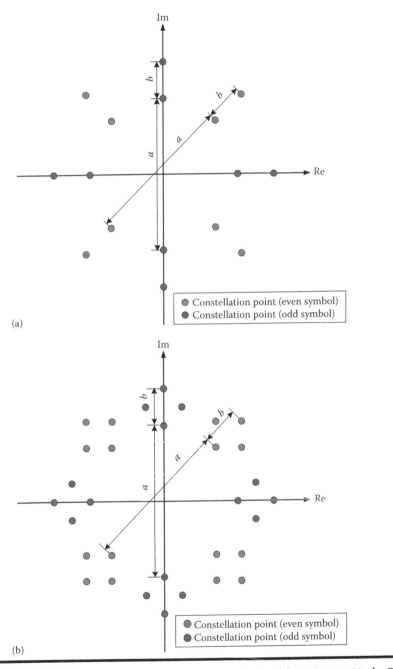

Figure 3.4   Constellation diagram of AT-DMB. Mode B with BPSK (a). Mode Q with QPSK (b).

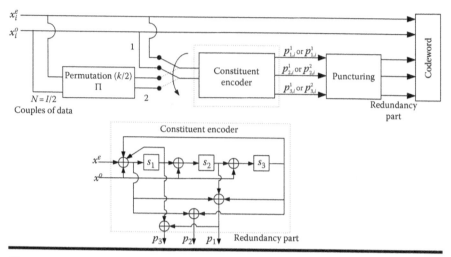

Figure 3.5 Structure of turbo encoder for the EL encoding.

encoder with a duo-binary circular recursive systematic convolutional (CRSC) code for channel encoding to the EL. This coding scheme allows only equal error protection (EEP).

As illustrated in the figure, this turbo encoder controls the code rate by puncturing from a mother code rate of 1/4. Four EEP protection levels are defined for bit rates in multiples of 32 kbps, and these correspond to the code rates 1/4, 1/3, 2/5, and 1/2, respectively.

3.3.3 Time Interleaving

The data throughput of the EL in mode B is half the size of mode Q. Therefore, the size of memory for time interleaving of the EL in mode B is half the size of mode Q. To increase robustness against burst errors and make good use of residual memory capacity for interleaving, the interleaving depth in mode B is doubled from 384 to 768 ms.

3.3.4 AT-DMB Encoder

Basically, T-DMB provides multimedia service including video, audio, and interactive data. T-DMB uses "ISO/IEC 11172-3 (MPEG-1 audio) layer II" and "13818-3 (MPEG-2 audio) layer II" for audio service. For video service, T-DMB adopted "ITU-T Rec. H.264 | ISO/IEC 14496–10 Advanced Video Coding" for video, "ISO/IEC 14496-3 MPEG-4 ER-BSAC or MPEG-4 HE AAC v2" for its associated audio, and "ISO/IEC 14496-11 MPEG-4 BIFS" for its interactive data. But T-DMB was optimally designed for audio service; it was not sufficient for video service. Therefore, T-DMB adopted Reed–Solomon code as outer channel coding only for video service to improve the performance of reception.

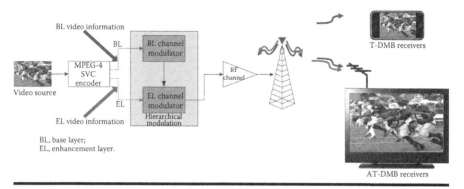

Figure 3.6   Concept of scalable video service in AT-DMB.

At the first stage of the AT-DMB development, high-quality video service called scalable video service guaranteeing backward compatibility with T-DMB video service was considered as one of killer applications. For AT-DMB scalable video service, AT-DMB adopted base line profile of Recommendation ITU-T H.264 | ISO/IEC 14496-10 amendment 3 for scalable video and "ISO/IEC 23023-1 MPEG Surround" for its scalable audio.

Figure 3.6 shows the concept of scalable video service in AT-DMB, which provides VGA video and 5.1 channel audio. The BL carries T-DMB video service with QVGA video and stereo audio, the EL carries the additional information that can make VGA video from QVGA video and surround audio from stereo audio. Basically, scalable video coding has various scalable coding technologies such as frame rate, picture resolution, bit rate, etc. AT-DMB adopted picture resolution scalable coding technology. It makes AT-DMB provide VGA video.

As described in Figure 3.6, the BL stream and the EL stream are modulated hierarchically at the AT-DMB modulator and then the modulated signal is sent to receivers. Thus, commercial T-DMB receivers get the conventional video service, but AT-DMB receivers get the BL and the EL signals together and reconstruct VGA video and 5.1 channel audio.

Figure 3.7 shows the conceptual architecture to split the AT-DMB SVC stream into the two-layer streams, synchronize the two-layer streams, and transmit them.

3.4 AT-DMB Performance

3.4.1 Simulation Results

We performed computer simulations to assess the performance of AT-DMB in contrast with T-DMB at various channel environments. The block diagram of the model used for the analysis and simulation of the AT-DMB system is shown in Figure 3.8. As shown in the figure, the two streams of the BL and the EL are input to the

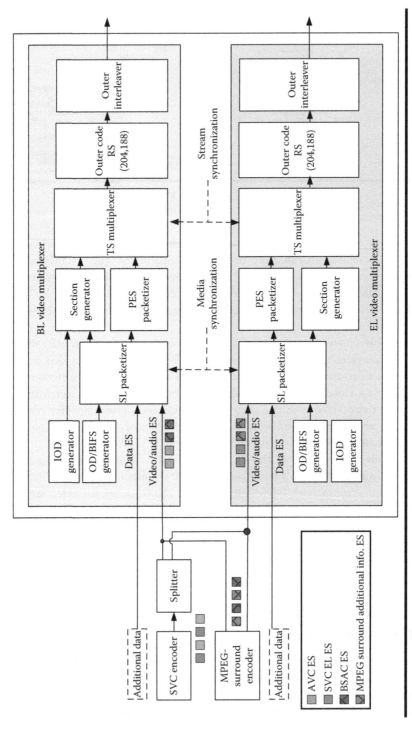

Figure 3.7   Conceptual architecture of the AT-DMB scalable video service encoder.

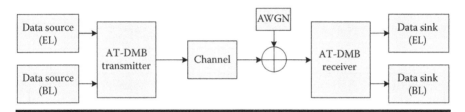

Figure 3.8   Block diagram of simulation.

AT-DMB transmitter, and then the AT-DMB signal is generated. Additive white Gaussian noise is added to the channel for emulating the channel environments.

The channels used in the simulations were a Rician channel for outdoor fixed channel model in which line of sight is secured, a Rayleigh channel for indoor fixed channel model, and the TU6 channel for the mobile. The receiver for the BL used the noncoherent detection as T-DMB/DAB typically did, while the EL signal detector used the decision-directed method [18].

The simulation parameters of AT-DMB are specified in Table 3.1. In the simulations, the carrier frequency was 200 MHz and the vehicle speed was set to be maximum 1200 km/h to reflect a fast fading effect. Perfect synchronization was assumed.

In evaluating the performance of AT-DMB, we used bit error rate (BER) as primary criterion. Two reference points within the device were used at which BER is measured. Here, BER after passing through either convolutional or turbo decoder was measured as the reference value. According to the T-DMB/DAB specification, the value of 10^{-4} is recommended as BER value for audio service at the end of Viterbi decoder, but the value of 10^{-3} is sufficient as BER value for video service at the end of Viterbi decoder. The reason is that the RS decoder is cascaded to Viterbi decoder in order to improve BER in case of video service.

Figure 3.9 shows the performance of mode B with the required SNR according to the Doppler frequency in the TU6 channel. The graphs show that the signal reception of the EL is much better than T-DMB for high Doppler. The reason is that the AT-DMB EL adds turbo coding in addition to the convolutional code in T-DMB, and it uses twice the depth of time interleaving in T-DMB in case of mode B. The required SNR of the BL decreases when the value of α increases, but the required SNR of the EL increases. That is because the maximum distance between the constellations of the EL diminishes when the value of α increases.

Tables 3.2 and 3.3 show performance results of AT-DMB in the fixed and mobile environment for a target BER 10^{-4} for audio service and target BER 10^{-3} for video service, respectively. As shown in Figure 3.9, if the transmitter transmits the AT-DMB signal instead of the T-DMB signal, the service coverage for T-DMB (BL) service may be reduced. When the value of α decreases, while the reception performance of the EL can be improved, the performance of T-DMB may be degraded. Therefore, in case of making the broadcast network plan of AT-DMB, the value of α should be carefully considered.

Table 3.1 Simulation Parameters of AT-DMB

Parameter	Specification	
	Base Layer	Enhancement Layer
Modulation	π/4-DQPSK	Mode B: BPSK Mode Q: QPSK
Forward error correction	Convolutional code (code rate: 1/2)	Turbo code (code rate: 1/2)
Time interleaving depth	384 ms	Mode B: 768 ms Mode Q: 384 ms
Frequency interleaving width	1.536 MHz	
Constellation ratio (α)	1.5, 2.0, 2.5, 3.0	
Nominal bandwidth	1.536 MHz	
FFT size	2,048	
Number of transmitted carriers	1,536	
Guard interval	504	
Sample time	1/2,048,000 s	

3.4.2 Field Trial

3.4.2.1 Measurement System and Environment

The AT-DMB transmission system used for field trials comprised a media encoder, a multiplexer, an exciter, and an RF transmitter. Two video contents were encoded and then transmitted to the BL and the EL separately. The total measurement route is about 21 km, and the maximum distance from the transmitting station is 4.4 km. The average speed of the van was about 60 km/h, and the maximum speed showed up was 85 km/h. Figure 3.10 shows the field trial area and the route that includes urban, open field, etc. The field trials were performed in Daejeon, Korea, and the antenna was vertically polarized and the radiation pattern was omnidirectional. The antenna gain was 3 dB and the height was 30 m, and the transmission power was 40 W.

The field trials were performed using a specially equipped mobile measurement system. Figure 3.11 illustrates the integrated system used. The system used a 2 dB gain omnidirectional receiving antenna placed on the van roof. The receiving

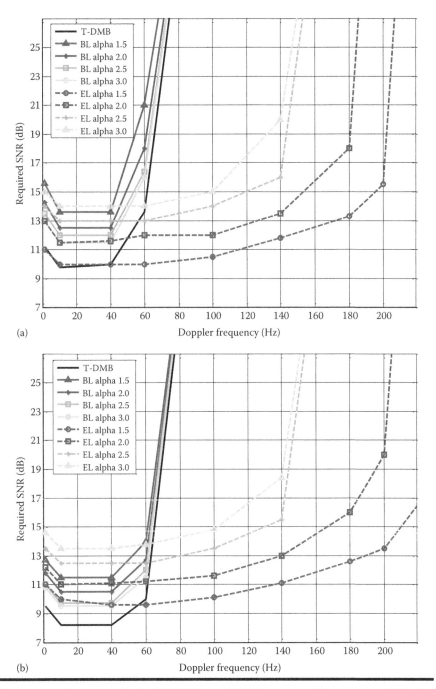

Figure 3.9 **Required SNR of T-DMB vs. AT-DMB mode B for a target BER 10⁻⁴ (a) and BER 10⁻³ (b). TU6 channel model.**

Table 3.2   Required SNR in dB for Audio Service

	α	Layer	AWGN	Rician	Rayleigh	TU6 (80 km/h)
T-DMB			5.2	5.6	8.3	9.8
AT-DMB (mode B)	1.5	BL	9.2	9.6	11.7	13.5
		EL	5.1	6.1	8.7	11
	2.0	BL	8	8.5	10.5	12
		EL	7.1	7.1	9.8	12
	2.5	BL	7.4	7.6	9.8	11.5
		EL	8.1	8.2	11.5	13
	3.0	BL	6.8	7	9.4	10.8
		EL	9.1	9.2	12.5	14
AT-DMB (mode Q)	1.5	BL	10.2	10.4	15	16.5
		EL	10	10.9	13	—
	2.0	BL	8.2	8.5	11.8	13.3
		EL	10.8	11	13.7	—
	2.5	BL	7.2	7.5	11	12
		EL	11.7	12	15	—
	3.0	BL	7	7.1	10	11.3
		EL	12	12.8	15	—

Note: Target BER 10^{-4}.

antenna height was 2 m. The mobile measurement system measured receiving success rate (RSR), packet error rate (PER), and location coordinates acquired using global positioning system (GPS) receiver. Receiving success means that there is no error in reception data for the duration of 1 s, and RSR is the rate of the receiving success time to the total receiving time as follows:

$$\text{RSR}(\%) = \frac{\text{Receiving success time}}{\text{Total receiving time}} \times 100. \tag{3.4}$$

Table 3.3 Required SNR in dB for Video Service

	α	*Layer*	*AWGN*	*Rician*	*Rayleigh*	*TU6 (80 km/h)*
T-DMB			4.5	4.8	7.2	8.2
AT-DMB (mode B)	1.5	BL	8.4	8.5	10.3	11.5
		EL	5	6	8.5	10
	2.0	BL	7.2	7.5	9.3	10.5
		EL	7.1	7.1	9.6	11
	2.5	BL	6.5	6.7	8.7	9.9
		EL	8	8	11	12.5
	3.0	BL	6	6.2	8.2	9.5
		EL	9	9	12.2	13.5
AT-DMB (mode Q)	1.5	BL	8.2	8.5	12	13
		EL	10	10.5	12.6	—
	2.0	BL	6.7	7	9.9	10.8
		EL	10.5	11	13.4	—
	2.5	BL	6	6.3	9	10
		EL	11.4	11.5	14.5	—
	3.0	BL	5.5	5.8	8.4	9.5
		EL	12	12.5	15	—

Note: Target BER 10^{-3}.

3.4.2.2 Results and Analysis

Figure 3.12 shows the analyzed results of the field trials plotting the RSR with respect to receiving signal strength. Using the curve-fitting technique with the polynomial of sixth order, we analyzed the performance based on the received signal power indicator at which the RSR value begins to exceed 95%. As shown in the figures, for the BL, the smaller the value of α is, the lower is the performance degradation relative to the reference case of the T-DMB signal. But the EL, on the other hand, experiences the opposite behavior.

Figure 3.10   Field trial area and route in Daejeon, Korea, during the first field trials.

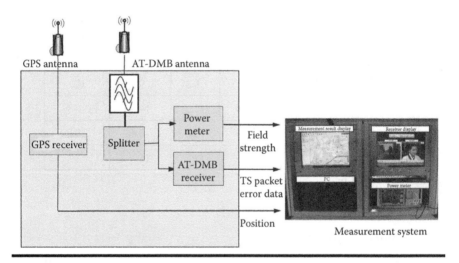

Figure 3.11   Mobile measurement system.

Table 3.4 shows the minimum required signal power for satisfying 95% of RSR obtained in the field trials as a function of the constellation ratio α and the turbo code rate for AT-DMB mode B. The data in the table show a similar tendency with the simulation results. As shown in the table, the value of α has to be set high in order to minimize reduction of the BL service coverage compared to T-DMB. But in order to increase the service coverage of the EL up to the level of the BL,

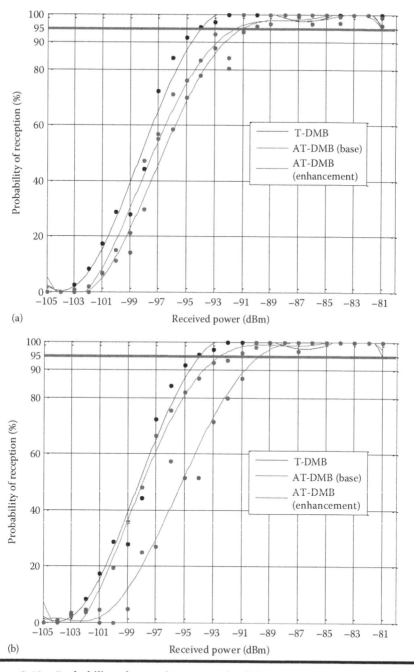

Figure 3.12 Probability of reception vs. received signal power. AT-DMB Mode B. BL code rate 1/2, EL code rate 1/2 (a) α = 1.5, (b) α = 2.0.

(continued)

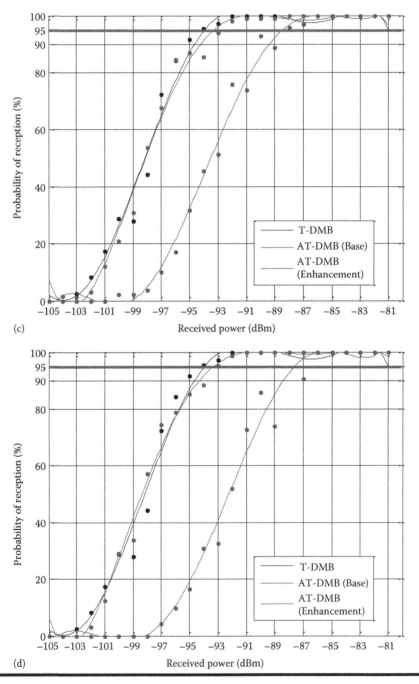

Figure 3.12 (continued) **Probability of reception vs. received signal power. AT-DMB Mode B. BL code rate 1/2, EL code rate 1/2, (c) α = 2.5, (d) α = 3.0.**

Table 3.4   Result Summary of the First Field Trials (AT-DMB Mode B)

	Constell-ation Ratio (α)	Layer	FEC (Code Rate)	Effective Data Rate (kbps)	Total Data Rate (kbps)	RSP (dBm)	RSP Difference between AT-DMB and T-DMB (dB)
T-DMB	∞		CC (1/2)	1152	1152	−93.9	0
AT-DMB	1.5	BL	CC (1/2)	1152	1152	−91.4	−2.5
		EL	Turbo (1/2)	576	1728	−90.9	−3.0
			Turbo (2/5)	448	1600	−91.7	−2.2
			Turbo (1/3)	384	1536	−95.1	1.2
			Turbo (1/4)	288	1440	−98.5	4.6
	2.0	BL	CC (1/2)	1152	1152	−92.5	−1.4
		EL	Turbo (1/2)	576	1728	−89.7	−4.2
			Turbo (2/5)	448	1600	−91.6	−2.3
			Turbo (1/3)	384	1536	−93.8	−0.1
			Turbo (1/4)	288	1440	−96.7	2.8
	2.5	BL	CC (1/2)	1152	1152	−93.1	−0.8
		EL	Turbo (1/2)	576	1728	−88.4	−5.5
			Turbo (2/5)	448	1600	−90.8	−3.1
			Turbo (1/3)	384	1536	−92.7	−1.2
			Turbo (1/4)	288	1440	−94.4	0.5
	3.0	BL	CC (1/2)	1152	1152	−93.2	−0.7
		EL	Turbo (1/2)	576	1728	−87.5	−6.4
			Turbo (2/5)	448	1600	−89.9	−4.0
			Turbo (1/3)	384	1536	−90.9	−3.0
			Turbo (1/4)	288	1440	−93.3	−0.6

the turbo code rate has to be low, reducing the payload of the EL. For example, in case the value of α is 3.0 and the turbo code rate is 1/4, the RSP of the BL and the EL compared to T-DMB decreases 0.7 and 0.6 dB each, but the payload of the EL is only 288 kbps.

In order to increase the payload of the EL, the value of α has to be lower and the turbo code rate higher, which increases the service coverage of the BL but reduces the coverage of the EL. For example, in case the value of α is 1.5 and the turbo code rate is 1/2, the RSP of the BL and the EL compared to T-DMB decreases 2.5 and 3.0 dB each, but the payload of the EL goes up to 576 kbps.

3.5 Conclusions

AT-DMB transmission and scalable video service technologies had been successfully developed from 2006 to 2009. These two technologies were standardized in Korea in 2009, and they were announced as ITU-R recommendations in 2011 and 2012. The field trial has been carried out in 2009 to check that these two technologies were fully backward-compatible with T-DMB. Through the field trial, AT-DMB was confirmed to provide more services than T-DMB with a little signal power loss against T-DMB.

AT-DMB can give differentiated services over T-DMB services to users exploiting its hierarchical structure. A variety of data services have been standardized after the domestic standardization of these two technologies in Korea. A few standards such as multichannel audio service backward- compatible with T-DMB stereo audio service and filecasting service were released as domestic standards in 2011. Other standards such as TPEG-RWI are now in the process of domestic standardization.

AT-DMB in Korea is expected to be launched in the near future after switch off from analog TV to digital TV. AT-DMB enables more valuable business models, since it allows for a more efficient and flexible provisioning of mobile broadcasting services.

References

1. ETSI TS 102 428 v1.1.1, Digital audio broadcasting (DAB); DMB video service; User application specification, June 2005.
2. S. Cho et al., System and services of terrestrial digital multimedia broadcasting (T-DMB), *IEEE Transactions on Broadcasting*, 53(1), 171–177, March 2007.
3. G. Lee, K. Yang, K. Kim, Y. Hahm, C. Ahn, and S. Lee, Design of middleware for interactive data services in the terrestrial DMB, *ETRI Journal*, 28(5), 652–655, October 2006.
4. ETSI TS 102 632 v1.1.1, Digital audio broadcasting (DAB); Voice applications, November 2008.
5. ETSI TS 102 635-1 v1.1.1, Digital audio broadcasting (DAB); Middleware; Part 1: System aspects, August 2009.

6. ETSI TS 102 635-2 v1.1.1, Digital audio broadcasting (DAB); Middleware; Part 2: DAB, August 2009.

7. Recommendation ITU-R BT.1833, Broadcasting of multimedia and data applications for mobile reception by handheld receivers, December 2007.

8. K. Kim, G. Lee, J. Lim, S. Lee, and D. Kim, Efficient generation of scalable transport stream for high quality service in T-DMB, *ETRI Journal*, 31(1), 65–67, February 2009.

9. TTAK.KO-07.0070, Specification of the advanced terrestrial digital multimedia broadcasting (AT-DMB) to mobile, portable, and fixed receivers, June 2009.

10. H. Choi, I. Shin, J. Lim, and J. Hon, SVC application in advanced T-DMB, *IEEE Transactions on Broadcasting*, 55(1), 51–61, March 2009.

11. TTAK.KO-07.0071, Advanced terrestrial digital multimedia broadcasting (AT-DMB) scalable video service, December 2009.

12. Report ITU-R BT.2049-3, Broadcasting of multimedia and data applications for mobile reception, May 2009.

13. Recommendation ITU-R BT.1833-1, Broadcasting of multimedia and data applications for mobile reception by handheld receivers, March 2011.

14. Recommendation ITU-R BT.2016, Error-correction, data framing, modulation and emission methods for terrestrial multimedia broadcasting for mobile reception using handheld receivers in VHF/UHF bands, March 2012.

15. V. Mignone and A. Morello, CD3-OFDM: A novel demodulation scheme for fixed and mobile receivers, *IEEE Transactions on Communications*, 44(9), 1144–1151, September 1996.

16. J. G. Proakis, *Digital Communications*, McGraw-Hill, New York, 2001.

17. C. Berrou, M. Jézéquel, C. Douillard, S. Kerouédan, and L. Conde, Duo-binary turbo codes associated with high-order modulations, *Proceedings of the ESA Workshop on Tracking Telemetry and Command Systems for Space Applications (TTC)*, Noordwijk, the Netherlands, 2001.

18. W. Kim, Y. Lee, H. Kim, H, Lim, and J. Lim, Coded decision-directed channel estimation for coherent detection in terrestrial DMB receivers, *IEEE Transactions on Consumer Electronics*, 53(2), 319–326, May 2007.

Chapter 4

An Overview of the North American ATSC M/H Mobile Broadcasting System and Its Next-Generation ATSC 3.0

Mike Simon and Mark A. Aitken

Contents

4.1 Introduction

For the purpose of this chapter, the use of the word "mobile," unless otherwise stated, is focused on broadcast television that has been enabled for mobile reception. In everyday terms, the use of the word "mobile" has become broadly used to define most every highly portable service other than broadcast television, with the wireless carriers at the top of the list. The following "truths" are stated to position the work in this chapter:

- Television broadcasters were late to recognize the need for a mobile platform.
- The United States developed Advanced Television System Committee 8 Level Vestigial Side Band (ATSC 8VSB) standard that was optimized solely for fixed reception.
- The standard was developed in a highly political environment with little input from the broadcast television business community.
- Limitations imposed to maintain backward compatibility have severely limited mobile capabilities.
- Understanding the "mobile business" will drive broadcasters to develop a new and more capable standard.

While the current ATSC standard may have specific limitations, it should be recognized that many of those limitations have their grounding in the politics that drove the adoption of the ATSC A/53 standard inside of a limited and somewhat technically arbitrary timeline.

When the United States embarked on exploring the opportunities for an "Advanced Television" standard, the world was still largely analog. The Advisory Committee on Advanced Television Service (ACATS) got its start in 1987. The IBM PC and the Space Shuttle had just celebrated their fifth birthday, and digital terrestrial TV existed nowhere else in the world. As a matter of fact, in 1987, the Moving Picture Experts Group (MPEG) did not even exist. It was an opportunity to develop and deploy the "first" commercial consumer high-definition television (HDTV) service.

The establishment of an "all digital" standard became a matter of U.S. industrial and political policy in 1990. The ability to convey HDTV inside of 6 MHz with a digital standard meant that there would be a "digital dividend" in the form of spectrum that could be auctioned after a period of transition instead of television broadcasters holding onto an additional 6 MHz (for a total of 12 MHz). Spectrum could be reclaimed from television broadcasters with a complex national plan that relied on channel moves and repacking into a reduced spectrum footprint. The Clinton administration was counting on $36B from auctions to help cover Washington's spending. Not much has changed.

Given that the establishment of an all-digital terrestrial television broadcast standard was of significant importance, and the United States was in the position of honor for developing and deploying it first, some "lines in the sand" became hard and required. When discussion arose in December 1993 about potential improvements (Coded Orthogonal Frequency Division Multiplexing) to the almost final "Grand Alliance" (GA) system that could extend ongoing work beyond mid-1995, such ideas were fairly quickly and quietly dropped from consideration, and the ATSC ended up ratifying and publishing the A/53 terrestrial standard in 1995.

It was 1998 before a U.S. Broadcaster, the Sinclair Broadcast Group, raised its hand and asked about the need to provide "ease of reception" and additional flexibility for over-the-air consumers including the need to provide future support of mobile DTV. Persistence over several years by a minority but growing and vocal part of the television Broadcast industry, and mounting awareness that mobile DTV might be an interesting way to increase viewership, forced the ATSC to begin investigations on enhanced reception improvements and ultimately mobility.

This fight became extremely visible and ultimately forced the Federal Communications Commission (FCC), the National Association of Broadcasters (NAB), and Association of Maximum Service Television (MSTV) to publicly address the performance limitations and ultimately to take a position to "stay the course." This served to bring the limitations of the ATSC A/53 8VSB standard to the forefront of the industry and forced manufacturers to address some of the shortcomings in the commercial implementations.

4.2 Development of Digital TV in the United States

In the United States, the development of the first digital TV (DTV) standard (ATSC A/53) was the outcome of a lengthy techno-political process. However, having insight into this chronicle is relevant today since the new ATSC A/153 M/H mobile handheld standard [1] is designed to interoperate and be backwardly compatible with the legacy A/53 fixed system [2]. In this spirit, Table 4.1 provides a brief chronology of some of the main techno-political events that shaped the development and adoption of the ATSC A/53 DTV (fixed service) and the A/153 M/H mobile standards. Also included are some recent events that may influence the development of a future ATSC (3.0) standard. The evolution from the analog (NTSC) TV system to a so-called Advanced TV system (ATV) in the United States originated approximately 25 years ago. It was in late 1989 that the first digital video platforms came to the forefront of this activity and activities soon evolved into a quest for the first DTV standard, with an emphasis on HDTV.

The GA system (1994) eventually became the U.S. DTV standard (ATSC A/53) and the politics of the day pushed for one main driving application: HDTV delivered to the living rooms of America via an outdoor antenna mounted 30 ft (9.1 m) in the air. The A/53 system was optimized for such a fixed-antenna-received DTV service and allowed each broadcaster use of a second channel for a simulcast DTV period to enable the United States to transition to DTV in the future.

Table 4.1   Brief Chronology of DTV in the United States

Date	Technical/Political Event
Aug. 1987	58 Broadcast groups petition FCC for an Advanced TV (ATV) service for the United States
Sept. 1987	ACATS formed to advise FCC
1988–1993	ACATS receives 23 proposals, selects 4 digital
May 1993	GA formed to merge (4 digital) into a single DTV system
Fall 1993	GA selects MPEG2 video, Dolby AC-3 audio, and MPEG2 transport
Feb. 1994	GA selects 8-VSB (single carrier) as the physical layer
Fall 1995	ATSC completes documentation of DTV (HDTV) standard
Dec. 1995	ACATS recommends the ATSC standard to FCC
April 1997	FCC table of allocations (a second channel) and rules (DTV transition starts)
1999–2000	Some broadcasters raised concerns of 8-VSB reception simple indoor antennas
2000–2001	ATSC forms task force on RF system performance to study COFDM vs. 8-VSB
Jan. 2001	FCC confirms support for keeping 8-VSB after consideration of COFDM
Feb. 2001	ATSC S9 issues RFP for A/53 backward compatible enhancements
2006–2007	Two viable proposals for mobile emerge and HW prototypes are demonstrated
April 2007	ATSC BOD authorizes work on a backward compatible mobile DTV standard
March 2008	ATSC S4 issues RFP ATSC M/H backward compatible mobile standard
June 2009	Transition to DTV completed in the United States, DTV Channels (2–51)
Oct. 2009	ATSC A/153 M/H standard is completed
April 2010	FCC Broadband Plan (Incentive auction of 120 MHz UHF Broadcast Spectrum)

Table 4.1 (continued)   Brief Chronology of DTV in the United States

Date	Technical/Political Event
April 2011	112 Economist send President Obama letter supporting Incentive Auctions
Sept. 2011	ATSC TG3 to develop nonbackward compatible Next-Gen standard (3.0)
Nov. 2011	Future of Broadcast Television Summit (FOBTV), Shanghai China
Feb. 2012	Congressional Budget Office (CBO) estimates 120 MHz Incentive Auction could raise $15 billion
Feb. 2012	U.S. Congress passes legislation, president signs law Incentive Auction Authority to FCC
March 2012	FCC forms Incentive Auction Task Force
March 2012	FCC retains world's leading experts in auction theory to design incentive auction

Simply stated, the FCC plan was for each broadcaster to operate both the old analog and a new digital channel for a period of time, switching off the analog service at some point in future to be defined by FCC (became June 2009). At the end of this transition period, all DTV services would be repacked into channels (2–51), recovering channels (52–69) for a future auction (this "digital dividend" was auctioned in 2008). To help maximize the DTV service area (reduce NTSC interference into DTV) during the simulcast period, the A/53 standard provided for the incorporation of an NTSC co-channel rejection filter in the DTV receiver, and precoding (to complement rejection filter) at the transmitter and distinct spectral shaping of the 8VSB waveform. The scheme was designed such that when co-channel NTSC interference was detected (receiver), the NTSC co-channel rejection filter would be activated in the DTV receiver to extend coverage. The trellis decoder (receiver) was then switched to a trellis code corresponding to the 8-VSB emitted trellis-coded modulation (TCM) with precoding to complement the filter, as will be explained. The TCM (precoding) is still emitted today after completion of DTV transition and can now only serve to constrain any effort of future extensibility. The 8VSB system inherently required only processing of the in-phase (I) channel signal. The receiver then only required a single A/D converter and a real (not complex) equalizer. These were all thought at the time (1994) to be desirable attributes to reduce overall system complexity and cost. One major requirement was to realize a usable HDTV video payload capacity (data rate) of ~18 Mbps, which was needed by the early prototype implementations of HDTV MPEG2 1080i video codecs. The exact maximum payload data rate

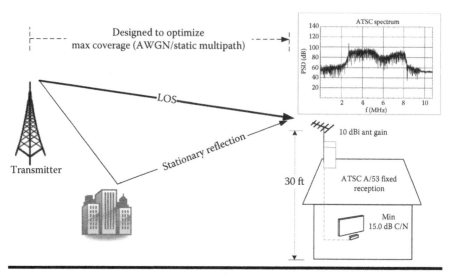

Figure 4.1   Fixed HDTV terrestrial channel model assumed in ATSC A/53 DTV.

supported including audio and auxiliary data is 19,392,658 bps over one 6 MHz terrestrial channel. The designers assumed a relatively undemanding channel model for the fixed HDTV terrestrial reception environment (30 ft outdoor antenna) being characterized mostly as line of sight or Rician channel with static multipath and considered to be limited by Gaussian noise (see Figure 4.1).

These assumptions and requirements drove the selection of the channel coding, time interleaving, training symbol sequences (support receiver equalizer) and the 8-level VSB TCM, resulting in a 15.0 dB carrier-to-noise ratio (CNR) in the AWGN channel, and a spectral efficiency of ~3.2 bps/Hz.

These (1994) decisions resulted in a monolithic HDTV system deprived of any physical layer scalability and or future extensibility. Despite the early objections raised in 1999 with respect to the severely limited "vision" of the capabilities, by March 2008, with perfect hindsight, the limitations of these decisions (driven by limited and poor assumptions) weighed heavily on engineers when the ATSC mobile standardization process began. This process was started when ATSC led work, and demonstrations by Samsung of increased reception technologies showed great promise in "unlocking" a usable degree of mobile reception capability. It was immediately realized and verified that the A/53 channel coding, time interleaving, training symbol sequences, and system signaling chosen in 1994 were completely the incorrect choices for a mobile fading environment. The real quagmire was that no fundamental changes could be made for an ATSC mobile extension. Any changes would impact the existing installed base of A/53 legacy receivers, and a firm stance was taken regarding the need for full "backward compatibility."

It is important to note that the ATSC M/H standard, constrained to be backward compatible with the ATSC A/53 standard, must support simultaneous HDTV/SDTV and mobile services in a 6 MHz channel, and is termed a dual stream system. The emitted A/153 waveform must transport both legacy MPEG2 transported MPEG2-encoded (HDTV/SDTV) services receivable by A/53 receivers and the new M/H mobile IP services. The M/H services use new robust channel coding, time interleaving, longer and more frequent training symbol sequences, and a robust system signaling. A novel mechanism was needed to construct and emit an A/153 M/H frame to guarantee that all A/153 mobile service (dual stream) waveforms remain undetectable and cause no performance degradation to legacy A/53 receivers while incorporating the performance improvements for a mobile fading environment and A/153 M/H receivers. The ATSC A/153 M/H system uses cross-layer design (CLD) techniques to introduce new technologies and meet these requirements; this has resulted in a very complex system design as will be explained in the following sections but was essential to meet the new mobile system performance as well as backward compatibility requirements.

Today, strong political and industrial economic forces in the United States are once again posturing to reduce the total amount of spectrum that will be allocated for DTV broadcasting in the future. Specifically, the FCC has received the statutory authority (Feb. 2012) from the U.S. Congress to conduct a voluntary incentive auction of spectrum as proposed under the FCC "National Broadband Plan." Part of this plan states that a need exists to recover large, continuous blocks of UHF broadcast spectrum, up to an additional 120 MHz. To accomplish this, the FCC is targeting UHF broadcast channels (31–51) as one option for recovery if future voluntary incentive spectrum auction schemes are massively successful. This recovery would then trigger a mandatory repacking of all remaining broadcasters (post-auction) into DTV channels (2–30). Any future planning on an ATSC (3.0) future standard will be significantly shaped by future political decisions in this area.

Later, in the section on ATSC 3.0, a brief review of some of the motives behind the FCC broadband plan will be discussed. The incentive auction model now being developed in the United States could be used as part of a second "digital dividend" scheme outside United States as well, at least in the minds of some regulators. To paraphrase the FCC chairman, "If successful, incentive auctions are envisioned to be the tool considered by other regulatory bodies around the world."

4.3 Overview of ATSC M/H A/153 Technologies and CLD Mechanisms

The ATSC A/153 mobile DTV system protocol stack is illustrated in Figure 4.2.

Only the lower M/H physical, Reed–Solomon (RS) frame, and signaling (TPC/FIC) layers shown are truly novel. These layers collectively enable the performance in a mobile fading environment while guaranteeing A/53

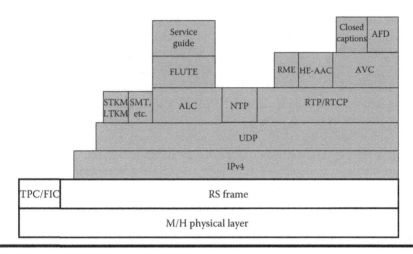

Figure 4.2   ATSC A/153 system protocol stack.

backward compatibility. Therefore, we will focus the discussion on these lower layers and the technology used to enable A/153 M/H in this chapter. The IPv4 and upper-layer technologies are mostly constrained versions from other standards. To gain insight into these layers, the reader should reference the ATSC A/154:2011 Recommended Practice [3].

Figure 4.3 shows the A/53 system block diagram; this system functionality must remain perpetual and will serve as the foundation on which the A/153 M/H dual stream system is built.

The A/53 system uses a fixed VSB data frame structure that is created downstream of the channel coding in the Mux block shown in Figure 4.3. A VSB frame contains 626 segments each composed of 832 symbols. The VSB data frame begins with one known data field sync (DFS) segment inserted in the Mux, followed by 312 data segments (channel coding blocks), then another known (DFS) segment, concluded with another 312 data segments. The DFS segments contain known training symbol sequences used by the A/53 receiver's equalizer and also low-level system signaling. These known DFS are inserted by the Mux with a periodicity of ~24 ms, for the A/53 fixed service channel model assumed. The studio (left side) HDTV/SDTV services are assembled in an emission Mux and sent as fixed (19,392,658 bps) 188 byte MPEG2 TS packets to the first block in the A/53 exciter. The randomizer block exclusive OR's all the incoming data bytes with a 16-bit maximum length pseudo random sequence reset at a DFS cadence and with the MPEG2 sync byte dropped resulting in 187 byte packets on the randomizer output. The next stage is a systematic RS encoder with a data block size of 187 bytes, with 20 RS parity bytes added at the end for error correction. Following this is a 52-data segment (intersegment) convolutional byte interleaver to improve the burst error correction performance of the RS code. Next, a 2/3 rate trellis encoder with one input bit

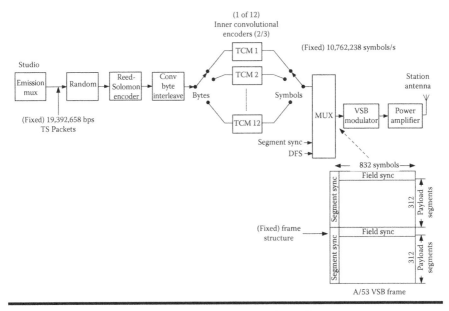

Figure 4.3 ATSC A/53 system block diagram and VSB frame structure.

encoded into two output bits using a 1/2 rate convolution code, while the other input bit is precoded to complement the comb filter in the receiver. Also to simplify the design of the trellis decoder in the A/53 receiver when the NTSC co-channel rejection filter is active, trellis code (intrasegment) interleaving using a group of 12 identical trellis encoders operating on the interleaved data symbols is used. These 12 groups were required since the creation of spectral nulls in the receiver required a comb filter with delays of 12 symbols. The resulting symbols are then mapped into an eight-level (one-dimensional) constellation. The Mux then assembles the VSB frame structure with a fixed symbol rate (10,762,238 symbols/s) locked to incoming TS packet rate (19,392,658 bps). Then, a known four-symbol pattern is inserted by Mux at the beginning of each of the 626 segments in VSB frame synchronous to the MPEG2 sync byte dropped by randomizer, and this is termed segment sync. The DFS symbol pattern is inserted by Mux with a periodicity of ~24 ms to finalize the construction of the VSB frame. In the final A/53 blocks, a pilot signal is inserted and the signal is D/A converted, filtered, and VSB modulated for emission in a manner that will help mitigate NTSC–DTV interference during the transition period.

In conclusion, the TCM scheme was optimized for the simulcast period and has good coding gain in an AWGN channel. However, 20 years later with an assessment in the direction of mobile, it was recognized that this coding scheme would act as a barrier to any possible operation in a mobile fading environment. The resolution for A/153 was the integration of a new FEC scheme including turbo coding. The basic

A/53 system functionality has been exposed so it can now be shown that some A/53 system attributes (hooks) can form the foundation to enable the technology enhancements listed as follows to provide the A/153 mobile system performance and assure the A/53 backward compatibility requirements.

The new technology enhancements to enable A/153 Mobile are as follows:

1. More frequent and longer training symbol sequences
2. Robust low-level system signaling
3. Robust FEC system
4. New reception paradigm

These may seem straightforward tasks, but when viewed through the constraints of maintaining A/53 backward compatibility with no legacy system degradation, the difficulty and system complexity grew exponentially and required CLD techniques to solve.

In Figure 4.4, on the left is shown the normal A/53 VSB field and two DFS training symbol sequences spaced ~24 ms apart. The length and periodicity of these A/53 DFS alone are insufficient to enable mobile channel acquisition including synchronization, channel estimation, and tracking in a mobile fading channel.

In Figure 4.4, on the right side is shown an A/153 M/H data field including new longer and more frequent training symbol sequences and robust low-level system signaling technology integrated. The new M/H training symbol sequences and robust system signaling symbols are placed at known symbol locations within the M/H VSB field to support the new M/H reception paradigm, to be discussed. To label or map these exact symbol locations in an M/H VSB field that is always delineated by a known A/53 DFS segment and 312 data segments (each 832 symbols), the M/H field can be considered as a (X–Y) symbol array

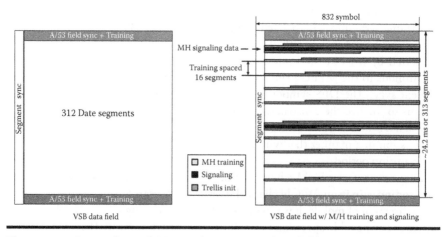

Figure 4.4   Comparison of A/53 and A/153 VSB data fields.

designated by (segment #, symbol #) that describes the precise symbol location in an M/H field. The new M/H receiver paradigm will first quickly locate and use the known M/H training symbol sequences for initial channel acquisition. Then, the robustly encoded symbols representing the low-level A/153 system information are decoded and the received signaling parameters are used to then locate and decode only the M/H content symbols of interest (called a parade). An M/H parade/s is (are) the symbol structure/s embodying a virtual M/H channel and usually is only a small percentage of the 10,762,238 symbols/s emitted in the A/153 dual stream system that simultaneously may also be broadcasting an A/53 HDTV channel.

However, one important implementation detail must be mentioned. If the same A/53 method shown in Figure 4.3 is used, where the Mux block (exciter) periodically inserts known DFS training symbol sequences (stored in memory), this will result in detrimental modifications to the emitted waveform and make all legacy A/53 receivers inoperable. Therefore, the data representing the known training symbol sequences must be encapsulated in transport stream (upstream) at studio, as part of the preprocessing steps for M/H to be discussed. The data that represent the known A/153 symbol training sequences is precalculated and stored for insertion in M/H Mux at studio. A prerequisite to correctly precalculate the training data is that all trellis encoders must be driven to a known "zero state" at the start of insertion of each training sequence in the exciter. The normal A/53 trellis encoder circuitry is modified for this purpose and is shown in Figure 4.5.

At the specific initialization instant, the trellis inputs (X2, X1) are switched from the normal (N) into the (I) initialization mode (modified dash lines), and the closed loop feedback positions the same logic level on both inputs of the exclusive OR gates shown. This logic produces a zero logic level on (D flip-flop) memory element output on the next symbol clock cycle. Therefore, by keeping the switch in the down (I) initialization mode for only two-symbol clock cycles, the trellis encoder is driven to an all-zero state independent of starting state as required, while maintaining the

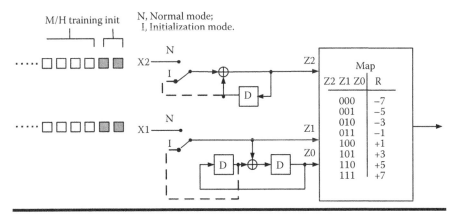

Figure 4.5   A/53 modified (1 of 12) trellis encoder.

normal trellis trajectory. The two-symbol initialization period is triggered in the exciter when the reserved initialization bits designed for this purpose appear at the (X2, X1) inputs of their respective trellis encoder. The initialization then occurs, but the initialization bits never enter trellis encoder; they serve only as placeholders. When initialization is complete, the switch moves up to (N) normal position, and the precalculated training data that immediately follow in the data stream produce the known training symbol sequence; the switch remains in the normal position until the next two-symbol cycle initialization period. These so-called initialization bits are shown at the start of each M/H training sequence in Figure 4.6. To complicate matters, the A/53 RS encoder shown in Figure 4.3 is located upstream of the trellis encoders; therefore, any initialization in trellis encoders will invalidate some of the already-calculated RS parity. Therefore, to mitigate these errors, nonsystematic RS re-encoding is used in exciter (M/H post-processing) to re-encode the RS parity on the fly downstream; this will be seen when we view the actual A/153 block and is an example of the complexity needed for A/53 compatibility.

The CLD that enables the M/H training, signaling, and turbo-coded data to be positioned at reserved symbol positions must be developed on the basic A/53 foundation. Referring back to Figure 4.3, some notable A/53 system attributes (hooks) are the *fixed* framing structure, *fixed* symbol, and *fixed* TS rates. These hooks can be leveraged in the CLD to enable the preprocessing of M/H data (studio), the sending of the M/H data to exciter, and the corresponding post-processing if some system constraints are applied. First, both the symbol clock (exciter) and TS clock (studio) must be locked to a GPS reference signal, and there must be a synchronous link (no rate adaptation or add/drop of null packets) between studio and exciter. Second, the signaling of which TS packet shall begin an M/H frame (symbol domain) is specified; this metric is currently not defined in A/53 but can be compatibly realized. These system hooks and system constraints are the foundation of the CLD that enables the A/153 system to function in a synchronous and deterministic manner; the CLD is discussed next.

Figure 4.6 shows an A/153 M/H conceptual block diagram with the blocks arranged in their effective positions to help visualize the key concepts of how the M/H system actually works more easily. The actual A/153 block diagram will then be presented. The A/153 time references used are either a GPS receiver or an NTP Stratum 1 Server (studio) and a GPS receiver (exciter). These are mandatory for the system timing and synchronization and will be discussed later. One input (top of Figure 4.6) is the normal A/53 transport stream from the existing ATSC emission Mux. The emission Mux data rate is reduced to allow for the inclusion of the M/H (preprocessed) data to form the total aggregate data rate of 19,392,658 bps. The common block is the A/53 byte interleaver, and when the switch shown is in the up position, the normal A/53 channel coding and modulation path are selected. The M/H IP streams are input into the M/H preprocessing blocks. The A/53 byte interleaver's input switch can select and combine (118) preprocessed M/H packets (termed an M/H group) along with (38) normal packets. The combined (156)

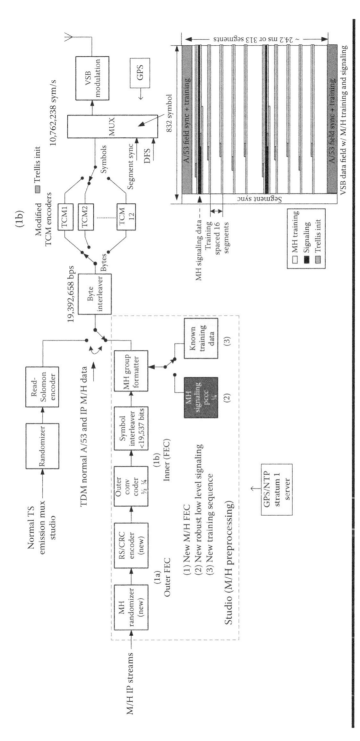

Figure 4.6   Conceptual A/153 M/H block diagram.

packet data structure is termed an M/H slot. In the M/H group formatter, all M/H training, signaling, and preprocessed data bytes are assembled and temporally formatted or prearranged. The formatting assures the M/H group block of data when processed later in time by the A/53 byte interleaver, and modified trellis encoders will produce M/H training sequences, robust signaling, and turbo-coded symbols at predetermined symbol locations within the 8-VSB data field. There is only one imaginable way this preprocessing and formatting of the M/H group data block (transport layer) can be successful. The upper M/H layers are designed to operate in a synergistic way that enhances the overall M/H system performance by given them knowledge of the lower physical layer processes. Therefore, the sharing of knowledge between layers to enhance the overall system performance is what constitutes the definition of a CLD in A/153 and is essential in meeting the goal of A/53 backward compatibility.

Now a brief walk-through of the M/H preprocessing blocks shown in Figure 4.6 is presented before we turn our attention to the actual A/153 block diagram and M/H frame structure.

The M/H IP stream input (H.264 A/V encoder output) is shown entering (blocks 1a + 1b), which forms the new M/H FEC chain. First, (1a) is a new outer FEC block comprising a randomizer and an RS/CRC block encoder. Then, (1b) comprises three blocks consisting of a new outer convolutional coder and symbol interleaver blocks that are effectively serially concatenated with the modified trellis encoders in the exciter. This serial concatenation process establishes the new inner FEC (turbo encoder). It should be realized the (1b) block is physically distributed with the outer convolutional coder and symbol interleaver (studio) and the modified trellis encoders in the M/H exciter (transmitter). This distributed architecture is possible with a synchronous CLD as will be explained. One reason this distributed architecture is used is that it enables the effective time sharing of the trellis encoders between normal and M/H turbo-coded data, and this improves the possible turbo coding gain realized in the M/H receiver. When viewed from the perspective of the M/H receiver (new reception paradigm), there exists only turbo-coded symbols that the M/H receiver will attempt to decode. No legacy A/53 symbols (based on their location) will be attempted to be turbo decoded; these symbols would only act as noise and reduce the possible turbo coding gain. Moreover, the A/53 receiver is unaware of any turbo-coded symbols and continues to decode (all) symbols as A/53 symbols and recovers a normal transport stream without any A/153-induced errors or degradation in A/53 system performance. The only system impact on A/53 is the reduced data rate given for the inclusion of A/153 M/H data.

The M/H group formatter block assembles the M/H signaling data (2) and M/H precalculated training data (3) along with the preprocessed data into a specific data block format using an A/53 *de-interleaver*. By formatting the assembled M/H group data block with an A/53 byte *de-interleaver* (inverse of A/53 byte interleaver) will effectively negate the effect of the A/53 byte *interleaver* shown and assure the known symbol locations in the physical layer. The (118) M/H group packets after formatting have a 4-byte MPEG2 header attached with an ATSC-reserved PID value of 0x1FF6. The ATSC-reserved PID 0x1FF6 is treated like null packets by A/53 receivers, and all emitted M/H packets are discarded to ensure backward compatibility. Note: Although all MHE packets are received at the A/53 receiver without any M/H-induced packet errors, the M/H payload data are useless to M/H receivers.

4.4 M/H Frame Structure and A/153 System Block Diagram

The M/H multiplexer is the entity at the studio location that accepts the dual stream inputs and contains all of the M/H preprocessing and then outputs an aggregated 19,392,658 bps packet stream structure termed an M/H frame that is only detectable by M/H receivers (see Figure 4.7).

In Figure 4.7, the basic element, the (156) packet M/H slot structure, is shown. Then, together, 16 M/H slots (2340 packets) form one M/H subframe shown. Then, five M/H subframes (12,480 packets) form the overall structure of one M/H frame that appears at the output of the M/H Mux with a periodicity of ~0.968 s, the equivalent of (20) A/53 VSB frames. Every M/H slot can be composed of either 118 M/H + 38 normal packets (termed mobile slot) or 156 normal packets (termed normal slot). The total data rate that can be allocated for mobile is variable from a minimum of 5 (mobile slots) per M/H frame to a maximum of 80 (mobile slots) per M/H frame. The M/H group (118 packets) in every mobile slot contains M/H signaling data (Figure 4.7) that when decoded by M/H receiver identifies the exact location of the M/H group (subframe # and slot #) within the M/H frame structure. The possible (5–80) M/H groups are equally distributed across each subframe using the allocation pattern shown in Figure 4.7. The spreading out

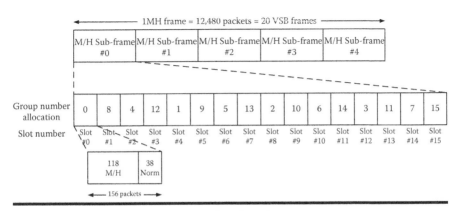

Figure 4.7   M/H frame structure packet domain.

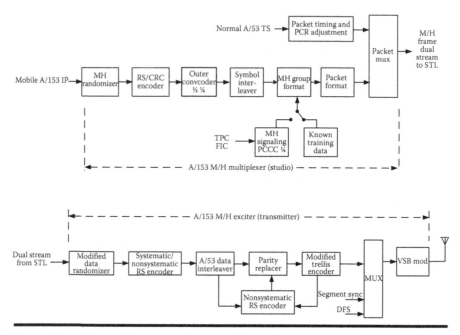

Figure 4.8   A/153 M/H Mux and exciter block diagrams.

in time of M/H groups over each subframe in an M/H frame ~0.968 s offers time diversity to help mitigate the mobile fading channel as is shown.

Figure 4.8 depicts the actual A/153 system block diagram. We will briefly discuss the M/H preprocessing blocks (M/H Mux) and then the post-processing blocks (M/H exciter) that will enable the ATSC mobile performance and ensure A/53 backward compatibility. Then, in the next section, the M/H receiver block and the M/H reception paradigm are discussed.

The M/H multiplexer is shown in the upper portion of Figure 4.8 and contains the M/H preprocessing, IP encapsulation, and multiplexing and outputs an M/H frame. First, the normal A/53 packet stream input must be reordered to open a placeholder time slot for the insertion of the 118 consecutive M/H packets (M/H group) by the packet Mux block shown. This requires that the original MPEG2 PCR values of the reordered packets shall be restamped to maintain backward compatibility with the A/53 transport stream buffer model for A/53 receivers. This functionality is provided by the packet timing and PCR adjustment block shown.

The input mobile IP packets are shown in Figure 4.8 entering the randomizer and then the RS/CRC encoder blocks that collectively form the M/H outer FEC function.

Figure 4.9 shows an expanded view of the RS frame structure generated in the RS/CRC encoder. A separate RS frame is generated for each M/H service in the M/H frame. The RS frame structure is a data block with M/H IP packets clocked in row by row. There are separate (IP addresses) used to transport the video and

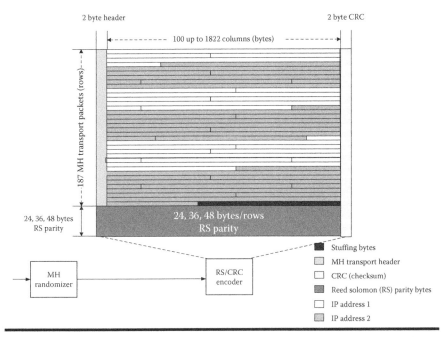

Figure 4.9 Expanded view RS/CRC frame encoder and resulting RS frame structure.

audio, etc. A 2-byte header is inserted at the beginning of each row and a 16-bit CRC of the row appended at the end. There are always exactly 187 rows, but the row length can vary from 100 to 1822 bytes depending on data rate allocated for M/H service. Finally there are 24, 36, or 48 RS parity bytes calculated (transverse direction) over each column of 187 bytes of the 187 rows. This places the 24, 36, or 48 rows of RS parity bytes with a 16-bit CRC appended at the bottom of RS frame. The RS frame structure is designed to enable RS erasure decoding in M/H receiver as an effective scheme for mitigating burst errors in a mobile fading channel.

The next blocks in Figure 4.8 are the outer convolutional encoder and symbol interleaver that are serially concatenated with modified trellis encoders in the M/H exciter by using the CLD to establish the M/H turbo encoder (inner FEC). The M/H group formatter then assembles the M/H (turbo-coded) IP data, pre-calculated training data, and the robustly encoded M/H signaling data and then A/53 byte de-interleaves to format the M/H group based on the knowledge of the lower physical layer processes. Then, each M/H packets has a 4-byte MPEG2 header inserted with ATSC-reserved PID value of (0x1FF6). The M/H packet with MPEG2 header is then termed a Mobile Handheld Enhancement (MHE) packet. The packet Mux block then assembles the composite dual stream output by multiplexing normal and or mobile slots to create M/H frames. The transport stream aggregate data rate is 19,392,658 bps, and this is output to STL for transport to M/H exciter.

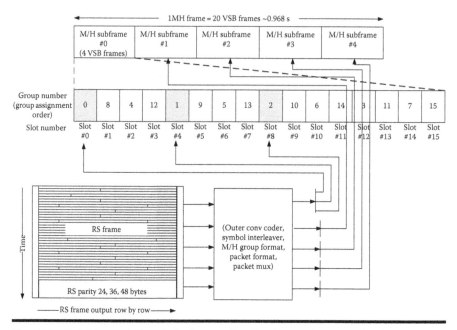

Figure 4.10 RS example of RS frame distribution over one M/H frame (time diversity).

Figure 4.10 is a high-level block showing an example of (1) M/H service establishing (1) RS frame being allocated (3) M/H groups per each subframe (0–4). This illustrates the M/H time diversity created by spreading the RS frame out in time over (1) M/H frame (~0.968 s) for transmission. Note: the M/H group assignment order is identical in each subframe (0–4).

In the M/H exciter, the first block is the modified data randomizer; this bypasses all MHE packets (PID 0x1FF6). If the MHE packets were randomized, this would destroy all preprocessing. However, all normal A/53 packets are still randomized as specified by the A/53 standard. The next block is the systematic/nonsystematic RS encoder. All normal A/53 packets are RS-encoded using the normal systematic method that appends all RS parity bytes at the end of the packet. The M/H receivers are unaware of any legacy RS encoding. In fact, they only have knowledge of the IP packet structure encapsulated at known M/H symbol locations. However, to maintain backward compatibility (no errors) for A/53 receivers, an RS nonsystematic encoding method is used to place the RS parity bytes at known byte locations within the payload area of MHE packets. The reason is that the normal systematic method of appending RS parity at end of packet will not work for MHE packets. The end positions in these packets are used in the creation of the long M/H training sequences. Fortunately, the A/53 receiver RS decoder will work with systematic or nonsystematic RS-encoded parity bytes, so backward compatibility is

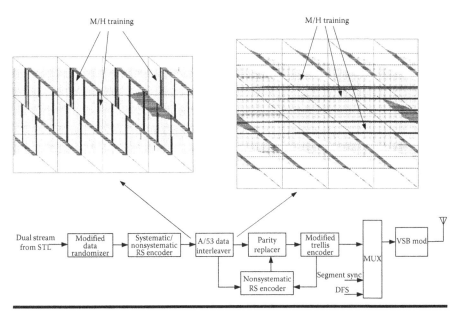

Figure 4.11   M/H group format before and after A/53 byte interleaver.

maintained. The A/53 byte interleaver block is unchanged from the A/53 standard, though its characteristic is known and used synergistically by (preprocessing) at upper M/H layer.

Figure 4.11 shows the M/H group before and after the A/53 byte interleaver in the M/H exciter. The M/H group format before A/53 byte interleaver in Figure 4.11 clearly displays evidence that the M/H training bytes have been synergistically preprocessed with knowledge of the A/53 interleaver. These M/H training bytes appear as vertical columns of bytes located at precalculated positions within the M/H group. After passing through the A/53 interleaver block, the M/H training bytes now appear horizontal and as continuous sequences of bytes that form the long M/H training sequences. These M/H training byte sequences are correctly positioned to develop the known training symbol sequences at the output of modified trellis encoders. It is important to observe that the 156 MHE packets (now termed data segments) are spread over exactly 170 data segments at the output of A/53 interleaver. These 170 data segments representing the M/H group after A/53 interleaver have (graphically) sawtooth regions of data over the beginning and ending segments, regions that are *void* of any M/H training bytes. These M/H training void regions form a serious M/H system constraints that will be discussed later in the section on M/H system performance. Next, trellis initialization is used by all the modified trellis encoder blocks to set the states to zero at the beginning of each M/H training sequence. However, the initialization process will result in RS parity errors to A/53 receivers unless mitigated. Therefore, the RS parity is recalculated after trellis initialization, and the RS parity bytes are replaced on the fly by the parity replacer block shown. The Mux block then inserts segment and DFS to construct A/53 VSB data fields. The DFS now have low-level signaling symbols that indicate M/H is being emitted. Finally, in Figure 4.8 is the 8-VSB modulation process that remains unchanged.

4.4.1 ATSC A/53 and A/153 M/H Receiver Blocks and Reception Paradigms

The basic A/53 and A/153 M/H receiver blocks are shown in Figure 4.12 and will briefly be discussed. The ATSC standards do not specify the receiver. However, these blocks can be considered representative for this discussion. First, it is assumed that an A/153 waveform is being transmitted. The reception process of the ATSC A/53 receiver is discussed first.

The M/H CLD results in the A/53 receiver being unaware of the presence of any M/H data in the emitted RF waveform. The A/53 receiver simply proceeds as normal to demodulate (equalize) Viterbi decode (100%) of symbols, de-interleave and RS decode with no resulting errors or system degradation being induced by the presence of M/H in the waveform. The A/53 de-randomizer outputs the recovered 19,392,658 bps TS to the de-multiplexer. All MHE packet PIDs (0x1FF6) are then ignored or discarded as if they were null packets (backward compatibility). The announcement and signaling metadata are provided by parsing the PSI/PSIP packets to select the (A/V) PIDs of interest and sending these to content decoders.

Next, a high-level walk-through of A/153 M/H reception is presented. The tuner selects the 6 MHz channel and down-converts the signal. The tuner output is sent to both the demodulator and to the known sequence detector block. The demodulator performs gain control, carrier recovery, and timing recovery on the signal.

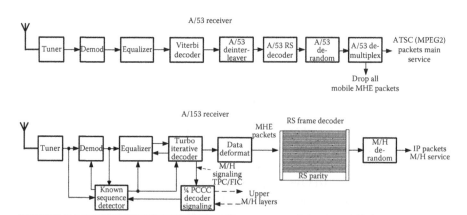

Figure 4.12   ATSC A/53 and A/153 receiver block diagrams.

The demodulator outputs the baseband signal to the equalizer and also feeds back signal to the known sequence detector. The known sequence detector now having inputs before and after demodulator starts searches for the known M/H training symbol sequences using correlation techniques. Once a sequence is detected, timing information is sent to the equalizer that makes use of the M/H training sequences to help acquire the channel. The timing information is also fed back to the demodulator to aide synchronization and carrier recovery. Next, the M/H signaling symbols from known locations in M/H group (mobile slot) are extracted in the turbo decoder and sent to the M/H signaling decoder. After decoding, the signaling information termed transmission parameter signaling (TPC) independently carried in each mobile slot is used. Some relevant TPC parameters are subframe #, slot #, parade ID, starting group #, total number of groups in parade, total number groups in subframe. These parameters enable M/H receiver to quickly identify and synchronize to the M/H frame symbol structure and identify all mobile slots in subframe associated with a parade ID. The TPC also contains parameters needed to decode the parade such as the RS frame and turbo code rates. A parade ID identifies the M/H service contained in an RS frame and assigned to mobile slots throughout the whole M/H frame. There may be multiple M/H services and parade IDs in an M/H Frame. The other M/H signaling information recovered is the fast information channel (FIC) sent in each M/H subframe, and this is used to enable a fast M/H service acquisition. The FIC contains signaling and announcement metadata and is collected and presented to the upper layers enabling a user to make selection of the M/H service wanted. The FIC primarily signaling function is the binding between M/H ensemble ID (M/H service) and the parade ID. Once this decision is made, the known sequence detector locates this parade ID in the M/H frame. The M/H turbo decoder then receives timing and signaling information of which slots need to be decoded and the turbo code rate needs to decoded, etc. The iterative turbo decoding process continues with all loops closed enabling the equalizer to track dynamically, etc. The turbo decoder then converges after several iterations and outputs a block of M/H data to the de-formatter. The turbo decoding then continues over all the mobile slots of the parade of interest until all M/H data are decoded. The RS/CRC decoder receives signaling information on the size and mode of M/H frame including number of RS parity bytes, etc. The M/H IP packets are then reassembled into the rows of the RS frame structure, and RS erasure decoding is performed. With 48-byte RS parity selected on each column, the CRC on each row in RS frame can detect an error and then declare a row erasure. The 48 bytes of RS parity can enable erasure decoding that can correct up to 48 row erasures in the RS frame. This RS erasure decoding is a powerful outer FEC technique that can mitigate burst errors following the inner turbo decoder and is effective in a mobile fading channel. Note: One constraint previously mentioned is the sawtooth region of an M/H slot that is void of M/H training symbols. These regions are more probable to burst errors, and this is one reason that 48-byte RS parity is being used by broadcasters in the field, and this will be discussed in the section on performance.

Group number (group assignment order)	0	8	4	12	1	9	5	13	2	10	6	14	3	11	7	15
Slot number	Slot #0	Slot #1	Slot #2	Slot #3	Slot #4	Slot #5	Slot #6	Slot #7	Slot #8	Slot #9	Slot #10	Slot #11	Slot #12	Slot #13	Slot #14	Slot #15

Figure 4.13 M/H receiver time-sliced reception.

The de-randomizer then outputs the M/H IP packets of the selected services to the content decoders. Figure 4.13 shows the M/H subframe from the example in Figure 4.10, which has one M/H parade being assigned (3) mobile slots (0, 4, 8) in each subframe of the M/H frame. The M/H training sequences and signaling in each mobile slot enable rapid synchronization and channel acquisition. These attributes can be used to conserve the battery energy by introducing a time-sliced reception procedure on the M/H receiver. The M/H receiver can be designed to power down nonessential circuits to realize energy savings during periods of time between the mobile slots of interest. In the subframe shown in Figure 4.13, the M/H receiver can switch to a reduced energy mode during slots (1, 2, 3), (5, 6, 7), (9, 10, 11, 12, 13, 14, 15) and only select to process and decode the M/H symbols having the content of interest during slots (0, 4, 8). When the parade selection decision is made, a command is given to the known symbol sequence detector (Figure 4.12), which synchronizes the reception process so that battery energy can be conserved on the M/H receiver.

4.5 ATSC A/153 M/H System Timing Requirements

This section explores the basic signaling and synchronization requirements between the M/H Mux (studio) and the M/H exciter (transmitter) to enable the CLD. There are four basic requirements that will be briefly discussed:

1. Synchronous transport link between M/H Mux and the M/H exciter
2. MHE packet PID (0x1FF6) used to identify M/H groups for processing in M/H exciter
3. First MHE packet of an M/H frame (packet domain) signaled to the M/H exciter
4. Phase of emitted M/H frame specified with respect to ATSC time (symbol domain)

The synchronous link requirement is established by deriving the TS_{clk} (M/H Mux) and $Symbol_{clk}$ (M/H exciter) from a common frequency reference (GPS) and by using the following equations:

$$TS_{clk} = 433,998/223,797 \times 10^7 \, \text{Hz} = 19,392,658 \, \text{bps} \tag{4.1}$$

$$Symbol_{clk} = 313/564 \times TS_{clk} = 10,762,238 \text{ Hz} \qquad (4.2)$$

The common frequency reference ensures that the input FIFO buffer (M/H exciter) will have no buffer management issues (under- or overflow) and remain synchronous by also specifying that no rate adaptation (add/drop null packets) is permitted as this would also destroy the mapping of packets to symbols, which is the basic requirement of the CLD.

Next, a point-to-point signaling channel must be established between M/H Mux and M/H exciter. The M/H exciter must identify all MHE packets (0x1FF6) and then process and or bypass them in various M/H post-processing blocks. The MHE PID (0x1FF6) is used for this signaling purpose. Next, the M/H frame is composed of exactly 12,480 packets, and a signaling mechanism is required for the M/H Mux to signal to the M/H exciter which packet in the incoming stream shall start an M/H frame (first packet of subframe #0, slot #0). Note: see ATSC A/110:2011 standard for A/153 detailed signaling specifications [4]. Finally, the last requirement is that the phase of the emitted M/H frame shall be timed with respect to ATSC time. The purpose of this last requirement is that the phase of the M/H frame of all M/H stations shall be time aligned at antenna air interface using GPS time. Since an RS frame spread over a whole M/H frame must be received ~0.968 s before starting to decode the RS frame, having a known M/H frame phase relationship between M/H stations may be desirable. This requirement is to enable future M/H receiver designs with improved channel change performance between M/H stations and/or improved handoff between M/H stations transmitting common content over large regional areas.

Figure 4.14 shows the timing relationships with respect to the phase of the M/H frame at the air interface of the antenna. First, the ATSC epoch (start of emission of first symbol of first M/H frame at the antenna air interface) is defined to be the same as the GPS epoch (January 6, 1980, 00:00:00 UTC). Then, the ATSC time equation (Equation 4.3) is used, given the GPS second count at next 1PPS Tick:

$$(\text{GPS seconds}) \times (4,809,375/4,654,936) = \# \text{M/H frames since epoch.} \qquad (4.3)$$

Given this factor (4,809,375/4,654,936) and GPS seconds count, the M/H Mux can calculate the phase of the M/H frame with respect to the epoch and release the start of the M/H frame with time stamps embedded to enable the M/H exciter to phase the emission of the M/H frame with respect to ATSC Time. This methodology can support either an MFN or SFN transmitter topology.

First, the M/H Mux in Figure 4.14 establishes an ATSC time reference; this happens in the ATSC time (cadence generator) block shown. The M/H Mux then assembles an entire M/H frame and waits for the correct time instant to release it into the STL. A precalculated time value called the max delay is established. The max delay time value selected must be longer than the actual transport delay time expected over

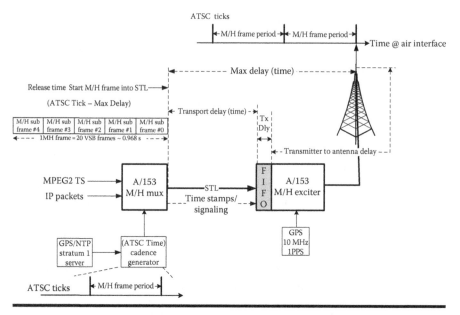

Figure 4.14   Basic timing relationships in A/153 M/H system.

STL to the antenna air interface. The M/H Mux then releases the start of the M/H frame into STL before or in advance of the actual emission time (ATSC "Tick") by a value of time equal to (ATSC Tick – Max Delay). The M/H Mux sends signaling including the max delay value and a time stamp indicating the actual instant the start of the M/H frame was actually released into STL. The M/H exciter then calculates the actual transport delay time by observing actual time of arrival of start of M/H frame (GPS), minus the time stamp value received from M/H Mux indicating the release time into STL. Also a static time value entered called transmitter antenna delay (TAD) represents the delay for the particular station (this is a calculated value by an engineer) used. The M/H exciter sets the FIFO input buffer size (TX DLY) equal to Max Delay – (Transport Delay + TAD). The FIFO buffer is then well regulated. By using this timing method, all M/H stations will emit their M/H frame at the air interface of their respective antenna in phase with the defined ATSC Time Tick.

4.6 ATSC A/153 M/H System Performance

This section discusses the main coding parameters available and basic M/H system performance. The M/H group in a mobile slot has (six fixed) training sequences, a selection of RS frame (24, 36, 48) RS parity, and a turbo code rate of ½ or ¼. Table 4.1 shows the results reported from the ATSC M/H lab tests with turbo code rates ¼ and ½; both had the RS parity fixed to 48 bytes.

Table 4.2   ATSC M/H Lab Measurements

	¼ Turbo	½ Turbo
CNR AWGN	3.2 dB	7.1 dB
CNR Pedestrian 3 km/h	11.2 dB	15.3 dB
CNR TU6 120 km/h	13 dB	17 dB
Payload	16%	33%
FEC Parity (Turbo + RS)	84%	67%

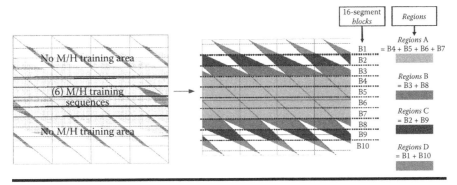

Figure 4.15   M/H group regions after the A/53 byte interleaver.

The ¼ rate turbo code provides the best overall performance, however, at an efficiency of 16%. The ½ rate turbo efficiency is 33%. However, there are some fundamental system constraints that are not obvious given the metrics in Table 4.2. There is an A/53 system attribute that proved to be very challenging and has significant limiting effects on the performance achievable at the ½ rate turbo code. The challenge is the A/53 byte interleaver output byte pattern. The best way to explain the phenomenon is by looking at Figure 4.15, which shows the output of A/53 interleaver.

The A/53 byte interleaver input is the 118-packet M/H group shown in Figure 4.11. The relevant observation is that in the interleaver output, these packets are spread over 170 segments. It was only possible to locate six long training signals in the middle region of these 170 segments. Graphically, the partition between the 118 M/H and 38 normal packets on the output appears as the sawtooth data pattern shown in Figure 4.15. The white area of the saw tooth is the normal A/53 packets. For M/H training to be effective, it must extend across the whole segment. The M/H receiver uses the M/H training symbol sequences at the physical layer to acquire and then track the mobile fading channel. The shared (sawtooth) regions must remain void of any long M/H training because any changes in this portioning would sacrifice A/53 backward compatibility. However, there are some mechanisms used that can help mitigate this inherent shortcoming. First, as shown in Figure 4.15, the 170 segments are divided into 10 blocks of segments and then 4 regions (A, B, C, D) are defined. The turbo code rate (½) or (¼) for these regions can be selected independently. The (A) region is in the middle and has the (6) M/H training signal sequences and majority of the M/H payload data. Empirical testing has shown that a ½ rate code performs well in the (A) region with the M/H training provided. The symmetric (B, C, D) regions are void of M/H training and provide poor performance in real-world portable and mobile conditions. Specifically, it was found that using a ¼ code rate in (B, C, D) and 48 bytes, RS parity (outer FEC) will clean up the errors resulting from poor equalizer tracking in these regions by M/H receiver. Therefore, to gain reasonable performance and better efficiency, it is common practice to find stations operating with mixed (½, ¼, ¼, ¼) code rates in regions (A, B, C, D) respectively and with 48 RS parity bytes at an approximate 25% efficiency.

In June 2011, a new Part 9 (Scalable Full Channel Mobile) was added to the A/153 M/H standard. This optional mode introduced by Part 9 enables a 156-packet M/H slot to contain more than 118 M/H packets, with the new M/H packets displacing some of the 38 normal packets. The end result was the introduction of more M/H training in all regions. Testing has confirmed that (½, ½, ½, ½) code rate and 48-byte RS parity can provide good performance and better overall efficiency. However, the trade-off is that the broadcaster must commit more data rate to M/H mobile and less to normal A/53 in the dual stream system to reap these benefits. There have been public demonstrations of Part 9 capable M/H receivers, but currently there is no M/H broadcasting using Part 9 options.

Finally, on the subject of a single-frequency network (SFN) and the ATSC M/H system. OFDM-based mobile DTV systems can usually benefit by using an SFN consisting of multiple coherent transmitters providing overlapping signals to provide higher, more uniform field strength and with multipath echoes being constructively combined at the receiver (so-called SFN gain). Moreover, the signals arriving from multiple spatial directions can also be used to help mitigate shadowing. The ATSC M/H system supports possibilities for operation of an SFN of multiple coherent and synchronously timed transmitters. However, there are several constraints that make benefiting from the use of an SFN a real challenge. First, the single-carrier ATSC M/H system must use an equalizer in the receiver to cancel multipath echoes. The combining of multipath echoes, possible in a well-designed and implemented OFDM system to achieve gain, is not achievable. There is always noise added when the echo is cancelled with an equalizer that always adds to the noise floor. Also the zero-dB echo condition in an SFN can be challenging and can add to CNR threshold required. The receiver complexity is always higher with an equalizer approach compared to OFDM. There are improvements in the ATSC M/H equalizer performance, but this is only part of the challenge in designing an ATSC M/H SFN. Second, the backward compatibility of legacy A/53 receivers can be a real concern when deploying an SFN. The multiple transmitters in an SFN will likely result in stronger and longer pre- and post-multipath echoes that can easily overwhelm the equalizer in legacy A/53 receivers in the field. Therefore, a careful SFN system design is needed to obtain benefits for M/H receivers while protecting A/53 legacy receivers deployed.

4.7 ATSC (3.0) "Next-Gen" Standard

In May 2010, ATSC announced the formation a "Next-Generation Broadcast Television" Team (PT-2), which explored potential technologies to be used to define a future terrestrial broadcast digital television standard. It was out of this initial activity that the ATSC Board authorized the formation of a Technical Standards Group (called TG3) specifically focused on special needs for future technologies and a series of voluntary technical standards and related recommended practices for the next-generation digital terrestrial television broadcast system. Such work is designed to serve the evolving needs of viewers and broadcasters alike for decades to come. ATSC 3.0 is a crucial long-term project that paves the way for futuristic terrestrial television broadcasting technologies and implementation topologies.

From a technology perspective, the good news is that ATSC 3.0 is likely to be incompatible with current broadcast systems. This will be driven by the fundamental underlying technical capabilities of the technologies employed. The hurdle for such a clean piece of paper design with no legacy constraints is that it must provide improvements in performance, functionality, and efficiency significant enough to warrant implementation.

In this context, the politics driving the possible recovery of an additional 120 MHz of UHF broadcast spectrum is worth understanding to have a proper perspective of the current situation. Any effort to develop an ATSC 3.0 standard will be shaped greatly by future political decisions and technology embraced.

There is a saying that any professional politician will never let an emergency or crisis go to waste. If there is thought to be some crisis now or in the near future, the American people become more willing to listen and go along with a politician's agenda to solve some imminent crisis. Also it is said that if there is no imminent crisis to move forward a political agenda, then create one. The crisis was given a name on October 5, 2009, when Julius Genachowski, FCC chairman, said that the biggest threat to the future of mobile TV in the United States was "the looming spectrum crisis." The FCC "National Broadband Plan" describes a means to avoid the before-then-unknown future spectrum crisis.

The wireless industry lobby has been openly engaged in signaling its willingness to pay for more wireless broadband spectrum to solve this supposed crisis. This crisis has also been termed the "spectrum crunch" by the FCC chairman. He describes a future that will limit mobile wireless operators from meeting future data capacity demands of consumers of 4G (long-term evolution, LTE)-enabled services by 2015, with a large percentage of the increased capacity projected to be video, without access to more "prime" spectrum. The FCC National Broadband Plan also speculates that the current broadcast spectrum is being utilized inefficiently today under legacy technology and regulations.

Ideally, a holistic approach would be for the FCC to seek input from technologists (engineers), lawyers (regulators), and economists (business communities) in a collaborative manner to bring additional spectrum into play quickly, being limited only by the laws of physics and economics and under new regulatory reform and with the goal to use new system architectures that use spectrum more efficiently. However, the debate has been dominated by the politicians, lawyers, and economists with little-welcomed technical input on spectrum efficiency. This can be observed in an April 6, 2011, letter sent to president Obama signed by 112 economists who specialize in telecommunications and auction theory design. "We understand that Congress is considering legislation that would give the FCC explicit authority to run 'incentive auctions' in which it would have the ability to distribute some portion of the auction proceeds to licensees who voluntarily give up their license rights," it reads. "We support such an effort and think it would increase spectrum efficiency in the United States." They believe their auction theory alone could increase spectrum inefficiency by selling the spectrum to the highest bidder at auction who *will then put the spectrum to its best economic use*.

If history is to teach us anything, we may see an equivalent repeat of the (2008) 700 MHz (Ch. 52–69) auction, where political expedience trumped physics, revenue, and consumer interests. The FCC imposed so many preconditions and allowed such a variety usages of the band; it prohibited the swift use of this spectrum in the market. The FCC is now (2012) opening a rulemaking to try to solve some of the technical interference and other issues it knew pre-auction (2008), which has seriously encumbered the use of spectrum relinquished at the close of the "digital transition." The largest mobile network operators still bid and won spectrum at this 2008 auction, some with no immediate intent to use. Their economic justification seems to have been driven by their want to protect their market share by prohibiting competitors from acquiring this spectrum.

The real driver today inside of Washington, DC that has overshadowed (trumped) everything is the large sum of money that incentive auctions might return, as high as $16 billion (net) as projected by the CBO. The CBO must score the future value or cost to the government on any new legislation proposed in Congress. The new legislation of interest, "Auction Authority," was granted with the passage of the "Middle Class Tax Relief and Job Creation Act 2012," which contained language authorizing the FCC to conduct voluntary incentive auctions. The legislation states that the projected incentive auction proceeds shall be used to fund portions of the middle-class tax relief bill and the construction of a nationwide LTE public safety network. The U.S. politicians like to stand in front of the TV cameras and claim responsibility for the passage of tax relief for all middle-class citizens or the building of a new public safety network to enable our first responders, something being planned ever since September 11. These are coveted messages that both democrat and republican politicians in Washington, DC can agree on in an election year (2012), since it will not increase the deficit and are presumably paid for by the proceeds of future incentive auctions. In early 2012, the politicians having received what they wanted (a "pay for" against spending) are satisfied. The U.S. Congress has already spent the money, and now it is the FCC's job to design and conduct the voluntary incentive auction to maximize proceeds with the oversight of congress.

In simple terms, what is a voluntary incentive auction? It is a combination of a *reverse* auction and a regular *forward* auction. The details of both the reverse and the forward auction will be worked out by the FCC and their hired economic consultants.

In the reverse auction, broadcast station licensees interested in voluntarily participating in the auction will submit a confidential bid of the amount of money they would accept for the return of their current broadcast licenses. The job of the FCC in the reverse auction is to convince more broadcasters to participate and to do so at the lowest asking price. The FCC may design several mechanisms to encourage broadcaster participation, such as the promise of future spectrum taxes or new creative regulatory policy for those who choose to remain in broadcasting as possible examples. Once the asking prices and the amount of all spectrum is determined (up to 120 MHz), the FCC will then assemble continuous blocks of spectrum; this will bring the most at auction. The FCC then conducts the normal forward auction process to maximize the proceeds by selling to highest bidders. Post-auction, the remaining broadcasters are then repacked by the FCC into channels (2–30).

The broadcast lobby, lead largely by the NAB, recognized the political forces in play and took the position not to oppose truly voluntary incentive auctions. This positioning was premised on the condition that such participation was truly voluntary and that those broadcasters that choose not to participate were held harmless. They worked to have specific language in the bill they believe would provide identifiable protection for broadcasters. The general assumption is that the laws of physics are still in effect, and such protections will limit the FCC from successfully repacking spectrum when maintaining the existing broadcaster coverage and or dealing with the border issues the language in the bill describes. Such details always seem to escape political expediency. Perhaps the first debate, and possibly litigation, may be on the definition of the word "voluntary." Does this means a broadcaster can choose not to participate in the auction without fear of wrath from the FCC? The broadband plan is full of talk about the need for technical innovation, but the FCC seems to have excluded broadcasters from such innovation. It would seem that they are the ones with spectrum, but with the possibility to be relevant with new and efficient technologies; they may well be major participants in a major way. By effectively wielding "broadcast" as an efficient means to off-load the consumption of mobile video services, they could be part of the solution to the "data crunch" driven by the inefficient architecture of the carrier infrastructure.

The ball is now in the court of the lawyers and regulators at the FCC. They may pursue their own agenda without attention to the physics and spectrum efficiency of TV broadcasters even though they must ask for comment in the upcoming proceedings. Perhaps a discussion of the real technical problems associated with the present carrier infrastructure that drives massive inefficiencies in the unicast distribution of common, "high-value" video entertainment content will emerge. Viewed simply, the future demand for capacity to serve the same popular video, large files, app updates, etc., to millions of wireless consumers in large populated areas by using only a cellular point-to-point architecture is at the pinnacle of inefficiency. Perhaps a more pragmatic or balanced approach that recognizes the value of broadcasting in meeting the objectives of the broadband plan will evolve. Broadcast, a point-to-multipoint means of distribution, excels in serving popular content. A critical rethink of future broadcast networks that can interoperate and off-load this type of traffic should be further investigated. Some broadcasters have already put forward proposal in this area that was immediately dismissed as folly because of the politics of needing a "pay for" spending and budget constraints that focus on immediate returns as opposed to long-term revenue gains. The authors believe that without a change in system architecture, there is no amount of spectrum that can solve the predicted future crisis. The FCC will likely find themselves having to keep auctioning spectrum in the future until they address the problem holistically.

The technologies and capabilities that could be part of an effective and efficient future TV broadcast system exist today. This is not the real barrier to a future, fully capable ATSC 3.0 standard. The real question should be "will the FCC permit the market to innovate now or in 10 years?" The voluntary incentive auction legislation states that the auctions must be complete in 10 years.

This section hopefully has given some insight into the techno-political process unfolding in the United States.

4.8 Conclusions

It should be clear to all that having the right technology and right solutions does not drive success by itself. The interrelated nature of the politics, the technical and business climate, and the diverse forces of extreme markets at work will do as much in shaping the outcome as will great technological prowess. It is in this environment that the present debate regarding spectrum usage has been born.

It will take patient and artfully skilled technical and political leadership to find and drive the correct balance against many needs—real needs driven by free-market economics, the perceived needs of a consuming public that wants its entertainment content when and where they desire to consume it, and the political need to balance business and political forces that often work at odds to the electability of any politician.

Ubiquity for TV broadcasters was once achieved by ensuring a television set existed in every living room, bringing news, weather, sports, traffic, entertainment, and emergency warnings into the homes of America. Times have changed. The dynamics driving mobile services today is still a question of being ubiquitous. The difference is that such ubiquity today is defined by a public that increasingly demands that their news, weather, sports, traffic, entertainment, and emergency warnings be delivered to them personally wherever they may be.

Crafting a wireless distribution network that interoperates and is capable of delivering that level of ubiquity with the spectrum available will be a challenge, but can be achieved today because of the capability to harmonize standards that make truly heterogeneous networks possible. Leadership will be required.

References

1. A/153: ATSC mobile DTV standard, available on www.atsc.org (accessed: March 2012).
2. A/53: ATSC digital television standard, available on www.atsc.org (accessed: March 2012).
3. A/154: 2011 ATSC mobile DTV recommended practice, available on www.atsc.org (accessed: March 2012).
4. A/110: 2011 ATSC standard for transmitter synchronization, available on www.atsc.org (accessed: March 2012).

Chapter 5

Overview of the Chinese Digital Terrestrial Multimedia Broadcasting System

Wenjun Zhang, Yunfeng Guan, and Dazhi He

Contents

5.1 Introduction

This chapter provides an overview of the Digital terrestrial television (DTT) standard of China, known as Digital Terrestrial Multimedia Broadcasting (DTMB), which is capable of providing high-quality multimedia services including mobile broadcasting. China has the largest TV market in the world, and its proportion of television audience using terrestrial method is pretty high (see Figure 5.1). The rest of the chapter is structured into five sections covering technical specifications

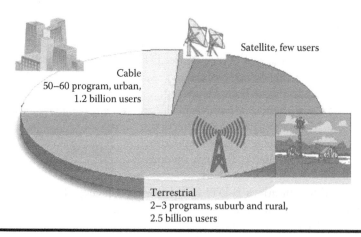

Figure 5.1   Proportion of television users in China.

of DTMB, specifications of DTMB single-frequency network (SFN), DTMB mobile application in SFN transmission network in Shanghai, technical difference compared to CMMB, and a conclusion of all.

5.2 Technical Specifications of DTMB

5.2.1 History and Roadmap

China started to develop its own digital television terrestrial broadcast standard in 1994. The following year, the central government founded the HDTV Technical Executive Experts Group (TEEG), with members from tens of domestic universities and research institutes. Three years later, the first HDTV prototype including high-definition encoder, multiplexer, modulator, demodulator, de-multiplexer, and high-definition decoder was developed, which covered all the key parts from transmitter to receiver. Its deployment in the live broadcasting of the celebration of the nation's 50th Anniversary on October 1, 1999, was an important success.

In 2001, China called for proposals for the Digital Terrestrial Television Broadcasting (DTTB) standard. Five proposals were submitted including the Advanced Digital Television Broadcasting–Terrestrial (ADTB-T), a single-carrier approach from HDTV TEEG, and the Digital Multimedia Broadcasting–Terrestrial (DMB-T), a multi-carrier approach from Tsinghua University. In 2003, another multi-carrier approach named as Terrestrial Interactive Multiservice Infrastructure (TiMi) was proposed by the Academy of Broadcasting Science. After lab tests, field trials, and intellectual property analyses, a united working group headed by the Chinese Academy of Engineering was founded in 2004 to incorporate the three proposals mentioned earlier.

Finally, in August 2006, the incorporated standard called "Frame structure, channel coding and modulation for digital television terrestrial broadcasting system" was issued [1]. It contains both single-carrier and multi-carrier options and supports various multiprogram SDTV/HDTV terrestrial broadcasting services. The announcement of this standard is indeed a milestone of Chinese digital TV industry. The standard was then made mandatory after August 1, 2007.

The commercialization of this standard can be gradually realized in three phases. First commercial trials were held in six cities including Beijing, Shanghai, Tianjin, Shenyang, Qingdao, and Qinhuangdao. During the Olympic Games in 2008, HDTV services were provided through terrestrial broadcasting. In the second phase, a larger-scale transition from analog to digital TV is being carried out. The last phase is to switch off existing analog TV permanently in 2020 (see Figure 5.2).

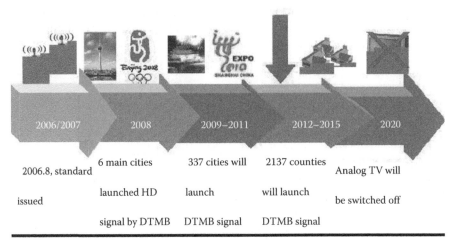

Figure 5.2   Roadmap of China's DTTB standard development.

5.2.2 Brief Description of the Chinese DTMB Standard

The block diagram of Chinese DTTB transmit system is shown in Figure 5.3. The signal is transmitted in frames. The functional descriptions of main blocks are listed later.

5.2.2.1 Frame Structure

Each frame contains a frame header (FH) and a frame body (FB). The FB consists of system information and coded data. Both FH and FB have the same symbol rate of 7.56 Msps. Three frame structures have been defined as shown in Table 5.1.

The FHs use pseudo-noise (PN) sequences. There are three different PN lengths: 420, 595, and 945. The three types of FH are abbreviated as PN420, PN595, and

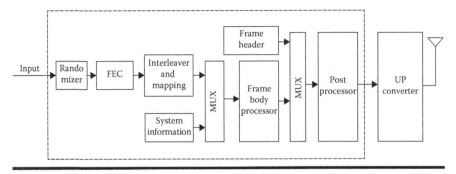

Figure 5.3   Block diagram of DTTB transmission system.

Table 5.1   Frame Structure of DTMB

	FH	*FB*
Frame structure 1	420 symbols (PN420), 55.6 μs	3780 symbols, 500 μs
Frame structure 2	595 symbols (PN595), 78.7 μs	3780 symbols, 500 μs
Frame structure 3	945 symbols (PN945), 125 μs	3780 symbols, 500 μs

PN945, respectively. PN420 and PN945 are made up of complete m-sequences (PN sequences) of lengths 255 and 511 with their cyclical extensions as preambles and postambles, while PN595 is made up of the first 595 symbols of an m-sequence of length 1023.

5.2.2.2 Frame Body Processing

As for the FB processing block, there are two options on the parameter C, which denotes the number of carriers. One is C1 mode indicating single-carrier modulation scheme, and the other is C3780 mode indicating multi-carrier modulation with 3780 subcarriers. In both options, the input data are coded with a low-density parity check (LDPC) with code rates of 0.4, 0.6, or 0.8. Constellation mapping scheme includes 64QAM, 32QAM, 16QAM, 4QAM, and 4QAM-NR (Nordstrom Robinson). It should be noted that in C3780 modes, a block interleaver in frequency domain is further applied within each FB.

5.2.2.3 System Information

Each FB contains 36 system information symbols that are BPSK mapped. The first four symbols are the indicator of C1 or C3780 modes, and the last 32 symbols are spread spectrum protected Walsh codes to indicate necessary system information including constellation mapping, LDPC code rate, and interleaving mode.

5.2.2.4 Forward Error Correction

LDPC code is used as part of the forward error correction (FEC) in the standard. There are three different LDPC code rates: 0.4, 0.6, and 0.8. The length of one code word is 7493 bits. BCH (762, 752) coding is concatenated outside the LDPC. The adopted BCH can correct only one bit error per LDPC code word. The BCH code does not contribute much for error correction, but it ensures that frames carry completely correct TS packets. A convolution symbol interleaver with two different depths is used with the LDPC. One is $M = 240$, and the other is $M = 720$. They both have the same number of interleaving branches, which are 52.

5.2.2.5 Square-Root-Raised-Cosine

A square-root-raised-cosine (SRRC) filter with a roll-off factor of 5% is used as shaping filter to limit the bandwidth of the transmitted signal to 8 MHz. Additionally, in C1 mode, dual pilots whose power is 16 dB lower than the total average power are optionally inserted at ±0.5 symbol rate onto the transmitted data.

5.2.2.6 Remarks and Explanations

There are some extra explanations to instruct the reader to understand the DTMB standard.

5.2.2.6.1 Single-Carrier C1 and Multi-Carrier C3780

It is interesting to point out that the option of multi-carrier modulation in the standard has no virtual subcarriers and pilots. Contrarily from classical orthogonal frequency division multiplexing (OFDM), DTMB does not make use of a cyclic prefix (CP), but introduces a PN sequence in the FHs. Figure 5.4 gives a graphic explanation.

Thus, the only difference between single carrier and multi-carrier in the standard is the iFFT of 3,780 points, which is included in the FB processor in Figure 5.3 and could be regarded as a data filter for FB. As a result, for both single carrier and multi-carrier, there are two types of channel estimation and compensation methods developed. One is an all-time domain processing approach (TD), which employs a time domain, code-enhanced, and data-directed adaptive decision feedback equalizer (DFE) with LMS algorithms. The other is a hybrid time- and frequency-domain processing approach (HTFD), which implements a channel estimator in time domain with the known FH and compensates the FB in frequency domain. The coexistence of single carrier and multi-carrier in the standard is designed to cope with different needs due to the various developing stages of radio and television systems in different areas of China and its vast population.

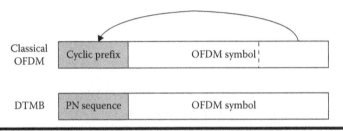

Figure 5.4   Replacement of the CP with a PN sequence in DTMB.

5.2.2.6.2 Dual Pilots

In the case of C1, two pilots can be inserted as an option at ±0.5 symbol rate to the baseband signal frame. The total transmission power of these two pilots is 16 dB lower than the total signal power. Pilots will be imposed upon each symbol of the calendar day frame (index starts from 0): under odd symbol indices, the real and imaginary parts of the symbol are 1 and 0, respectively; otherwise, the real and imaginary parts of the symbol are –1 and 0, respectively.

The purpose of dual pilots is to help the receiver's synchronization. By using the dual pilots, the receiver can quickly detect the carrier frequency offset (CFO) and timing frequency offset (TFO). The CFO will shift the position of dual pilot and the TFO will enlarge or shrink the position difference of dual pilots.

5.2.2.6.3 PN Phase

Among three kinds of FH, PN420 and PN945 can be a cyclic extended PN sequence, shifted with every different signal frame, while PN595 is the fixed part of PN1023 and has no cyclic extended. Figure 5.5 describes the detailed composition of three FHs.

The PN phase of PN420 and PN945 will be different from each frame. The purpose of this design for the shift PN phase is to help the channel estimation of receiver. When in a quasi-static channel, the combination of FHs with different PN phase can help the channel estimation to be more precise.

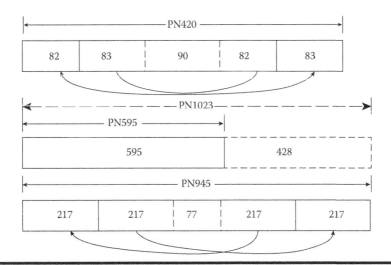

Figure 5.5 Frame structure of DTMB.

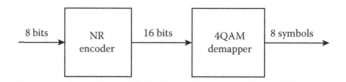

Figure 5.6   Block diagram of the 4QAM-NR.

5.2.2.6.4 4QAM-NR

In DTMB standard, there is one working mode named 4QAM-NR, which is the NR encoder concatenated with a 4QAM demapper. Figure 5.6 gives diagram for this mode.

The NR code is designed for the time domain equalizer of receiver. The equalizer can combine the NR decoder to greatly enhance the ability of equalizer, especially for the dynamic channel conditions [2].

5.2.3 Novelty of This Standard

The main technical innovations for the DTMB standard include the following:

Time-domain PN sequences are used as FHs. They can serve as training sequence for the equalizer at the receivers, and they may be also used as guard interval to mitigate inter-symbol interference. Choosing a known PN sequence as FH also provides higher spectrum efficiency, because the sequence can also be used for the channel estimation, instead of inserting both continuous and scattered pilot in traditional COFDM approach. Furthermore, it achieves faster channel acquisition since this can be done directly in time domain, compared with the European digital terrestrial TV standard Digital Video Broadcasting–Terrestrial (DVB-T).

The FEC code employed is an LDPC. Up to now, performance of LDPC is regarded as the best in coding theory. Compared with the concatenated code used in the first-generation digital terrestrial TV systems, e.g., DVB-T or the American ATSC, the LDPCs employed in DTMB have superior performances.* LDPC can provide superior error correction capability for better sensitivity, resulting in larger coverage with the same transmit power. The performance of the LDPC codes for AWGN channel adopted in DTMB is shown in Table 5.2.

The system information is protected with spread spectrum technology. DTTB defines many options such as single-carrier modulation (C1), multi-carrier modulation (C3780), three FH options, three FEC code rates, five constellation mappings, two interleaving depths, fixed or rotated PN in FH, etc. Different combinations result in hundreds of operation modes. The adopted spread spectrum technique

---

* The recent progress in RA (repeated accumulate) LDPC adopted in DVB-T2 (Digital Video Broadcasting–Second-Generation Terrestrial) has shown a better performance.

Table 5.2   LDPC Performance for Tolerance of Vulnerable (ToV)

Code Rate	Block Length (Bits)	Information Bits	E_b/N_0 (dB)	Performance away from Shannon Limit (dB)
0.4	7488	3008	2.1	2.3
0.6	7488	4512	2.3	1.6
0.8	7488	6016	3.3	1.2

for the system information allows receivers to identify correct working mode in all kinds of severely distorted channels.

5.3 Technical Specifications of DTMB for Single-Frequency Networks

An SFN is a network where all transmitters in different locations are synchronized to send the same signal simultaneously through the same frequency channel to realize the reliable coverage across a given territory. In DVB-T SFN, the transport stream (TS) from the multiplexer is transported first to the SFN adapter for adaptation. Then, the TS including mega frame initialization packets (MIPs) is transported to all transmitters through the distribution network and transmitted after being converted to radio frequency by synchronizing handling.

While in a DTMB SFN, the second frame (SF) structure ensures that there are integer numbers of TS packet in one second. So a second initialization packet (SIP) in one second is mixed with TS by the DTMB SFN adaptor to guarantee the synchronizing of different transmitters. The sketch map of DTMB SFN architecture is given in Figure 5.7.

The DTMB SFN adapter [3] includes two function modules: SIP insertion and bitrate adaptation. It is a key equipment to construct the SFN and realize the format conversion from input TS to SFN adaptation TS. The block diagram of the DTMB SFN adapter is illustrated in Figure 5.8.

5.3.1 SIP Insertion Function

5.3.1.1 SF Definition

SF is the basic data frame for the DTMB SFN adapter, and the time duration of one SF is 1 s, including all the information bits to modulate eight superframes. The quantities of TS packets included in one SF are illustrated in Table 5.3 with different modulation orders, frame head modes, and code rates. It should be noted that the 4QAM-NR mode actually means that NR coding is performed before 4QAM mapping. From the view of mapping side, the NR coding is a kind of quasi-orthogonal premapping.

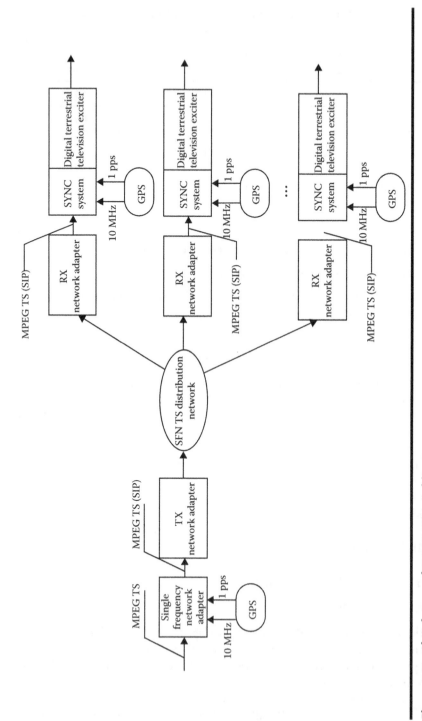

Figure 5.7 Sketch map of DTMB SFN architecture.

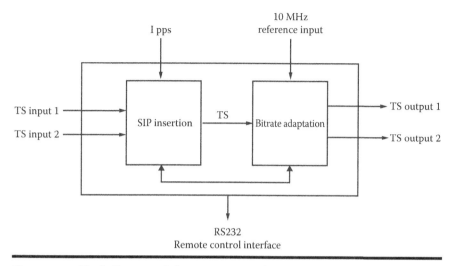

Figure 5.8 Block diagram of the DTMB SFN adapter.

5.3.1.2 SIP Insertion and Handling

The DTMB SFN adapter inserts one SIP into the input TS every 1 s. The inserting moments are in alignment with 1 pps from a GPS clock, as is shown in the sketch map of SIP insertion depicted in Figure 5.9.

The transmitters in the SFN detect and receive the SIPs in the TS, which include the maximum delay $T_{delay\text{-}max}$. The relation between the maximum delay $T_{delay\text{-}max}$, added delay of exciter $T_{delay\text{-}add}$, and $T_{delay\text{-}transmitted}$ is expressed by Equation 5.1:

$$T_{delay\text{-}max} = T_{delay\text{-}add} + T_{delay\text{-}transmitted}, \tag{5.1}$$

where

$T_{delay\text{-}max}$ is the maximum delay indicating the uniform transmission time of TS of several transmitters relative to 1 pps of GPS

$T_{delay\text{-}transmitted}$ is the distribution network transmission delay indicating the transmission time of TS through distribution network after being transported by the SFN adapter

$T_{delay\text{-}add}$ is the added delay of exciter indicating the excess delay handled by each exciter so that several transmitters can simultaneously send

The value of maximum delay will not exceed 0.9999999 s. See the sketch map of SIP handling in Figure 5.10.

The format of the SIP inserted into the TS is the same as that of MPEG2-TS packet, made up of a 4-byte header and a 184-byte data field. The organization of the SIP is shown in Figure 5.11.

Table 5.3 Sketch Map of SIP Insertion

Modulation Order	Type of Frame Header	Code Rate	Payload Data Rate (bps)	TS Packet Rate (Packet/s)
4QAM-NR	420	0.8	5,414,400	3,600
	595	0.8	5,197,824	3,456
	945	0.8	4,812,800	3,200
4QAM	420	0.4	5,414,400	3,600
		0.6	8,121,600	5,400
		0.8	10,828,800	7,200
	595	0.4	5,197,824	3,456
		0.6	7,796,736	5,184
		0.8	10,395,648	6,912
	945	0.4	4,812,800	3,200
		0.6	7,219,200	4,800
		0.8	9,625,600	6,400
16QAM	420	0.4	10,828,800	7,200
		0.6	16,243,200	10,800
		0.8	21,657,600	14,400
	595	0.4	10,395,648	6,912
		0.6	15,593,472	10,368
		0.8	20,791,296	13,824
	945	0.4	9,625,600	6,400
		0.6	14,438,400	9,600
		0.8	19,251,200	12,800
32QAM	420	0.8	27,072,000	18,000
	595	0.8	25,989,120	17,280
	945	0.8	24,064,000	16,000

Table 5.3 (continued) Sketch Map of SIP Insertion

Modulation Order	*Type of Frame Header*	*Code Rate*	*Payload Data Rate (bps)*	*TS Packet Rate (Packet/s)*
64QAM	420	0.4	16,243,200	10,800
		0.6	24,364,800	16,200
		0.8	32,486,400	21,600
	595	0.4	15,593,472	10,368
		0.6	23,390,208	15,552
		0.8	31,186,944	20,736
	945	0.4	14,438,400	9,600
		0.6	21,657,600	14,400
		0.8	28,876,800	19,200

1 pps signal of GPS
1 pps signal of GPS
1 pps signal of GPS
1 s
1 s
18–19
20–188
Power control
Filled bytes
t
SIP packet insertion
SIP packet insertion
SIP packet insertion

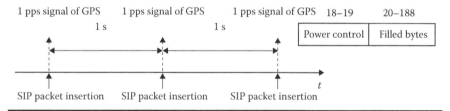

Figure 5.9 Sketch map of SIP insertion.

1 pps signal of GPS
1 pps signal of GPS
1 pps signal of GPS
1 s
1 s
18–19
20–188
Power control
Filled bytes
Transmission delay
Added delay
Transmission delay
Added delay
SIP packet
SIP packet
t
Maximum delay
Maximum delay

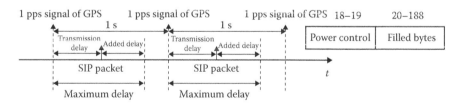

Figure 5.10 Sketch map of SIP handling.

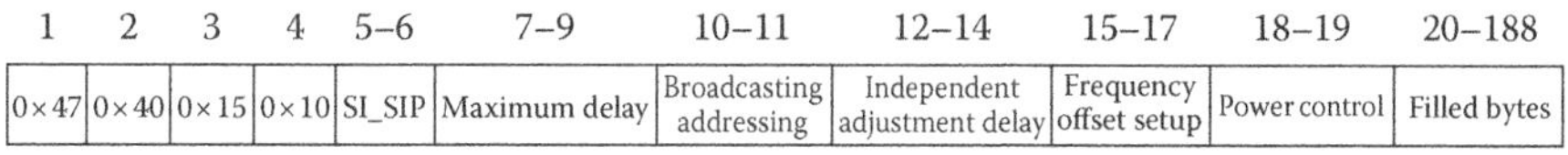

1
2
3
4
5–6
7–9
10–11
12–14
15–17
18–19
20–188
0×47
0×40
0×15
0×10
SI_SIP
Maximum delay
Broadcasting addressing
Independent adjustment delay
Frequency offset setup
Power control
Filled bytes

Figure 5.11 Sketch map of SIP organization.

The data sections of SIP are defined as follows:

■ SIP header: made up of 32 bits in compliance with the requirement in GB/T 17975.1-2000, herein PID number is 0x0015.
■ SI_SIP: made up of 16 bits P0–P15 and all the bits are defined in Table 5.4.

Table 5.4   Definition of SI_SIP

Bits	Purpose	Definition
[P0: P1]	FH mode	00: PN420
		01: PN595
		10: PN945
		11: reserved
[P2]	Number of carriers	0: C = 1
		1: C = 3780
[P3: P5]	Constellation mapping	000: 4QAM-NR
		001: 4QAM
		010: 16QAM
		011: 32QAM
		100: 64QAM
		101~111: reserved
[P6: P7]	Code rate	00: 0.4
		01: 0.6
		10: 0.8
		11: reserved
[P8]	Interleaving mode	0: 240
		1: 720
[P9]	Double pilot	0: used
		1: not used
[P10]	PN phase	0: no rotation
		1: rotation
[P11: P15]	Null	Reserved

Figure 5.12   Format of null packets for SFN adaptation.

■ Maximum delay: made up of 24 bits, the unit is 100 ns, and the range of maximum delay is 0x000000–0x98967F, corresponding to the time range of maximum delay of 0–0.9999999 s. The value of maximum delay shall be larger than the transmission delay in the program distribution network.

■ Broadcasting addressing: made up of 16 bits to address some transmitter in SFN, and the range of addressing is 0x0000–0xFFFF, herein 0x0000 indicates addressing all the transmitters.

■ Independent adjustment delay: made up of 24 bits. Based on maximum delay, add independent adjustment delay to a given transmitter by broadcasting addressing to ensure that all the transmission signals of transmitters in SFN are in compliance with the specific relations of delay.

■ Frequency offset setup: made up of 24 bits to add frequency offset value to a given transmitter by broadcasting addressing. The deliberate frequency offset value is relative to the center frequency of the RF channel in use. The unit is 1 Hz, and the range is $[-8388,608, 8388,607] \times 1$ Hz.

■ Power control: made up of 16 bits to control the transmitting power of a given transmitter by broadcasting addressing. The highest bit is power switching, bit "1" indicating on, bit "0" indicating off; the lower 15 bits are power control word, with the unit of 0.1 dBm and the range of $[0, 32,767] \times 0.1$ dBm.

■ Filled bytes: totally 169 bytes, each byte being 0xFF.

5.3.2 Bitrate Adaptation Function

According to system output bitrate clock, the DTMB SFN adapter inserts one SIP into the TS at the rising edge of each 1 pps from a GPS clock and reads data from front-end buffer at other locations. If the data are not enough to fill in one TS packet, new null packets for SFN adaptation are then inserted to produce bitrate adaptation of TS, the format of which is illustrated in Figure 5.12. The output TS bitrate of the DTMB SFN adapter is completely equal to the payload data rate for transmitter operation mode required by the SFN adapter, locked at the 10 MHz clock reference from a GPS clock.

5.4 DTMB Mobile Application in Shanghai SFN Transmission Network

Since 2002, Shanghai has used RF channel 39 (718–726 MHz) for mobile DTV service to public transportation vehicles and has achieved great commercial success.

Commuters can enjoy real-time mobile DTV service on the bus, while broadcasters see their advertisement income growing, thanks to a larger audience. Initially, the European terrestrial broadcast standard DVB-T was chosen to be applied to the building of the SFN in Shanghai, with transmission mode 2K FFT, 1/4 guard interval, QPSK, and 2/3 code rate. An SFN network was built up step by step over 3 years, with four SFN transmitters being built, emission time being fine-tuned and adjusted, and diversity receivers being introduced. Information about DVB-T SFN transmitters on CH-39 is listed in Table 5.5.

The field testing results of DVB-T is shown in Figure 5.13.

In 2006, the DTMB terrestrial digital TV standard was officially released. The corresponding laboratory test and field trials report showed that some of its

Table 5.5 Technical Parameters of Stations in the Channel 39

Transmitter Name	Antenna Height (m)	Power (W)
Oriental Pearl	420	1200
Radio and Television Bureau	130	1100
Oriental TV	140	700
Education TV	92	1100

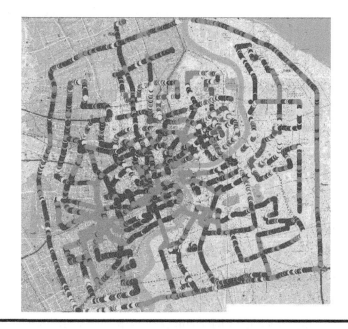

Figure 5.13 Coverage map with Oriental Pearl transmitter only.

Table 5.6   Parameters of One Mode of DTMB

Carrier mode	Single carrier
Modulation	4QAM
PN Mode	PN595
Code rate	0.8
Interleaving length	Int720 [720 ms]
Net transmission capacity	10,396 Mbps

Table 5.7   Performance of One Mode of DTMB

Channel	Specification Requirement (dB)	Receiver A	Receiver B
AWGN	6	5.5	6.4
Rayleigh	9.5	8.7	8
Rice	6.5	6	7.4
0 dB 30 μs	11	10.5	9.5
0 dB MaxLen	60	289	78
BU 70 Hz	17	14.9	15.5

performance parameters and data capacity outperformed DVB-T that was already in operation in Shanghai. Parameters of one mode of DTMB with single-carrier QPSK modulation are listed in Table 5.6.

This mode is recommended by specification of DTMB receiver [4], released in 2011. The specification has listed some mandatory performance requirements for receivers sold in China. Table 5.7 shows performance requirements of this mode and some receivers' test results.

From Table 5.7, some parameters, i.e., 0 dB MaxLen and BU70 Hz [4], have very promising results, which are critical for SFN implementation, especially for the application based on existing DVB-T infrastructure.

5.4.1 Implementation of SFN Transmit Network of DTMB

To fully utilize the existing infrastructure, the SFN transmitter network of DTMB will be based on existing transmitter sites. The network construction has three stages.

5.4.1.1 Stage 1

With two transmitters: Oriental Pearl and HongQiao. Field tests were conducted to obtain the coverage of the two stations. Based on field test results, a DTMB coverage planning tool was designed for SFN and used to compute the overlapping coverage area. The Oriental Pearl station (420 m), located in the eastern part of urban Shanghai, is the city's highest TV tower, while HongQiao station is the highest TV tower (150 m) in western urban Shanghai. These two stations can cover a majority part of urban Shanghai.

By using software planning tool and field test results, overlapping area can be found that the difference of RSL between the two stations is less than 5 dB. Figure 5.14a shows the overlapping area. The receiving conditions of SFN with the two stations are shown in Figure 5.14b.

From Figure 5.14b, it is clear that some receiving problems can be resolved by the SFN, while some new receiving spots may occur in the overlapping areas. Stage 2 will try to overcome these problems.

5.4.1.2 Stage 2

Field testing of overlapping areas is conducted with both transmitters on. With the lab test results listed in Table 5.7 and analysis of planning tool, some adjustments of time delay and powers are made to achieve better receiving effect. The SFN network of two stations must have the SFN adaptor, which must comply with the draft of "Technical specifications and methods of measurement for Digital Terrestrial Television Broadcasting Modulator" [3]. This specification supports the adjustment of time delay of each station respectively. The aim of such adjustment is to reduce time delay between signals transmitted from two stations so that the channel profile does not exceed receiver capability. Adjusting transmitting power and direction is another effective method to mitigate multipath distortion. After some adjustment, the improved receiving conditions of SFN with Oriental Pearl and HongQiao stations are shown in Figure 5.15. It should be noted that the failure spots marked by rectangle labels have been reduced substantially.

5.4.1.3 Stage 3

Based on the experience of the first two SFN stations, the network planning algorithm needs to take the following three matters into consideration when building more stations. First is traditional radio power coverage, which is the base of SFN network and should be as large as possible. Second is overlapping area of strong signal between adjacent stations, which should be as smaller as possible. Last is average time delay in the overlapping area, which is a key factor in network's optimization. With the help of planning tools and road testing, the complete DTMB SFN network with six transmitters is successfully implemented. In stage 3, another

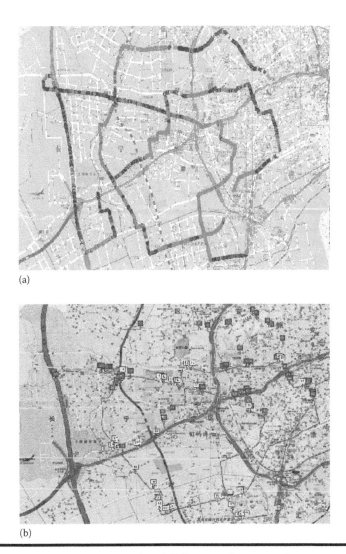

(a)

(b)

Figure 5.14 (a) Coverage map—signal differences between the two transmitters. Light color indicates there is not much contrast between the differences of RSL while the saturated color represent the difference of RSL is little. (b) Receiving effect with two transmitters: Oriental Pearl and HongQiao. Labels with saturated color represent the failure spots with Oriental Pearl transmitter only; labels with light color represent the failure spots with both transmitters.

Figure 5.15 Receiving effect with the broadcasting of Oriental Pearl and HongQiao station after adjusting transmitter power and antenna directivity. Labels with saturated color represent the failure spots with Oriental Pearl transmitter only; labels with light color represent the failure spots with both transmitters.

four stations is set up to achieve full signal coverage of Shanghai. Table 5.8 lists the information of SFN with six stations. The improved DTMB SFN coverage and reception will fully satisfy the need of mobility TV in Shanghai.

The complete DTMB SFN is shown in Figure 5.16. Figure 5.16a is the RSL coverage of the DTMB SFN for the city's downtown area. Figure 5.16b reflects corresponding receiving signal conditions.

Table 5.8 Information of SFN with Six Stations

Name	Power (W)	Signal Delay (μs)	Antenna Height (m)
Oriental Pearl	1000	0	420
HongQiao	700	14	150
Orient TV station	1000	2	140
DaNing	500	9	100
Education TV station	600	5	92
Radio and Television Bureau	400	5	130

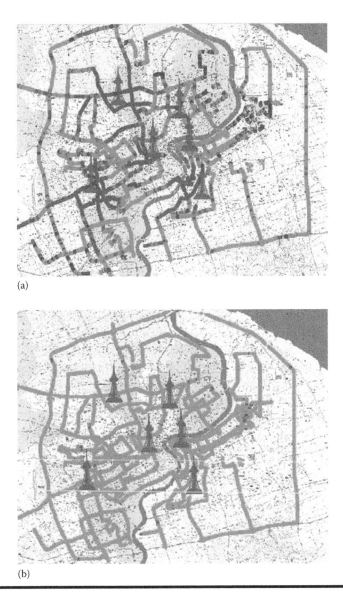

(a)

(b)

Figure 5.16 **(a) Coverage map with six SFN stations. Light color represents the RSL between –70 and 20 dBm; Saturated color represents the RSL lower than –70 dBm. (b) Receiving conditions with six SFN stations. Light color represents the successful reception and saturated color is for the failure of reception.**

As shown in Figure 5.16, the SFN of DTMB in Shanghai achieved good coverage and much improved reception. In our tests, the successful reception rate achieved over 97.76% when RSL is greater than −80 dBm.

5.4.2 Technical Challenges and Solutions

Reviewing the implementation and testing process, some technical challenges to the design of SFN in Shanghai are summarized from both transmitter side and receiver side.

5.4.2.1 Challenges on the Transmitter Side

The challenge in the design of transmitters is how to balance the SFN coverage and reception. There are two concerns. First is to achieve full signal coverage with sufficient RSL. Second is to have good reception for the overlapping areas of SFN network.

The solution to the first concern is to apply wide coverage pattern to obtain the best coverage of the six stations, with the use of the RF propagation network planning tool, geographical database, and corresponding 3D building with demographics database. Besides, the RF propagation network planning tool should be applied to reach the coverage requirement to such extent that the overlapping areas between different stations are sufficiently small.

The solution to second concern is time delay adjustment, which is proved to be an effective method to reduce the multipath impact to receiver in the overlapping areas of SFN. Reduction of time delay of arrived signals between two adjacent stations will bring about a less harmful channel profile, which is a useful and essential way for the design and evaluation of the SFN coverage and transmitter design.

5.4.2.2 Challenges on the Receiver Side

There are two concerns in the receiver side: how to deal with the severe multipath distortion caused by SFN and how to track the Doppler spread of dynamic fading channel.

The solution to the first issue is using time domain DFE with a virtual center. At the beginning of DTMB deployment, only one station from Oriental Pearl broadcasted the DTMB signal. Some DTMB set-top boxes equipped with early generation of chipset worked well. However, the reception problems occurred when the HongQiao station started broadcasting, especially in those overlapping areas. Test results showed that most of the reception problems can be categorized as SFN reception difficulties caused by strong pre-echo and long post-echo. Tests and simulations have confirmed that DFE with a virtual center is able to deal with the SFN channel in most cases. This technology can refer to [5].

Another issue's solution is the use of diversity DFE. Detailed information can be found in [6], which will use the maximal ratio algorithm to combine the signals from different kinds of DFE. A typical application of diversity DFE is the electronics display board used in the city's bus stations. The SFN network deteriorated the channel profile in the overlapping area, and the bypassing vehicles led to the spreading of channel Doppler, which impacted the channel estimation and equalizer status. Receivers with diversity DFE had shown a good performance for the dynamic channel in these overlapping areas.

5.5 Comparison between DTMB and CMMB

5.5.1 Overall Structure of CMMB

CMMB is a mixed satellite/terrestrial wireless broadcasting system designed to provide audio, video, and data service for handheld receivers with less than 7 in. wide LCD display. This system is backed by the State Administration of Radio, Film and Television and has been specified as industry standard GY/T 200.1-2006 and GY/T 220-2006. The overall structure of CMMB can be illustrated in Figure 5.17.

A scheme of broadcasting from satellite and terrene simultaneously is designed in CMMB. Signal from satellite in S band cover the whole nation while signal from the network on the terrene cover urban region with dense population, with

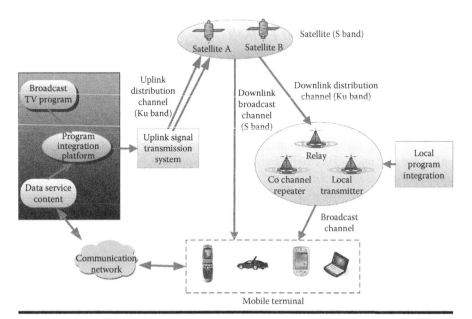

Figure 5.17   Overall structure of CMMB.

better quality of service. By the way of satellite and terrene, the CMMB try to be a seamless cover system. However, until now, only the terrene broadcasting has been constructed and applied.

5.5.2 Main Features of CMMB

Different with the DTMB, CMMB is designed for provide television and radio broadcasting services to car television and handheld terminals with small screen such as mobile phone, PDA, etc [7]. The most important features of CMMB [8] are as follows:

1. Capability of providing digital television broadcasting program, integrated information and emergency broadcasting, capability of incorporated seamless coverage with satellite and terrestrial network, and capability of supporting public services.
2. Supporting mobile phone, PDA, MP3, MP4, digital camera, notebook, and small terminals in car, train, ships, and plane to receive multimedia service such as video, audio, and data.
3. Supporting the management system with encryption, authorization, and controlling, supporting unified standard and unified operation, and supporting national roaming of the subscribers.

5.5.3 Frame Structure, Channel Code, and Modulation of CMMB

The frame structure, channel code, and modulation of CMMB is named as Satellite–Terrestrial Interactive Multi-service Infrastructure (STiMi), which is collectively referred to a kind of coding and modulation technology based on satellite and terrestrial channel. STiMi is the core transmission technology of CMMB. The functional structure is shown in Figure 5.18.

From Figure 5.18, the input data stream from upper layer is processed by FEC, interleaving, and constellation and then multiplexed with scatter pilots and continual pilots. Afterward, the data are processed by OFDM modulation, and then the FH is inserted to form the frame in the physical layer. Finally, after up-converting, the signal is transmitted to the aerial. The main characters of STiMi are as follows:

1. Adopting RS code and highly structured LDPC code to improve reception sensitivity, reducing hardware complexity of encoder and decoder, suitable for ASIC implementation.
2. Adopting BPSK, QPSK, and 16QAM mapping, suitable for service transmission with different quality of service.

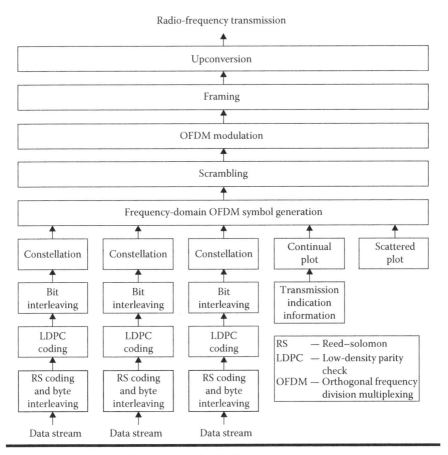

Figure 5.18 Functional structure of STiMi in CMMB.

3. Adopting OFDM modulation technology to improve capacity against multipath fading.

4. Adopting timeslot technology, to reduce power consumption in terminal.

5. Adopting spreading beacon technology in time domain, to reduce capture time and to improve capacity against carrier offset and multipath delay spread.

6. Adopting pilot technology to reduce complexity of demodulation and to guarantee channel estimation and equalization in complex wireless transmission conditions.

5.5.4 Technical Summary of STiMi

STiMi is designed for mobile handheld reception, high-sensitivity, mobility, and battery supply. This system is based on timeslot frame structure and OFDM

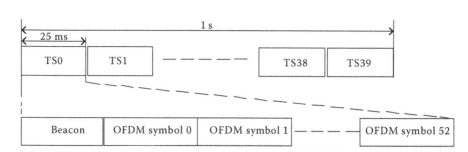

Figure 5.19 Frame structure of STiMi in CMMB.

modulation technology, logic channel technology, and synchronization beacon technology. The frame structure of STiMi is described in Figure 5.19.

The physical layer of STiMi includes one control logical channel (CLCH) and 1–39 service logic channel (SLCH). The CLCH is used to broadcast system control information to terminals and occupies only timeslot 0. The SLCH can be configured to use one or more several timeslots to provide the difference in transmission capacities for broadcasting services. The allocation of CLCH and PLCH is illustrated in Figure 5.20.

The channel coding and modulation scheme for each PLCH can be configured respectively, while the configuration of CLCH is fixed with RS(240, 240), LDPC 1/2, BPSK, scrambler initial value 0. The channel coding and modulation scheme of SLCH can be configured to different modes to provide different throughputs.

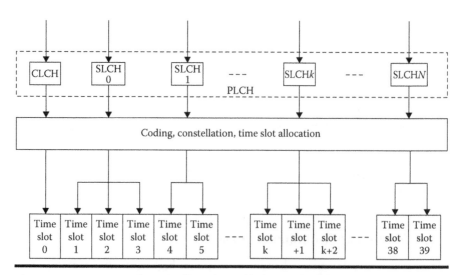

Figure 5.20   CLCH and PLCH allocation of STiMi in CMMB.

One timeslot consists of one beacon and 53 OFDM symbols, which is described in Figure 5.19.

The frame structure of STiMi is one of the key technologies of CMMB. Difference with the continual transmission with one mode in DTMB, CMMB with timeslot technology is especially suitable for the power-saving reception. Mobile terminals commonly powered by battery only need to decode the CLCH in timeslot 0 and then receive the signal of several timeslots related to one SLCH. Within a second time, the mobile terminal isn't work in most of time.

5.6 Conclusions

This chapter has provided a technical review on Chinese terrestrial digital television system, known as DTMB. Furthermore, it has introduced the SFN specification of DTMB, which is very important for the application of this standard. Finally, a six-transmitter SFN for DTMB system, successfully built in Shanghai, has been elaborated. It is the first SFN implementation of single-carrier QPSK-modulated system. Valuable experience was gained from different technical approaches, i.e., time delay adjustment, antenna direction, and power adjustment. DFE with virtual center and diversity DFE are very effective in terms of mitigating the strong pre-echo and long delay echo in the dynamic channel condition. Some reception algorithms for SFN of DTMB are promising techniques to be investigated.

The system of CMMB is overviewed, and some technical features of CMMB are summarized. The physical layer technology, named as STiMi in CMMB, is introduced. The timeslot frame structure of STiMi is also compared with DTMB.

References

1. GB 20600-2006, *Framing Structure, Channel Coding and Modulation for Digital Television Terrestrial Broadcasting System*, June 2006.
2. D. He, Y. Guan, W. Liang, F. Ju, and W. Zhang, Combined NR decoding with decision feedback equalizer for Chinese DTTB receiver, *IEEE Transactions on Broadcasting*, 54(3), 477–481, September 2008.
3. GY/T 229.1-2008, *Technical Specification and Methods of Measurement for Digital Terrestrial Television Broadcasting Single Frequency Network Adapter*, March 2008.
4. GB/T 26686-2011, *General Specification for Digital Terrestrial Television Receiver*, December 2011.
5. Patent, A kind of time domain decision feedback equalizer with a virtual center, application number in China, 200510065448.2.
6. Patent, Time domain adaptive equalizer and equalizing method, application number in China, 200910047642.6.
7. J. Zhou, Z. Ou, M. Rautiainen, T. Koskela, and M. Ylianttila, Digital television for mobile devices, *IEEE Multimedia*, 16(1), 60–70, January 2009.
8. State Administration of Radio, Film, and Television (SARFT) of China, Chinese mobile multimedia broadcasting technical white paper, June 2008.

Chapter 6

DVB-T2 for Mobile and Mobile DVB-T2 (T2-Lite)

David Gozálvez, David Gómez-Barquero,
Iñaki Eizmendi, Gorka Berjón-Eriz, and Manuel Vélez

Contents

6.1 Introduction

The first-generation European standard for the provision of digital terrestrial TV (DTT), known as Digital Video Broadcasting–Terrestrial (DVB-T), presents two important limitations for the provision of mobile services. On the one hand, it does not feature time slicing, and receivers cannot accomplish power saving by switching off their front ends during some portions of time. On the other hand, the interleaving duration is limited to few milliseconds due to the lack of time interleaving (TI) at the physical layer. DVB-H (handheld) reuses the physical layer of DVB-T and incorporates additional features at the link layer in order to adapt the transmission to the handheld environment, i.e., time slicing for power saving, and multi-protocol encapsulation–forward error correction (MPE-FEC) for time diversity. Although it is possible to build shared DVB-T/H networks, stationary and mobile services cannot coexist in the same frequency channels and, thus, dedicated DVB-H networks are eventually required to achieve good coverage levels. The Japanese system Integrated Services Digital Broadcasting–Terrestrial (ISDB-T) supports in-band transmission of mobile services with one segment (see Chapter 2). In this case, mobile services in ISDB-T are transmitted in a narrow portion of the entire bandwidth, enabling very low power consumption in mobile receivers, while the rest of the bandwidth is dedicated to the transmission of stationary services.

The second-generation European standard for the provision of DTT is known as DVB-T2 (second-generation terrestrial) and incorporates some important additions for the provision of mobile services, including a very flexible time interleaver at the physical layer, which can provide interleaving durations ranging from hundreds of milliseconds up to several seconds. DVB-T2 also introduces the utilization of physical layer pipes (PLPs) for the transport of data. Each PLP is a virtual container that can be configured with a particular set of transmission parameters, including the constellation, the code rate, and the TI. By means of multiple PLPs, it is possible to allocate multiple use cases, e.g., fixed, portable, and mobile, in the same frequency channel. However, the simultaneous provision of stationary and mobile services in DVB-T2 with multiple PLPs is limited by the fact that the fast Fourier transform (FFT) mode and the pilot pattern must be common for all the

PLPs in a T2 multiplex.* The T2-Lite profile has been recently incorporated into the DVB-T2 specification as a manner to improve the coexistence of stationary and mobile services in DVB-T2 networks. The objective is to combine in the same T2 multiplex a DVB-T2 signal optimized in terms of FFT size and pilot pattern for high throughput, together with a T2-Lite signal optimized for high robustness. T2-Lite signals can also be transmitted as a stand-alone signal that occupies an entire frequency channel. The new profile is based on a limited subset of the existing DVB-T2 standard aimed at mobile broadcasting and designed to allow simpler receiver implementation by avoiding the most complex configurations and by establishing further restrictions in terms of memory and data rate. In particular, it is stated that the chip area of a T2-Lite demodulator is less than half the chip area of a DVB-T2 demodulator. The introduction of additional profiles was foreseen in DVB-T2 with the inclusion of future extension frames (FEFs). Thanks to this feature, it is possible to efficiently share the capacity of one frequency channel in a time division manner between different T2 signals, with each signal having specific time slots. The T2-Lite profile will be automatically implemented in the next generation of DVB-T2 receivers since it has been introduced as part of the DVB-T2 standard in the version 1.3.1 of the specification [1]. This will allow chip manufacturers and network operators to benefit from economies of scale and reduce the costs associated with T2-Lite. Furthermore, T2-Lite-only receivers with significantly reduced chip size and costs can appear in the future to address the mobile/portable market.

This chapter investigates the capabilities of DVB-T2 for the provision of mobile services. The two techniques available in the standard for the coexistence of stationary and mobile services, i.e., multiple PLPs and T2-Lite, are studied and compared. The performance is evaluated by means of computer simulations together with laboratory measurements.

The rest of the chapter is structured as follows. Section 6.2 introduces the provision of mobile services in DVB-T2 and reviews the new transmission techniques included in the standard. Section 6.3 describes the T2-Lite profile: the restrictions that have been applied to reduce the complexity of receivers, the elements that have been removed from the specification, and the new elements that have been added for higher robustness and flexibility. Section 6.4 presents some simulation results that illustrate and compare the performance of DVB-T2 and T2-Lite in mobile scenarios. Section 6.5 is dedicated to the laboratory tests carried out to study the influence of different configuration parameters in the mobile performance of DVB-T2. Section 6.6 explains the advantages of diversity techniques in DVB-T2 and T2-lite from a network planning point of view. Finally, the chapter is concluded in Section 6.7.

---

* Stationary services are generally transmitted with large FFTs and sparse pilot patterns in order to achieve a high spectral efficiency with sufficient robustness for rooftop antennas. On the other hand, reception in mobile scenarios requires the utilization of smaller FFTs and more dense pilot patterns to follow the rapid variations in the time and frequency domain and to cope with the inter-carrier interference that is caused by the Doppler spread.

6.2 Mobile Reception in DVB-T2

The DVB-T standard [2] was originally designed for fixed and portable reception, and it does not incorporate TI. Despite a penalization in terms of carrier-to-noise ratio (CNR), mobile reception is still possible, and the wide availability of deployed DVB-T networks has motivated the provision of vehicular DVB-T services in several countries such as Germany or Taiwan [3]. These networks are planned for portable indoor reception without directive rooftop antennas, and they also enable mobile reception with vehicle mounted antennas. In contrast with DVB-T, the Korean standard Terrestrial–Digital Multimedia Broadcasting (T-DMB) (see Chapter 3) and the Japanese standard ISDB-T feature TI spanning hundreds of milliseconds, which provides better robustness against impulsive noise and fast fading in mobile scenarios.

DVB-T2 introduces several improvements over DVB-T that enable a higher robustness, including better FEC, rotated constellations, TI, and distributed multiple-input single-output (MISO). The convolutional and Reed–Solomon (RS) codes used in DVB-T for channel coding have been replaced by a concatenation of low-density parity-check (LDPC) and Bose–Chaudhuri–Hocquenghem (BCH) codes, achieving a performance close to the channel capacity in additive white Gaussian noise (AWGN). In addition, DVB-T2 features rotated constellations as a manner to improve the performance of the system in fading channels. A time interleaver has been included at the physical layer in order to combat impulsive noise and benefit from time diversity in mobile scenarios. It can be configured on a service basis and provide interleaving durations ranging from few milliseconds up to several seconds. The TI is also very flexible and offers different trade-offs in terms of transmission robustness (time diversity), latency, and power saving. For the first time in terrestrial broadcasting systems, DVB-T2 includes a distributed MISO technique, based on the Alamouti code [4], for improving the reception in single frequency networks (SFNs). Distributed MISO reuses the existing network infrastructure by operating across the antennas of different transmitters and, therefore, can be used without any additional cost for network operators.

6.2.1 Forward Error Correction

DVB-T2 inherits the FEC coding scheme from the DVB-S2 (second-generation satellite) standard based on the concatenation of LDPC and BCH codes. The inclusion of LDPC codes achieves a performance close to the channel capacity in AWGN, whereas the BCH code removes the presence of error floors. Generally, it is stated that the combination of LDPC and BCH codes outperforms the combination of convolutional and RS codes in about 3 dB in AWGN. The utilization of irregular LDPC codes in DVB-T2 results in a nonuniform error protection of each code bit, as it depends on the column weight of the parity check matrix [5]. In a similar manner, the reliability of the bits in the case of 16QAM, 64QAM, and

256QAM is not uniform either. For the purpose of controlling the correspondence between code bits and constellation points, a bit interleaver is placed between the LDPC encoder and the time interleaver.

6.2.2 Rotated Constellations

Rotated constellations were introduced in DVB-T2 for improving the system performance in fading scenarios [6] by means of signal-space diversity [7]. When rotated constellations are used, a rotation is applied to the constellation symbols in such a manner that the two components in the imaginary plane, in-phase I and quadrature Q, carry all the binary information. Then, the real and imaginary parts of each symbol are cyclically delayed and end up being transmitted in different subcarriers. This increases the reception robustness in the presence of deep fades but requires a high demodulation complexity (see Chapter 11).

6.2.3 Time Interleaving

The time interleaver in DVB-T2 consists of a block interleaver that operates on a cell level. Each cell represents a complex symbol that is transmitted in a different subcarrier within the orthogonal frequency division multiplexing (OFDM) symbols. The time interleaver takes as input groups of cells referred to as TI blocks that are interleaved by the time interleaver one after the other. After TI, the cells are mapped to T2 frames for transmission. DVB-T2 does not support fast zapping, and receivers have to wait until the complete reception of one TI block before they can de-interleave and process the FEC blocks. The longer the interleaving duration, the longer the receivers must wait prior to the de-interleaving of the TI blocks. The average channel change time is approximately 1.5 times the interleaving duration and, therefore, long TI in DVB-T2 cannot be provided at the physical layer with tolerable channel change times.

If there is more than one PLP, the data channels are multiplexed in time slices within the T2 frames, with each PLP generally carried in one time slice per frame. In this case, subslicing, frame skipping, and multi-frame interleaving can be used when mapping the PLPs to the T2 frames for transmission [8]. With subslicing, the PLPs are carried in multiple subslices per frame. This accomplishes a more uniform distribution of information over time and improves time diversity, although it reduces the power saving in receivers. Frame skipping is a mechanism by which the PLPs are transmitted only in a subset of T2 frames regularly spaced over time, which gives better power saving at the expense of worsened time diversity. Lastly, multi-frame interleaving extends the transmission of each TI block beyond one T2 frame (<250 ms) and across multiple frames. This results in better diversity especially against shadowing, but increases the channel change time. It must be noted that the interleaving duration in DVB-T2 is limited by the amount of time de-interleaving (TDI) memory in receivers and also by the Modulation and Coding

(MODCOD) configuration. Broadly speaking, interleaving durations in the order of seconds are possible only for low data rate services. For more details about TI in DVB-T2, please refer to Chapter 12.

6.2.4 Channel Estimation

As with DVB-T, DVB-T2 relies on the transmission of scattered pilots for channel estimation. Scattered pilots are transmitted with reference information that is known to the receiver and are placed regularly inside the OFDM symbols according to the pilot pattern. Channel estimates are computed in the subcarriers with pilot information, whereas interpolation is used to obtain the estimates in the rest of subcarriers. In particular, time interpolation is performed first between scattered pilots and across OFDM symbols. Then, frequency interpolation is performed on each OFDM symbol and across all subcarriers [9]. The separation between pilots defines the Nyquist limit of channel estimation, i.e., the maximum channel variations that can be followed in the receiver. Denser pilot patterns increase the Nyquist limit at the expense of capacity, as a reduced number of subcarriers can be used for the carrying of data. DVB-T2 supports eight pilot patterns (from PP1 to PP8) with different separations in time and frequency between scattered pilots. One special case of channel estimation in DVB-T2 is pilot pattern PP8, which has been designed to be used together with the so-called CD3-OFDM technique [10]. In this case, estimates are computed also in data subcarriers after decoding the information.

The choice of the FFT size is related to the spectral efficiency of the system and the tolerance against inter-carrier interference (ICI). For the same absolute value of guard interval, larger FFTs allow the deployment of SFNs with a lower fraction of the transmission time assigned to the guard interval. For the same fraction of guard interval, larger FFTs allow a longer distance between transmitters. On the other hand, the robustness of the transmitted signal against ICI is proportional to the separation between subcarriers and inversely proportional to the carrier frequency. This means that a larger FFT size in the same channel bandwidth presents a lower tolerance against Doppler, and, therefore, the maximum velocity that is supported by the system is lower.

6.2.5 Distributed MISO

The arrival of similar-strength signals from different transmitters in line-of-sight (LoS) scenarios can cause deep notches in the frequency response of the channel. These notches can erase a significant percentage of subcarriers and degrade the quality of service (QoS) in an important manner. The objective of distributed MISO is to combine the signals from different transmitters in an optimum way and remove the presence of notches from the channel. In non-line-of-sight (nLoS) scenarios, the utilization of distributed MISO can also improve the reception as a result of better diversity.

The distributed MISO technique of DVB-T2 is based on a modified version of the Alamouti code [4], which is a very simple space-frequency block code designed for increasing the diversity in systems with two transmit antennas. In DVB-T2, the code is applied across pairs of transmitters in order to improve the reception in SFNs without any modification in the existent network infrastructure. Assuming that x_1 and x_2 are the information symbols to be transmitted in the first and second subcarriers, the modified Alamouti code included in DVB-T2 can be represented with the following transmission matrix:

$$G_{ALAM} = \begin{bmatrix} x_1 & x_2 \\ -x_2^* & x_1^* \end{bmatrix}, \tag{6.1}$$

where the rows correspond to transmitters and the columns correspond to subcarriers. From Equation 6.1, we can see that one transmitter (first row) outputs the information symbols without any modification, while the other transmitter (second row) outputs an encoded version of the original symbols based on complex conjugates. The modified version of the Alamouti code included in DVB-T2 possesses the same properties of the original code, i.e., full diversity with reduced (linear) complexity required at the receiver side to demodulate the signal [4]. In order to retain the orthogonality of the Alamouti code, it is necessary that the channel remains approximately constant during the transmission of symbols from the same pair. Due to the long duration of the OFDM symbols, the channel is likely to experiment significant variations between two consecutive OFDM symbols, especially in mobile scenarios. For this reason, the Alamouti code in DVB-T2 is applied on consecutive subcarriers in the frequency domain instead of consecutive OFDM symbols on the time domain.

In order to use distributed MISO in DVB-T2, the density of pilots needs to be doubled for the same performance of channel estimation. This overhead increase should be taken into account when evaluating the improvement of distributed MISO in terms of spectral efficiency. For example, the use of pilot pattern PP1 in MISO provides the same performance as pilot pattern PP3 in single-input single-output (SISO), but the overhead due to channel sampling increases from 4.2% up to 8.3%. It should be pointed out that MISO transmissions in DVB-T2 cannot match the channel sampling of SISO with pilot pattern PP1.

6.2.6 *Physical Layer Signaling*

The physical layer signaling in DVB-T2 is transmitted inside preamble symbols known as the P1 and P2 symbols at the beginning of each T2 frame. The P1 symbol is the first OFDM symbol transmitted in the T2 frames and is intended for fast identification of available T2 signals. At the same time, it also enables the reception of the P2 symbols in a very robust way even on mobile channels. P2 symbols are transmitted right after the P1 symbol and carry the Layer 1 (L1) signaling, divided in L1 presignaling and L1 postsignaling. The L1 presignaling consists of very basic information (200 bits) and is always transmitted with the most robust configuration available (BPSK and code rate 1/5). The L1 postsignaling enables the reception of the actual data and is transmitted with constellations BPSK, QPSK, 16QAM, or 64QAM and code rate 1/2. This way, it is possible to adjust the transmission of the L1 postsignaling so that it is more robust than the data (i.e., 3 dB) and, at the same time, results in the lowest possible overhead. The L1 postsignaling can be divided into the configurable part and the dynamic part. The configurable part signals the configuration of the PLPs and is seldom changed during the transmission of T2 signals. The dynamic part signals the position of the PLPs in each frame and changes from frame to frame.

L1 signaling is protected by the same BCH and LDPC codes used for the data path, although, in this case, only the 16K LDPC codes (i.e., 16,200 bits) can be selected. Shortening and puncturing are used to adjust the amount of L1 postsignaling to the LDPC code. In order to maximize the time diversity, the codewords with the L1 signaling information are uniformly distributed over all the P2 symbols of one T2 frame (e.g., 2 P2 symbols are used in the 8K FFT mode). This accounts for an interleaving duration of just several milliseconds, which may not be sufficient in mobile scenarios [11]. For instance, the interleaving duration of the L1 signaling with 8K OFDM symbols in 8 MHz channels is around 2 ms (1 ms per OFDM symbol).

DVB-T2 includes two mechanisms for increasing the robustness of the L1 signaling known as L1 repetition and in-band signaling. The former increases the robustness of the L1 signaling by transmitting in each T2 frame the L1 dynamic part that corresponds to the current and the next T2 frame. The latter transmits the L1 dynamic part embedded in the data path so that it possesses the same robustness as the data. Although the utilization of L1 repetition and in-band signaling introduces a delay in transmission of one T2 frame and one interleaving frame, respectively, it does not result in an increase in the channel change time. In addition to L1 repetition and in-band signaling, it is also possible to improve the robustness of the L1 signaling by reusing when decoding the codewords the information that has been already decoded successfully in previous frames, and that is also transmitted in the current frame. This is the case of the L1 configurable part, which is seldom changed during the transmission of T2 signals and generally remains the same from frame to frame. If L1 repetition is used, this technique can also be applied to the L1 dynamic part corresponding to the current frame, provided that the L1 signaling was successfully decoded in the previous frame.

6.3 T2-Lite Profile

The T2-Lite profile targets exclusively mobile and handheld receivers, and thus it contains only the transmission modes that are better suited for mobile reception while minimizing the amount of complexity that is required in receivers.

For example, it establishes restrictions in terms of time interleaver memory, service data rate, and FEC processing rate in order to simplify the cost of handheld devices. It also removes from the specification those modes that are not well suited for the provision of mobile services in the ultra-high frequency (UHF) band. On the other hand, the number of new elements in T2-Lite has been restricted in order to retain the maximum compatibility. In particular, the new profile incorporates more robust code rates to enable the reception at lower CNR thresholds and also allows a higher flexibility for the multiplexing of different T2 signals in the same frequency channel. These additions make not possible for current DVB-T2 receivers to demodulate T2-Lite signals, although future receivers based on the 1.3.1 version of the standard are expected to incorporate these features. On the other hand, mobile and handheld receivers based on the T2-Lite profile might be able to demodulate certain DVB-T2 transmissions as long as these are compatible with the limitations established in the profile.

6.3.1 Restrictions of T2-Lite

The TI memory in T2-Lite has been reduced from $2^{19} + 2^{15}$ cells down to 2^{18} cells (see Chapter 12), as this has an important impact on the cost of receivers due to the required silicon. Although the restriction in terms of TI memory roughly halves the maximum interleaving duration that can be provided for any given data rate, it is acceptable in the context of T2-Lite, as mobile services are generally transmitted with much lower data rates than fixed services. Table 6.1 displays the maximum interleaving duration that is achievable for all possible MODCOD configurations in the case of a mobile service that is transmitted at 375 kbps (320 kbps H.264 video, 48 kbps High-Efficiency Advanced Audio Codec (HE AAC) v2 audio, and 7 kbps signaling overhead). As it is shown in the table, the maximum interleaving duration for this type of service varies from 0.4 to 3.3 s.

Table 6.1 Maximum Interleaving Duration (s) in T2-Lite for a Service Data Rate of 375 kbps

MODCOD	TI	MODCOD	TI	MODCOD	TI	MODCOD	TI
QPSK 1/3	0.4	16QAM 1/3	0.9	64QAM 1/3	1.3	256QAM 1/3	1.8
QPSK 2/5	0.5	16QAM 2/5	1.1	64QAM 2/5	1.6	256QAM 2/5	2.1
QPSK 1/2	0.6	16QAM 1/2	1.2	64QAM 1/2	1.8	256QAM 1/2	2.4
QPSK 3/5	0.8	16QAM 3/5	1.6	64QAM 3/5	2.5	256QAM 3/5	3.3
QPSK 2/3	0.9	16QAM 2/3	1.8	64QAM 2/3	2.7	256QAM 2/3	NA
QPSK 3/4	1	16QAM 3/4	2	64QAM 3/4	3	256QAM 3/4	NA

In T2-Lite, the aggregated data rate of any data PLP and if present, its common PLP, has been limited to 4 Mbps. In addition, the rate at which the FEC chain can process cells in the receiver buffer model of T2-Lite has been scaled down from 7.6×10^6 cells/s according to the constellation. This facilitates a simpler implementation of the FEC decoder chain. For example, in the case of one-dimensional demapping (i.e., without rotated constellations) of log-likelihood ratios (LLRs), where the number of operations is proportional to the number of bits carried per complex symbol, the soft demapping can be performed with the same complexity regardless of the constellation.

The FEC in T2-lite has been also restricted to the utilization of short FEC frames, so that only the less complex decoding of 16K LDPC codes must be implemented in receivers. DVB-T2 supports both short and long FEC frames, with the former using 16K LDPC codes with 16,200 bits per codeword, and the latter using 64K LDPC codes with 64,800 bits per codeword. Compared to the 16K LDPC codes, the utilization of 64K LDPC codes provides a performance advantage in the order of a fraction of dB at the expense of some additional latency, although this can be considered negligible in the context of broadcasting.

6.3.2 Elements Removed in T2-Lite

T2-Lite supports a limited subset of transmission modes compared to DVB-T2. The 32K FFT mode is not well suited for mobile reception at UHF frequencies due to the low separation between subcarriers and has been removed from the profile. The 1K FFT mode has been also removed since the transmission at L-band is not expected with the new specification. It should be noted that while the transmission of 2K signals is feasible in the 1.5 MHz channels of the very-high-frequency (VHF) band, at high-frequency bands such as the L-band, the lower separation between subcarriers in the 2K FFT mode compared to the 1K mode may limit the performance at high velocities. Pilot pattern PP8, which relies on CD3-OFDM for channel estimation in mobile scenarios, is not supported in T2-Lite, as the combination of the CD3-OFDM technique for channel estimation with TI or multiple PLPs may result in poor performance. In addition, three combinations of FFT size, pilot pattern, and guard interval (two in SISO mode and one in MISO mode) are not allowed in order to reduce the number of combinations and simplify the implementation of receivers.

The highest code rates in DVB-T2, which are 4/5 and 5/6, have been removed from T2-Lite. For the same spectral efficiency, the use of higher-order constellations and more robust code rates is preferable in mobile scenarios due to the fact that lower code rates provide better diversity against fading. On the other hand, code rates 2/3 and 3/4 are not allowed to be used together with the 256QAM constellation, as these configurations do not provide sufficient robustness in the mobile scenario. In a similar manner, the use of rotated constellations is not allowed for the 256QAM constellation, as in this case, the gain does not compensate the extra complexity that is required for the demapping of LLRs.

6.3.3 Elements Added in T2-Lite

The new elements added for T2-Lite are related to the FEC chain and the configuration of FEF parts. Regarding the FEC chain, two code rates 1/3 and 2/5 have been added together with their respective bit to cell demultiplexers. The new code rates have been inherited from DVB-S2 in order to enable the reception at lower CNR thresholds. This way it is possible to extend the coverage of T2-Lite services at the expense of lower capacity. In addition, T2-Lite uses the parity interleaver that is included in DVB-T2 for 16QAM, 64QAM, and 256QAM constellations for the new code rates and for all the constellations, including QPSK. The parity interleaver is designed to improve the performance of LDPC decoding in fading channels by ensuring that the bits connecting to the same check node end up being transmitted in different cells. Otherwise, the erasure of one cell due to fading may result in the loss of multiple bits connecting to the same check node, which is detrimental to the error-correcting capabilities of the code. In addition to more robust code rates and the use of the parity interleaver, the duration of the FEF part in T2-Lite has been increased from 250 ms up to 1 s. This allows a higher flexibility when multiplexing T2 signals in the same frequency channel, as the DVB-T2 signal is expected to occupy the major part of the transmission time.

6.3.4 Combination of T2 Signals

A T2-Lite signal can occupy an entire frequency channel or can be multiplexed with other signals by means of FEF parts. The combination of T2-Lite with DVB-T2 transmissions is expected to be the first manner in which commercial T2-Lite services will be transmitted over the air. For example, it would be possible to dedicate the 80% of the transmission time to DVB-T2 and the 20% to T2-Lite. Assuming that the T2-Lite signal is transmitted with FFT size 8K (with extended carrier mode), QPSK1/2, and pilot pattern PP1, the total capacity for T2-Lite services is 1.5 Mbps per channel (8 MHz bandwidth). This would allow up to four services at 375 Kbps to be carried in the T2-Lite signal.

The previous example could be accomplished by different configurations. For example, it is possible to alternate DVB-T2 frames of 200 ms with T2-Lite frames of 50 ms. In this case, the interleaving duration is about 50 ms (with subslicing), whereas the average channel change time is approximately 175 ms. In order to extend the TI, a multi-frame interleaving of four frames can be performed to provide an interleaving duration of 800 ms with an average channel change time of 1.3 s. On the other hand, it is also possible to alternate DVB-T2 transmission periods of 800 ms with T2-Lite frames of 200 ms. This configuration provides slightly better power saving and reduced signaling overhead as the receiver has to synchronize with the T2-Lite signal in fewer occasions. However, it does not allow multi-frame interleaving, as the resulting channel change time would be higher than 2 s on average. Therefore, the TI in this case is limited to intra-frame interleaving, resulting in an interleaving duration of 200 ms (with subslicing), and an average channel change time of approximately 0.7 s.

6.3.5 Provision of Radio Services in T2-Lite

The T2-Lite profile is very well suited for the provision of digital radio services. T2-Lite-only receivers can be implemented with a low level of complexity while providing good robustness in all types of reception environments. For example, by using the 10% of the transmission time for T2-Lite in a combined DVB-T2/T2-Lite multiplex, it is possible to accommodate about 18 radio services at 64 kbps with HE-AAC v2. Alternatively, four mobile TV channels can be carried by using the 20% of the transmission time. Nowadays, the reference standard for digital radio is the DAB standard and its evolution, the DAB+ specification [12]. Compared to DAB+, which implements a concatenation of convolutional and RS coding similar to DVB-T, the T2-Lite profile features better FEC protection, thanks to the use of LDPC codes. This enables the reception at lower CNR values or, alternatively, allows more radio services to be carried in the same multiplex. Generally speaking, the improvement of LDPC codes compared to convolutional codes for low CNR values in AWGN is stated to be around 3 dB in terms of CNR and around 0.5 bits per symbol in terms of spectral efficiency [5]. T2-Lite can also provide better time diversity by means of longer TI. For low data rate services such as the case of radio, DVB-T2 can provide interleaving durations in the order of 1 s with tolerable channel change times, whereas the TI in DAB+ is about 360 ms. The use of distributed MISO in T2-Lite also provides better performance in SFNs. It must be noted that the transmission of T2-Lite services at higher frequency bands than UHF and below 3 GHz is limited by the selection of FFT modes adopted in the profile. In particular, the lowest FFT mode in T2-Lite is 2K, whereas DAB+ includes 2K, 1K, 512, and 256 FFT modes. Since the amount of ICI caused by the Doppler shift is proportional to the subcarrier separation and inversely proportional to the carrier frequency, a 2K FFT mode may not achieve sufficient protection against Doppler effects in the 1.712 MHz channels of the L-band (1.5 GHz).

6.4 Physical Layer Simulations

In order to evaluate the performance of DVB-T2 and T2-lite in mobile channels, we have first employed simulations in the TU6 channel. This channel is made of six taps having wide dispersion in delay and relatively strong power and is representative of mobile reception for Doppler frequencies above 10 Hz. For multiple receive antennas (single-input multiple-output, SIMO), we have modeled each propagation path between the transmit and the receive antennas as an independent TU6 profile, so that no spatial correlation exists between antennas. The results have been

Table 6.2 Simulation Parameters

Parameter	Value	Parameter	Value
Bandwidth	8 MHz	FFT mode	8K
Guard interval	1/4	MODCOD	QPSK 1/2
TI memory	2^{18} cells	QoS criterion	BB FER 1%
Channel model	TU6, PI, PO, PI	Doppler frequency	1.67, 10, 33, and 80 Hz

obtained for 10 Hz, 33 Hz, and 80 Hz of Doppler, which corresponds to a user velocity of 18 km/h, 60 km/h, and 144 km/h at a frequency carrier of 600 MHz. In addition to the TU6 channel, we have also considered the pedestrian channels developed during the Wing TV project [13] for the evaluation and validation of DVB-H. In particular, a pedestrian outdoor (PO) channel and a pedestrian indoor (PI) channel were defined for a user velocity of 3 km/h (1.67 Hz of Doppler at 600 MHz). The parameters used for the simulations are summarized in Table 6.2. Robust configurations of MODCOD are better suited for mobile reception, as they can provide better protection in mobile channels with sufficient capacity for low data rate services. Regarding the utilization of TI, the simulations assume the transmission of 2^{18} cells per interleaving frame, which represents half of the TI memory in DVB-T2, and the entire TI memory in T2-Lite. The results presented in this section have been obtained under the assumption of ideal channel estimation. In real receivers, however, some loss can be expected due to the effect of ICI, imperfect interpolation, and the presence of noise in the pilot cells.

6.4.1 Time Interleaving

The performance of subslicing in the TU6 channel model is represented in Figure 6.1. According to the figure, the gain of subslicing is about 1 dB with 80 Hz of Doppler and about 1.5 dB with 10 Hz. If subslicing is not enabled, the interleaving duration is limited by the number of cells that are transmitted per frame (approx. 50 ms in this case), whereas with subslicing, the interleaving duration is given by the frame length (up to 250 ms). It must be noted that the utilization of subslicing may provide higher gains if a lower number of cells than 2^{18} are transmitted per frame (e.g., low data rate services). From the results, we can see that a high number of subslices are required to exploit the time diversity of the TU6 channel with 80 Hz of Doppler, whereas the number of subslices does not impact the system performance in the case of 10 Hz of Doppler. The explanation for this is simple: a higher number of subslices result in a more uniform distribution over time, achieving a better averaging of fading in fast varying channels. If the channel varies slowly, the interleaving duration rather than the number of subslices is more important for time diversity.

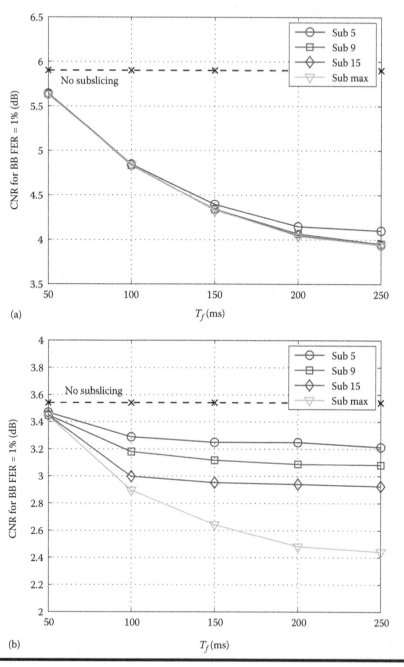

Figure 6.1   Performance of subslicing in DVB-T2 in the TU6 channel with Doppler 80 Hz (a) and 10 Hz (b).

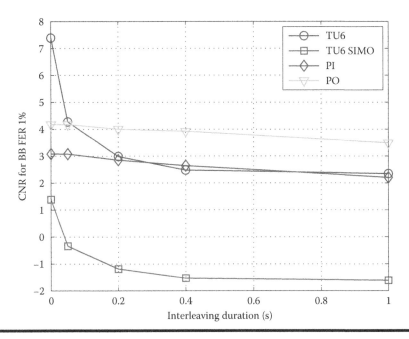

Figure 6.2   Performance of DVB-T2 TI in the TU6 channel (60 km/h) and in the PI/PO channels (3 km/h).

Figure 6.2 illustrates the performance of DVB-T2 in different channel models according to the interleaving duration. We can see that the gain of TI in both pedestrian channels is approximately 1 dB. The reduced gain of TI in this case is motivated by the low user velocity, which requires very long TI in order to achieve significant gains. In contrast, the gain of TI in the TU6 channel is approximately 5 dB and saturates with an interleaving duration of around 200 ms in such a way that longer interleaving durations do not achieve much larger gains. If two receiving antennas are used (SIMO), the gain of TI in the TU6 channel is reduced down to 3 dB. For interleaving durations lower than 1 s, both TI and SIMO counter the presence of multipath fading in the received signal, and, thus, the use of one technique diminishes the gain that can be achieved with the other.

6.4.2 Distributed MISO

In Figure 6.3, we represent the gain of distributed MISO in the SFN TU6 channel with two transmitters. Each transmitter in the SFN is modeled as an independent TU6 profile so that the received signal results from the combination of two TU6 profiles, each having a unique power level and delay. According to the figure, the gain of distributed MISO is largest when there is no delay or power imbalance between transmitters. The presence of power imbalance is detrimental for diversity,

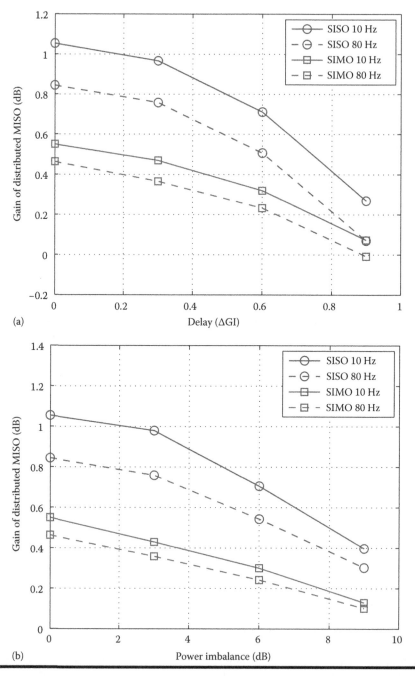

(a)

(b)

Figure 6.3 Gain of distributed MISO in DVB-T2 with two transmitters. Influence of the time delay between the two contributions (a). Influence of the power imbalance (b).

as it reduces the contribution of one transmitter to the received signal. On the other hand, the reception of delayed signals in SFNs is similar to the utilization of DD, providing better frequency diversity as a result of increased delay spread. However, this is disadvantageous for distributed MISO, as the rapid variations between subcarriers in the frequency domain can affect the performance of the Alamouti code. Because of this, the gain of distributed MISO diminishes when the delay between the signals from different transmitters increases. Reception at higher Doppler frequencies or the implementation of SIMO also reduces the improvement of distributed MISO as a result of better diversity in the time and space domain. As it is shown in Figure 6.3, the gain of distributed MISO in some cases is negligible, and it may actually become negative, resulting in a performance loss compared to SISO.

In Figure 6.4, we show the capacity curves in the SFN TU6 channel with and without distributed MISO. The curves have been obtained for a Doppler of 33 Hz, whereas the effect of the power imbalance and the delay between transmitters has been averaged over the simulations. In the left side of the figure, the capacity curves are represented without considering the loss due to the pilot overhead, whereas on the right side, the loss has been taken into account. We have assumed the pilot pattern PP3 for SISO and the pilot pattern PP1 for distributed MISO. Both patterns provide the same channel estimation and result in an overhead of 4% and 8%, respectively. From the figure, we can see that the gain of distributed MISO is lower than 0.3 dB. With the pilot overhead, the use distributed MISO lowers the overall spectral efficiency, as the limited gain in robustness does not compensate for the lower capacity.

6.4.3 Signaling Path

The main objective in this section is to investigate the robustness of the signaling path in mobile channels. As it has been explained before, it is important that the signaling path is transmitted at least 3 dB more robust than the data path in order to ensure the reception of services. In Figure 6.5, we illustrate the capacity curves of the L1 postsignaling when repetition and the advanced decoding technique described in Section 6.2.6 are used in the TU6 channel. Each point in the curves corresponds to the four transmission modes that are possible for the L1 postsignaling; BPSK 1/2 (0.5 bpc), QPSK 1/2 (1 bpc), 16QAM 1/2 (2 bpc), and 64QAM 1/2 (3 bpc). For the sake of comparison, we have also represented the capacity curves of the data path with maximum subslicing and maximum frame length, resulting in 250 ms of interleaving duration. In this case, each point in the curve corresponds to QPSK 1/2 (1 bpc), 16QAM 1/2 (2 bpc), 64QAM 1/2 (3 bpc), and 256QAM 1/2 (4 bpc). By looking at these curves, it is possible to identify for each MODCOD configuration of the data path, the transmission mode that is necessary in the L1 signaling to achieve a robustness of 3 dB better or more. According to the results, the use of repetition is not sufficient to ensure a 3 dB advantage for the L1 postsignaling when the data path is transmitted with QPSK, and, in this case,

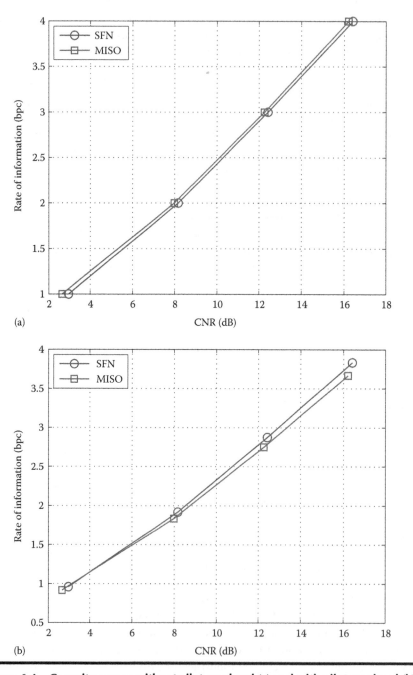

(a)

(b)

Figure 6.4 **Capacity curves without pilot overhead (a) and with pilot overhead (b).**

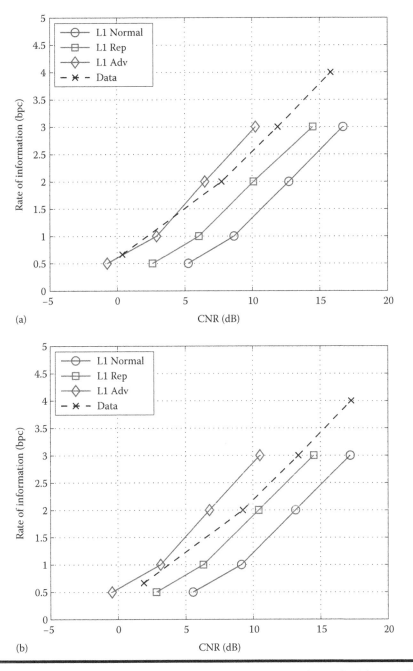

Figure 6.5 **Performance of L1 postsignaling in the TU6 channel with Doppler 80 Hz (a) and 10 Hz (b).**

advanced decoding is needed. For the rest of constellations, repetition can provide the adequate robustness. It must be noted, however, that advanced decoding allows the L1 postsignaling to be transmitted with a higher-order constellation (e.g., from BPSK to QPSK) while maintaining the same level of robustness. This reduces the overhead caused by the physical layer signaling and leaves more capacity for the transmission of data in the P2 symbols.

6.5 Laboratory Measurements

Laboratory tests have been carried in order to study the influence of the FFT size, the pilot pattern, and the TI in the mobile performance of DVB-T2. To perform these tests, the DVB-T2 signal is generated in a PC card-based DTT multi-standard modulator. This card also includes a channel simulator module, so different CNR levels and multipath propagation models can be adjusted in the tests, including Doppler effects. The generated signal is received and decoded by a professional DVB-T2 receiver, which provides several signal-quality measurements. The reception of 30 consecutive seconds without errors has been selected as the QoS criterion in the tests. Because of the random characteristics of the simulated channels, the measurements have been repeated several times and the performance results are presented as the average value. Figure 6.6 shows the measurement setup used in these tests.

The main simulation parameters are shown in Table 6.3, including the DVB-T2 signal configuration and the simulated channel. The channel model is a standard TU6 profile with different Doppler frequencies. The TU6 channel with no Doppler and the Gaussian channel have also been measured in order to investigate the degradation caused by mobile reception. All the combinations of FFT size, modulation scheme, code rate, and pilot pattern (if allowed by the standard) have been analyzed. In total, 174 different configurations have been measured in the laboratory tests.

Table 6.4 shows the guard intervals selected for each combination of FFT size and pilot pattern. PP8 pilot pattern has not been considered in the laboratory tests since it is intended for stationary reception. In a similar manner, the 256QAM constellation has been excluded from the study due to the fact that its CNR requirements are too high to be achieved in mobile reception conditions.

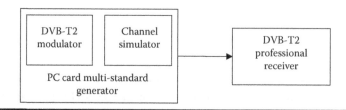

Figure 6.6 Laboratory tests setup.

Table 6.3   Laboratory Measurement Parameters

Parameter	Value	Parameter	Value
Bandwidth	8 MHz	Extended mode	No
FFT size	2K, 4K, 8K, 16K, 32K	Guard interval	1/128, 1/32, 1/16, 1/8, 1/4
Modulation scheme	QPSK, 16 QAM, 64 QAM	Rotated constellations	On
Code rate	1/2, 2/3	FEC size	16,200, 64,800 bits
Pilot pattern	PP1, PP2, PP3, PP4, PP5, PP6, PP7	TI	From 0 to 90 ms
PAPR reduction	Off	HEM	On
L1 modulation	BPSK	L1 repetition	Off
Channel model	TU6	Doppler frequency	2.5, 5, 10, 33, 40, 80 Hz
Number of PLPs	1		

Table 6.4   Guard Interval Values Selected for Each FFT Size and Pilot Pattern Combination

	32K	*16K*	*8K*	*4K*	*2K*
PP7	1/128	1/32	1/32	1/32	1/32
PP6	1/32	1/32	—	—	—
PP5	—	1/16	1/16	1/16	1/16
PP4	1/16	1/16	1/16	1/16	1/16
PP3	—	1/8	1/8	1/8	1/8
PP2	1/8	1/8	1/8	1/8	1/8
PP1	—	1/4	1/4	1/4	1/4

6.5.1 CNR Performance in Mobile Reception

This section focuses on the CNR degradation caused by the Doppler frequency, i.e., the receiver speed. In the tests, the FEC size is 64K and the TI duration is between 70 and 80 ms. Figure 6.7a shows the results for different pilot patterns with the 4K FFT size, which can be considered as a compromise value between

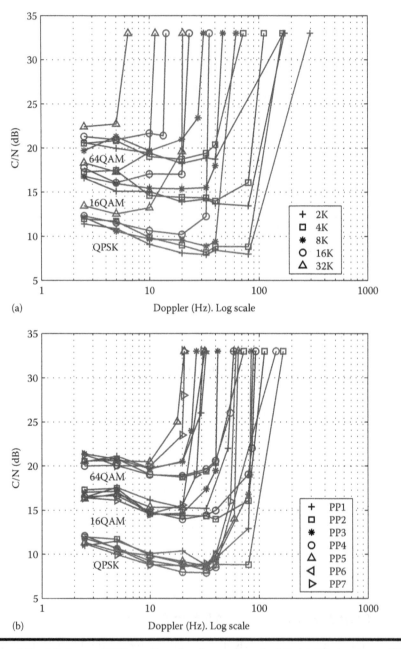

Figure 6.7 Minimum CNR vs. Doppler for code rate 1/2 (a) Pilot pattern PP2. (b) FFT size 4K.

capacity and robustness against ICI. In Figure 6.7b, the study focuses on the influence of the FFT size with pilot pattern PP2, which provides the best performance according to Figure 6.7a. The highest CNR value obtained in the figures for each configuration corresponds to the maximum tolerated Doppler when no external noise is added in the channel simulator. In that situation, the internal noise of the receiver is the main noise source, and the CNR value has been estimated to be around 33 dB. The CNR threshold values are also calculated for the TU6 channel without Doppler and for the Gaussian channel (Table 6.5). It should be pointed out that the values corresponding to the Gaussian channel are higher than the simulation results provided in the Implementation Guidelines [9], and therefore, they can be used as a reference to define the implementation loss of the hardware setup used in the tests.

It can be seen in Figure 6.7 that lower CNR values are obtained as the Doppler increases. This is caused by the lower correlation of the fast fading channel in combination with the TI. However, the CNR drastically increases near the maximum allowed Doppler. The chosen pilot pattern has less influence in the CNR results than the FFT size for medium Doppler values, but both are determinant for the maximum Doppler that is supported by the system (see Section 6.5.2). On the other hand, there is no noticeable influence of the pilot pattern and the FFT size for stationary channels (Table 6.5), even in the case of the configuration with 2K, PP7, and guard interval 1/32, where the duration of the guard interval (7 µs) is close to the delay spread of TU6 channel. Finally, there is a significant increase in the CNR values when comparing the TU6 channel with no Doppler and the TU6 channel with 2.5 Hz of Doppler.

6.5.2 FFT Size and Pilot Pattern Influence

The maximum Doppler spread that is supported for different combinations of FFT size, pilot patterns, and MODCOD has been measured in order to identify the influence of these parameters in the mobile performance of DVB-T2 (Figure 6.8). In these tests, the FEC size is 64K and the TI duration is between 70 and 80 ms.

The results show that even pilot patterns (PP2, PP4, and PP6), which contain a pilot carrier in every second OFDM symbol, achieve better performance than odd pilot patterns (PP1, PP3, PP5, PP7), in which the pilot carrier is present in one out of four symbols. For example, comparing pilot patterns PP1 and PP2, which introduce the same overhead (8.33%), the maximum Doppler with PP2 pattern doubles the one with PP1. Similar results are obtained when comparing FFT sizes. As the FFT size is increased by a factor of two, the maximum Doppler is reduced by the same factor. The overhead added by the FFT size is related to the guard interval duration. If the guard interval duration is fixed, then the overhead is doubled when the FFT size is reduced by a factor of two. Therefore, these parameters should be selected as a trade-off between capacity and robustness against Doppler effects.

Table 6.5   Minimum CNR (dB) for Gaussian and TU6 (No Doppler) Channels for Code Rate 1/2

		Gaussian Channel					TU6 0 Hz				
		32K	16K	8K	4K	2K	32K	16K	8K	4K	2K
QPSK	PP7	1.4	1.4	1.5	1.6	1.7	3.3	3.2	3.3	3.5	3.6
QPSK	PP6	1.5	1.3				3.5	3.0			
QPSK	PP5		1.4	1.5	1.7	1.7		3.1	3.2	3.4	3.6
QPSK	PP4	1.4	1.4	1.5	1.5	1.6	3.4	3.2	3.2	3.4	3.4
QPSK	PP3		1.5	1.5	1.7	1.7		3.4	3.4	3.4	3.6
QPSK	PP2	1.4	1.5	1.5	1.5	1.7	3.4	3.3	3.5	3.5	3.6
QPSK	PP1		1.5	1.8	1.7	1.9		3.6	3.7	3.7	3.8
16QAM	PP7	6.5	6.5	6.5	6.5	6.6	8.6	8.7	8.8	8.9	9.0
16QAM	PP6	6.5	6.6				8.7	8.7			
16QAM	PP5		6.5	6.5	6.6	6.6		8.7	8.8	8.9	9.1
16QAM	PP4	6.6	6.5	6.6	6.6	6.6	8.7	8.6	8.7	8.9	8.8
16QAM	PP3		6.5	6.6	6.6	6.6		8.8	8.9	8.9	8.9
16QAM	PP2	6.5	6.6	6.6	6.6	6.6	8.7	8.8	8.8	8.9	8.9
16QAM	PP1		6.6	6.6	6.6	6.6		9.0	8.9	9.0	9.0
64QAM	PP7	10.9	10.9	11.0	11.0	11.0	13.1	13.1	13.2	13.2	13.4
64QAM	PP6	10.9	10.9				13.0	13.2			
64QAM	PP5		11.0	10.9	10.9	11.0		13.1	13.1	13.3	13.3
64QAM	PP4	10.9	10.9	11.0	11.0	10.9	13.1	13.1	13.1	13.2	13.2
64QAM	PP3		11.0	10.9	11.0	11.0		13.2	13.3	13.5	13.5
64QAM	PP2	11.0	10.9	11.0	10.9	11.0	13.1	13.2	13.2	13.2	13.3
64QAM	PP1		11.0	10.9	11.0	11.0		13.2	13.3	13.3	13.3

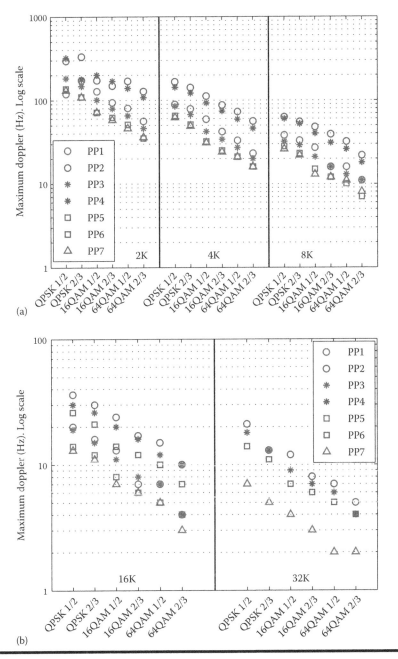

Figure 6.8   Maximum tolerable Doppler. (a) FFT 2K, 4K, and 8K. (b) FFT 16K and 32K.

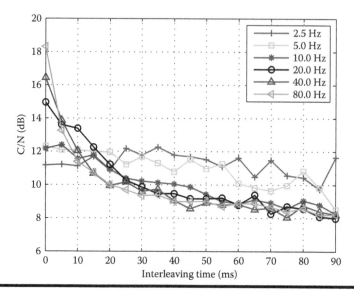

Figure 6.9 Minimum CNR vs. interleaving time for 4K, PP2, QPSK, 1/2.

6.5.3 Time Interleaving

Figure 6.9 shows the influence of TI in the minimum CNR that is required in six different Doppler scenarios. These measurements have been carried out using a DVB-T2 mode with 4K FFT, PP2, QPSK 1/2, and guard interval 1/8. In these tests, an FEC size of 16K has been selected and the TI varies from 0 to 90 ms. The different interleaving values have been obtained by dividing the T2 frames in several interleaving blocks of desired interleaving length. Longer interleaving times are not allowed for this configuration with a single PLP due to the limited amount of TDI memory in receivers (see Sections 6.2.3 and 6.4.1).

The results in this case show that mobile reception improves with longer values of TI. Moreover, it can be observed that, as the interleaving duration increases, the performance gain of TI is larger in high Doppler scenarios. In order to achieve larger gains in the low Doppler scenarios, the interleaving duration should be increased beyond 90 ms by means of subslicing and multi-frame interleaving.

6.6 Network Planning Discussion

6.6.1 Mobile Reception in DVB-T2 Networks

Besides fading, mobile reception is challenged by the use of low-gain antennas at ground level in outdoor and indoor scenarios. Table 6.6 shows the link margin that is necessary in DVB-T2 networks for fixed, portable indoor, vehicular, and

Table 6.6   Link Budget of DVB-T2 Networks in the UHF Band

	Fixed (dB)	Portable Indoor (dB)	Vehicular (dB)	Handheld Indoor (dB)
Multipath fading loss[a]	0.2	2	7	2
Height loss[b]	0	23.5	23.5	23.5
Feeder loss	4	0	0	0
Building penetration loss	0	11	0	11
Antenna gain	11	0	0	−9.5
Location correction factor[c]	9	13	13	13
Link margin total	2.6	49.9	41.5	59.4
Distance to fixed	—	−47.3	−41.3	−56.8
Distance to portable indoor	—	—	6	−9.5

[a] Compared to AWGN according to the values presented in [9]. The MODCOD is assumed to be QPSK 1/2 in portable indoor, vehicular, and handheld indoor reception, and 256QAM 3/5 in fixed reception.
[b] Height loss corresponds to the urban scenario.
[c] Location correction factor corresponds to good coverage quality (i.e., 95% in fixed, portable indoor, and handheld indoor, and 99% in vehicular).

handheld indoor reception in the UHF band. The values listed in the table correspond to those presented in [14]. Compared to fixed reception, which is typically performed with rooftop antennas at 10 m, there is a height loss resulting from the low antenna height of about 1.5 m, a loss in the antenna gain caused by the utilization of omnidirectional and integrated antennas, and a building penetration loss in the case of indoor scenarios. On the other hand, the presence of multipath fading in the received signal causes a loss of 0.2 dB for fixed reception (Rice F1 channel), a loss of 2 dB for portable and handheld indoor reception (Rayleigh P1 channel), and a loss of 7 dB for vehicular reception (TU6 channel) compared to AWGN without TI. It should be pointed out that the table does not include the loss caused by the Doppler spread at high velocities. As it is shown in Section 6.5, this degradation limits the maximum velocity of users and depends on the transmission mode, particularly the FFT size and the pilot pattern, and also on the quality of the channel estimation performed at the receiver side.

The presence of shadowing in the service area makes coverage predictions statistical, such that the coverage probability rather than the signal level is used for network planning purposes. As a result, a location correction factor is needed in order to ensure a minimum coverage probability within the service area. This is generally computed by assuming that shadowing follows a lognormal distribution with zero mean and standard deviation σ. For σ = 5.5 dB (outdoor reception), location correction factors of 3, 7, 9, and 13 dB achieve a coverage probability of 70%, 90%, 95%, and 99%, respectively. Typical targets in terrestrial DVB systems are 95% for good portable and pedestrian reception, and 99% for good vehicular reception.

Apart from diversity techniques, terrestrial broadcasting networks can employ low code rates and small constellations to improve the coverage of mobile services. Mobile services in DVB-T2 networks can be transmitted in dedicated RF channels or can be multiplexed together with fixed services. The utilization of multiple PLPs or the T2-Lite profile in DVB-T2 allows fixed and mobile services to be multiplexed in time without interfering with each other, such that there is no performance degradation compared to using dedicated RF channels. In DVB-T2 networks planned for fixed reception, using a configuration based on QPSK 1/2 instead of 256QAM 3/5 provides a gain of 15.1 dB (Rice F1 channel). In networks planned for portable indoor reception, a similar approach would be to use QPSK 1/2 instead of 64QAM 1/2, resulting in a gain of about 9.9 dB (Rayleigh P1 channel) [9]. In T2-Lite, using the more robust code rates added to the specification, 2/5 and 1/3, provides an additional gain of around 1 and 2 dB, respectively.

As shown in Table 6.6, the distance to the link margin of fixed reception is −47.3 dB for portable indoor, −41.3 dB for vehicular, and −56.8 dB for handheld indoor. The utilization of more robust MODCODs together with diversity techniques cannot compensate for such high values, and therefore, mobile services can only be provided in a best effort manner. Table 6.7 shows the coverage probability of

Table 6.7 Coverage Probability of Mobile Services in DVB-T2 Networks

Diversity Techniques	Network Planned for Fixed Reception[a]		Network Planned for Portable Indoor[b]	
	Vehicular	Handheld Indoor	Vehicular	Handheld Indoor
—	1%	0%	100%	95%
SIMO	10%	—	100%	—
TI 250 ms	1%–7%	0%	100%	95%–97%
SIMO + TI 250 ms	10%–22%	—	100%	—

[a] The values have been obtained assuming 256QAM 3/5 for fixed reception and QPSK 1/2 for mobile and handheld indoor reception.

[b] The values have been obtained assuming 64QAM 1/2 for fixed reception and QPSK 1/2 for mobile and handheld indoor reception.

vehicular and handheld indoor reception in DVB-T2 networks planned for fixed and portable indoor reception (coverage probability 95%). We can see that the coverage of vehicular reception in networks planned for fixed reception is below acceptable values (90%), whereas handheld indoor reception is not possible across the entire coverage area. It is important to notice that the height loss shown in Table 6.6 corresponds to the urban scenario and that lower values are used in the case of suburban and rural scenarios (height loss of –17.5 and –16.5 dB, respectively). For example, in rural scenarios, DVB-T2 networks planned for fixed reception can achieve a coverage probability of around 70% for vehicular reception with SIMO and TI. On the other hand, networks planned for portable indoor reception take into account the loss caused by the reception at ground level and the use of lower gain antennas. As a result, they provide full coverage for vehicular reception and achieve good coverage levels for handheld indoor reception (95%). Nevertheless, it must be noted that the building penetration loss depends on several factors, including the materials and the construction of the buildings, and it can be higher than the values listed in Table 6.6. In this case, the more robust code rates included in T2-Lite (1/3 and 2/5) can be necessary to enable good coverage levels for handheld indoor reception.

In order to illustrate the advantage of diversity techniques in terms of network planning, it is possible to translate the link margin gains computed in this chapter into a coverage extension of mobile services in terrestrial broadcasting networks. In Table 6.4, we represent the link margin gain obtained by diversity techniques and the corresponding coverage extension for a typical urban scenario (propagation exponent $\alpha = 3.5$ [14]).

These values can be also interpreted as a reduction of the network infrastructure that is required for the provision of mobile services. In Table 6.8, we can see that the incorporation of two receive antennas in vehicles can multiply by a factor higher than two, the coverage area of mobile services in DVB-T networks. It must be noted that the gains of TI depend on the user velocity and route across the service area. The values listed on the table vary between the case when the user is not

Table 6.8 Coverage Extension of Mobile Services in DVB-T2 Networks[a]

Diversity Technique	Link Margin Gain (dB)	Coverage Extension
SIMO	6	120%
TI 250 ms	0–5	0%–93%
SIMO + TI 250 ms	6–9	120%–227%

[a] The gains of SIMO correspond to an implementation based on two co-polar antennas with sufficient separation between the antennas. The gains of time interleaving depend on the user velocity and the trajectory across the service area. The highest value corresponds to the case when the user is moving at a constant velocity of 144 km/h.

moving (no gain) and the case in which the user is traveling at a constant velocity of 144 km/h (largest gain). In order to quantify the coverage extension of TI in a precise manner, it is necessary to perform dynamic system-level simulations that take into account the mobility of the users in realistic scenarios [15].

6.6.2 Multiplexing of Mobile Services in DVB-T2 Networks

Network operators can decide the use of multiple PLPs or the T2-Lite profile in order to multiplex fixed and mobile services in the same RF channels. In the case of multiple PLPs, the FFT mode and the pilot pattern must be common for all the PLPs transmitted in the same multiplex, whereas the T2-Lite profile allows alternating different configurations in the time domain by means of FEF parts. For the discussion regarding the utilization of multiple PLPs and T2-Lite, we compare two different DVB-T2 configurations in terms of spectral efficiency. The first configuration is based on multiple PLPs and employs a robust combination of FFT size and pilot pattern for the entire multiplex. In contrast, the second configuration is based on T2-Lite and alternates between different combinations for the transmission of fixed and mobile services. In Table 6.9, we can see that both configurations provide the same robustness for fixed and mobile services and also the same duration of guard interval.

The total spectral efficiency in one RF channel that combines fixed and mobile services by means of multiple PLPs, S_{PLP}, can be approximated by [16]

$$S_{PLP} = \frac{N_d}{N_c} \cdot (1 - G) \cdot \left(T_f \cdot B_f + T_m \cdot B_m \right), \tag{6.2}$$

where

N_c is the total number of subcarriers per OFDM symbol
N_d is the number of subcarriers with data information per OFDM symbol

Table 6.9   Configuration for the Combination of Fixed and Mobile Services in DVB-T2 Networks

Configuration Parameter	Configuration Based on Multiple PLPS		Configuration Based on T2-Lite	
	Fixed	*Mobile*	*Fixed*	*Mobile*
FFT size	8K Ext	8K Ext	32K Ext	8K Ext
Guard interval	1/4	1/4	1/16	1/4
Pilot pattern	PP1	PP1	PP4	PP1
Constellation	256QAM	QPSK	256QAM	QPSK
Code rate	3/5	1/2	3/5	1/2

G is the fraction of guard interval

B_f is the number of bits per cell in the case of fixed services

B_m is the number of bits per cell in the case of mobile services

T_f is the portion of time dedicated to the transmission of fixed services

T_m is the portion of time dedicated to the transmission of mobile services

On the contrary, if the T2-Lite profile is used instead, the total spectral efficiency, S_{LITE}, can be computed as follows:

$$S_{LITE} = \frac{N_d^f}{N_c^f} \cdot \left(1 - G_f\right) \cdot \left(T_f \cdot B_f\right) + \frac{N_d^m}{N_c^m} \cdot \left(1 - G_m\right) \cdot \left(T_n \cdot B_n\right), \qquad (6.3)$$

where N_c^f, N_d^f, G_f, N_m^f, N_n^f, and G_m are, respectively, the number of subcarriers per OFDM symbol, the number of subcarriers with data information per OFDM symbol, and the guard interval in the case of fixed and mobile services.

It is important to note that although we have not included the percentage of time that is needed for the transmission of L1 signaling, this represents less than 1% of the total transmission time, and thus, it has a minimal impact in the calculations. Assuming that the 80% of the transmission time is reserved to the transmission of fixed services ($T_f = 0.8$) and that the 20% is dedicated to the transmission of mobile services ($T_m = 0.2$), Equations 6.2 and 6.3 reveal that the total spectral efficiency for the configuration using multiple PLPs is approximately 2.3 bpc, whereas for the configuration using T2-Lite, it is approx. 3 bpc. This represents an advantage for T2-Lite in the order of 30%.

In Figure 6.10, we illustrate the overall spectral efficiency for DVB-T2 networks using multiple PLPs or T2-Lite according to the percentage of time that is dedicated to the transmission of mobile services. We can see that the advantage of T2-Lite diminishes when the transmission of mobile services occupies a higher percentage of the transmission time up to the point at which there is no time reserved for the transmission of fixed services ($T_m = 1$), and the T2-Lite profile provides no gain.

6.7 Conclusions

In this chapter, we have reviewed the main characteristics of DVB-T2 for the provision of mobile services including the T2-Lite profile. This is based on a subset of the DVB-T2 standard aimed at mobile receivers with reduced complexity and power consumption. The simulation results have shown that the utilization of TI at the physical layer and SIMO is very important to improve the reception robustness of DVB-T2 services in mobile scenarios. In contrast, the influence of the distributed MISO technique included in the standard is much less significant depending on the power imbalance and the delay between the signals from different transmitters. On the other hand, the lack of TI in the

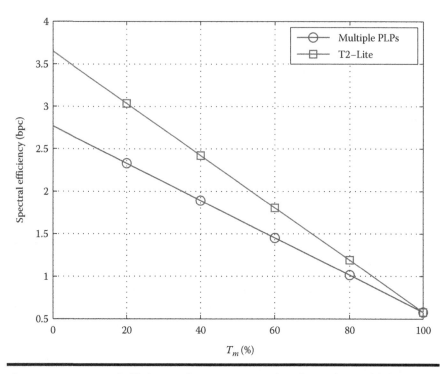

Figure 6.10   Spectral efficiency of DVB-T2 networks when using multiple PLPs or the T2-Lite profile for the combination of fixed and mobile services.

signaling information may limit the mobile performance of DVB-T2. The results have shown that the utilization of repetition and advanced decoding techniques is needed in order to ensure sufficient robustness for the signaling information compared to the data. The laboratory tests have shown the influence of the FFT size and the pilot pattern in the mobile performance of DVB-T2 and, particularly, in the maximum velocity that is supported by the system. Small FFTs and dense pilot patterns increase the robustness against Doppler at the expense of less capacity for the transmission of services.

Despite the increased reception robustness, DVB-T2 networks planned for fixed reception cannot provide mobile services with good coverage levels in urban scenarios, and thus, networks targeting mobile reception must be deployed with a larger infrastructure. On the contrary, networks planned for portable indoor reception can reach handheld terminals by using a more robust MODCOD. Mobile services can be introduced in DVB-T2 transmissions by means of PLPs or by using the T2-Lite profile. For the same reception robustness, the T2-Lite profile can achieve a 30% increase in terms of capacity compared to multiple PLPs when the 20% of the transmission time is dedicated to mobile services.

References

1. ETSI EN 302 755 v.1.3.1, *Frame Structure Channel Coding and Modulation for a Second Generation Digital Terrestrial Television Broadcasting System (DVB-T2)*, April 2012.
2. ETSI EN 300 744 v.1.6.1, *Framing Structure, Channel Coding and Modulation for Digital Terrestrial Television*, January 2009.
3. G. Pousset, Y. Lostanlen, and Y. Corre, Positioning mobile TV standards DVB-T vs. DVB-H, White Paper, Dibcom, Paris, France, 2008.
4. S. M. Alamouti, A simple transmit diversity technique for wireless communications, *IEEE Journal on Selected Areas in Communications*, 16(8), 1451–1458, October 1998.
5. M. Eroz, F. W. Sun, and L. N. Lee, An innovative low-density parity-check code design with near-Shannon-limit performance and simple implementation, *IEEE Transactions on Communications*, 54(1), 13–17, January 2006.
6. C. A. Nour and C. Douillard, Rotated QAM constellations to improve BICM performance for DVB-T2, *Proceedings of IEEE ISSSTA*, Bologna, Italy, 2008.
7. J. Boutros and E. Viterbo, Signal space diversity: A power and bandwidth efficient diversity technique for the Rayleigh fading channel, *IEEE Transactions on Information Theory*, 44(4), 1453–1467, July 1998.
8. D. Gozálvez, D. Gómez-Barquero, D. Vargas, and N. Cardona, Time diversity in mobile DVB-T2 systems, *IEEE Transactions on Broadcasting*, 57(3), 617–628, September 2011.
9. ETSI TR 102 831 v1.2.1, *Implementation Guidelines for a Second Generation Digital Terrestrial Television Broadcasting System (DVB-T2)*, August 2012.
10. V. Mignone, A. Morello, and M. Visintin, CD3-OFDM: A novel demodulation scheme for fixed and mobile services, *IEEE Transactions on Communications*, 44(9), 1144–1151, September 1996.
11. T. Jokela, M. Tupala, and J. Paavola, Analysis of physical layer signaling transmission in DVB-T2 systems, *IEEE Transactions on Broadcasting*, 56(3), 410–417, September 2010.
12. ETSI EN 300 401 v1.4.1, *Radio Broadcasting Systems; Digital Audio Broadcasting (DAB) to Mobile, Portable and Fixed Receivers*, June 2006.
13. European Celtic Wing TV Project, Deliverable D15b: Simulation report, 2006.
14. Frequency and network planning aspects of DVB-T2, Technical Report, EBU Tech 3348, May 2012.
15. D. Gómez-Barquero, P. Unger, T. Kürner, and N. Cardona, Coverage estimation for multiburst FEC mobile TV services in DVB-H systems, *IEEE Transactions on Vehicular Technology*, 59(7), 3491–3500, September 2010.
16. D. Gozálvez, Combined time, frequency and space diversity in multimedia mobile broadcasting systems, PhD dissertation, Universitat Politèctica de València, Valencia. Spain, 2012.

Chapter 7

An Overview of the Next-Generation Mobile Digital Video Broadcasting Standard DVB-NGH

David Gómez-Barquero

Contents

185

7.1 Introduction

Digital Video Broadcasting—Next-Generation Handheld (DVB-NGH) is the handheld evolution of the second-generation digital terrestrial TV (DTT) technology, DVB-T2 (Terrestrial Second Generation). The first release of the DVB-T2 specification was officially published in 2009 with the purpose of improving the efficiency of DTT networks to cope with the increasing demand for high-quality TV services (especially HDTV) [1]. DVB-T2 succeeds in achieving a capacity increase between 50% and 70% over its predecessor DVB-T (Terrestrial) [2].

DVB-NGH was created with the objective of being the reference mobile multimedia broadcasting standard, outperforming in terms of capacity and coverage existing technologies, like the first-generation mobile TV terrestrial standard DVB-H (Handheld) [3], and the hybrid satellite–terrestrial mobile TV standard DVB-SH (Satellite to Handheld) [4]. DVB-NGH is the first broadcasting system that incorporates multiple-input multiple-output (MIMO) antenna schemes. MIMO overcomes the information-theoretic limits of single-input single-output (SISO) wireless communication systems without any additional bandwidth or increased transmit power, and makes DVB-NGH a third-generation broadcasting standard. MIMO allows multiplexing information streams across the antennas, forming a spatial radiation pattern per stream, which can result in a significant improvement of the system performance, especially at high signal-to-noise ratio (SNR) values [5]. The main drawback is that existing DTT network infrastructure needs to be upgraded at both transmitter (e.g., an additional transmit antenna, RF feedings, power combiners and amplifiers, etc.) and receiver sides (two antennas and RF front ends are required to demodulate the signal). This is the reason why MIMO was excluded from the DVB-T2 standard [2].

The standardization process of DVB-NGH started at the end of February 2010 and is currently at its final stages, and it is expected to be completed and delivered to ETSI in January 2013. The main issues that prevail in mobile broadcasting are related to improvements in coverage to cope with the more severe propagation conditions and reduce the network infrastructure investments, and improvements in the power consumption of the receivers to increase the battery life. Hence, whereas the main driver for DVB-T2 was increased capacity, the main goal of DVB-NGH is improved coverage, especially for pedestrian indoor reception. The main commercial requirements of DVB-NGH [6] were as follows:

■ A minimum 50% spectrum efficiency improvement compared to DVB-H working under the same conditions

■ The possibility of combining DVB-NGH and DVB-T2 signals in one radio frequency (RF) channel

■ The possibility of configuring the system either as a unidirectional broadcasting system or as a bidirectional system with an interaction channel (e.g., provided by a cellular network)

■ The minimization of the overhead such as packet headers and metadata without losing any functionality

■ The possibility of offering an NGH service in different qualities with specific robustness and graceful degradation in fringe areas of the network

■ The possibility of transmitting location-based services within single-frequency networks (SFNs) with a minimum increase to network overhead

■ The possibility of deploying a satellite component as a complement of the terrestrial network

The DVB-NGH specification defines four so-called profiles: the base (sheer-terrestrial) profile, the MIMO terrestrial profile, the hybrid terrestrial–satellite profile, and the hybrid MIMO profile. The base profile is mandatory, whereas the rest of the profiles are optional. The base profile is based on the DVB-T2 and T2-lite standards. It provides full description of all details. The MIMO and the hybrid profiles describe only the differences with reference to the base profile. The hybrid MIMO profile outlines only the differences with reference to the MIMO and hybrid profiles.

The main technical elements introduced in the base profile of DVB-NGH with respect to DVB-T2 are as follows: scalable video coding (SVC) [7] with multiple physical layer pipes (PLPs), time-frequency slicing (TFS), full support of an Internet Protocol (IP) transport layer with a dedicated protocol stack, header compression mechanisms for both IP and transport stream (TS) packets, new forward error correction (FEC) coding rates for the data path (down to 1/5), nonhomogeneous constellations for 64-QAM and 256-QAM, four-dimensional rotated constellations

(4D-RC) for QPSK, improved time interleaving in terms of zapping time, end-to-end latency and memory consumption, improved physical layer signaling in terms of robustness, capacity and overhead, a novel distributed multiple-input single-output (MISO) transmit diversity scheme for SFNs, and efficient provisioning of local content in SFNs.

The DVB-NGH MIMO profile is best suited for outdoor medium/high-signal use cases such as portable outdoor or vehicular reception. Unfortunately, the low SNRs of portable indoor reception drastically reduce the MIMO gain. DVB-NGH has adopted a novel transmission scheme known as enhanced spatial multiplexing with phase hopping (eSM-PH) for improved robustness in the presence of spatial correlation [8]. The MIMO profile of DVB-NGH has also adopted a new bit interleaver that eases the implementation of iterative structures that can provide significant gains on the top of the MIMO gain [9]. The eSM-PH code and the bit interleaver have been optimized for the case of deliberated power imbalance between the transmit antennas, which may be useful in some deployment scenarios where not all the receiver population supports MIMO.

The hybrid profile of DVB-NGH allows providing coverage to rural areas where the installation of terrestrial networks could be uneconomical. A hybrid terrestrial–satellite network is probably the most cost-effective network topology for mobile broadcasting to wide areas, including not densely populated areas. DVB-NGH enables to complement the terrestrial coverage (generally in the ultrahigh frequency [UHF] band) with an optional satellite component in the L- or S-frequency bands. In addition to the multifrequency network (MFN) configuration, an SFN configuration with both networks operating at the same frequency in the L or S bands is also possible. The satellite component of DVB-NGH has been designed with the goal of keeping the maximal commonality with the terrestrial component to ease its implementation at the receiver side. Specific solutions have been specified in the standard to deal with the characteristic problems of mobile satellite reception and to seamlessly combine the signals coming from the two networks. In particular, the hybrid profile of DVB-NGH allows for long time interleaving (e.g., around 10 s) at the physical layer with fast zapping support in order to cope with long signal outages due to the blockage of the line-of-sight (LoS) with the satellite, and single-carrier OFDM (SC-OFDM) for hybrid MFN networks in order to reduce the peak-to-average power ratio (PAPR) of the transmitted signal by the satellite.

The hybrid MIMO profile allows the use of MIMO in the terrestrial and/or the satellite components within a hybrid transmission scenario.

This chapter provides an overview of the main technical solutions adopted in DVB-NGH. Section 7.2 reviews the main commercial requirements of DVB-NGH and its standardization strategy. The rest of the chapter has a similar structure as the second part of the book. It has been structured in four sections, one for each profile. Section 7.3 describes the base profile. Section 7.4 is devoted to MIMO terrestrial profile. Section 7.5 deals with the hybrid terrestrial–satellite profile. Section 7.6 describes the hybrid MIMO profile. Finally, the chapter is concluded with Section 7.7.

7.2 Standardization Process of DVB-NGH

7.2.1 Analysis of the Commercial Requirements

Despite the superior technical performance of DVB-NGH, one of its main advantages compared to first-generation mobile broadcasting DVB systems is the possibility of transmitting DVB-NGH services in-band within a DVB-T2 multiplex in the same RF channel. This feature alleviates the investment required to start providing NGH services, since it is possible to reuse the existing DVB-T2 infrastructure without deploying a dedicated DVB-NGH network [6]. And it may encourage broadcasters to launch services gradually based on the demand of the local market.

The combination of DVB-NGH and DVB-T2 in the same multiplex is possible, thanks to the future extension frame (FEF) concept of DVB-T2. With the definition FEFs, the DVB-T2 standard let the door open to future enhancements. Thanks to these frames, it is possible to accommodate different technologies in the same multiplex in a time division manner. In DVB-T2, all frames start with a preamble OFDM symbol known as P1, which identifies the type of frame. The position in time and the duration of the FEFs are signaled in the physical layer-1 (L1) signaling in the T2 frames. This way, DVB-T2 legacy receivers, not able to decode a FEF, simply ignore the transmission during that time. During the transmission of FEFs, terminals can switch off their RF front-end saving power, like in a stand-alone DVB-T2 discontinuous transmission. Although a DVB-T2 FEF can contain any technology, the existence of combined DVB-T2/NGH receivers finally pushed the elaboration of a DVB-NGH physical layer strongly inspired in DVB-T2. Figure 7.1 illustrates the combined transmission of DVB-T2 with FEFs in the same multiplex. It should be pointed out that the technologies transmitted in the FEFs see the T2 frames as "FEFs."

DVB-NGH is the third type of specified FEF of DVB-T2. The other two are for identifying transmitters in SFNs [10] and for the mobile profile of DVB-T2 known as T2-Lite [1]. A typical deployment scenario for DVB-NGH or T2-Lite in a shared DVB-T2 multiplex is the one depicted in Figure 7.1: one FEF of reduced

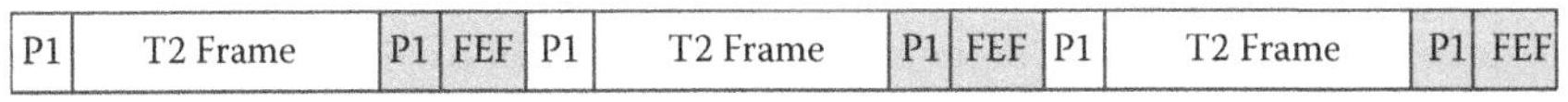

P1	T2 Frame	P1	FEF	P1	T2 Frame	P1	FEF	P1	T2 Frame	P1	FEF

Figure 7.1 Coexistence of T2 frames and FEFs in a single multiplex. Each T2 frame and FEF starts with a preamble P1 OFDM symbol that identifies the type of frame.

size (e.g., 50 ms duration) after every T2 frame with a considerably longer duration (e.g., 200 ms). This configuration devotes most of transmission time to DVB-T2 services, which are supposed to be the main services for the broadcast operator, but it allows introducing some mobile services. Moreover, since FEFs are rather short, the zapping time of the DVB-T2 services is practically not affected by the introduction of the new technology.

The combination of DVB-NGH with a cellular network requires a full end-to-end IP system, such that it is possible to deliver the same audio/visual content over the two networks in a bearer-agnostic way [6]. It should be pointed out that there is no convergence in networks, since both broadcasting and cellular networks are used separately and independently. Convergence is realized in services, platforms, and multimode terminals that support both radio access technologies. DVB-NGH includes full support of an IP transport layer, in addition to the most common transport protocol used in broadcast systems: MPEG-2 TS. DVB-NGH supports two independent transport protocol profiles for TS and IP, each one with a dedicated protocol stack. Note that this approach is possible because the physical layer packet unit in DVB-NGH, known as base band frame (BB frame), is content agnostic. This is a big difference compared to first-generation DVB standards, whose physical layer unit is MPEG-2 TS packets. In order to transmit IP over MPEG-2 TS, it is necessary to introduce an additional encapsulation protocol known as Multi Protocol Encapsulation (MPE). In the second-generation DVB standards, IP datagrams can also be transmitted using the Generic Stream Encapsulation (GSE) protocol [11]. GSE was designed to carry IP datagrams and provides efficient IP datagram encapsulation over variable-length link layer packets, which are then directly scheduled on the physical layer BB fames, rather than transmitted over MPEG2-TS\MPE. Additionally, GSE has less overhead. The relative overhead depends on the average PDU size, but approximately up to 70% overhead reduction can be achieved.

One of the most important conclusions of the study item performed prior to the development of DVB-NGH was that a significant data net throughput is today wasted in signaling overheads such as packet headers and metadata, and an optimization of the signaling and packet encapsulation was recommended to yield an important throughput improvements without losing any functionality [12]. This recommendation was subsequently adopted as part of the DVB-NGH commercial requirements [6]. Also the latency in the service reception and hence the end user experience is emphasized in the commercial requirements. Compared to DVB-T2, DVB-NGH has considerably improved the bandwidth utilization efficiency without any compromise to the functionality of the system. In particular, it has adopted new TS and IP packet header compression mechanisms, and it has improved the physical layer L1 signaling and the physical layer adaptation in terms of reduced signaling overhead.

The two transport protocol profiles of DVB-NGH have been also designed with the commercial requirement of transmitting layered video, such as SVC [7], with multiple PLPs.* Layered video codecs allow for extracting different video representations from a single bit stream, where the different sub-streams are referred to as layers. SVC is the scalable video coding version of H.264/AVC (Advanced Video Coding). It provides efficient scalability on top of the high coding efficiency of H.264/AVC. SVC encodes the video information into a base layer, and one or several enhancement layers. The base layer constitutes the lowest quality, and it is a H.264/AVC-compliant bit stream, which ensures backward compatibility with H.264/AVC receivers. Each additional enhancement layer improves the video quality in a certain dimension: temporal, spatial, and quality scalability.

The combination of layered video codec such as SVC with multiple PLPs presents a great potential to achieve a very efficient and flexible provisioning of mobile TV services in DVB-NGH systems. By transmitting the SVC base layer using a heavily protected PLP and the enhancement layer in one PLP with moderate/high spectral efficiency, it is possible to cost-efficiently provide a reduced quality service with a very robust transmission, while providing a standard/high quality for users in good reception conditions. The benefits of using SVC compared to simulcasting the same content with different video qualities in different PLPs with different robustness are twofold [13]. First of all, SVC has reduced bandwidth requirements compared to simulcasting. Compared to single-layer AVC coding, the SVC coding penalty for the enhancement layer can be as little as 10% [7]. Second, with SVC, it is possible to provide a graceful degradation of the received service quality when suffering strong channel impairments with seamless switching between the different video qualities.

The DVB-T2 specification states that DVB-T2 receivers are only expected to decode one single data PLP at a time [1]. But they must be able to decode up to two PLPs simultaneously when receiving a single service: one data PLP and its associated common PLP, which is normally used to transmit the information shared by a group of input data streams. This feature cannot be used to deliver SVC with multiple PLPs with differentiated protection in DVB-T2 because the behavior of the receivers when only one PLP is correctly received is not specified [13]. Compared to DVB-T2, DVB-NGH has enhanced the handling of multiple PLPs belonging to the same service.

Following the commercial requirements, DVB-NGH has also adopted two complementary techniques to insert local content in SFNs. The two most important local content types are probably local news and local advertising. Both are already widely implemented using temporal windows in the national DTT service in many countries. But the exponential growth of digital content platforms has also fostered the development of multimedia content and interactive services targeted at

---

* A PLP is a logical channel at the physical layer that may carry one or multiple services, or service components. Each PLP can have different bit rates and error protection parameters (modulation, coding rate, and time interleaving configuration, MODCODTI).

specific geographic communities. These services have the potential to create significant citizen and consumer benefits, as well as new revenue streams for broadcasters. Such local content may be potentially relevant for future broadcast services, serving audience needs that are not fully met by the current national and local broadcasting. However, the technical infrastructure of broadcast networks and the economic viability of local services have not been clearly established, and thus the audience demand for them has not been adequately assessed. DVB-NGH has adopted two complementary techniques, one based on hierarchical modulation and the other based on orthogonal insertion. Each technique addresses different use cases with different coverage-capacity performance trade-off, such that DVB-NGH will allow exploring the viability of inserting local services in SFNs in a way that has not been possible before. For example, DVB-NGH will allow inserting local content from a single transmitter in the network or providing local services with the same coverage as global services.

The NGH study mission also recommended targeting both sheer terrestrial networks and hybrid terrestrial–satellite networks for DVB-NGH in order to avoid market fragmentation, as happened with DVB-H and DVB-SH [12]. Although DVB-NGH technically outperforms DVB-SH in terms of capacity and coverage, the main advantage of DVB-NGH is that there is no need to deploy a dedicated terrestrial network in the S or L bands. Thanks to the FEF feature of DVB-T2, it is possible to reuse the existing DVB-T2 infrastructure in the UHF band to start providing DVB-NGH services in-band of a T2 multiplex and add the satellite component in the L or S frequency bands forming a hybrid MFN.

7.2.2 Standardization Strategy

The DVB-NGH specification was initially targeted to be completed by the end of 2011 [6]. The goal was to have commercial devices available on the market already in 2013 in order to be successful in the marketplace. The NGH study mission predicted that by 2015 time frame "rich media" content consumption would drastically increase with a variety of new devices due to the digital switch over and the convergence of fixed and mobile services along with telecommunication services [12]. However, the standardization process of DVB-NGH was delayed due to the development of the mobile profile of DVB-T2, known as T2-Lite, and an attempt of cooperation with the cellular standardization body 3GPP.

7.2.2.1 T2-Lite

In DVB-T2, it is possible to allocate multiple use cases, e.g., stationary, portable, and mobile, in the same frequency channel using multiple PLPs. Each PLP is a virtual container that can be configured with a particular set of transmission parameters. However, the simultaneous provision of stationary and mobile services is limited by the fact that the fast Fourier transform (FFT) mode and the pilot pattern must be common for all the PLPs in the T2 multiplex.*

The T2-Lite profile was developed in order to improve the coexistence of fixed and mobile services in DVB-T2 networks. With T2-Lite, it is possible to combine in the same multiplex a DVB-T2 signal optimized for high throughput and fixed reception and a T2-Lite signal optimized for high robustness and mobile reception in terms of modulation, coding rate, time interleaving, FFT size, and pilot pattern. It should be pointed out that T2-Lite signals can also be transmitted as a stand-alone signal.

The T2-Lite profile is based on a limited subset of DVB-T2 aimed at mobile broadcasting and designed to allow simpler receiver implementation by avoiding the most complex configurations and by limiting the required memory and the maximum service data rate. The profile has been introduced as a mandatory annex of the DVB-T2 specification in release 1.3.1 [1]. A detailed overview of the *DVB-T2* and *T2-Lite* mobile broadcasting technologies can be found in Chapter 6. It should be pointed out that DVB-NGH receivers have to support T2-Lite.

7.2.2.2 Common Broadcast Specification between 3GPP and DVB

During the standardization process of DVB-NGH, the standardization forum DVB contacted 3GPP organization asking to consider a potential collaboration in the area of mobile broadcasting in order to maximize the commercial success of DVB-NGH. A joint workshop took place for mutual presentations of 3GPP and DVB standardization activities, and the creation of a study item was proposed to the 3GPP technical subgroup dealing with services. The objective was to "study the feasibility of, and creating common service requirements and use cases for, a common broadcast specification which can be used in a 3GPP mobile communications network and a broadcasting network that is based on DVB or other similar standards." This proposal introduced, for the first time, the concept of Common Broadcasting Specifications (CBS), but it was not accepted by 3GPP due to lack of support from mobile operators. DVB decided then to complete the development of DVB-NGH and to address again the "CBS" topic after the finalization of its specifications.

Chapter 9 presents preliminary studies about a *common broadcast specification of* state-of-the-art *3GPP and DVB standards* to provide a broadcast overlay optimized for mobile and operated in conjunction with a broadband unicast access.

---

* Stationary services are generally transmitted with large FFTs and sparse pilot patterns in order to achieve a high spectral efficiency with sufficient robustness for rooftop antennas. On the other hand, reception in mobile scenarios requires the utilization of smaller FFTs and more dense pilot patterns to follow the rapid variations in the time and frequency domain, and to cope with the Inter-Carrier Interference (ICI) that is caused by the Doppler spread.

7.3 Base Profile of DVB-NGH

7.3.1 Bit-Interleaved Coding and Modulation

The BICM module of DVB-NGH is based on the module of DVB-T2 with a number of modifications that have been introduced in order to improve robustness, increase spectral efficiency, and optimize existing blocks while limiting the receiver complexity. The BICM module of DVB-NGH is at the cutting edge of coded modulation technologies. It spans an SNR range of more than 20 dB and supports spectral efficiencies ranging from 0.67 to 5.87 bits per constellation symbol for the single-antenna mode with a gap to Shannon capacity ranging from 2 to 3 dB at quasi-error-free point in fading static channels.

Figure 7.2 shows the BICM module of the base profile of DVB-NGH and illustrates the main differences with respect to DVB-T2. As in DVB-T2, the FEC scheme is based on a serial concatenation of a Bose–Chaudhuri–Hocquenghem (BCH) code and a Low-Density Parity Check (LDPC) code. The gain achieved compared to the FEC scheme used in first-generation DVB systems based on the concatenation of a convolutional and Reed–Solomon codes ranges from

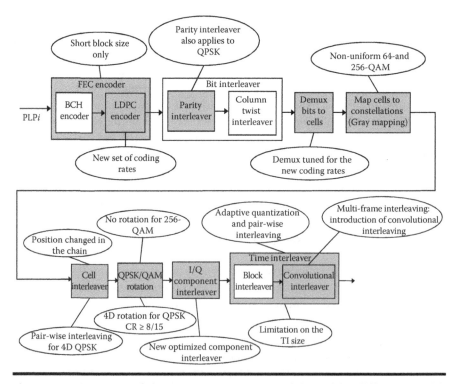

Figure 7.2   Structure of the DVB-NGH BICM module and its differences with DVB-T2.

3 dB for low-bit-rate services to more than 5 dB for high-bit-rate services in stationary channels.

The restrictions aimed to reduce the receiver complexity are as follows: only the short LDPC codeword size of 16,200 bits (16k) is allowed, the size of the time interleaver (TI) memory is halved compared to DVB-T2, the use of rotated constellations is prohibited in 256-QAM, and the maximum service bit rate is limited to 12 Mbps including both source and parity data. These restrictions are the same as the ones adopted in T2-Lite, except for the fact that in T2-Lite, the maximum service bit rate is limited to 4 Mbps taking into account only source data.

The BICM modifications introduced to improve the transmission robustness and coverage are the following. At the FEC code level, new more robust LDPC coding rates have been adopted extending the operating SNR range. Coding rate values uniformly distributed over the range 1/5–11/15 have been adopted (the most robust coding rate in DVB-T2 is 1/2. In T2-Lite, it is 1/3). At the constellation level, nonuniform constellations for 64-QAM and 256-QAM, and four-dimensional rotated QPSK constellations have been introduced in order to achieve a shaping gain and increased robustness to deep fade events, respectively. At the interleaving level, the DVB-T2 parity bit interleaver is used for all constellation sizes (it is not used for QPSK in DVB-T2). A new I/Q component interleaving has replaced the cyclic Q delay of DVB-T2 to exploit the signal-space diversity provided by rotated constellations, especially for multi-frame interleaving and TFS. The main advantage of rotated constellations when they are used in combination with multi-frame interleaving and TFS is that they can reduce the loss of information against signal outages in the time and frequency domain. The component interleaver has been designed in such a way that the components of each rotated symbol end up being transmitted with the maximum possible separation in the time domain and in different RF channels. Finally, the TI has introduced a convolutional interleaver (CI) for multi-frame interleaving (i.e., across multiple frames), which provides important advantages in terms of latency and memory utilization, and the use of adaptive cell* quantization for a more efficient use of the TI memory with low-order constellations.

More details on the *BICM module of DVB-NGH* can be found in Chapter 11. The chapter also provides some illustrative performance results and a comparison with DVB-H, such as the one illustrated in Figure 7.3. In the figure, we can see that DVB-NGH outperforms DVB-H in the AWGN channel by around 2.0–2.5 dB. In a Rayleigh fading channel, the gain ranges from 3.0 to 7.0 dB.

Chapter 12 is specifically devoted to the new *time interleaver of DVB-NGH*. The TI adopted in DVB-NGH is very flexible and supports multiple trade-offs in terms of time diversity, latency, and power saving. It consists of a combination of a block interleaver (BI) and a CI. The BI is inherited from DVB-T2 and performs intra-frame interleaving (i.e., within the frame) in the order of hundreds of milliseconds

---

* The TI in DVB-NGH (and DVB-T2) operates with cells (constellation symbols).

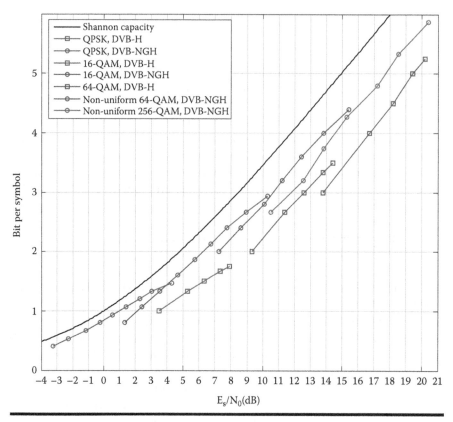

Figure 7.3 Required SNR to achieve an FER 10^{-4} after LDPC decoding over AWGN channel. Comparison with the Shannon limit and DVB-H.

[14], whereas the CI performs multi-frame interleaving and provides interleaving durations up to several seconds. The CI is introduced in such a way that it is possible to combine both de-interleavers into a single one at the receivers, such that each interleaver does not need dedicated memory. The benefits of the CI are that it requires around half the memory than a BI to provide the same TI duration and that it reduces the average zapping time by about 33%. It should be noted that these benefits apply only for multi-frame interleaving, when the CI is employed. The adaptive cell quantization technique allows a higher number of cells to be stored within a given physical time de-interleaving (TDI) memory for robust transmission modes that tolerate a higher quantization noise. DVB-NGH has the same TDI memory size as T2-Lite for 64-QAM and 256-QAM (i.e., 2^{18} cells or constellation symbols), but for QPSK and 16-QAM, it is possible to interleave up to 2^{19} cells in time. This optimization of the TDI memory allows longer TI durations for a given service data rate or higher service data rates for a given TI duration with both intra-frame and multi-frame interleaving.

Although DVB-NGH has the same TI memory size as T2-Lite, it makes a much more efficient use of the available memory. All in all, for QPSK and 16-QAM, DVB-NGH multiplies by four the maximum interleaving duration compared to T2-Lite.

7.3.2 Time-Frequency Slicing

TFS is a novel digital TV wireless transmission technique that consists in transmitting the services across several RF channels, breaking the existing paradigm of transmitting TV services in a single RF channel (see Figure 7.4). With TFS, the reception of a particular service is performed following dynamically over time the frequency hops among the different RF channels.

TFS was originally proposed in the standardization process of DVB-T2, but it was finally made an informative part of the standard due to the need of implementing two tuners in the receivers to receive a single service.* TFS can provide important gains in terms of both capacity and coverage [15]. The combination of many RF channels into a single TFS multiplex allows for an almost ideal statistical multiplexing (StatMux) gain for variable bit rate services. The frequency diversity provided by TFS can be very significant, since services can be potentially spread over the whole UHF frequency band. In general, the signal of each RF channel is affected by different propagation conditions that cause imbalances in the received strength at each location although the same power is transmitted in all channels [16]. The imbalances depend not only on the characteristics of the particular propagation scenario, but also on the frequency-dependent behavior of some physical elements of the transmission chain and the presence of interferences from other networks.

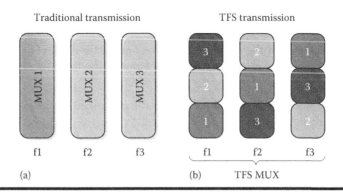

Figure 7.4 Traditional digital TV transmission (a) and TFS (b) over three RF channels.

---

* It was not possible to guarantee always a time interval between successive frames of the same service long enough for frequency hopping among RF channels with a single tuner.

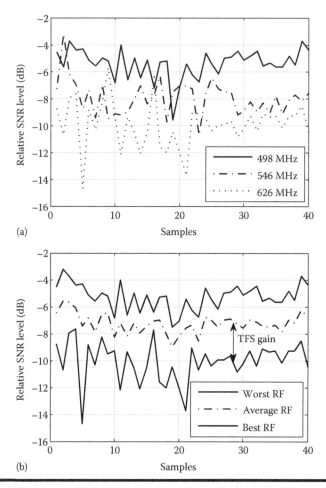

Figure 7.5 **Samples of the outdoor measurements of three RF channels (a) and maximum, average, and minimum signal level among the three RF channels (b).**

Figure 7.5a shows a set of outdoor measurements performed by the Swedish DTT operator Teracom. In a traditional DTT network, the perceived coverage at a given location is determined by the channel with the worst signal level in each location. This assures correct reception of all the multiplexes, which is a natural commercial requirement. With TFS, on the contrary, the coverage is determined by the average signal strength among all RF channels used by the TFS transmission. Hence, an indication of the TFS coverage gain at a particular location may be approximately computed as the difference between the average SNR value of the RF channels and the minimum instantaneous SNR among all RF channels (see Figure 7.5b).

The TFS coverage depends on the number of RF channels involved in the transmission and the frequency spacing among them. In general, the gain increases with the number of RF channels and the frequency spacing. The potential gain is very high because with appropriate coding and interleaving, even a fully lost RF channel may still allow correct reception of all services. Rotated constellations can further improve the performance of TFS by means of increasing diversity [17]. DVB-NGH has adopted TFS because it can be operated with a single* tuner without adding excessive complexity at the receivers. It has been included in all four profiles of DVB-NGH. The main benefit of TFS for DVB-NGH is in terms of improved coverage and transmission robustness. The additional frequency diversity is especially important for pedestrian reception conditions, where the time diversity is very little or inexistent. Moreover, for mobile reception, the increased frequency diversity can reduce the requirements for time interleaving. Another important benefit of TFS is the possibility to find spectrum for DVB-NGH services more easily and in a more flexible way, being possible to combine several RF channels with different percentages of utilization allocated to DVB-NGH, thanks to the logical frame structure defined in DVB-NGH.

Chapter 13 provides a detailed overview of *Time-Frequency Slicing* (*TFS*), describes its implementation in DVB-NGH, and provides illustrative performance evaluation results about the coverage gain with field measurements and physical layer simulations calibrated within the NGH standardization process.

7.3.3 Logical Frame Structure of DVB-NGH and Bundling DVB-T2 FEFs

The main envisaged use case for DVB-NGH is to share one multiplex with DVB-T2 in a time division manner using FEFs. In the case of T2-Lite, there is a one-to-one relation between FEFs and T2-Lite frames. That is, each DVB-T2 FEF is a T2-Lite frame. DVB-NGH allows for a more flexible and efficient allocation between FEFs and logical frames.

DVB-NGH has defined a new logical frame structure suited to the transmission using DVB-T2 FEFs, which has been especially designed to be compatible with TFS, and it may be seen as a generalization of TFS. The logical frame structure provides a lot of flexibility to TFS, because it relaxes constraints such as having the same length and allocation for the FEFs in all RF channels, or synchronizing the different T2 multiplexes. Furthermore, it allows the combination of FEFs with different transmission modes, frequency bands, etc., enabling hybrid terrestrial–satellite terminals with a single demodulator (tuner).

The logical frame structure defined in DVB-NGH is formed by the following elements: logical super-frame, logical channel group, logical channels, logical frames, and NGH frames. Depending how the logical frame is mapped into the

---

* The limitation in the coded service data rate and the guard interval provided by T2 frames relax the time constraints to correctly follow the frequency variations of the transmitted TFS signal.

NGH frames and the number of RF channels used, it is possible to distinguish four types of logical channels. Logical channels (LCs) type A and B use a single RF channel (i.e., time domain bundling).

For LC type A, each logical frame is mapped to one NGH frame (like in T2-Lite), whereas for LC type B, each logical frame is mapped to multiple NGH frames. Thus, each NGH frame may contain data from several logical frames of the same logical channel. Compared to the reference LC type A, this configuration has a reduced L1 signaling overhead because there is no need to transmit all physical signaling information in each NGH frame. For very short FEFs, which may be the most representative use case for the initial transmission of DVB-NGH services (like the example shown in Figure 7.1), the physical layer signaling could become relevant. It should be pointed out that the adopted solution does not degrade the zapping time and handover time, provided that the FEFs are transmitted with a sufficient high frequency.

LCs of type C and type D employ several RF channels with TFS. In both cases, logical frames may span several NGH frames. For LCs of type C, logical frames are mapped to multiple NGH frames that are transmitted over a set of RF channels. The NGH frames should be separated from channel to channel in order to allow reception with a single tuner. For LCs of type D, each logical frame is mapped to multiple NGH frames that are time-synchronized over the different RF channels.

A detailed description of the logical frame structure of DVB-NGH can be found in Chapter 14. The chapter also reviews the frame structure in DVB-T2 and details the operation of a DVB-T2 multiplex with FEFs.

7.3.4 Physical Layer Signaling

The physical layer signaling in DVB-NGH has two main functions as in DVB-T2. First of all, it provides a means for fast signal detection, enabling fast signal scanning. Second, it provides the required information for accessing upper layers, i.e., the L2 signaling and the services themselves. DVB-NGH has enhanced the physical layer signaling of DVB-T2 in three different aspects:

- Higher signaling capacity
- Improved transmission robustness
- Reduced signaling overhead

Compared to DVB-T2, DVB-NGH has increased the capacity of the signaling preamble and the L1 signaling. In DVB-T2, the physical layer signaling is transmitted in preamble OFDM symbols at the beginning of each frame, known as P1 and P2 symbols. The preamble P1 symbol is the first OFDM symbol of each frame and provides seven signaling bits, which, among other basic signaling information, identifies the type of frame. The frame type is signaled with three bits. These three bits are not sufficient to signal all profiles of DVB-T2, T2-Lite, and DVB-NGH. Therefore, DVB-NGH has introduced an additional preamble P1 (aP1) symbol to identify the terrestrial MIMO and the hybrid terrestrial–satellite SISO and MIMO profiles. The presence of the aP1 symbol is signaled in the P1 symbol. For the basic profiles of DVB-NGH, as well as for DVB-T2 and T2-Lite, there is no aP1 symbol. Furthermore, the new logical frame structure of DVB-NGH avoids any limitation in the maximum number of PLPs that can be used in the system due to L1 signaling constraints. The L1 signaling capacity has been increased because it is not constrained to the preamble P2 symbol(s).

DVB-NGH adopts for L1 signaling new mini LDPC codes of size 4320 bits (4k) with a coding rate 1/2. Although 4k LDPC codes have a worse performance than the 16k LDPC codes used in DVB-T2 for L1 signaling, the reduced size of 4k LDPC codes is more suitable for the L1 signaling because it reduces the amount of shortening and puncturing. In DVB-T2, L1 LPDC codewords are shortened (i.e., padded with zeros to fill the LDPC information codeword) and punctured (i.e., not all the generated parity bits are transmitted), which decrease the LDPC decoding performance. The adopted 4k LDPC codes have the same parity check matrix structure as the 16k LDPC codes used for data protection. This allows for efficient hardware implementations at the transmitter and receiver side efficiently sharing the same logic.

DVB-NGH has also adopted two new mechanisms to improve the robustness of the L1 signaling known as incremental redundancy (IR) and additional parity (AP). AP consists of transmitting punctured bits in the previous frame. In case there is need for more parity bits, the IR mechanism extends the original 4k LDPC code into an 8k LDPC code of 8640 bits. The overall coding rate is thus reduced from 1/2 down to 1/4. L1-repetition as in DVB-T2 can be optionally used to further improve the robustness of the L1 signaling as a complement of AP and IR.

Figure 7.6 compares the robustness of the L1-post signaling in DVB-T2 (with and without L1-repetition) and DVB-NGH (for different configurations of AP and L1-repetition). The performance has been evaluated with physical layer simulations for the mobile channel TU6 (6-Tap Typical Urban) for a Doppler frequency of 80 Hz. The size of the L1-pre is representative for eight PLPs. In the figure, it can be noted that DVB-NGH considerably improves the robustness of the L1-post compared to DVB-T2. The gain due to 4k LPDC without repetition and AP is around 1 dB. The difference between the most robust configurations in DVB-T2 and DVB-NGH is around 4 dB.

The robustness improvement of the L1 signaling in DVB-NGH can be translated into a reduction of the signaling overhead for the same coverage. Moreover, DVB-NGH has restructured the L1 signaling structure of DVB-T2 in order to further reduce the signaling overhead. Instead of signaling the configuration of each

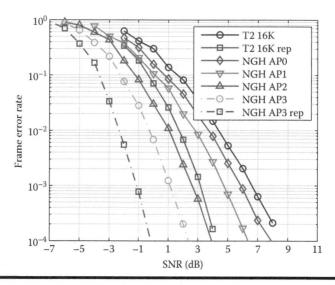

Figure 7.6 **Robustness comparison of the L1-post in DVB-T2 and DVB-NGH. Modulation is QPSK. TU6 channel with 33.3 Hz Doppler.**

PLP, PLPs are associated in groups with the same settings, reducing the required signaling information. Furthermore, it is possible to split in several frames signaling parameters that are in practice static and that are transmitted in DVB-T2 in every frame. The new logical frame structure of DVB-NGH also avoids transmitting all L1 signaling information in each NGH frame.

Chapter 15 provides an overview of the *physical layer signaling in DVB-T2 and DVB-NGH*. Illustrative performance evaluation results are presented comparing the robustness of the signaling of DVB-T2 and DVB-NGH, and the robustness of the data and the signaling paths in DVB-NGH.

7.3.5 System and Upper Layers

DVB-NGH is a physical layer standard that supports two independent transport protocol profiles for TS and IP, each one with a dedicated protocol stack (see Figure 7.7). The profiles have been independently designed in order to improve the bandwidth utilization. Two header compression mechanisms have been adopted for both TS and IP packets. For the TS profile, for PLPs that carry only one service component, the TS header can be reduced from three bytes to only one byte. For the IP profile, the unidirectional mode of the Robust Header Compression (ROHC) protocol has been adopted [19]. Both profiles also allow the transmission of SVC with multiple PLPs. Compared to DVB-T2, DVB-NGH has enhanced the handling of multiple PLPs belonging to the same

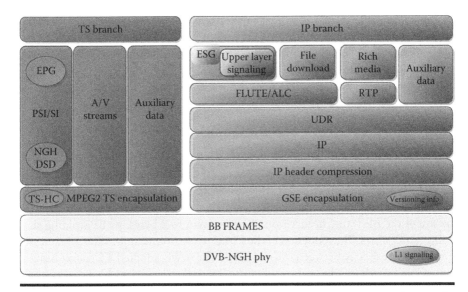

Figure 7.7   TS and IP profiles of DVB-NGH, including header compression mechanisms. Both profiles may coexist in the same multiplex. The TS signaling and service discovery are based on the PSI/SI tables with a new NGH delivery system descriptor. The upper layers IP signaling is carried within the ESG for better convergence with IP-based systems (cellular networks) and is compatible with OMA-BCAST.

service, being possible to receive up to three data PLPs simultaneously plus the common PLP. The receiver buffer model, which determines the necessary time delays at the transmitter in order to avoid buffering overflow or underflow in the receivers, has been updated, and a new signaling to provide mapping between service components and PLPs and the scheduling of the PLPs at the physical layer has been defined.

The upper layer solution for the IP profile is based on Open Mobile Alliance Mobile Broadcast Services Enabler Suite (OMA-BCAST). However, it should be noted that the specification allows using other upper-layer solutions, which may be defined in the future. OMA-BCAST is an open global specification for mobile TV and on-demand video services that can be adapted to any IP-based mobile delivery technology. OMA-BCAST specifies a variety of features including content delivery protocols for streaming and file download services, electronic service guide (ESG) for service discovery, service and content purchase and protection, terminal and service provisioning (e.g., firmware updates), interactivity, notifications, etc. OMA-BCAST specifies separate set of specifications for the different physical layer bearers, such as 3GPP, DVB-H, and, now DVB-NGH. In order to minimize the redundancy and latency on the signaling, and hence also the complexity and signaling overhead, the OMA-BCAST adaptation for DVB-NGH has taken radical approach to split the majority of the signaling between physical layer and on the top of IP, inside the OMA-BCAST service guide structures. Only minimal signaling is defined in the L2, for optimizing the receiver latency and power saving.

Chapter 16 provides an overview of the *system and upper layers of DVB-NGH*.

7.3.6 Overhead Reduction Methods

Compared to DVB-T2,* DVB-NGH reduces the overhead at the physical layer, thanks to its improved signaling and physical layer adaptation, as well as at the network layer and above through the introduction of packet-header compression schemes for both TS and IP profiles. The improved physical layer adaptation reduces the size of the BB frame header from 8 bytes down to 3 bytes, while avoiding signaling ambiguities. The reduction of the L1 signaling overhead provides an increase in the system capacity between 1% and 1.5%. The newly adopted TS packet header compression method allows the size of the TS packet header to be reduced from 4 bytes to only 1 byte, providing a 1.1% system capacity increase. With ROHC, the IP packet overhead can be reduced to approximately 1% of the transmitted data, yielding a capacity increase between 2.5% and 3.5%.† Both packet header compression mechanisms are optional.

7.3.6.1 Physical Layer Adaptation

DVB-NGH has defined a new physical layer adaptation mode in order to reduce the overhead and prevent ambiguities. The signaling has been reorganized separating the PLP-specific information (only present in the L1 signaling) from the BB frame information (only present in the BB frame headers) avoiding duplication and improving consistency.

7.3.6.2 TS Packet Header Compression

DVB-NGH introduces an optional header compression method for TS packets, which allows reducing the header size from 4 bytes to only 1 byte. The compression is performed on the transmitter side, and the information needed for restoring the header in the receiver is signaled in the BB frame header and the L1 signaling, such that the compression and decompression process is transparent. This technique is applicable only to PLPs that carry one program component,

---

* Compared to first-generation mobile broadcasting systems, the encapsulation overhead for IP delivery is reduced by up to 70% in DVB-NGH, thanks to the use of GSE as link layer encapsulation protocol instead of the older MPE over MPEG-2 TS.

† Assuming an average payload data unit of 1000 bytes, typical overhead values for IPv4 and IPv6 packets are 4% and 6%, respectively.

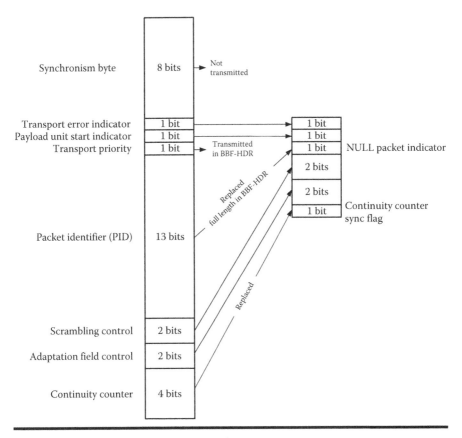

Figure 7.8   TS packet header compression.

identified by one program identifier (PID). The TS packet header is compressed as follows (see Figure 7.8):

■ The synchronism byte is removed as in DVB-T2.
■ The 1 bit transport priority indicator is removed and transmitted in the BB frame header.
■ The 13 bit PID field is replaced by a single bit to signal null packets. The PID value is signaled in the BB frame header.
■ The 4 bit continuity counter is replaced with a 1 bit duplication indicator.

7.3.6.3 IP Packet Header Compression

One of the new features of DVB-NGH is the use of the ROHC protocol [19] to reduce the overhead of IP packets. ROHC, however, introduces inter-packet dependencies in the transmitted stream that can increase the zapping time and introduce packet error propagations. Hence, an NGH adaptation layer for ROHC has been introduced in order to reduce the negative impact of ROHC on the zapping time and improve the robustness of the compressed flow (see Figure 7.9). Moreover, the NGH adaptation layer for ROHC guarantees backward compatibility with the standalone ROHC framework, which allows the reuse of exiting software implementations of the ROHC protocol.

ROHC can be modeled as an interaction between two state machines, one compressor machine and one decompressor machine. The ROHC framework defines the state machine transitions and describes procedures for start of transmission and error recovery. ROHC classifies the protocol header fields depending on their changing pattern between consecutive packets. A simplified view of this classification consists in dividing the fields into three types: inferred, static, and dynamic. The inferred fields are the ones that contain values that can be inferred from other protocol header fields or from lower-level protocols and do not need to be transmitted. The static fields are expected to be constant throughout the lifetime of the packet flow (e.g., IP destination address) and therefore must be communicated to the receiver only once, or expected to have well-known values (e.g., IP version), and therefore do not need to be communicated at all. Dynamic are the fields that are expected to vary during the transmission of the packet flow. The efficiency of the ROHC scheme depends on the setup of the compressor, the characteristic of the transmitted IP flow, and whether or not the NGH adaptation layer for ROHC is used.

Chapter 17 provides an overview of the different *overhead reduction methods of DVB-NGH*. It describes the technical and operational issues relevant to each overhead reduction mechanism and presents header compression results for ROHC.

7.3.7 Local Service Insertion in Single-Frequency Networks

DVB-NGH has adopted two complementary techniques to transmit local content in SFNs, known as hierarchical and orthogonal local service insertion (H-LSI and O-LSI, respectively). Both techniques provide very important capacity gains compared to the classical SFN approach where the local content is transmitted across the whole service area, but each technique addresses different use cases with different coverage-capacity performance trade-off, such that the optimum transmission technique depends on the target use case and the particular scenario (location and power of the transmitters, distribution of the local service areas, etc.) considered. Figure 7.10 illustrates the coverage areas for global and local services in an SFN for both techniques. For H-LSI, the coverage of the local services is limited to the areas surrounding the transmitters, whereas with O-LSI, local services have basically the same coverage as the global services. For both techniques, the transmission of local content through the whole SFN network can be scheduled in a way that different local areas do not interfere with each other.

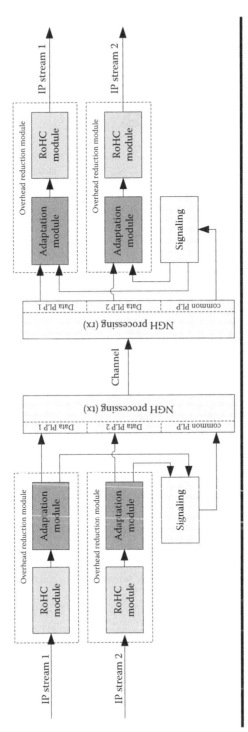

Figure 7.9 **Block diagram of DVB-NGH including header compression stage.**

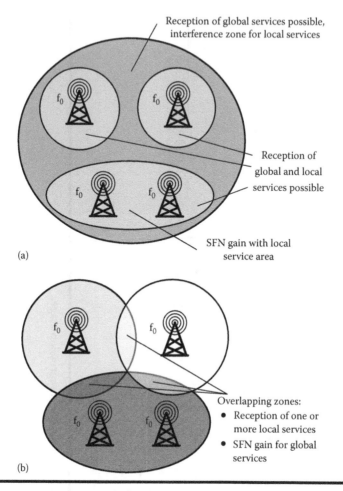

Figure 7.10 Coverage areas for global and local services in an SFN with H-LSI (a) and O-LSI (b).

7.3.7.1 Hierarchical Local Service Insertion

H-LSI uses hierarchical modulation [20], which allows combining two independent data streams, known as high priority (HP) and low priority (LP) streams, into a single stream with different robustness. With H-LSI, local services are transmitted in the LP stream on top of the global services in the HP stream with a hierarchical modulation. Transmitters inserting local content add an additional QPSK constellation on top of the QAM constellation used for global services. DVB-NGH supports hierarchical 16-QAM and 64-QAM. H-LSI increases the available data rate in the SFN. When the global PLP employs QPSK modulation, and the same coding rate is used for the global and local PLPs, it is possible to achieve up to 100% capacity gain when all carriers are used to transmit local content (the maximum percentage of local services is 50%).

H-LSI can be used to transmit local services on top of the global services in areas close to the transmitters. As depicted in Figure 7.10, the hierarchical modulation cannot assure the same coverage for local and global services. This means that in some areas it is not possible to receive any local service at all. Another drawback of the hierarchical modulation is that it suffers from interlayer interference, because each stream acts as noise to the other. That means that the coverage of the global services is reduced when local services are transmitted on top of them. The susceptibility to noise of the HP stream can be reduced by increasing the spacing between HP constellation symbols or using a lower coding rate, at the expense of degrading the performance of the LP stream and reducing the available total capacity, respectively.

7.3.7.2 Orthogonal Local Service Insertion

The O-LSI scheme specifies groups of OFDM carriers in specific OFDM symbols for the exclusive use of particular transmitters to transmit local services. The main benefits of O-LSI are twofold. First, the coverage of the global services is not affected by the local services. Second, the coverage of the local services is potentially the same as the coverage of the global services (see Figure 7.10). It should be noted that local services do not fully benefit from the SFN gain (except within an LSA containing several transmitters), and hence there is no power gain. But there is still a statistical network gain. In the overlapping zones between adjacent transmitters, receivers can decode more than one local service in addition to the global service because they do not interfere with each other.

With O-LSI, the transmission capacity depends on the number of local service areas employing a specific set of carriers. The transmission capacity can be increased using a MODCOD with a higher spectral efficiency, since it is possible to transmit the carriers devoted to local services with higher power, because for each transmitter, only some carriers within one OFDM symbol are active. It should be noted that the total transmission power of each OFDM symbol is the same, but at each transmitter, only one set of local carriers is active. The capacity gain of O-LSI compared to simulcasting strongly depends on the percentage of local services and the number of local service areas, but it can take values up to 100%.

Chapter 18 describes the complementary *local service insertion techniques of DVB-NGH* for providing global and local contents in SFN. The chapter also identifies the main use cases of local content in mobile broadcast networks and the corresponding DVB-NGH system configurations. Furthermore, the applicability of these approaches in terms of network topologies, implementation issues, and performance evaluation is analyzed.

7.4 MIMO Terrestrial Profile of DVB-NGH

The DVB-NGH specification defines the implementation of MIMO techniques as an optional profile in order to exploit the diversity and capacity advantages made possible by the use of multiple transmission elements at the transmitter and receiver. The use of the MIMO profile is signaled in preamble P1 symbol, which is followed by an additional preamble aP1 symbol that provides information about the FFT size and guard interval used.

During the standardization process of DVB-NGH, two types of MIMO techniques, known as rate-1 and rate-2, were distinguished according to their multiplexing capabilities* and compatibility with single-antenna receivers.

The MIMO rate-1 codes exploit the spatial diversity of the MIMO channel without the need of multiple antennas at the receiver side (i.e., they can also be referred as MISO schemes). They can be applied across the transmitters of SFNs reusing the existing DTT network infrastructure. DVB-NGH has adopted the distributed MISO scheme of DVB-T2 based on Alamouti coding [21] and adopted a novel transmit diversity scheme known as enhanced single-frequency network (eSFN). Chapter 20 describes in detail the two *MIMO rate-1 schemes adopted in DVB-NGH*. It should be pointed out that the two schemes are part of the base profile, since they are compatible with single-antenna receivers.

MIMO rate-2 codes exploit the diversity and multiplexing capabilities of the MIMO channel. They mandatorily require at least two antenna aerials at both receiver and transmitter sides. Hence, existing DTT network infrastructure must be upgraded (among others, second transmit antenna, cooling systems, RF feedings, power combiners, etc.). The MIMO rate-2 code adopted in DVB-NGH is intended for a 2×2 MIMO system and is known as eSM-PH. A cross-polar configuration has been selected because of the excessive antenna separation in UHF frequency range for co-polar antennas, and the increased robustness against LoS condition for the cross-polar setup. A detailed description of *eSM-PH* is presented in Chapter 21.

Chapter 19 introduces *basic MIMO concepts* focusing on issues central to its use in DVB-NGH, describes the channel sounding campaign and resulting *UHF cross-polar MIMO channel* developed within the standardization process and used for performance evaluation, and discusses *MIMO implementation issues* into new and existing networks that continue to also support SISO transmission.

---

* MIMO allows the multiplexing rate of information across the antennas to be increased. The most simple way of increasing the data rate consists in dividing the information symbols between the transmit antennas. This is referred to as pure spatial multiplexing (SM).

7.4.1 MIMO Rate-1 Codes

7.4.1.1 Alamouti

The Alamouti code is a very simple space-frequency block code designed for increasing the diversity in systems with two transmit antennas [21]. It achieves full diversity with reduced (linear) receiver complexity. In DVB-T2, the Alamouti code is applied across pairs of transmitters in order to improve the reception in SFN. One transmitter sends the information symbols without any modification, while the other transmitter sends an encoded version of the original symbols (it is applied by encoding in pairs the information symbols of adjacent subcarriers). By using the Alamouti code, it is possible to combine the signals from different transmitters in an optimum way and remove the presence of notches from the channel. But in order to use the Alamouti code, it is necessary to double the number of pilot patterns so that receivers can estimate the channel response from each transmit antenna.

7.4.1.2 Enhanced Single-Frequency Network

eSFN is a cyclic delay diversity [22] scheme that consists on applying a linear predistortion function to each antenna in such a way that it does not affect the channel estimation in receivers. This technique increases the frequency diversity of the channel without the need of specific pilot patterns or signal processing to demodulate the signal. eSFN is also well suited for its utilization as a distributed MISO technique in SFN. The randomization performed in each transmitter can avoid the negative effects caused by LoS components in this kind of networks. In addition, by using a different predistortion function at each transmitter, it is possible to allow for unique transmitter identification within the network, which can be used, e.g., for monitoring applications.

7.4.2 Enhanced Spatial Multiplexing with Phase Hopping

The presence of correlation in the MIMO channel due to the lack of scattering or insufficient antenna separation is especially detrimental for SM [23]. eSM-PH retains the multiplexing capabilities of SM and, at the same time, increases the robustness against spatial correlation. The conceptual block diagram of eSM-PH is illustrated in Figure 7.11. The transmission matrix of eSM can be represented as the concatenation of the regular SM transmission matrix with a precoding matrix such that the information symbols are weighted and combined before their transmission across the antennas. The weighting of the information symbols depends on a rotation angle, which has been specifically tuned for each constellation. In addition, a phase hopping term is added to the second antenna in order to randomize the code structure and avoid the negative effect of certain channel realizations. The phase hopping term changes periodically within each FEC codeword.

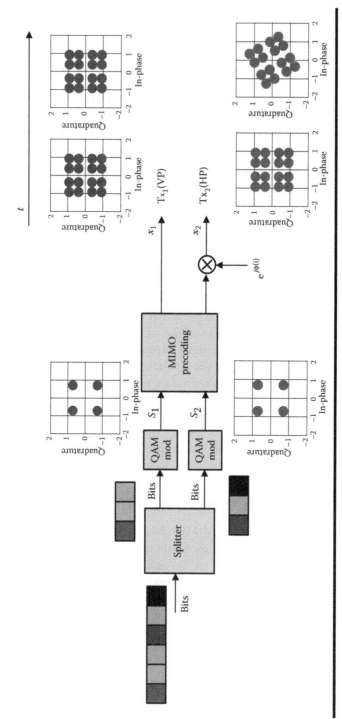

Figure 7.11 Block diagram of eSM-PH with gray labeled 4-QAM symbol constellation on each antenna.

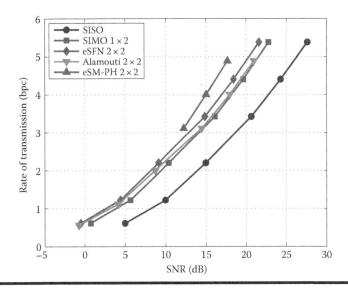

Figure 7.12   Performance of the different DVB-NGH MIMO schemes in the NGH outdoor MIMO channel with 60 km/h speed.

Figure 7.12 compares the performance of eSM-PH with SISO, SIMO with two receive antennas, and eSFN and Alamouti with two transmit and two receive antennas. The results include the effect of pilot overhead. A pilot density of 1/12 has been assumed for SISO, SIMO, and eSFN, whereas for Alamouti and eSM-PH, the pilot density is 1/6 (i.e., double, the pilot patterns used for eSM-PH are the same as for MISO Alamouti in DVB-T2). The figure highlights the significant gains achieved by the different DVB-NGH MIMO schemes over SISO. The performance of MIMO Alamouti lies between SIMO and eSFN due to the effect of increased pilot overhead. eSM-PH provides interesting gains only in favorable reception conditions, such as portable outdoor or vehicular reception. Nevertheless, it should be pointed out that the MIMO profile has also adopted a new bit interleaver that exploits the quasi-cyclic structure of the adopted LDPC codes. It exhibits a low complexity, low latency, and fully parallel design that ease the implementation of iterative structures that can provide significant gains on the top of the MIMO gain [9].

eSM-PH can be transmitted with power imbalance between the antennas to ease its introduction and coexistence with SISO transmissions. A deliberated transmitted power imbalance avoids envelope power fluctuations at the transmitter and provides a reasonable coverage reduction for SISO/SIMO terminals by lowering the existing transmit antenna power slightly, while eSM-PH maintains good performance due to its optimized performance to overcome this situation. The power imbalances considered in the standard are 3 and 6 dB. For these cases, the eSM-PH rotation angle has been optimized to reduce the performance loss due to transmit power imbalance.

7.5 Hybrid Terrestrial: Satellite Profile of DVB-NGH

The DVB-NGH specification allows for the deployment of an optional satellite component complementing the coverage provided by a terrestrial network. The hybrid profile includes several technical solutions specifically developed to deal with the characteristic problems of mobile satellite reception and to receive the signals coming from the two networks with a single tuner and to combine them seamlessly. It has been designed with the goal of keeping the maximal commonality with the terrestrial component to ease its implementation at the receiver side. The main elements introduced are the following:

- Extended convolutional inter-frame time interleaving (up to approx. 10 s) with fast zapping support with a uniform-late CI profile
- SC-OFDM for the satellite component in MFN hybrid networks in order to reduce the PAPR of the transmitted signal

DVB-NGH has also specified the scheduling of the terrestrial and satellite transmissions in MFN such that parallel reception of both signals is possible for terminals with a single demodulator. This has been made possible, thanks to the introduction of the concept of the logical channel and the logical channel group.

Compared to a sheer terrestrial receiver, a satellite receiver requires at least an additional external TDI memory to account for the long time interleaving requirements at the physical layer, SC-OFDM demodulation, and a tuner covering the satellite frequency bands (L and S) (see Figure 7.13).

Long time interleaving with fast zapping support is a key feature for mobile TV satellite transmissions. The LMS channel is characterized by long signal outages due to the blockage of the LoS with the satellite caused by tunnels, buildings, trees, etc., which can be compensated only with a long time interleaving duration [24]. However, time interleaving increases the end-to-end latency and the channel change (zapping) time, which is a crucial quality of service parameter, especially for TV use. Generally, it is considered that zapping times around 1 s are satisfactory, whereas more than 2 s are felt as annoying [25].

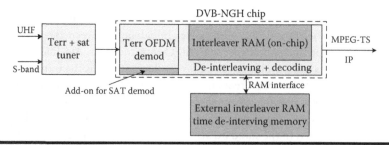

Figure 7.13 Block diagram of a DVB-NGH receiver supporting both the base sheer terrestrial profile and the hybrid satellite–terrestrial profile.

The hybrid profile makes use of an external TDI memory as a complement of the on-chip TDI memory, as shown in Figure 7.13. The size of the on-chip TDI memory is the same as that of the basic terrestrial profile, and it is only used for intra-frame block interleaving, whereas the external TDI memory is used for inter-frame convolutional interleaving. Fast zapping is supported using a uniform-late profile of the CI, like in DVB-SH [4] (see Figure 7.14). The uniform-late profile introduces a trade-off between overall performance in mobile channels and performance after zapping.

The reduction of the PAPR of the transmitted signal is also key in satellite communications in order to maximize the efficiency of the high-power amplifier on board the satellite. Time Division Multiplexing (TDM) has been the reference scheme because of its reduced power fluctuations that allow driving the amplifiers with small input back-offs, optimizing the amplifier power efficiency. But DVB-NGH has also adopted SC-OFDM for the satellite component of hybrid MFN networks (for hybrid SFN networks, OFDM is used). The reason is that it provides an important PAPR gain compared to OFDM while preserving many commonalities with OFDM. The main difference at the receiver is the introduction of a de-spreading function (see Figure 7.15). The spreading at the transmitter can be viewed as a way of spreading each symbol over the entire spectrum. Hence, the implementation cost of including SC-OFDM in the chips is markedly lower than TDM modulation.

SC-OFDM provides an approximate gain in the order of 2.5 dB in terms of reduced PAPR for high-power satellite amplifiers, which can be directly translated into an increase in the coverage provided by the satellite, achieving a gain of about 1.5 dB in the link budget at low input back-offs.

The use of the hybrid profile is signaled in preamble P1 symbol, which is followed by an additional preamble aP1 symbol that provides information about the use of SC-OFDM, and the FFT size and guard interval used.

Chapter 22 provides an overview of the *hybrid profile of DVB-NGH* and describes the main technical solutions specifically developed to deal with the characteristic problems of mobile satellite reception and to seamlessly combine the signals coming from the two networks with a single tuner.

Chapter 23 describes and evaluates the SC-OFDM modulation that was selected along with pure OFDM to implement the transmissions on the DVB-NGH satellite link.

7.6 Hybrid MIMO Profile of DVB-NGH

The hybrid MIMO profile allows the use of MIMO on the terrestrial and/or satellite elements within a hybrid terrestrial–satellite network. Cases included are one or two (cross-polar, linear polarization) terrestrial antennas in combination with one or two (cross-polar, counter-rotating circular polarization) satellite antennas, being

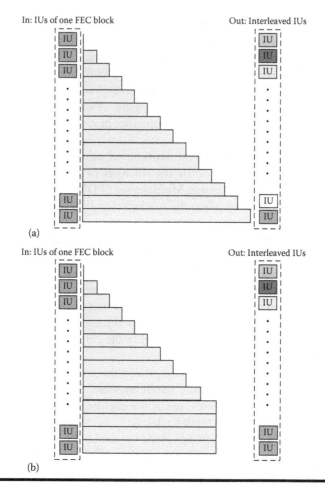

Figure 7.14 Convolutional interleaving profiles: Uniform (a) and Uniform-Late (b).

thus possible to have up to four transmit antennas (see Figure 7.16). At least one of the transmission elements (i.e., terrestrial or satellite) must employ multiple antennas; otherwise, the use case lies within the hybrid profile. Both MFN and SFN network configurations are possible.

For hybrid MFN configurations, in the case that satellite waveform is SC-OFDM, spatial multiplexing encoding for rate-2 MIMO is simple SM instead of eSM-PH (i.e., neither MIMO precoding nor phase hopping is applied).

For hybrid SFN configurations, the satellite simply repeats the transmission of the terrestrial transmitters. An important motivation for this is to avoid increasing the amount of pilot symbols required for channel estimation (e.g., in the four transmitter case, if channel responses for all channels are required, the pilot density is four times the density required for the same estimation accuracy to the

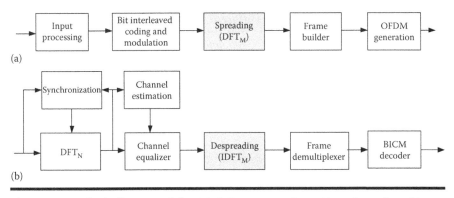

(a)

(b)

Figure 7.15   Block diagram of the SC-OFDM transmitter (a) and receiver (b).

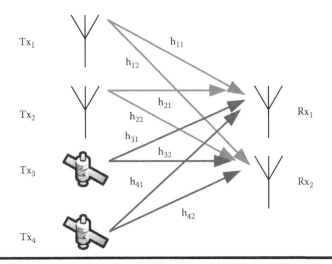

Figure 7.16   4 × 2 hybrid MIMO antenna configuration.

one-transmitter SISO case). Another consideration motivating this choice is that the need to guarantee a satisfactory receiver performance is possible even if one of the streams is lost (a highly probable event in such mobile systems). The selected codes for SFN scenarios are presented in Table 7.1.

When single polarization is available at both satellite and terrestrial site, Alamouti coding or eSFN can be applied. When the terrestrial transmissions are dual polarized and the satellite utilizes only a single polarization, the satellite can simply transmit the same symbols as one terrestrial polarization (this is called "Alamouti + QAM"). eSFN can also be used. To achieve a rate-2 transmission, virtual MIMO (VMIMO) can be used, where the single polarized transmitter emulates at the transmitter side an optimized 2×1 channel while the dual polarized transmitter emits rate-2 MIMO. When the terrestrial transmissions use

Table 7.1 MIMO Schemes for Different Hybrid MIMO SFN Scenarios

#Tx	Terrestrial	Satellite	Schemes
2	Single polarization	Single polarization	Rate-1: eSFN, Alamouti
3	Dual polarization	Single polarization	Rate-1: eSFN, Alamouti + QAM
			Rate-2: VMIMO
3	Single polarization	Dual polarization	Rate-1: eSFN, Alamouti + QAM
			Rate-2: VMIMO
4	Dual polarization	Dual polarization	Rate-1: eSFN, Alamouti + Alamouti
			Rate-2: eSM-PH (terr.), eSM-PH + eSFN (sat.)

single polarization and satellite transmissions are dual polarized, Alamouti + QAM can be utilized. Now only the roles of terrestrial and satellite transmitters are reversed as compared to the code presented for the case of dual terrestrial and single satellite polarization. eSFN and VMIMO are also possible. In this case, both terrestrial and satellite transmissions utilize dual polarization. For rate-1, the same Alamouti blocks are transmitted from both satellite and terrestrial transmitters. Further, eSFN can be used for satellite to enhance the operation. This code is called "Alamouti + Alamouti." For rate-2, eSM-PH is used in both terrestrial and satellite components. The satellite transmits the same signal except for the eSFN predistortion.

The use of the hybrid MIMO profile is signaled in preamble P1 symbol, which is followed by an additional preamble aP1 symbol that provides information about the use of SC-OFDM, the FFT size and guard interval used. The hybrid MIMO profile allows combining long time interleaving with MIMO [26].

7.7 Conclusions

DVB-NGH is the first third-generation DVB standard to include MIMO antenna techniques. DVB-NGH is also the first mobile broadcast system that exploits the diversity of the mobile channel in all dimensions: frequency diversity among several RF channels with TFS, time diversity with long time interleaving (in the order of 10 s) at the physical layer with fast zapping support, and space diversity with MIMO. Furthermore, DVB-NGH employs two-dimensional and four-dimensional rotated constellations (also known as signal-space diversity) to increase the diversity order. All the new technological solutions adopted in DVB-NGH, together with the high performance of DVB-T2, make DVB-NGH a technology with superior capacity and coverage performance compared to existing mobile multimedia broadcasting standards.

However, questions remain regarding the commercial prospects of the standard given the limited success of similar standards in the past. Despite the superior technical performance of DVB-NGH, one of its main advantages compared to first-generation mobile broadcasting DVB systems is the possibility of transmitting DVB-NGH services in-band within a T2 multiplex in the same RF channel without constraints from coexistence, enabling to reuse the existing DVB-T2 infrastructure and avoiding to deploy a dedicated DVB-NGH network. However, the mobile profile of DVB-T2 known as T2-Lite has also this feature, and first commercial devices are starting to be available already in 2012. In this respect, the hybrid terrestrial–satellite profile of DVB-NGH may be the one with a higher commercial potential initially, and the terrestrial profile could, e.g., reach the market as an evolution of T2-Lite if it becomes successful. DVB-NGH could also take advantage of the introduction of the new High-Efficiency Video Coding standard, which is expected to provide significantly improved coding efficiency as compared with the current state-of-the-art MPEG-4 AVC video coding. But in any case, the progress beyond prior state-of-the-art in digital video broadcasting achieved with DVB-NGH makes it the reference point for future/upcoming generations of digital broadcasting technologies. For example, the MIMO techniques developed within DVB-NGH provide a starting point for the possible use of MIMO in the context of fixed rooftop reception.

Acknowledgments

The author would like to thank all his colleagues from the DVB TM-H working group that have contributed to the development of the DVB-NGH specification.

References

1. ETSI EN 302 755 v.1.3.1, Frame structure channel coding and modulation for a second generation digital terrestrial television broadcasting system (DVB-T2), April 2012.
2. L. Vangelista et al., Key technologies for next-generation terrestrial digital television standard DVB-T2, *IEEE Communications Magazine*, 47(10), 146–153, 2009.
3. G. Faria, J. A. Henriksson, E. Stare, and P. Talmola, DVB-H: Digital broadcast services to handheld devices, *Proceedings of the IEEE*, 94(1), 194–209, 2006.

4. I. Andrikopoulos et al., An overview of digital video broadcasting via satellite services to handhelds (DVB-SH) technology, in B. Furht and S. Ahson, eds., *Handbook of Mobile Broadcasting: DVB-H, DMB, ISDB-T and MediaFLO*, CRC Press, Boca Raton, FL, 2008.

5. J. Paulraj, D. A. Gore, R. U. Nabar, and H. Bölcskei, An overview of MIMO communications—A key to Gigabit wireless, *Proceedings of the IEEE*, 92(2), 198–218, 2004.

6. DVB Commercial Module sub-group on Next Generation Handheld, Commercial requirements for DVB-NGH, CM-1062R2, June 2009.

7. H. Schwarz, D. Marpe, and T. Wiegand, Overview of the scalable extension of the H.264/MPEG-4 AVC video coding standard, *IEEE Transactions on Circuits and Systems for Video Technology*, 17(9), 1103–1120, 2007.

8. S. Moon et al., Enhanced spatial multiplexing for rate-2 MIMO of DVB-NGH system, in *Proceedings of the IEEE ITC*, Jounieh, Lebanon, 2012.

9. D. Vargas, D. Gozálvez, D. Gómez-Barquero, and N. Cardona, MIMO for DVB-NGH, The next generation mobile TV broadcasting, *IEEE Communications Magazine*, in press.

10. ETSI TS 102 992 v1.1.1, Structure and modulation of optional transmitter signatures (T2-TX-SIG) for use with the DVB-T2, September 2010.

11. ETSI TS 102 606 v1.1.1, Digital video broadcasting (DVB): Generic stream encapsulation (GSE) protocol, October 2007.

12. DVB Technical Module sub-ground on Next Generation Handheld, NGH study mission report, TM-H 411, June 2008.

13. C. Hellge, E. Guinea, D. Gómez-Barquero, T. Schierl, and T. Wiegand, HDTV and 3DTV services over DVB-T2 using multiple PLPs with SVC and MVC, in *Proceedings of the IEEE Broadcast Symposium*, Alexandria, VA, 2012.

14. D. Gozálvez, D. Gómez-Barquero, D. Vargas, and N. Cardona, Time diversity in mobile DVB-T2 systems, *IEEE Transactions on Broadcasting*, 57(3), 617–628, 2011.

15. EBU Technical Report 3348, Frequency and network planning aspects of DVB-T2, May 2012.

16. J. J. Giménez, D. Gozálvez, D. Gómez-Barquero, and N. Cardona, A statistical model of the signal strength imbalance between RF channels in a DTT network, *IET Electronic Letters*, 48(12), 731–732, 2012.

17. C. Abdel Nour and C. Douillard, Rotated QAM constellations to improve BICM performance for DVB-T2, in *Proceedings of the IEEE ISSSTA*, Bologna, Italy, 2008.

18. T. Jokela, M. Tupala, and J. Paavola, Analysis of physical layer signaling transmission in DVB-T2 systems, *IEEE Transactions on Broadcasting*, 56(3), 410–417, 2010.

19. C. Bormann et al., RObust Header Compression (ROHC): Framework and four profiles: RTP, UDP, ESP, and uncompressed, ITEF RFC 3095, July 2001.

20. H. Jiang, P. A. Wilford, A hierarchical modulation for upgrading digital broadcast systems, *IEEE Transactions on Broadcasting*, 51(2), 223–229, 2005.

21. S. M. Alamouti, A simple transmit diversity technique for wireless communications, *IEEE Journal on Selected Areas in Communications*, 16(8), 1451–1458, 1998.

22. S. Kaiser, Spatial transmit diversity techniques for broadband OFDM systems, in *Proceedings of the IEEE GLOBECOM*, San Francisco, CA, 2000.

23. R. U. Nabar, H. Bolcskei, and A. J. Paulraj, Transmit optimization for spatial multiplexing in the presence of spatial fading correlation, in *Proceedings of the IEEE GLOBECOM*, San Antonio, TX, 2001.

24. F. Pérez-Fontán et al., Statistical modelling of the LMS channel, *IEEE Transactions on Vehicular Technology*, 50(6), 1549–1567, 2001.
25. H. Fuchs and N. Färber, Optimizing channel change time in IPTV applications, in *Proceedings of the IEEE BMSB*, Las Vegas, NV, 2008.
26. D. Gozálvez, Combined time, frequency and space diversity in multimedia mobile broadcasting systems, PhD dissertation, Universitat Politèctica de València, Valencia, Spain, 2012.

Chapter 8

An Overview of the Cellular Broadcasting Technology eMBMS in LTE

Jörg Huschke and Mai-Anh Phan

Contents

8.1 Introduction

Mobile networks have emerged from voice telephony networks to multimedia delivery networks. With the advent of smartphones and tablet PCs, mobile TV services have become quite popular during the past 2 years. Apart from live TV channels, the offerings often include special mobile editions with highlights from the weekly TV program, such as series and comedies, delivered as looped channels.

The user data rates available by current 3G and 4G networks, including LTE, enable high-quality video transmission. Due to the enormous number of 3G terminals in use today, outnumbering by far mobile broadcast terminals of any broadcasting standard, the majority of today's mobile TV services are delivered over existing 3G networks [1,2]. Existing 3G operators have enough capacity in 3G networks to scale up for a mass market of mobile TV services. Latest 3G technology such as high-speed data packet access (HSDPA) [3,4] gives room for several steps of capacity increases. This allows for more users while benefiting from both the diversity and the quality of mobile TV services. The underlying technology is called packet-switched streaming (PSS) [5–7]. PSS is nowadays supported by all UMTS terminal vendors and offers high-quality streaming services for live or on-demand services. Further quality improvements have been achieved by the introduction of the advanced H.264 video codec [8] and by the introduction of streaming bearers with specific quality-of-service (QoS) support.

Nevertheless, the increasing popularity of mobile TV and similar services may lead to situations in which many users want to watch the same content at the same time. Examples are live events of high interest such as soccer matches, game shows, etc. For those cases, multicasting as known from the Internet or broadcasting is clearly more appropriate technology.

The work on adding broadcast/multicast support to 3G networks started back in 2002 when 3GPP created a work item for broadcast/multicast services in GSM/UMTS, respectively, called multimedia broadcast and multicast service (MBMS). 3GPP release 6 specifications included the first version of MBMS. An overview on the functionality and performance is provided in [9]. MBMS introduced only small changes to the existing radio and core network protocols. Therefore, MBMS can be introduced by a pure software upgrade in general as long as the underlying hardware provides sufficient processing power, which is the case for all state-of-the-art network nodes. This reduces the implementation costs both in terminals and in the network. It makes cellular broadcast a relatively cheap technology if compared to noncellular broadcast technologies [10–12], which require new receiver hardware in the terminal and significant investments into a new network infrastructure.

MBMS focuses on the transport aspects of broadcast and multicast services. It provides transport bearers over which IP multicast [13] packets can be delivered. MBMS also provides protocols and codecs for the delivery of multimedia files and streams on top of the IP multicast bearers [14].

Service layer functionality related to multicast/broadcast transport services has been defined by the Open Mobile Alliance (OMA) in the Mobile Broadcast Services 1.0 (BCAST 1.0) specification, which was finalized in 2007 [15,16]. BCAST 1.0 addresses features like content protection, service and program guides, transmission scheduling, notifications, and service and terminal provisioning. It enables, besides linear TV services, also podcast services. OMA BCAST 1.0 is agnostic with respect to the underlying broadcast/multicast distribution scheme and applies to MBMS and other noncellular broadcast systems like DVB-H.

The rest of this chapter is organized as follows: We will start with introducing the streaming and download delivery methods. After that, we describe the benefits of MBMS single-frequency networks (MBSFNs) and briefly address how it is supported in WCDMA. After giving an overview of the LTE downlink physical layer, we explain how it implements MBSFN. We describe further aspects of eMBMS in LTE, like the architecture and service and MBSFN area concept and briefly present performance results. In the subsequent section, we elaborate on the resource allocation and scheduling for eMBMS. Before we conclude the chapter, we give an overview of enhancements to eMBMS in LTE added in recent 3GPP releases.

8.2 MBMS User Services: Streaming and Download

The MBMS user service addresses service layer protocols and procedures above the IP layer. The MBMS user service is a toolbox, which includes a streaming and a download delivery method. These delivery methods do not differ between or depend on the MBMS bearer services. Figure 8.1 shows that the MBMS user services "Download" and "Streaming" sit on top of the MBMS bearer services that were explained in the previous section. Both the "Download" and the "Streaming" user services deliver media data encoded in various formats, e.g., video in H.264 and audio in AMR or AAC format.

The MBMS download delivery method is intended to increase the efficiency of file distributions, including messaging services such as MMS. The download delivery method allows for error-free transmission of files via the unidirectional MBMS bearer services. The files are "downloaded" and stored in the local files-system of the mobile phone. The streaming delivery method is intended for continuous reception and play-out used, e.g., in mobile TV applications.

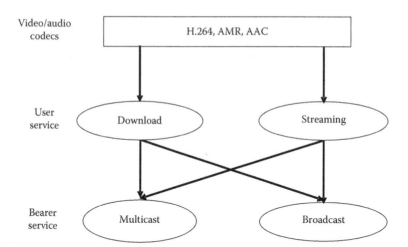

Figure 8.1   MBMS bearer and user services.

Figure 8.2 shows the protocol stacks that are used for MBMS as specified in [14] (gray boxes are defined therein by reference). The left side depicts the part of the protocol stack that requires an IP unicast bearer. The right side shows the part of the protocol stack that was designed for multicast/broadcast bearers built over UDP. Since UDP packets can also be sent over unicast bearers, the right side of the protocol stack can also be implemented on top of a unicast bearer.

It can be seen that service announcements and other metadata can be delivered over both unicast and broadcast/multicast connections. This means that a client can, for instance, download service announcement-related information from a web page or it receives the information via a broadcast/multicast bearer. Unicast and broadcast/multicast delivery of service announcement information can also be combined. For the associated delivery procedures, certain procedures such as point-to-point (PTP) file repair and reception reporting require a unicast connection whereas other procedures such as point-to-multipoint (PTM) file repair (e.g., multicasting of missing packets) can be executed over a broadcast/multicast bearer. Similar holds for the security functions. Some of them such as registration and MBMS service keys distribution require a unicast connection, whereas others such as MBMS traffic keys distribution work over broadcast/multicast bearers. The actual transfer of media data, i.e., streaming and download, was designed for broadcast/multicast bearers, but as already said earlier, it can also be done over a unicast bearer.

8.2.1 Download Delivery Method

The download delivery method is intended for file distribution services, which store the received data locally in a terminal. It can be used to deliver arbitrary files from one source to many receivers efficiently. Some recent video services on the

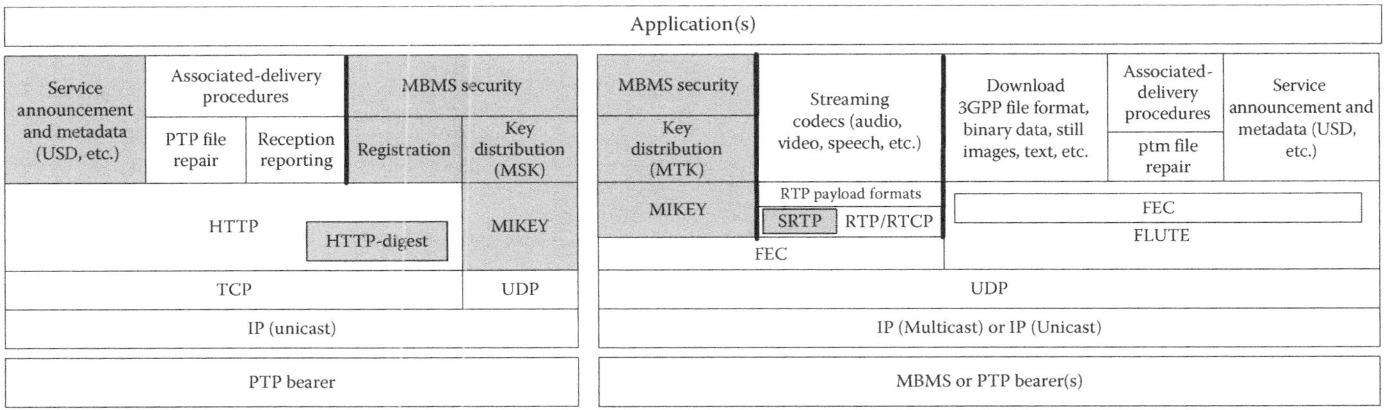

Figure 8.2 Protocol stacks used for MBMS user services.

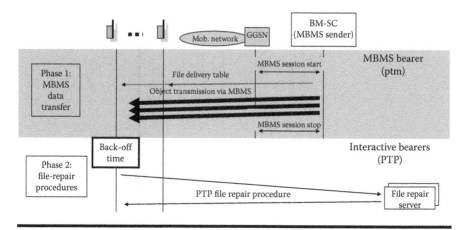

Figure 8.3 MBMS download procedure principle with PTP file repair.

Internet appear to the end user as streaming services, but actually, they use file-based transmission where the entire media file is divided into fragment files that are transmitted sequentially, using, e.g., the DASH protocol [17]. Existing content-to-person MMS services, which deliver short video clips related to live events like a soccer match via MMS, will greatly benefit from this feature. Today, those services rely on PTP connections for MMS delivery. In the future, the existing MMS subsystem can be easily interfaced with a BM-SC that then distributes the clip via MBMS download. The principle of an MBMS download delivery is shown in Figure 8.3.

The files are delivered during the MBMS data transfer phase (see phase 1 in Figure 8.3). The MBMS transmission bearer is activated with the MBMS session start message. This message triggers the paging process in the radio access network (RAN), which in turn informs the MBMS receivers about an upcoming transmission. After the MBMS bearer is successfully established, the BM-SC starts sending the actual MBMS download data. Included in the download data is a file containing the file delivery table. The FLUTE [18] protocol is used to send the files via UDP and allows the FEC protection of the files using the IETF FEC framework [19]. After the MBMS data transmission, bearer resources are released via the MBMS session stop message.

During the MBMS data transfer phase, certain mobile phones may experience packet losses due to fading conditions or handovers. Naturally, full reliability cannot be offered in a pure unidirectional distribution scheme because the packet loss rate can be excessive for some users. Therefore, three packet error recovery schemes are foreseen for the download delivery method. The most important one is the use of application layer forward error correction (AL-FEC) code, which allows recovery of lost packets without any server interaction already during the MBMS data transfer phase. The other two recovery schemes use file repair procedures (see phase 2 in Figure 8.3), where the first scheme is a PTP repair mechanism using interactive bearers and the other one is a PTM repair mechanism using MBMS bearers.

The Raptor AL-FEC code [20] was chosen as a basis for FEC protection of the files [21]. The Raptor AL-FEC code generates a number of redundant FEC symbols for each source block. One major advantage of the Raptor code is its rateless property, i.e., the possibility to generate a large number of FEC redundant symbols out of one source block. Generally, Raptor codes can handle even large files as one source block. But since mobile phones have a limited amount of fast memory for decoding, a single-source block may contain only up to 4100 kB of data. Thus, larger files have to be subdivided into multiple source blocks. A broadcast of newly created FEC packets during the MBMS data transfer (phase 1) is of benefit for all receivers, which have not successfully reconstructed the original source block.

Currently, there is a study item ongoing in 3GPP on enhanced FEC. Several coding schemes have been proposed, e.g., RaptorQ. The main benefit of RaptorQ is further reduced encoding and decoding complexity as well as slightly reduced redundancy overhead required for successful decoding. The latter is more significant for very small FEC block sizes.

During a file repair procedure, further Raptor AL-FEC packets are transmitted to the receivers. If an interactive bearer is used, the repair data are independently sent to different receivers and can even be tailored to the actual losses of that receiver. On the contrary, if the MBMS bearer is used, the same repair data are sent only once to multiple receivers and the repair data should be useful for all receivers with losses. Therefore, the rateless property of the Raptor code is very beneficial for the PTM repair mechanism.

If a file repair procedure is used (phase 2), the MBMS client waits until the end of the transmission of files or sessions and then identifies the missing data from the MBMS download. Afterward, it calculates a random back-off time and selects a file repair server randomly out of a list. Then, a repair request message is sent to the selected file repair server at the calculated time. The file repair server responds with a repair response message either containing the requested data (interactive bearer), redirecting the client to an MBMS download session (MBMS bearer), redirecting the client to another server, or, alternatively, describing an error case. The BM-SC may also send the repair data on an MBMS bearer (possibly the same MBMS bearer as the original download) as a function of the repair process. The performance of the post-delivery file repair procedures described earlier has been analyzed in [22–24].

8.2.2 Streaming Delivery Method

The streaming delivery method is intended for the continuous reception and play-out of continuous media like video, audio or speech. A typical example of an application using the streaming delivery method is the transfer of live channels in a mobile TV service. RTP [25] is used as transport protocol for MBMS streaming delivery. RTP provides means for sending real-time or streaming data over UDP (see Figure 8.2). As previously mentioned, packet losses could occur during the streaming data transfer that would result in distortions of the received video and audio quality at the receiver side. In case of streaming with its real-time constraints, file repair procedures are not suitable. However, the Raptor AL-FEC can also be used with the streaming delivery method to reduce the residual error rate to the acceptable level.

AL-FEC increases the time diversity at the application layer, as IP packet burst errors that are significantly shorter than the AL-FEC block duration can be recovered by choosing sufficiently low AL-FEC code rate. On the downside, the use of AL-FEC in principle increases the zapping time as well. The performance of AL-FEC is worse than having a single FEC at the physical layer, but it requires less memory. AL-FEC is also suitable for achieving very low error rates below 10^{-4} for which the physical layer codes have not been designed due to the availability of retransmissions in the case of unicast.

AL-FEC performance for streaming delivery over MBMS and other broadcast technologies has been investigated in [26]. For eMBMS, Huschke [27] evaluates streaming performance with AL-FEC. The service criterion is that 95% of users achieve AL-FEC decoding failure probability below 10^{-5}. Code block size was adapted to cover 2 s of streaming data. The best combination of AL-FEC code rate and physical layer modulation and coding scheme (MCS) has been searched for each of a set of inter-site-distance (ISD) values. The larger the ISD, the more robust the MBMS transmission must be made. The more robust the transmission, the lower is also the achievable spectral efficiency. The radio network simulations showed that a spectral efficiency of 2.5 bps/Hz can be achieved up to an ISD of 1 km even when serving deep indoor users.

The FEC stream bundling concept shown in Figure 8.4 allows the protection of the actual audio/video data together with synchronization information (RTCP) and possibly decryption information. Packets of one or more UDP flows may be used to construct the source blocks for the FEC protection. The FEC redundancy information is transmitted in one separate traffic flow. Since the Raptor code is a systematic FEC code, the receiver can simply ignore the FEC flow, if no transmission errors occur. The advantage of the FEC stream bundling concept, as shown in Figure 8.4, is that the FEC efficiency is increased when protecting several data flows together, because the FEC code works on a larger portion of data.

Figure 8.5 depicts how one or more out of several possible packet flows of different types (audio, video, text RTP and RTCP flows, MIKEY flow) are sent to the FEC layer for protection.

The source packets are modified to carry the FEC payload ID, and a new flow with repair data is generated. The receiver takes the source and repair packets and buffers them to perform, if necessary, the FEC decoding. After appropriate buffering, received and recovered source packets are forwarded to the higher layers.

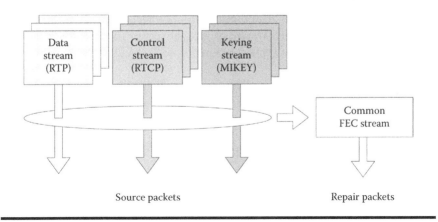

Figure 8.4 FEC stream bundling concept.

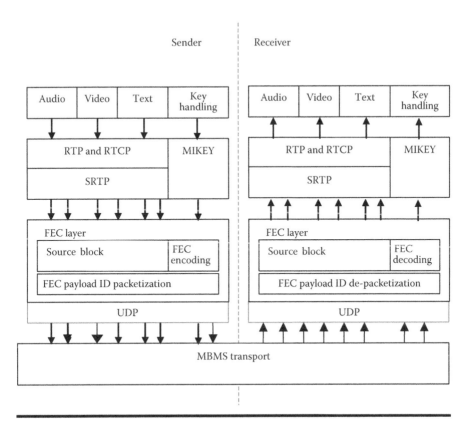

Figure 8.5 FEC mechanism for the streaming delivery method interaction diagram.

MBMS streaming delivery may also use reception reporting, i.e., reports about received parts of the service that are sent back to the server. The reception reporting procedure allows an operator to collect reception statistics and usage patterns of the service from actual end users.

The MBMS user service framework is harmonized with OMA BCAST and DVB CBMS specifications. In particular, the same set of audio/video codecs is supported.

8.3 Major Standardization Releases

MBMS evolution in 3GPP 3G networks from Release 6 is part of the two main tracks of standard developments, as depicted in Figure 8.6:

- Evolution of WCDMA
- Long-term evolution (LTE) of UTRA (denoted E-UTRA in the 3GPP specifications)

The evolution of WCDMA comprises improvements for MBMS for the PTM radio bearers, mainly the introduction of so-called MBMS single-frequency networks

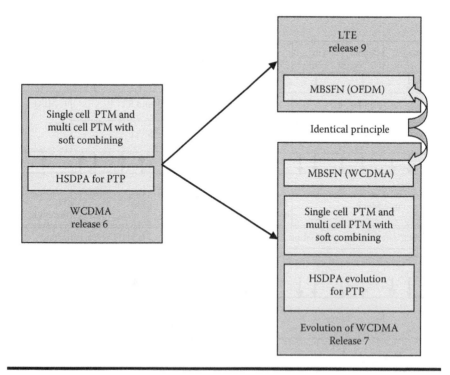

Figure 8.6 MBMS evolution tracks toward MBSFN. Releases refer to 3GPP releases.

(MBSFNs), and HSPA evolution that allows for higher capacity using unicast radio bearers. In 3GPP Release 7, also support of unicast bearers for streaming [5,6] or OMA Push [28] as part of an MBMS user service has been added. The use case that has driven the integration of unicast bearers into MBMS is access to an MBMS service in the home network from a visited network that does not support the same MBMS services as in the home network or that does not support MBMS at all.

In a parallel track, 3GPP has standardized LTE. MBMS has been introduced to LTE in 3GPP Release 9 as of March 2010. The technical specifications of E-UTRA are given in the 36-series of 3GPP set of specifications. They include an overall description in [29]. One of the main goals for LTE was efficient support of MBMS [30].

The most important innovations in LTE with respect to MBMS are the support of flexible carrier bandwidths of up to 20 MHz, which is four times the fixed bandwidth of WCDMA, use of OFDM modulation in the physical layer, and efficient and flexible support for MBSFN, including time-division multiplexing between MBSFN and unicast traffic on the same carrier. Before we further elaborate on MBMS in LTE, we provide a brief excursion into the MBSFN support of WCDMA.

8.4 MBMS in WCDMA Releases 7 and 8

For WCDMA, 3GPP has introduced MBSFNs in Release 7 for FDD and TDD in 2007. In Release 8, this has been further developed into Integrated Mobile Broadcast (IMB), which is designed to operate in unpaired spectrum. IMB principles are identical to those for MBSFN FDD, except that a new synchronization code has been added to resolve potential cell search issues of legacy TDD terminals.

In the conventional WCDMA downlink, all the physical channels of a cell, except for the synchronization channels, are scrambled by cell-specific codes. While scrambling codes are used to differentiate cells, orthogonal spreading code is used to separate multiple code-division multiplexed channels within a cell. With orthogonal spreading, a receiver does not experience own-cell interference when the base station signal travels through a flat channel. State-of-the-art receivers are capable of removing own-cell interference even in the case of non-flat channel, using, e.g., linear MMSE chip equalizer or G-RAKE techniques. Due to lack of signal orthogonality, however, these receivers do not suppress other-cell interference as effectively.

In contrast, in the MBSFN transmission mode, the same scrambling is used for all participating cells. When different base station uses the same waveform to simultaneously send a common set of MBMS content channels, the received signal is then the same as that for a single-source transmitted signal traveling through a heavily dispersive channel, where each path corresponds to a signal path between a base station transmitter and the receiver. In this case, other-cell interference shares the same orthogonal properties as the own-cell signals. As a result, an advanced receiver (e.g., zero-forcing equalizer) not only collects the signal energy contributed by the multiple base station signals, but also gets rid of interference arising due to multipaths and transmission from multiple base stations. In order to exploit the resulting very high signal to interference plus noise ratio (SINR), Release 7 adds support for 16QAM for multicast transmission. This boosts the spectral efficiency of MBMS in a WCDMA network significantly. In order to limit the receiver complexity, the delay spread needs to be limited, and therefore the cells in the MBSFN need to transmit the same signal at the same time. Synchronization in the order of a few microseconds is required. The synchronization of the base stations, called NodeBs, connected to one radio network controller (RNC) is ensured by the RNC. As a consequence, synchronization is not ensured across RNCs, and, thus, MBSFN across RNCs is not supported.

In a proposed application of the MBSFN mode, a NodeB provides multiple carriers of which some operate in the MBSFN mode, which are referred to as dedicated MBSFN carriers and use cell common scrambling, and some carriers operate in the normal 3GPP Release 6 mode and use cell-specific scrambling. On the dedicated MBSFN carriers, multiple MBMS physical channels can be code-division multiplexed. However, the same channelization code is used in all the NodeBs that belong to the same MBSFN to transmit the same MBMS content channel. Of course, the usual 3GPP Release 6 time multiplexing of MBMS radio bearers is possible.

Table 8.1 summarizes MBSFN spectral efficiency results from [31] obtained from radio network simulations with 2800 m ISD, 3 sectors per site, urban area propagation environment, with channel model "Vehicular A," and users moving at 3 km/h. The performance criterion is the spectral efficiency achievable with 95% coverage and 1% BLER. Results are shown for receivers of 3GPP Type-2 (implementing G-RAKE or equivalent techniques) and 3GPP Type-3 (like Type-2 plus implementing antenna diversity). With Type-3 advanced receivers that can equalize, i.e., take advantage of, the signals from the seven closest NodeBs, spectral

Table 8.1 MBMS Capacity and Spectrum Efficiency Achieved by SFN Operation (Using 90% of the Cell Power) for Receivers of Type-2 (Implementing G-RAKE or Equivalent Techniques) and Type-3 (Like Type-2 plus Implementing Antenna Diversity)

Receiver	Receiver Capable of Equalizing Signals from 3 NodeBs	Receiver Capable of Equalizing Signals from 7 NodeBs
Type-2	1.54 Mbps	2.62 Mbps
	0.31 bps/Hz	0.53 bps/Hz
Type-3	3.01 Mbps	5.38 Mbps
	0.60 bps/Hz	1.08 bps/Hz

efficiency in excess of 1 bps/Hz can be achieved if 16QAM is employed. For less advanced Type-3 receivers, or in case the ISD is significantly larger, so that only the signals from 3 NodeBs can be equalized, the spectral efficiency is still 0.6 bps/Hz.

The higher cell capacity also allows for higher bearer data rates, thus the maximum supported bearer data rate has been increased to 512 kbps. As the cell capacity is several times higher than 512 kbps, time-sliced transmission of services is enabled, where each service is mapped to one 2 ms subframe in a radio frame. This significantly improves the battery saving in the receiver.

Finally, 3GPP has standardized an optional flat architecture for WCDMA in Release 8. The RNC functionality can be integrated into the NodeB, and IP multicasting in the core network is supported down to the NodeBs. Functionality and signaling via the Iur interface have been added to enable the necessary coordination between NodeBs, in particular for the synchronization between NodeBs in order to perform MBSFN transmission.

8.5 MBMS in LTE: eMBMS

MBMS in LTE uses an evolved architecture in order to support MBSFNs with high flexibility, which was an important design goal of LTE from the start. Furthermore, for LTE, it is desired to support MBSFN transmission and user individual services on the same carrier. The architecture needs to support the coordinated allocation of radio resources within the carrier for MBSFN transmission across all cells participating in the particular MBSFN. Figure 8.7 shows the eMBMS architecture,

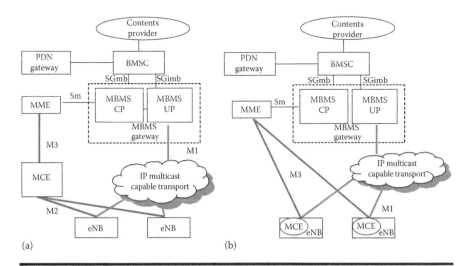

Figure 8.7   MBMS architecture in SAE/LTE. (a) Separate MCE. (b) MCE function-ality integrated into eNB.

which is based on enhancements to the LTE Release 8 architecture. The default architecture is shown to the left.

The following logical entities are defined as follows:

- *BM-SC.* The broadcast/multicast service center (BM-SC) controls MBMS sessions and corresponding MBMS bearers.
- *MBMS GW.* The MBMS gateway (GW) is an entity that is located between the content provider and the evolved base stations (eNode Bs or eNBs). The control plane (CP) of the MBMS GW is involved in the MBMS session start/setup toward the LTE RAN via the mobility management entity (MME). The user plane (UP) is responsible for delivering the user data over the IP multicast capable transport network to the eNBs and participates in the content synchronization for MBMS services using MBSFN. The MBMS GW is part of the evolved packet core.
- *MME.* In the context of MBMS, the MME is responsible for session control signaling.
- *MCE.* The multicell/multicast coordination entity (MCE) is an entity responsible for coordinating the usage of MBSFN transmission within the same MBSFN area in the LTE RAN. Therefore, in the architecture alternative shown to the right of Figure 8.7, where the MCE is integrated to each eNB, the resource configuration in all MCEs must be configured consistently by O&M since there is no interface for coordination between MCEs. Otherwise, an MBSFN area can only cover the cells served by the respective eNB.
- *eNB.* The eNB is the evolved base station in LTE responsible for multiplexing, framing, channel coding, modulation, and transmission.

The following logical interfaces are defined as follows:

- *M1.* Logical interface between the MBMS GW and the eNBs. The transport on this interface will be based on IP multicast. The MBMS content is transported in a frame or tunnel protocol, in order to support content synchronization and other functionalities. IP multicast signaling is supported in the transport network layer in order to allow the eNBs to join an IP multicast group.
- *M2.* Logical control interface between the MCE and the eNBs. This interface is used to coordinate the setting up of an MBMS service in the eNBs for MBSFN operation. The signaling transport layer is based on IP.
- *M3.* Interface between MME and MCE. Supports MBMS session control signaling, including the QoS attributes of each service (does not convey radio configuration data). The procedures comprise, e.g., MBMS Session Start and Stop. Stream control transmission protocol (SCTP) is used as signaling transport, i.e., PTP signaling is applied.
- It is not precluded that M3 interface can be terminated in eNBs. In this case, MCE is considered as being part of eNB. Therefore, M2 does not exist in this scenario. This is depicted in Figure 8.7, which depicts two envisaged

deployment alternatives. In the scenario depicted on the left, MCE is deployed in a separate node. In the scenario on the right, MCE is part of the eNBs.

■ *Sm*. The reference point for the CP between MME and MBMS GW.
■ *SGmb*. The reference point for the CP between BM-SC and MBMS GW.
■ *SGi-mb*. The reference point for the UP between the BM-SC and MBMS GW.

8.5.1 LTE Downlink Physical Layer

In the following, we provide a brief introduction to the E-UTRA downlink physical layer [32,33]. The E-UTRA downlink uses OFDM, because it efficiently supports flexible carrier bandwidth, allows frequency domain scheduling, is resilient to propagation delays, which is particularly beneficial for SFN broadcasting configurations, and is well suited for multiple-input multiple-output (MIMO) processing.

The possibility to operate in vastly different spectrum allocations is essential. Different bandwidths are realized by varying the number of subcarriers used for transmission, while the subcarrier spacing remains unchanged. In this way, operation in spectrum allocations of 1.4, 3, 5, 10, 15, and 20 MHz can be supported. Due to the fine frequency granularity offered by OFDM, a smooth migration of, for example, 2G spectrum is possible. Frequency-division duplex (FDD), time-division duplex (TDD), and combined FDD/TDD are supported to allow for operation in paired as well as unpaired spectrum.

To minimize delays, the transmit time interval is only 1 ms, where one subframe can be submitted. Figure 8.8 outlines the time-domain structure for LTE downlink

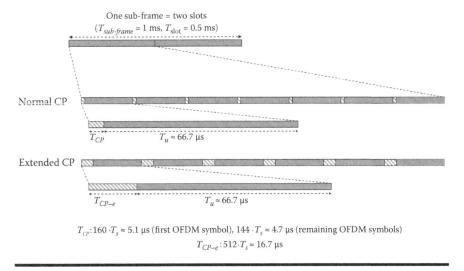

Figure 8.8   LTE downlink subframe and slot structure. (From Dahlman, E., *3G Evolution: HSPA and LTE for Mobile Broadband*, **Academic Press, Amsterdam, the Netherlands, 2008.)**

transmission. Each 1 ms subframe consists of two slots of length $T_{slot} = 0.5$ ms. Each slot consists of several OFDM symbols. A subcarrier spacing $\Delta f = 15$ kHz corresponds to a useful symbol time $T_u = 1/\Delta f \approx 66.7$ μs. The overall OFDM symbol time is then the sum of the useful symbol time and the cyclic prefix (CP) length T_{CP}. Signals from eNBs arriving within the CP duration of the user equipment (UE) synchronization point contribute useful signal energy and thereby improve the coverage. Signals arriving outside the CP produce interference. Since the CP does not contain user data, its length is a trade-off between the time fraction available for user data and the SINR value achievable with the desired error probability. In order to cope with different propagation delays caused by different cell sizes, LTE defines two CP lengths for a typical subcarrier spacing of $\Delta f = 15$ kHz, the normal CP and an extended CP, corresponding to seven and six OFDM symbols per slot, respectively, as illustrated in Figure 8.8. The exact CP lengths can also be obtained from Figure 8.8. It should be noted that in case of the normal CP, the CP length for the first OFDM symbol of a slot is somewhat larger, compared to the remaining OFDM symbols, in order to fill the 0.5 ms slot.

By extending the CP from 4.7 to 16.7 μs, it is possible to handle very high delay spreads that can occur in a large cell with a radius of up to and exceeding 120 km. For larger distances, the extended CP can even be increased by a factor of two resulting in 33.3 μs. In order to limit the relative overhead imposed by this extended long CP, the OFDM useful symbol time is also doubled for the configuration with the long extended CP of 33 μs. In order to maintain the same capacity with an unchanged carrier bandwidth, the subcarrier spacing is also reduced by a factor of two, resulting in $\Delta f_{low} = 7.5$ kHz. The extended CPs are mainly intended for MBSFN transmission. Further details will be provided in the next chapter.

A resource block is defined as a two-dimensional time-frequency resource that has a time duration of one 0.5 ms slot and a frequency bandwidth of 180 kHz. For a normal CP, a resource block consists of 12 subcarriers during a 0.5 ms slot, as illustrated in Figure 8.9. In case of normal CP, each resource block thus consists

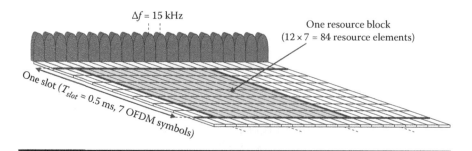

Figure 8.9 Downlink resource blocks assuming normal CP. With extended CP, there are six OFDM symbols per slot and, consequently, 72 resource elements per resource block. (From Dahlman, E., *3G Evolution: HSPA and LTE for Mobile Broadband*, Academic Press, Amsterdam, the Netherlands, 2008.)

Table 8.2 Resource Block Parameters

Configuration			# Subcarriers	# OFDM Symbols
Normal CP	T_{CP}=4.7 µs	Δf=15 kHz	12	7
Extended CP	T_{CP}=16.7 µs	Δf=15 kHz		6
Extended CP	T_{CP}=33.3 µs	Δf_{low}=7.5 kHz	24	3

of $12 \times 7 = 84$ resource elements, and in case of extended CP, the resource block consists of 72 resource elements for a subcarrier spacing of Δf= 15 kHz (12·6) and also for Δf_{low} = 7.5 kHz (24·3). The numerology is summaraized in Table 8.2. Each subcarrier of each OFDM symbol can be modulated by quadrature phase shift keying (QPSK), 16-QAM, or 64-QAM modulation schemes.

A subframe can be used for MBMS transmission or unicast transmission. The downlink shared channel (DL-SCH), which is a transport channel, uses the physical layer to provide unicast radio bearers with feedback from the UEs. The feedback enables the use of hybrid automatic repeat request (HARQ) retransmission and also the UE individual adaptation of transmission parameters to the particular radio conditions, including MIMO transmission. The unicast transmission is more efficient than MBSFN transmission only for services with a very small number of users per cell, for small cells even for less than one user per cell on average. The reason for the superior efficiency of MBSFNs is the elimination of intercell interference.

8.5.2 MBSFN

If a larger number of users of a particular MBMS are present in a cell, broadcast radio transmission in the cell is more suitable, which can be used either in single-cell or multicell transmission mode. For PTM transmission in both single-cell and multicell modes, a new transport channel, the multicast channel (MCH) was defined. The MCH can be time-multiplexed on a subframe granularity of 1 ms with other transport channels such as the DL-SCH. Thus, in contrast to MBSFN configured in WCDMA, it is not necessary to use an MBSFN dedicated carrier in LTE.

A multicell transmission essentially means that the cells transmitting the MBMS service are configured to form an MBSFN (see also Section 8.3). If an MBSFN with multiple cells is established using a particular MCH, then the same MCH information is transmitted time aligned from these cells using identical transport formats, identical resource allocations, and identical scrambling (cell-group-specific scrambling, see earlier). From a UE point of view, such multicell MCH transmission will appear as a single MCH transmission. However, it is a channel aggregated from all cells involved in the MBSFN transmission and will typically have a large delay spread due to the differences in the propagation delay as well as residual transmit-timing differences. In order to be able to properly demodulate the multicell

MCH transmission, the UE needs an estimate of the aggregated channel. For this to be possible, MCH specific reference signals are needed that are identical for all cells involved in the MBSFN, i.e., identical time/frequency locations and identical reference signal sequences are used. If more than a few cells are configured for the same MBSFN, the CP of 16.7 μs should be used. In case "high-power high-tower" sites are available or of deployments in low-frequency bands, good coverage can be achieved with even higher distance between sites, and in this case, the extended long CP of 33 μs should be used.

8.5.3 Synchronization

Within a so-called MBSFN area, all eNBs need to be synchronized with a microsecond tolerance, and the radio frames need to be aligned. The MCE is responsible for configuring identical MBSFN subframe allocations and MCH scheduling periods (MSPs) in all cells of an MBSFN area, as well as the MCH MCS, satisfying the guaranteed bit rate of the MBMS bearer. The MCE also defines the common order in which services are scheduled in all eNBs of an MBSFN area.

Finally, content synchronization needs to ensure that the IP packet multiplexing and mapping to transport blocks in MBSFN subframes are identical in all these cells, taking into account that IP packets have varying size and packet losses can occur between the BM-SC and the eNB. This is achieved by the SYNC protocol [34], where the packet flow is grouped into synchronization sequences. A separate instance of the SYNC protocol is associated with each MBMS bearer. For each synchronization sequence, the BM-SC tries to ensure that it does not send more packets to the eNB than allowed by the guaranteed bit rate of the MBMS bearer, discarding packets if necessary. The BM-SC labels all packets of a synchronization sequence with an identical time stamp telling the eNB when to start the transmission of the first packet of that synchronization sequence. The time stamp has to cover transfer delays between the BM-SC and all eNBs in the MBSFN area to ensure that all of them have received and buffered the packets of an MSP before any of the eNBs is allowed to transmit the first packet. The MSP is configured by the MCE, but must be an integer multiple of the synchronization sequence duration to make this concept work.

In order to limit the value range of timestamps, they are measured relative to the start of a synchronization period which can be up to 600 s, rather than e.g., relative to the start of a service session. In other words, the PDU flow is grouped into synchronization sequences which are in turn grouped into synchronization periods. The end of a synchronization period can be detected by the eNBs by receiving a PDU with signifcantly lower timestamp then PDUs received recently.

The BM-SC maps each IP packet to one SYNC protocol data unit (PDU) of "Type 1," which carries data without IP header compression ("Type 2" has been defined for PDUs with IP header compression, but is only supported in WCDMA, but not LTE). Figure 8.10 shows a sequence chart of a typical PDU transfer in a synchronization sequence.

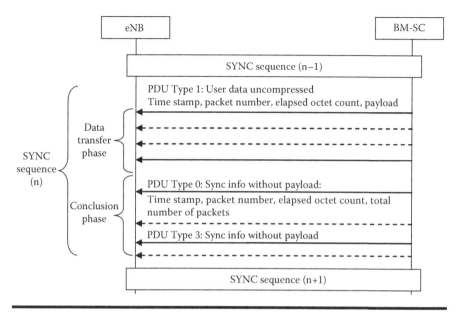

Figure 8.10   Illustration of a typical SYNC PDU sequence.

In addition to the time stamp, "Type 1" PDUs include a packet number that is reset at the start of the synchronization sequence, and increases for every subsequent packet, and an elapsed octet counter that indicates the number of octets transferred since the start of the synchronization sequence. In order to indicate the end of the synchronization sequence and thus conclude whether any of the "Type 1" PDUs got lost, the BM-SC may complete a synchronization sequence by sending a PDU "Type 0," which basically contains the same control information as "Type 1" and also indicates the total number of transmitted packets as well as the total number of octets, or "Type 3," which additionally contains a list of all packets and corresponding packet sizes. "Type 0" and "Type 3" PDUs can be sent several times during the conclusion phase to increase the reliability.

The eNB has a buffer corresponding to one MSP, inserting the IP packets extracted from the received "Type 1" PDUs, concatenating the packets of each service and putting the services in the order as configured by the MCE. In case of lost "Type 1" PDUs, the "Type 0" and "Type 3" PDUs enable the eNB to place subsequently received IP packets at the correct positions in the buffer. This has the advantage that an eNB can selectively mute only those subframes where some of the lost IP packets had to be mapped. By muting such subframes, instead of transmitting data in arbitrary order, it is prevented that the eNB creates interference in the MBSFN area. The layer two header size that is added by the eNB depends on the size of the received IP packets and whether multiple IP packets can be multiplexed in the same transport block. Therefore, if multiple subsequent SYNC "Type 1" PDUs are lost, it might not be possible for the eNB to determine—only based on information obtained from PDU "Type 0"—to which subframes the lost IP packets have to be mapped. This issue is solved by the PDU "Type 3," which informs the eNB about the sizes of the lost IP packets.

The varying layer two header size is one reason why the BM-SC does not know the exact number of user data bits fitting into an MSP. If the BM-SC has sent too many packets for an MSP so that the eNB buffers would overflow, the eNBs discard the excess packets starting from the tail of the packet sequence.

8.5.4 Channel Mapping

The protocol architecture of LTE consists of the physical, transport, and logical channel layers. Figure 8.11 introduces the MBMS channels on each layer as well as the unicast data channels (in gray), and their mapping.

Each MCH is tied to an MBSFN area. Each PMCH can have a different MCS, therefore all MTCHs multiplexed into the same (P)MCH use the same MCS. The subframe of the PMCH carrying the MCCH, however, can have a different MCS, in order to enable the configuration of a more robust MCS for the MCCH.

One MBMS service is mapped on exactly one MTCH, and multiple MTCHs can be multiplexed in the same transport block, which uses the complete bandwidth within a subframe. Thus, a terminal that is interested in a specific MBMS service must "filter" its service of interest from a transport block that contains data from multiple MTCHs. In order to distinguish these data units, each MTCH will be mapped to a specific logical channel identifier (LCID). LCID = 0 is reserved for the MCCH. Since each MBSFN area transmits only one MCCH, and only one MCH carries the MCCH, MTCHs that are mapped to another MCH may also use LCID = 0.

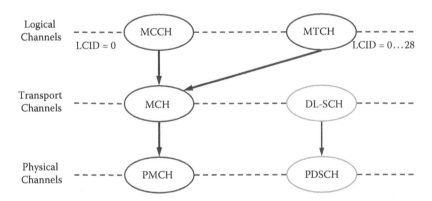

Figure 8.11 eMBMS channels and layer mapping. MCCH, Multicast control channel; MTCH, Multicast traffic channel; MCH, Multicast channel; PMCH, Physical multicast channel.

8.5.5 MBMS Area Concept

For the MBMS service provisioning, the MBSFN area and MBMS service area need to be distinguished. The MBMS service area, which is typically provided to the service provider, defines a geographic area where a service shall be broadcasted. Within the network, the operator identifies each MBMS service area by one or more MBMS service area identities (MBMS SAIs), and each MBMS SAI defines a group of cells. A cell can belong to and is therefore addressable by one or more MBMS SAIs.

An MBSFN area defines the set of cells participating in the transmission of signals for one or more services in MBSFN mode. An MBMS SAI may comprise one or more complete MBSFN areas. Overlap between MBSFN areas, as well as between MBMB service areas, is supported. This also enables a smaller MBSFN area to overlap a large MBSFN area, so that, e.g., regional and nationwide MBSFN areas can coexist.

The relationship between MBMS Service Areas, MBMS SAIs, and MBSFN areas is illustrated in Figure 8.12. MBMS Service Area A consists of MBMS SAI #1 and MBMS SAI#2. MBMS SAI#1 covers MBSFN area #1, #2, and #3. The MBMS services that are provided in MBSFN area #1 and #2, belonging to the MBMS SAI #1, do not have to schedule the MBMS data synchronously. The synchronization requirement is valid only within the same MBSFN area. MBSFN area #1 operates, e.g., on carrier frequency f1, and MBSFN area#4 could cover carrier frequency f2. Both MBSFN areas cover the same geographic area.

Within an MBSFN area, there can be reserved cells that do not carry the MBMS subframe allocation information. These cells may nevertheless transmit the MBMS signals in MBMS subframes, in order to support neighbor cells that are not classified as reserved.

UEs in cells at the border of an MBSFN area will suffer from a high level of interference if the neighbor cells not belonging to the MBSFN area transmit different signals in the subframe used by the MBSFN area. Such border cells inside the MBSFN area may therefore be configured as reserved cells, such that UEs located in these cells

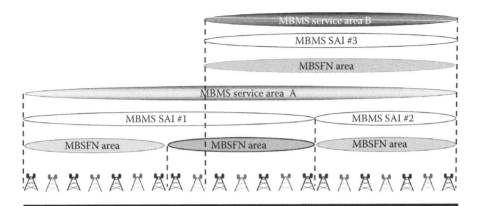

Figure 8.12 MBMS-related area concept.

do not expect the availability of the MBMS service, due to the lack of essential MBMS signaling. Reserved cells at the border can thereby serve as an interference guard zone. Thanks to the low height of the eNB antenna towers and the low transmit power, typically only a few rings of cells around an MBSFN area are needed for the guard zone, depending on the used MCS. Therefore, the subframes can be reused for another MBSFN area or unicast traffic already a few kilometer or few tens of kilometers away.

8.5.6 Performance

Based on the assumptions made by 3GPP, initial radio network simulations have been carried out. Results are presented in [35]. The key parameters are shown in Table 8.3.

Figure 8.13 shows the achieved broadcast capacity versus ISD for the case of multicell broadcasting. A large cluster of cells transmits the same broadcast content synchronously, thereby achieving signal aggregation gains and avoiding strong intercell interference. It can be seen that the spectrum efficiency requirement of 1 bps/Hz, i.e., 5 Mbps in the assumed 5 MHz bandwidth, is achieved for ISDs of up to approximately 1850 m ISD for the 3GPP simulation scenario "Case 1," i.e., in the 2 GHz band and with 20 dB indoor penetration loss, and up to 4700 m for the "Case 4," i.e., in the 900 MHz band and with 10 dB indoor loss [32]. Naturally, the capacity decreases with increasing separation between transmitters as the power per transmitter is assumed to be fixed and the proportion of cells that are so far away that they cause interference rather than contribute useful signal increases.

In the single-cell broadcasting case, different cells may transmit different signals on the same time-frequency resource. Therefore, signal aggregation is not possible and the spectral efficiency levels significantly lower. However, the throughput in the single-cell case can be further increased by introducing a kind of frequency reuse scheme where the same time-frequency resource is used not at all or only with limited power in adjacent cells.

Table 8.3 Radio Network Simulation Parameters

Spectrum allocation	10 MHz
Base station power	40 W
Propagation	$L = 35.3 + 37.6 * \log(d)$, d is distance in meters (includes $P_L = 20$ dB penetration loss)
	14 dB_i transmitter antenna gain
	8 dB lognormal shadowing
ISD	500–4000 m (varied)
MCSs	Depending on ISD
	QPSK, 16QAM, or 64QAM, and turbo coding

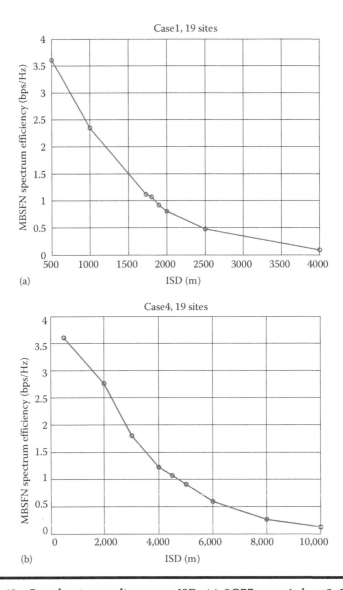

Figure 8.13 **Broadcast capacity versus ISD. (a) 3GPP case 1, i.e., 2 GHz with 20 dB indoor loss; (b) 3 GPP case 4, i.e., 900 MHz with 10 dB indoor loss.**

8.6 Resource Allocation and Scheduling of MBMS Services

MBMS data transmission in MBSFN mode is time-multiplexed with other LTE unicast traffic. This is an advantage over WCDMA-based MBMS transmission, where the use of MBSFN was confined to a dedicated carrier. Up to 6 of the

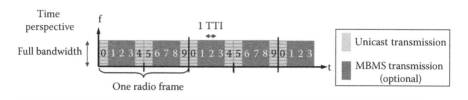

Figure 8.14   Time multiplexing of MBMS and unicast transmissions.

10 subframes of a radio frame are configurable for MBMS in the FDD mode and up to 5 in the TDD mode. Figure 8.14 shows which of the subframes can be used for MBMS transmission and which are reserved for unicast.

8.6.1 MCH Subframe Allocation

The subframes actually used for MBMS transmission are defined as a pattern covering either one radio frame or four radio frames and repeats after a configurable number of radio frames. All MCHs within one MBSFN area use a common subframe allocation (CSA) pattern. For the largest repetition period, i.e., 2.56 s, the smallest possible allocation of subframes is 0.3%, which can be sufficient for very low-bit-rate services, audio streaming, or road traffic messages. Among the subframes defined by the CSA, consecutive subframes are allocated to the same MCH. Figure 8.15 illustrates an example of the allocation within one CSA period.

8.6.2 Scheduling of MBMS Services

Within the subframe allocation of one MCH, the data of different MTCHs are multiplexed in time and grouped to MSPs as illustrated in Figure 8.16. Different MCHs can use different MSPs. All data of one MTCH scheduled for a particular MSP are transmitted consecutively. An MSP begins with an MCH scheduling information (MSI) that defines the last subframe used by each MTCH within the MSP. The MSI is not necessary to be received but enables the UE to enter power-saving mode until the subframe where the MTCH of interest to the UE starts in the current MSP.

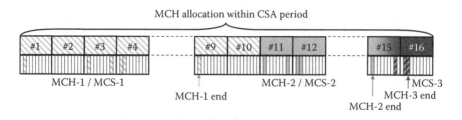

Figure 8.15   Illustration of MCH to subframe allocation within one CSA period.

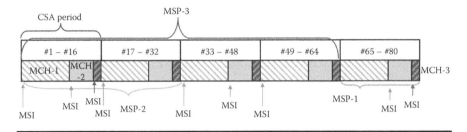

Figure 8.16 MSP and MSI.

8.6.3 Signaling Flow for MBMS Subframe Identification by the UE

The following sequence represents the signaling information a UE needs to receive from the point where a user has chosen a service of interest down to the point where the UE can start receiving the first subframe with payload data. On the service layer, the user service description (USD) [14] provides the terminal with relevant information for MBMS service reception, such as service and session ID and session start times.

1. USD: choose services of interest
   a. Obtain service and session ID
2. Read system information relevant for MBMS reception
3. MBMS-specific system information contains MCCH and notification configuration
4. MCCH
   b. MCH configuration
   c. Session list per MCH
      i. Service and session ID to LCID mapping
      ii. If service not listed, monitor notification
5. For reduced UE battery consumption
   d. MSI
   e. Scheduling information for services (MTCHs)
      iii. Duration if scheduled
      iv. Special codepoint if unscheduled

8.6.4 MCCH Updating

The MCCH carries essential eMBMS service information. It is transmitted every MCCH repetition period configurable to 320 ms, 640 ms, 1.28 s, or 2.56 s, as illustrated in Figure 8.17. Information on the MCCH can change only at the beginning of an MCCH modification period, configurable to 5.12 or 10.24 s.

A UE not yet receiving an MBMS session is not required to monitor the MCCH continuously, in order to save battery power. When information changes on an

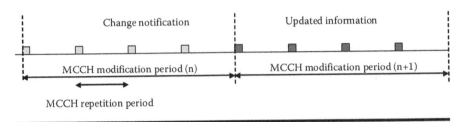

Figure 8.17 MCCH transmission timing.

MCCH, a change notification is transmitted a configurable number of times per modification period, using a message on the physical downlink control channel (PDCCH, mainly used for unicast control signaling). This message indicates which of the configured MCCHs will change. Since the PDCCH is transmitted only in the small control region of a subframe, whereas the MCCH occupies the larger data part of a subframe, having to receive only the PDCCH in order to detect MCCH changes is more power efficient for the UE. Once the UE has detected the change notification message, it will also read the MCCH.

8.7 Recent Enhancements

In the work on LTE Release 10, 3GPP has added MBMS user counting functionality, and for LTE Release 11, 3GPP has included support for MBMS service continuity in multifrequency deployments. Release 11 is scheduled to be finalized in September 2012.

8.7.1 User Counting for MBSFN Activation

A user counting procedure enables the MCE to activate/deactivate MBMS services in a predefined MBSFN area depending on user interest. The counting procedure is triggered by MCE if allowed by the BM-SC. Counting is possible for Release 10 MBMS UEs in RRC_CONNECTED mode, not in RRC_IDLE mode. Counting is supported for services already provided in an MBSFN area, as well as for services not yet provided in an MBSFN area. The counting request is transmitted on the MCCH, for both ongoing and non-ongoing services. All Release 10 UEs in RRC_CONNECTED mode will respond to the counting request.

If the number of UEs responding to be interested in a service exceeds an operator-set threshold, the MCE can enable the preconfigured MBSFN area so that the service gets broadcasted. If the UE is switched from RRC_CONNECTED to RRC_IDLE mode, it can continue MBMS service reception.

If the threshold is not exceeded, then the number of users interested in the service is considered to be so low that delivering the service in unicast mode only in the cells with interested users is more efficient from a radio resource perspective.

If the MBMS service is also available via unicast and not provided in broadcast mode, the interested UEs can access it via unicast and thereby automatically enter RRC_CONNECTED mode. If the counting is then initiated, all interested UEs would be in RRC_CONNECTED mode. Otherwise, if UEs remain in RRC_IDLE, the MCE would receive no counting response from these UEs and therefore no reliable indication whether to enable the broadcast or not.

8.7.2 Service Continuity in Multifrequency Networks

An MBMS service that is provided via MBSFN is generally provided only on one frequency, while multiple frequencies can be deployed in a geographic area to cope with increasing traffic, i.e., identical content is not duplicated on other frequencies. In order to support service acquisition and service continuity in multifrequency networks within the same MBSFN area, the four scenarios as depicted in Figure 8.18 should be considered. For convenience, the following description considers a UE that can only receive MBMS on a serving frequency, i.e., on a frequency where it is camping (for RRC_IDLE) or where it is served (for RRC_CONNECTED). We discuss four service acquisition and continuity scenarios. In this example, we assume that MBMS is provided on frequency f2, the UE's service of interest is only provided in MBSFN area #1, and the MBMS session is ongoing in all four scenarios. If the UE moves

1. Into the coverage area of MBSFN area #1, the UE should be moved to the MBMS frequency, if it is not already on the MBMS frequency (service acquisition)
2. Within the MBSFN area, the UE should be kept on MBMS frequency layer (service continuity)
3. From MBSFN area #1 to non-MBSFN area or adjacent MBSFN area #2, there is no need to keep the UE on the MBMS frequency
4. Within non-MBSFN area, there are no MBMS-related requirements and the UE should follow normal mobility procedures

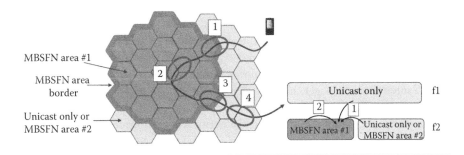

Figure 8.18 Scope of MBMS service continuity.

The inter-frequency cell reselection/handover to the MBMS frequency f2 can also be triggered when the UE is already within the MBSFN area, but on the wrong frequency, and the service is about to start. Further details are described later.

In LTE Releases 9/10, service acquisition and service continuity are feasible only for UEs in RRC_IDLE, which may prioritize the MBMS frequency carrying the service when the session is ongoing. Information about the frequency and the session start and stop can be provided to the UE via the USD [14]. However, for UEs in RRC_CONNECTED without a separate receiver to allow any reception on a different frequency, service acquisition/continuity is more challenging. First, if the service is ongoing according to the USD, the UE has to find out whether it is available in the UE's current location and on which frequency the service of interest is provided. This is more challenging for UEs in RRC_CONNECTED because they can only tune to other frequencies when they are not receiving anything on the serving frequency. Second, since the network controls handovers and serving cell configurations, the UE in RRC_CONNECTED needs to indicate its MBMS reception status to the network that can thus initiate appropriate action.

In order to provide the UE with sufficient information in LTE Release 11, so-called MBMS assistance information is provided to the UE by both the USD (from the service layer) and the network. The USD of a service contains information in which MBMS SAIs and on which frequencies the service is provided. Each cell in the network broadcasts MBMS SAIs of cells for its own frequency and for its neighboring frequencies in its system information. If the UE finds a match between the MBMS SAI in the USD of the service of interest and in the system information, it can derive the frequency where the service is provided from the network's system information. Based on the MBMS SAI information, the UE in IDLE_MODE can prioritize this frequency, and the UE in RRC_CONNECTED mode can send its MBMS interest indication to the network. This interest indication contains a list of one or more MBMS frequencies according to the UE's interest and capability of parallel MBMS reception on different frequencies, and also a priority bit, which can be used by the network in case of congestion.

If MBMS reception is prioritized, the network can thus release a unicast bearer to keep the UE on a congested MBMS frequency. In order to minimize the signaling load, the UE indicates its interest on a frequency level rather than on a service level and only when the content of the signaled information has changed, e.g., when the user interest has changed or when the service availability has changed in terms of time or location. In handover scenarios, the source cell forwards the MBMS interest to the target cell, such that the signaling load on the radio interface is further reduced.

For backward compatibility, i.e., if the network does not provide MBMS SAIs in its system information, or if there are Rel-9/10 UEs, the MBMS SAI information from the network or the USD, respectively, can be ignored, and the UE falls back into Rel-9/10 behavior.

8.8 Conclusions

eMBMS enables LTE networks to offer true broadcast services. Thereby LTE networks are able to handle even mass-delivery of media—live or prerecorded—that is of interest to a wide audience. We have shown that SFN-based broadcasting strongly benefits from the prevailing small cell deployment of LTE networks. The eMBMS service and SFN area size can be tailored geographically from a single cell up to nationwide and areas of different size can overlap, all on the same carrier frequency. eMBMS is part of the continued further evolution of LTE standardization in 3GPP. The short cycles of issuing new standard releases ensure timely implementation of technology opportunities and market demands.

References

1. M. Eriksson et al., The GSM/EDGE Radio Access Network—GERAN; system overview and performance evaluation, *Proceedings of IEEE Vehicular Technology Conference (VTC) Spring*, Tokyo, Japan, 2000.
2. H. Holma and A. Toskala, *WCDMA for UMTS—Radio Access for Third Generation Mobile Communications*, Wiley, New York, July 2004.
3. J. Sköld and M. Lundevall, S. Parkvall, and M. Sundelin, Broadband data performance of third-generation mobile systems, *Ericsson Review*, 82(1), 14–23, 2005.
4. E. Dahlman, S. Parkvall, J. Sköld, and P. Beming, *3G Evolution: HSPA and LTE for Mobile Broadband*, Academic Press, Amsterdam, the Netherlands, 2008.
5. 3GPP TS 26.234 v11.1.0, Transparent end-to-end packet-switched streaming service (PSS). Release 11; September 2012.
6. I. Elsen, F. Hartung, U. Horn, M. Kampmann, and L. Peters, Streaming technology in 3G mobile communication systems, *IEEE Computer*, 34(9), 46–53, September 2001.
7. 3GPP TS 26.245 v11.0.0, Transparent end-to-end packet switched streaming service (PSS); Timed text format. Release 11; September 2012.
8. ITU-T Recommendation H.264 (03/05), Advanced video coding for generic audiovisual services " |ISO/IEC 14496-10:2005: " Information technology—Coding of audio-visual objects—Part 10: Advanced video coding.
9. F. Hartung U. Horn, J. Huschke, M. Kampmann, T. Lohmar, and M. Lundevall, Delivery of broadcast services in 3G networks, *IEEE Transactions on Broadcasting*, (1), 188–199, March 2007.
10. M. Kornfeld and G. May, DVB-H and IP datacast—Broadcast to handheld devices *IEEE Transactions on Broadcasting*, 53(1), 161–170, March 2007.
11. S. Cho, G. Lee, B. Bae, K. Yang, C. Ahn, S. Lee, and C. Ahn, System and services of terrestrial digital multimedia broadcasting (T-DMB), *IEEE Transactions on Broadcasting*, 53(1), 171–178, March 2007.
12. M. Chari, F. Ling, A. Mantravadi, R. Krishnamoorthi, R. Vijayan, G. Walker, R. Chandhok, FLO physical layer: An overview, *IEEE Transactions on Broadcasting*, 53(1), 145–160, March 2007.
13. S. McCanne, Scalable multimedia communication using IP multicast and lightweight sessions, *IEEE Internet Computing*, 3(2), 33–45, March–April 1999.
14. 3GPP TS 26.346, Multimedia broadcast/multicast service (MBMS); Protocols and codecs.

15. Open Mobile Alliance, Mobile broadcast services, www.openmobilealliance.org (accessed 31.12.2007).

16. Open Mobile Alliance, Service guide for mobile broadcast services, www.openmobilealliance.org

17. ISO/IEC DIS 23009–1.2, Dynamic adaptive streaming over HTTP (DASH), November 2011.

18. T. Paila, M. Luby, R. Lehtonen, V. Roca, and R. Walsh, FLUTE—File delivery over unidirectional transport, IETF RFC 3926.

19. M. Watson, FECFRAME requirements, IETF Draft-Ietf-Fecframe-Req-00, work in progress.

20. A. Shokrollahi, Raptor codes, *IEEE Transactions on Information Theory*, 52(6), 2551–2567, June 2006.

21. M. Luby, T. Gasiba, T. Stockhammer, and M. Watson, Reliable multimedia download delivery in cellular broadcast networks, *IEEE Transactions on Broadcasting*, 53(1), 235–246, March 2007.

22. T. Lohmar and M. Elisova, Evaluation of the file repair operations for multicast/broadcast download deliveries, *Proceedings of the European Wireless*, Nicosia, Cyprus, 2005.

23. T. Lohmar, Z. Peng, and P. Mähönen, Performance evaluation of a file repair procedure based on a combination of MBMS and unicast bearers, *Proceedings of the IEEE WOWMOM*, Niagara Falls, NY, 2006.

24. T. Lohmar and J. Huschke, Radio resource optimization for MBMS file transmissions, *Proceedings of the IEEE International Symposium on Broadband Multimedia Systems and Broadcasting (BMSB)*, Bilbao, Spain, 2009.

25. H. Schulzrinne et al., RTP: A transport protocol for real-time applications, IETF RFC 3550, July 2003.

26. T. Stockhammer, A. Shokrollahi, M. Watson, M. Luby, and T. Gasiba, Application layer forward error correction for mobile multimedia broadcasting, in *Handbook of Mobile Broadcasting: DVB-H, DMB, ISDB-T and MediaFLO*, CRC Press, Boca Raton, FL, 2008.

27. J. Huschke, Facilitating convergence between broadcasting and mobile services using LTE networks, *Proceedings of the ITU Telecom World 2011 Technical Symposium*, Geneva, Switzerland, 2011.

28. OMA, Push OTA protocol, WAP-235-PushOTA-20010425-a, April 2005.

29. 3GPP TS 36.300 v11.1.0, Evolved universal terrestrial radio access (E-UTRA) and evolved universal terrestrial radio access network (E-UTRAN); Overall description; Stage 2, March 2012.

30. 3GPP TR 25.913 v.9.0.0, Requirements for evolved UTRA and evolved UTRAN, December 2009.

31. 3GPP TR 25.905 v7.2.0, Improvement of the multimedia broadcast multicast service (MBMS) in UTRAN, January 2008.

32. 3GPP TR 25.814 v.7.1.0, Physical layer aspects for evolved UTRA, October 2010.

33. E. Dahlman, S. Parkvall, and J. Skold, *4G: LTE/LTE-Advanced for Mobile Broadband*, Academic Press, Amsterdam, the Netherlands, 2011.

34. 3GPP TS 25.446 v.10.2.0, MBMS synchronisation protocol (SYNC), December 2011.

35. Ericsson, E-UTRA performance checkpoint: MBSFN, contribution R1-071958 to 3GPP RAN1, April 2007.

Chapter 9

Universal DVB-3GPP Broadcast Layer: An Enabler for New Business in Mobile Broadcasting Landscape

Christian Gallard, Jean-Luc Sicre,
Matthieu Crussière, Catherine Douillard,
Gérard Faria, and Sylvaine Kerboeuf

Contents

9.1 Introduction

In February 2010, as a response to the Digital Video Broadcasting (DVB) Call for Technologies related to definition of the second-generation mobile broadcasting system DVB-NGH (next-generation handheld), the mobile network operator Orange proposed 3GPP Evolved—Multimedia Broadcast Multicast Service (E-MBMS) waveform as a candidate for DVB-NGH. DVB standardization group did not warmly welcome this proposal. However, understanding the opportunity it offered to reach the world of mobile devices (smartphones and tablets), and to provide a solution to absorb the ever-increasing video traffic on mobile networks, the digital TV standardization forum contacted 3GPP organization in November 2010, asking to consider a potential collaboration in the area of mobile broadcasting. A joint workshop took place in March 2011 in Kansas City (United States) for mutual presentations of 3GPP and DVB standardization activities. In May 2011, in Xian (China), the creation of a study item (SI) was proposed to the 3GPP technical subgroup dealing with services (3GPP SA1). The objective was to *study the feasibility of, and creating common service requirements and use cases for, a common broadcast specification which can be used in a 3GPP mobile communications network and a broadcasting network that is based on DVB or other similar standards* [1]. This proposal, introducing, for the first time, the concept of *Common Broadcasting Specifications* (CBS), was discussed but not accepted by 3GPP SA1, due to lack of support from mobile operators. Nevertheless, on the DVB side and after the finalization of the DVB-NGH specifications (June 2012), the "CBS" topic should be addressed again.

This chapter aims at promoting broadband–broadcast cooperation, the broadcast part being based on a clever blend of state-of-the-art 3GPP and DVB standards, the outcome being presented here in details. An overlay optimized mobile broadcasting mode operated in conjunction with a unicast access could be a good mix to cope with the announced mobile data tsunami. This chapter also elaborates why such cooperation could be a win-win situation for each actor of the multimedia value chain.

Section 9.2 explains why, in an endless increasing mobile video consumption and due to its intrinsic characteristics (e.g., high spectral efficiency dedicated to service throughput with minimal "radio housekeeping" overhead), mobile broadcasting could efficiently contribute to absorb part of the mobile data traffic (e.g., via off-loading, similarly to 3G-WiFi cooperation at home). Moreover, contextual elements are provided to highlight the "win-win" opportunity for both broadband and broadcast ecosystems and the new businesses enabled by the broadband/broadcast cooperation.

To become commercially successful, mobile broadcasting technology must be integrated in smartphones and tablets and in broadband base stations. It is quite sensible then to assume that the long-term evolution (LTE) broadcast mode has to be used as a starting point for broadband/broadcast cooperation. Accordingly, an enhanced version of E-MBMS taking into account typical broadcast requirements (e.g., large coverage area, long time interleaving, etc.) could provide a strong basis for a unified physical layer standard. This assumption guided the design of the E-MBMS enhancements proposed in Section 9.3.

Initially, CBS was assumed to be a downlink-only specification, operated on a dedicated carrier. However, CBS could also be transmitted "in band" with LTE or LTE-advanced unicast (as is the case in the current LTE/E-MBMS specifications) or "in band" with DVB-T2 broadcast (as is the case with DVB-T2 "base" and "Lite" or DVB-NGH, which benefits of the future extension frames (FEFs) feature).

9.2 Multimedia Traffic Explosion and Broadband/Broadcast Cooperation

9.2.1 Predicted Explosion of Mobile Data Traffic on Mobile Operator Networks

Since the availability of the mobile TV service on smartphones and tablets, consumers have been experiencing multimedia applications delivery in mobile conditions over 3G unicast transmissions with better experience than ever. But this service is suitable for only a reduced number of users per cell as it is not strongly spectrum efficient: two different users, watching the same content at the same time, consume two spectral resources.

According to CISCO forecast [2], mobile data traffic on wireless mobile networks will dramatically increase in the coming years. Mobile video traffic was 52%

of traffic by the end of 2011. As pictured in Figure 9.1, *Global mobile data traffic will increase 18-fold between 2011 and 2016… Mobile Video Will Generate Over 70 Percent of Mobile Data Traffic by 2016* ([2], p. 3 and p. 10 Fig. 6). Furthermore, it is estimated that the percentage of time spent using Internet at work or "on the move" represents, on average, approximately 60% of the total mobile data use.

Ericsson's report [3] observes a similar trend (see Figure 9.2): in Q4 2009, mobile data traffic surpassed voice traffic, and this trend seems to be confirmed in the forthcoming years. At least one-third of the traffic for each type of device (mobile PC, tablet, or smartphone) is already online video.

Byte Mobile's report [4] confirms that around 50% of the mobile data traffic is already video (see Figure 9.3, based on the feedback of four operators).

Therefore, mobile data consumption explosion is clearly a main challenge for telecom operators, which must avoid the congestion of mobile networks having the finite capacity that results from the finite amount of spectrum available to mobile telecommunications.

9.2.2 Off-Loading of Mobile Data Traffic from Mobile Networks to Broadcast Networks

To avoid congestion of mobile networks, traffic off-loading is a crucial and key objective.

In CISCO's analysis [2], it is assumed that "mobile offload increases from 11% (72 petabytes/month) in 2011 to 22% (3.1 exabytes/month) in 2016," as depicted in Figure 9.4.

The current operated solution is to off-load traffic on an IEEE-WiFi connection: WiFi is widely available at home, at work, on public hotspots and IEEE 802.11-n or 802.11-ac standards perform well with the high data rates required for video transmission, using MIMO and/or the unlicensed 5 GHz spectrum. Such a configuration is considered with interest by European mobile operators [5].

In 3GPP specifications, since Release 8, the defined access network discovery and selection functions help a device to discover non-3GPP available systems, provide to the end-user operators' policies for preferred connection, and then constitute very helpful features to off-load mobile broadband traffic on WiFi connection.

When the terminal is "on the move," WiFi is probably not anymore the appropriate off-loading. Instead, a broadcasting technique will enhance delivery, particularly at peak time in crowded areas. In this context, popular contents could be broadcast over a given area, with a guaranteed quality of service (QoS), regardless of the number of active users.

Broadcasting technology was dedicated in the delivery of live TV exclusively, but nowadays, video consumption is driven by on-demand requests. Hopefully, broadcast delivery method is not incompatible with both consumption modes!

Filecasting, as planned, for instance, in the Japanese ISDB-Tmm (see Chapter 2), can be used to fill in the cache of a terminal, with, e.g., e-news or video-clips,

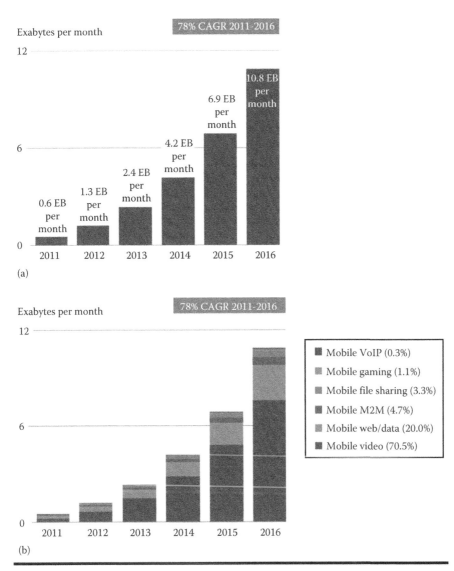

Figure 9.1 **CISCO mobile data traffic forecasts. (a) CISCO forecasts 10.8 exa-bytes per month of mobile data traffic by 2016. (b) Mobile video will gener-ate over 70% of mobile data traffic by 2016. (From CISCO white paper, Cisco visual networking index: Global mobile data traffic forecast update, 2011–2016, February 2012, http://www.cisco.com/en/US/solutions/collateral/ns341/ns525/ns537/ns705/ns827/white_paper_c11-520862.pdf, pp. 5, 10.)**

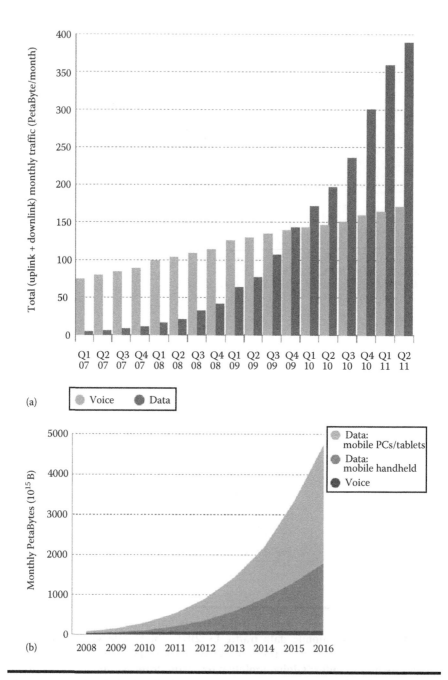

Figure 9.2 Ericsson mobile data traffic forecasts. (a) Global total traffic in mobile networks, 2007–2011. (b) Mobile traffic: voice and data, 2008-2016. (From Ericsson, Traffic and market data report, November 2011, http://hugin. info/1061/R/1561267/483187.pdf)

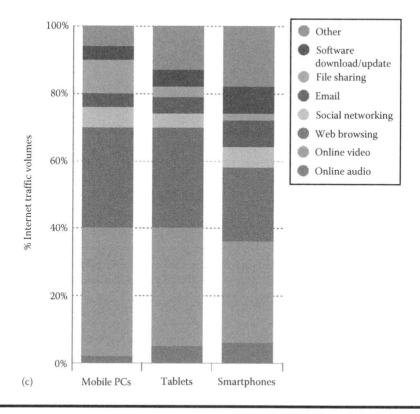

Figure 9.2 (continued) Ericsson mobile data traffic forecasts. (c) Application internet traffic volumes by device type. (From Ericsson, Traffic and market data report, November 2011, http://hugin.info/1061/R/1561267/483187.pdf)

anticipating the request of an end-user. Indeed, broadcasting is very efficient to download popular or bandwidth (BW)-hungry contents into the "profiled" end-user terminals' cache. This technique, also known as "predictive datacasting" or "preemptive downloading," is able to prevent daily network congestion by smoothing efficiently the BW required at peak time or during a particular event. A side effect of "predictive datacasting," just like "podcasting," is the capacity of this method to deliver a service with a perfect QoS, whatever the access network situation at particular location or at a particular instant.

Broadcasting standards are defined in dedicated standardization forum like DVB, ARIB, or ATSC in order to translate, in the digital era, traditional broadcast delivery of "TV" or "radio" services. Mobile stakeholders have nevertheless seen the interest of a broadcast mode embedded in broadband systems, to ease video consumption on not-dedicated but connected terminals: Integrated Mobile Broadcasting (IMB) system based on 3G technology has been specified in 3GPP Release 8. This standard employs the TDD bands (available between FDD bands,

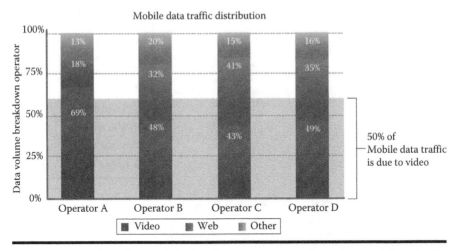

Figure 9.3 Mobile data traffic distribution. (Extract from Bytemobile, Mobile analytics report, February 2012, Copyright 2012 Bytemobile, Inc.—Q1 2012.)

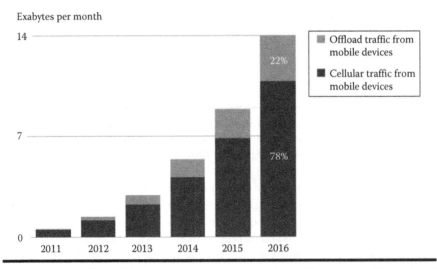

Figure 9.4 Traffic off-load forecast. (From CISCO white paper, Cisco visual networking index: Global mobile data traffic forecast update, 2011–2016, February 2012, http://www.cisco.com/en/US/solutions/collateral/ns341/ns525/ns537/ns705/ns827/white_paper_c11-520862.pdf, p. 12.)

but not used by Western mobile operators) and is compatible with unicast 3G technology, but as it requires a dedicated network, it still looks for supporters inside 3GPP ecosystem. 3GPP has also specified E-MBMS standard within LTE system (more details on Chapter 8), to embed a broadcasting mode in the mobile broadband system of the future: Broadcast and unicast delivery methods are multiplexed in time, within LTE spectrum without requiring any dedicated network. Even if better spectrum efficiency is expected from E-MBMS when compared to IMB (i.e., OFDM versus CDMA waveform), E-MBMS will achieve a throughput lower than IMB, as it will not constitute anymore a 100% downlink system and throughput will be "waste" for unicast delivery.

Overlay solutions exist and are defined outside the mobile ecosystem, like DVB-T2 Lite (see Chapter 6) and DVB-NGH (see Chapter 7). They have some advantages like to be 100% downlink systems, to exhibit high performance in terms of spectral efficiency (i.e., no latency and memory constraints), but present the main drawback that the 100% downlink system is shared with digital TV services.

DVB started in early 2009 the definition of DVB-NGH specifications. A "DVB-T2 Lite" specification has been completed in 2010, allowing to multiplex in time DVB-T2 frames having parameter sets optimized for "stationary" and "mobile" reception, thus providing the opportunity to insert "in-band" (i.e., using the existing infrastructure) TV services targeting mobile TV. An enhanced-version "DVB-NGH," including advanced features (e.g., MIMO scheme, hybrid terrestrial and satellite coverage, etc.), is announced for mid-2012.

9.2.3 Mobile Broadband/Broadcast Convergence: An Enabler for New Business

The M^3 project [6] has made clear that multimedia business cases relying on mobile broadband–broadcast hybrid access appear as a "win-win" opportunity for mobile network operators both to preserve and to monetize the value of their network infrastructures.

The mobile broadband–broadcast hybrid access generates business-to-business (B2B) value in the mobile multimedia chain and constitutes a relevant key enabler leading to the industrial and commercial successes of business-to-consumer (B2C) models. In this perspective, a set of guidelines is proposed with the aim to help solving the various bottlenecks related to mobile broadband–broadcast last mile access hybridization.

Wireless broadband–broadcast hybridization access is on the agenda, but without any direct route to success!

The so-called convergence of today's (2012) landscape is rather a chaotic world made of very few success stories, as business chains of both broadcast and broadband are experimenting a crunch period where old value chains are confronted to new business models (e.g., Over The Top or OTT services). On the one side, the value chain of telco/broadband world is still driven by the access fees, as the average revenue per user (ARPU) constitutes major incomes of the mobile operators. On the other side, the audience hunting and the linearly programmed multimedia contents stay the relentless drivers of the broadcast value chains. These two traditional business chains are stressed by the Internet *way of consumption*, with a shift from a passive to an interactive and on-demand media consumption (from linear broadcast TV to catch-up TV) along with the proliferation of many Internet content providers. Moreover, the first wave of Internet content services was mainly based on free contents, which prevented the services distributors from developing a sustainable business. The example of the Application Store business model illustrates successfully the second wave of mobile Internet content services and endangers the business models of the traditional stakeholders. On the Internet today, the "contents" business model is based on painless micro-payments rather than premium services involving monthly fees.

In such a context, using cooperatively the existing resources provided by broadcast and broadband networks should permit to decrease globally the operating cost of distributing the ever-increasing video traffic induced by the proliferation of smart terminals and Internet-like usage of multimedia contents, without requiring additional investment in network infrastructure and then providing a way to balance a subtle business plan.

Vertical business of linear broadcast services is endangered.

In the "old analogue times," some ideal value chains used to exist for TV and radio broadcasting: a vertical business model starting at the contents production and ending at the consumer level with a dedicated terminal (either a fixed or a mobile one). The old-guard broadcast players with well-established expertise are endangered by a range of new entrants like Apple, Google, and Hulu. The traditional linear TV model evolves today toward a nonlinear, personalized, on-demand TV model as viewing habits evolve. This transformation is the trigger point of a large set of complex combinations of vertical and horizontal value chains. Such complexity and diversity of value chains are not in favor of additional/complementary network investments (i.e., increased capital expenditures or CAPEX) devoted to a unique service (i.e., mobile TV) and based on a unique and indirect source of revenues (i.e., advertisement).

In the multimedia business, the value is moving, leaving the access infrastructure to focus on services and contents.

An important aspect to consider today is the market trend to concentrate the value in the services and not in the network access: it constitutes a severe disabling factor in any attempt to invest on a broadcast downlink access, cooperating or not with the mobile broadband access. On the one side, mobile broadband network operators are not confident enough in the perspective of return on investment resulting on the rollout of a dedicated mobile broadcast network enlarging the capacity of their native broadband network, even if broadcast used to be the basic access of in-vehicle telematics and entertainment. On the other side, asking the users to pay for a better QoS through a "broadcast enhanced option" might be commercially hazardous. Nevertheless, the road toward success for mobile broadband operators is, definitely, to combine traditional flat-fee (access-driven) business models with more flexible and service-driven (individual) pricing models by making services differentiation and QoE monetization. This calls for primordial coupling between the services and the network assets, like a range of QoS guarantees, intelligent traffic control, and enforcement policies. Operating a complementary broadcast access will give without any doubt an additional degree in service differentiation to the mobile network operators.

Economic value of mobile broadcast access exists, yet it cannot be directly monetized.

The past years have shown that the value of the mobile broadcast is not in a direct B2C model. Indeed, the mobile broadcast infrastructure required to serve mobile terminals cannot be economically balanced with a unique "TV on the move" service. The B2C models of the mobile multimedia chain are mainly driven by the content sellers, being an OTT content provider, a Telco operator, or a broadcaster. With the evolution of customer habits in media consumption, the relevant business cases are mainly "OTT content" driven and "Telco" driven, as money for investment exists in both (OTT) Contents and Telco businesses. A broadcast downlink (terrestrial or satellite) does have an economic B2B value as broadcast offers good and guaranteed downlink QoS, easy mass access, and an optimal spectral efficiency. The value of a mobile broadcast access leaned to a mobile broadband access actually lies in a "Network as a Service" (NaaS) capacity that will foster the industrial and commercial successes of B2C models. "Broadcast"-driven business cases do not seem easy to consider, especially when noticing the failure of mobile broadcast rolling out everywhere, at the exception of Korea and Japan, who benefited from governmental active policies.

Broadband–broadcast hybrid access appears as a "win-win" opportunity for mobile network operators in front of the OTT players, to preserve and to monetize the value of the network infrastructures they operate.

By offering an improved QoS over a given area, broadcast access, operated as a complement to broadband mobile access, acts as a smart mobile access network for both individual interactive and on-demand services while allowing for global push of contents (such as live TV and datacasting) to a massive amount of users. By deploying such a smart hybrid coverage, a network operator is in an excellent position to monetize its hybrid network usage, to optimize frequency band utilization and to contract with (OTT) service operators in a win-win agreement, particularly in areas where mobile broadband access is saturated. Moreover, operating a broadcast downlink together with personalization and interactive features constitutes a potential source of new revenues coming from advertisers.

Spectrum regulation and public authorities' decisions are of utmost importance.

In the wireless domain, indeed, the governmental policies have a very structuring impact: frequency access policy is confirmed as a strong enabling (or disabling) factor. In this domain, the broadcast and broadband wireless worlds, depending on history, political, and national rules, are in opposition. The condition in which the spectral BW is monetized is of utmost importance when preparing hybridization and convergence of mobile access. In such a context and considering the forthcoming frequency shortage, the efficiency of the broadcast delivery expressed in terms of "bit/second per Hertz per square kilometer" appears as a major asset with regard to both spectral and area coverage, especially because the broadcast paradigm allows to escape to the "frequency reuse" mechanism that governs the planning of broadband networks. Probably, some metrics integrating bit-rate/area/power have to be defined to rank network infrastructures.

More convergence for a worldwide endorsed broadcast physical layer is highly desirable.

The diversity of digital terrestrial broadcast standards operated around the world is a serious and severe disabling factor regarding the availability and cost of "universal receivers." The terminal impact in building the business chain is far from being incidental: smart mobile devices are identified as strong enabling factors for broadband–broadcast service convergence (e.g., iPhone distribution and increased live TV consumption). A unified mobile terrestrial broadcast downlink would boost the broadband downlink access, and used as a complement to the indoor IEEE Wi-Fi access, it would constitute an overlay network able to address terminals in mobile conditions, both outdoor and indoor.

9.3 First Thoughts about a Common Physical Layer

9.3.1 Convergence Levels

The business models built on a combination of mobile broadband and mobile broadcast accesses are strongly linked to the degree of cooperation or convergence between the two networks. From the end-user service side, three convergence modes or levels can be identified:

- *Terminal level:* convergence starts when a terminal is able to consume multimedia services, delivered on a broadband and/or a broadcast access, under the control of the user or the application or the network.
- *Network level:* convergence comes from network operators having a cooperation strategy established over a given area, to improve QoS to entitled users. Network convergence uses a "network switch server" managing the last mile access with the handover and/or overlay strategy under various conditions including period of the day and the location of the users.*
- *Service level:* interactive TV services requiring both broadband and broadcast accesses and sometimes designed at the contents production stage (e.g., the-movie-you-decide-the-end) constitute a good example of convergence at service level. HbbTV is another example of service convergence, where the predominant broadcast access benefits from ancillary data delivered over the broadband access used in OTT mode. In the future, providing that production cost will not be a disabling factor, such a service convergence could lead to very sophisticated cooperative platforms using multi-access to support "enhanced interactive services."

---

* IMB broadcast standard, integrated into 3GPP portfolio, is an early example of such network convergence.

These three basic convergence levels can be combined to offer the best QoS in a given context of constraints (service type, access, user location and density, time of the day, etc.)

The "physical layer convergence" constitutes a crucial enabling factor, as any upper-level convergence scenario implies a terminal's front end having simultaneous access to both broadband and broadcast networks. One of the major objectives of the unified broadcast physical layer is to ease the design of cost-effective front end; it also implies harmonization of mobile broadcast specifications to make them as universal as IEEE-WiFi and 3GPP-LTE broadband ones.

For the time being in the broadcast world, standardization, regulation framework, and frequency access rules are fragmented by continent and even by countries. This fragmentation prevents the deployment of nomadic device (handhelds, tablets, laptops) equipped with a broadcast-capable chipset in addition to the classical broadband-modem.

A converging mobile broadcast downlink would greatly help any type of cooperation between mobile broadband and mobile broadcast network operators, including coverage collaboration or RAN-sharing-like strategy. It is pointed out that if the lack of a "universal" broadcast access persists, IEEE-WiFi even with its quite poor QoS will become the dominant means to provide the searched "universal last mile wireless access" for mobile services and devices.

9.3.2 Concepts for a Common Broadcasting Specification

In the framework of DVB-NGH standardization process emerged the idea of a broadcast system able to deliver "Direct to Nomadic (DTN)" services over hybrid network topologies involving satellite, broadcast towers, cellular base stations, or hot spots then using a wide range of frequency bands (from 100 MHz to 3 GHz). This "broadcast" approach leads to consider the specification of a "family" of DVB-NGH broadcast frames having commonalities but sufficiently flexible to satisfy the constraints of the network in which they are used. This network-centric approach is well adapted to the one-to-many delivery system, for which all receiving situations could occur simultaneously even if it is orthogonal to the terminal-centric approach of the one-to-one broadband system that adapts and tailors the transmission for the specific situation of each and every terminal.

Even if broadband network operators are able to optimize individually the transmission efficiency of each link according to the location of each user, such a strategy leads to increase the volume of delivered contents when several users request simultaneously the same data. From a global network point of view, the transmission efficiency within a cell can hence rapidly fall down considering popular multimedia services. Worst, network congestions can occur over certain periods of time with a direct negative impact on the QoS.

In that sense, the "broadcast" approach embeds, by nature, considerable advantages in terms of efficiency, since the downlink BW is not shared between, but rather jointly offered to terminals. As a consequence, the transmission efficiency for a given service does not depend on the number of demands, the QoS support is guaranteed over the network whatever the number of users, and even the cost of delivery (per user) decreases in proportion with the number of users. Moreover, as the one-to-many delivery does not require management of the receiving ends, the quasi-totality of the capacity offered by the physical layer is available for multimedia service and data: as the broadband and the broadcast physical layers offer each near-to-Shannon capacity, the broadcast usage of the available capacity is always more efficient than the broadband unicast one.

These considerations constitute the rationale for a "Common Broadcast Specifications" (CBS) able to supply either broadcast type or broadband type of multimedia contents to nomadic terminals, using a broadcast transmission method over a dedicated 100% downlink radio frequency channel. Otherwise, the CBS should be available on various networks (satellite, broadcast, cellular, hot spots) operated at different frequency bands in order to permit the delivery of multimedia services to terminals equipped with CBS demodulator and broadband modem, whatever the network the nomadic terminals have access to while "on the move."

In that perspective and from the general considerations already introduced, the following few principles have been adopted when designing the CBS:

1. CBS will include "MAC" and "PHY" layers.
2. CBS-PHY will define a family of frames, as follows:
   a. The waveform will allow various network topologies (e.g., satellite, cellular, broadcast).
   b. The transmission parameters allow wide frequency band operation (e.g., 30…3000 MHz).
   c. The frame syntax will distinguish broadband or broadcast service types.
3. CBS frames will start with a preamble giving the required information to identify and demodulate the payload part of the frame.
4. Several types of frames multiplexed in time could be used in a given instantiation of the CBS.
5. CBS should offer logical structures to ease mapping of the services and their recovery by the receivers. Such logical containers should be spread within the frame in both time and frequency domains to offer every type of diversities.

The following sections give further insights and analyses supporting these principles.

9.3.2.1 Frame Syntax Considerations

CBS should be used to deliver either pure broadcast services (i.e., TV) or pure broadband services (i.e., streamed contents) that have not the same structure and have different "service" channel requirements. Multiplexes of linearly programmed TV services are self-documented (i.e., electronic program guides giving information on multiplexed services and programmed contents), allowing the receivers to self-discover the available TV services. Broadband services are intermittently using the transmission resources and are sustained by a procedure (i.e., broadcast session announcement, session opening, session closing) implemented through a control channel requiring a bidirectional wireless connection. In other words, broadcast services can be operated using only CBS, while CBS has to be considered as a capacity extension of the broadband infrastructure that must manage broadband services and instruct broadband terminals on the way to access contents over the CBS infrastructure.

Broadcast and broadband services requiring two modes for accessing services, it implies different syntaxes of the frames as auxiliary channels sustaining such services are of different natures: self-documentation for broadcast, session identification for broadband.

9.3.2.2 Resource Allocation Considerations

Regarding the organization of the data payload within the frames, this data payload being shared between several contents (i.e., TV programs for broadcast, streamed data for broadband), it shall be split in subparts, each one possibly experiencing a specific channel encoding process (i.e., code rate, modulation, etc.). Various methods could be used to do so, but 3GPP-LTE provides an interesting two-step way to achieve it: first, map individual services to a set of "resource blocks" (RBs) and then map such a set of RBs onto the transmitted frame while insuring time and frequency diversity. CBS should probably use the same approach for the broadcast contents.

9.3.2.3 Waveform and Parameter Choices

As far as waveform to be used within a given category of frame (i.e., broadband or broadcast categories then frame syntax) is concerned, its selection must be done in relation with the network topology and frequency in use. For terrestrial implementation of the CBS, OFDM modulation is the obvious choice, but for a satellite implementation of the CBS, single-carrier modulation seems a better option to avoid the high peak-to-average power ratio issue of the multi-carrier modulation, and then single-carrier OFDM (SC-OFDM) may be a very valuable option (see Chapter 23).

In both cases, the modulation parameters (FFT and GI size as the SISO/MISO/MIMO options) must be selected in relation with the topology of the transmitter network: cell size (i.e., the maximum delay spread of the multipath channel) and frequency band (i.e., expected resilience against the maximum Doppler occurring at the maximum receiver's motion speed).

At a first step, it is instructive to summarize and comment the DVB-NGH and 3GPP E-MBMS parameters and it is the purpose of the study presented in the next section.

As a result of this study, an E-MBMS frame is proposed as the first family member of the "Common Broadcast/Broadband Specification."

9.3.3 DVB-NGH and 3GPP E-MBMS Parameters Survey

DVB and 3GPP constitute two different standardization ecosystems devoted to complementary telecommunication systems: "TV" digital broadcast for DVB and mobile communications for 3GPP. The mobile broadcast component is then the common border between both groups.

Due to their respective origins, details of the physical layer specifications have many differences as, for instance, the fields where optimum performances are reached.

Table 9.1 gives the main parameters defined in DVB-T2/-NGH and 3GPP E-MBMS systems.

A first fundamental difference between both systems is the maximum coverage that can be achieved, which is proportional to the guard interval* (GI) range defined for each system, without any specific processing on receiver side. For E-MBMS, the maximum GI is 16.67 μs, which leads to a maximum cell radius of 5 km. In comparison, the DVB-NGH maximum GI (1/4) in an 8 MHz channel and an 8K FFT (likely to be used) leads to a maximum cell radius of more than 67 km. It is also worth mentioning that DVB-H roll-out studies in France had envisaged a maximum coverage of 33.6 km, which remains larger than what could be offered by E-MBMS. These values are of high importance since the cost of the network (i.e., CAPEX) is closely related to the number of transmission sites required to reach optimum coverage of a given area.

The inter-carrier spacing values are also much contrasted between the systems. For E-MBMS, the typical standardized value is 15 kHz, which is much larger than the maximal value in DVB-NGH (around 1 kHz). E-MBMS specifications have to, actually, support high-mobility scenarios in which the Doppler effect becomes one of the most limiting factors, especially for the highest frequency bands. In addition, larger values of the inter-carrier spacing are more convenient to limit the uplink synchronization constraints. From DVB-NGH point of view, uplink constraints are naturally relaxed and the Doppler spread to consider is substantially reduced due to lower frequency band exploitation. It has to be reminded that the choice of the inter-carrier spacing directly impacts the overhead proportion arising from GI and is thus also a key parameter regarding the coverage capacity.

The 3GPP E-MBMS parameters presented here correspond to the extended cyclic prefix mode, which is the only mode to be fully specified in the current 3GPP release (R10—broadcast signal transmitted in a time division multiplex with

---

* Note that the guard interval is also commonly referred to as cyclic prefix (CP), especially in 3GPP literature. In the following, the used terminology regarding both DVB and 3GPP systems will be GI.

Table 9.1 Comparison of DVB-T2/-NGH and 3GPP E-MBMS Main Parameters

		DVB-T2/-NGH [7,8]		3GPP E-MBMS [9–12]	
Frequency bands	[MHz]	Band III	VHF [174–300] UHF [300–446]		
		Band IV	[470–606]		
		Band V	[606–790]	Digital dividend	[790, 862]
		L-Band	[1,452–1,492]	TDD	[1,900–1,920]
		MSS-Band	UL [1,980–2,010] DL [2,170–2,200]	FDD	DL [1,920–1,980] UL [2,110–2,170]
Channel BW	[MHz]	1.75 6 7 8 10[a]		1.4 3 5 10 15 20	
Typical channel BW	[MHz]	8		5	
FFT sizes		From 2 to 16 K		From 128 to 2 K	
Inter-carrier spacing (ΔF)	[kHz]	5 MHz 2,790–0,349	8 MHz 4,464–0,558	15,0	
GI sizes		5 MHz	8 MHz	5 MHz	8 MHz
Duration	[μs]	11–716	14–448	16,67	
Maximum cell radius (Terr.)	[km]	214	133	5	
Typical cell radius (Terr.)	[km]	54	33	5	
Typical cell radius (Sat.)	[km]	~1,000	~1,000	N/A	

(continued)

Table 9.1 (continued) Comparison of DVB-T2/-NGH and 3GPP E-MBMS Main Parameters

		DVB-T2/-NGH [7,8]		*3GPP E-MBMS [9–12]*	
Channel encoding					
FEC		LDPC code		Binary turbo code	
Codeword size		$N^b = 16,200$		$K^c = \{40..6,144\}$	
Code rate		From 1/3 to 11/15		From 0.1172 to 0.9258	
Physical time interleaver (intraframe)	[ms]	250		1^d	
Physical time interleaver (interframe) (for 4 Mbps service—maximum memory for satellite component to be confirmed)	[ms]	1,400		N/A	
Typical rates		5 MHz	8 MHz	5 MHz	8 MHz
Terrestrial — FFT size		8 K extended mode^e		512	1,024
Guard interval (proportion)		1/32		1/4	
Guard interval (duration)	[µs]	45	28	16.67	16.67
Modulation and coding scheme		16QAM 4/9		16QAM 1/2	
Pilot pattern		PP4 (pilot density 1/24)		Sect. 6.10.2 [10]	
Service bit rate	[Mbps]	**7.74**	**12.34**	**2.981^f**	**4.795^g**
Spectral efficiency	[bits/s/Hz]	**1.541**	**1.543**	**0.596**	**0.599**

Satellite		2 K normal mode		N/A
	FFT size			
	Guard interval	1/32		
	Modulation and coding scheme	QPSK 1/3		
	Pilot pattern	PP2 (pilot density 1/12)		
	Service bit rate [Mbps]	**2.74**	**4.38**	
	Spectral efficiency [bits/s/Hz]	**0.548**	**0.547**	

a The 10 MHz bandwidth is intended only for professional equipment and not for domestic receivers.

b Fixed coded block size.

c Information block side defined within a range specified in Reference 11, Table 5.1.3-3.

d Application-layer FEC based on Raptor Codes [13] is not taken into account here.

e Signaling and synchronization overhead not removed but negligible impact in a 250 ms long frame.

f Number of MBMS subframes 6/10. Number of OFDM symbols reserved for unicast per MBMS subframe 2/12. Transport blocks size 4968 and 1 code block per transport block.

g Number of MBMS sub-frames 6/10. Number of OFDM symbols reserved for unicast per MBMS subframe 2/12. Transport blocks size 7992. Two code blocks per transport block.

unicast signal). It can be noticed that a longer extended cyclic prefix mode is antici-pated by 3GPP for the MBMS dedicated cells, although not supported yet, which involves a larger cyclic prefix and a smaller inter-carrier spacing (cf. [10]). Anyway, even in this mode, the 3GPP inter-carrier spacing is still very large (7.5 kHz), and the cell radius (10 km) is still very small, when compared with DVB values.

Another major difference is related to the maximum time interleaving. In E-MBMS, it is restricted to 1 ms at physical layer level (Raptor codes [13] applied on application layer are not taken into account here), while it could reach up to 250 ms for DVB-NGH for intra-frame interleaving. Therefore, as DVB-NGH does not suffer from the very low latency restrictions imposed to E-MBMS (e.g., gaming over LTE), the larger time diversity implemented in DVB-NGH should provide better performance in mobility at the expense of a longer access delay (i.e., longer channel switching time).

Finally, it must be kept in mind that both systems address different (channel) BWs. E-MBMS is specified for BW values that are multiples of 5 MHz (except for the lower ones, 1.7 and 3 MHz), whereas DVB-NGH is operated in channels hav-ing typically a BW of 6, 7, or 8 MHz. This BW variety, implying specific "system clock" (i.e., sampling frequency), makes the system specifications, as they are today, strongly incompatible.

9.3.4 First Tracks Proposed

As already stated, LTE broadcast mode is used as the starting point in this study. Hence in the sequel, an enhanced version of E-MBMS integrating typical broad-cast requirements is investigated and proposed as a possible unified standard for broadcast applications.

9.3.4.1 Frame Structure

The 3GPP E-MBMS is specified in a time division multiplex context where broad-cast is multiplexed with unicast transmissions, and accordingly, the synchroniza-tion and signaling procedures rely on information delivered in the unicast part of the transmission. In the context of a carrier dedicated to downlink-only trans-mission, synchronization and signaling features have to be totally reconsidered. DVB-NGH offers a nice concept with the so-called P1 preamble symbol, which is used for quick synchronization and is also carrying basic signaling about FFT and GI size. It is proposed to apply the same "P1" concept to CBS, and interested readers could refer to Chapter 15 for details about the physical layer signaling of DVB-NGH.

Another assumption is that the radio frame structure inherited from LTE is reused here. Radio frame duration of 10 ms is kept unchanged in a first step even if this could be reconsidered in a second step especially to increase the intra-frame time interleaving depth. The slot structure still exists for LTE broadcast mode even

P1	DVB-T2	P1	DVB-T2	P1	CBS	P1	DVB-T2

Figure 9.5   TDM transmission of CBS in DVB-T2 FEF.

though data are transmitted using the full BW of one subframe. In the enhanced version of E-MBMS proposed here, subframe duration of 1 ms is kept unchanged while the slot concept may not be kept.

From broadcasters' point of view, the starting scenario is to multiplex, in a TDM, CBS frames with DVB-T2 frames. This can be realized simply by embedding the CBS signal in the FEF defined in DVB-T2 standard, as depicted in Figure 9.5.

In that case, the sampling frequency for CBS, if based on usual 3GPP frequencies (submultiple of 30.72 MHz), is different from DVB-T2 case (including P1 symbol). Fortunately, there is a nice relationship between 3GPP and DVB sampling frequencies. In a typical DVB case (8 MHz BW*), the sampling frequency is

$$F_{S\_DVB} = \frac{1}{T_{S\_DVB}} = \frac{1}{(7/64)\,\mu s} = 9.14\,\text{MHz.} \tag{9.1}$$

On 3GPP side, the sampling frequency is a submultiple of 30.72 MHz:

$$F_{S\_3GPP} = (1/k) \times (15,000 \cdot 2,048) = (1/k) \times 30.72\,\text{MHz.} \tag{9.2}$$

The ratio between these two values can then be derived:

$$\frac{F_{S\_3GPP}}{F_{S\_DVB}} = \frac{84}{25}. \tag{9.3}$$

So any required switch from DVB frequency to 3GPP frequency, when changing from DVB-T2 to CBS, should be made easier.[†] This could probably be achieved in a couple of tens of nanoseconds. What could also make synchronization and other receiver tasks easier, although not imperative, would be to define CBS length (in terms of samples) as a multiple of DVB sample duration; this would be the case if the upper equation were followed. Such a constraint is especially welcome for DVB receivers that do not demodulate the CBS (FEF) waveform. A complete number of F_{S_3GPP} and F_{S_DVB} periods would potentially mean that at the end of the CBS

---

[*] Same calculations could be derived for BW = {5, 6, 7} MHz, reminding that on DVB side, $F_{S\_DVB} = (8/7) * BW$.

[†] Some application scenarios could even not require such frequency switch if the receiver only needs to demodulate CBS frames or native DVB-T2 frames, and not both.

frame a very small guard time might be inserted so that the duration of the CBS frame is an integer number of the T_{S_DVB} period. The guard time, if needed, would be strictly smaller than $T_{S\_DVB} \approx 0.11$ µs and thus will have a negligible impact on the overall capacity. Note that these considerations are not the core of the proposal but are implementation issues.

9.3.4.2 Bit-Interleaved Coding and Modulation (BICM) Scheme

Discussions on channel coding in standardization bodies are always sensible. Comparison of performance and related complexities between supporters of turbo codes and promoters of LDPC codes often hide intellectual property rights and expected future royalties. For instance, after many contributions in the course of DVB-NGH standardization, it was agreed that in a context of mobile broadcasting, turbo codes were strongly adapted due to their huge flexibility, better performance, and reduced complexity. Nevertheless, LDPC codes have been finally adopted due to their compatibility with the existing DVB-S2, DVB-T2, and DVB-C2 standards.

This section does not intend to propose a particular coding scheme for a future common broadcasting standard, but it focuses on the general requirements of such a system in terms of signal-to-noise ratio (SNR) and spectral efficiency ranges, and presents the different BICM parameters that can be exploited in its design.

As far as the first generation of DVB mobile broadcasting standard DVB-H is concerned, SNR values provided in the implementation guidelines range from 4 to 22 dB for fixed and portable stationary channels, and from 8 to 27 dB in mobile channels. These SNR level ranges are obtained with a concatenation of an inner 64-state convolutional code and an outer Reed–Solomon code at the physical level, and an upper level Multiprotocol Encapsulation—Forward Error Correction code (MPE-FEC).

These types of figures are not—yet—available in the DVB-T2-Lite and DVB-NGH implementation guidelines; however, with the set of coding rates adopted for NGH (i.e., 1/3–11/15), the range of SNR covered is –1 to 20 dB for a Gaussian channel and 0–22 dB for a Rayleigh fading channel. The corresponding range of spectral efficiencies varies from 0.7 to 5.8 bps/Hz. The adoption of coding rates down to 1/5 for the satellite path would also allow SNR values 2 dB lower.

Whatever the exact SNR range and spectral efficiency range targeted by a future common standard, a very flexible coded modulation scheme will be required. This goes through the adoption of several constellations with different robustness levels and of an FEC code offering high flexibility in terms of coding rates. In order to cover spectral efficiency values ranging from less than 1 bps/Hz to more than 5 bps/Hz, at least on the order of –4 to –64 constellations are required. As far as the FEC code is concerned, classical constructions of turbo codes using puncturing provide the best performance for coding rates in the range 1/5 to 5/6 or 6/7. For coding rates lower than 1/5, each component code is required to compute more than two parity-check bits simultaneously, introducing correlation between the redundancy bits and thus worsening the decoder performance while increasing its complexity. For very high coding rates, finding internal interleavers able to provide low error floors is a tricky exercise, and the addition of an outer BCH code for coding rate higher than 6/7 is then usually required when low error rates are sought. The constructions of LDPC codes adopted in the second generation of DVB standards also suffer the same drawbacks. Below coding rate 1/3, they perform less efficiently since their parity-check matrix becomes denser when the coding rate decreases. Moreover, although coding rates as high as 9/10 have been adopted in DVB-S2 and DVB-C2, an outer BCH code is required to maintain acceptable error floors. An advantage of the LDPC decoders is that the decoder architectures provide an excellent throughput/complexity trade-off for high coding rates.

For an FEC code flexible enough in terms of coding rates, a given spectral efficiency value can be obtained with several combinations of constellation orders and coding rates. The relative performance of these different combinations depends on the type of channel under consideration. For example, in the Gaussian channel, it is better to use a low-order constellation with a high coding rate, whereas in high diversity fading channels choosing high-order constellations with low coding rates usually yields better performance, but at the price of increased complexity. For mobile channels, the best combination choice highly depends on the diversity available in the channel. Consequently, the choice of the constellation/coding rate association greatly depends on the type of channel that has to be considered as a priority.

Another flexibility level is the FEC block size. On DVB side, the coded block size is fixed and is equal to 16,200 bits; the information block size thus depends on the coding rate. On 3GPP side, 188 information block sizes (before coding) ranging from 40 to 6144 bits are available, thus offering more flexibility with respect to block size. Whatever the code family, the FEC block size has a direct impact on the correction performance. From a sheer error correction performance point of view, the longer the block size, the closer the performance to the Shannon capacity. Besides, for short block sizes, less diversity is available at the FEC block level. However, diversity can still be exploited by an appropriate design of the channel interleaver guaranteeing the good spreading of data in time and frequency. Furthermore, in a context of mobility, the use of shorter FEC block sizes can help to reduce decoding latency and memory requirements in the receiver. Finally, for a given framing, the use of shorter coded blocks makes filling up of the available data carriers easier and allows padding to be avoided.

9.3.4.3 Resource Allocation and Interleaving Schemes

The time interleaving depth defined inside 3GPP is clearly smaller than the one defined in DVB standards due to latency constraints. Two main strategies detailed hereafter are proposed to circumvent this limitation on physical layer (note here that application-layer FEC is standardized as well on 3GPP side): inter-frame and intra-frame interleaving schemes based on enlarged transmission time interval (TTI) and retransmission-based schemes reusing the hybrid automatic repeat request (HARQ) concept. Note that both approaches would require additional soft-buffer resources at the terminal.

9.3.4.3.1 Inter-Frame and Intra-Frame Interleaving with Enlarged TTI

As a starting point to this approach, it is proposed to increase the TTI value (set to 1 ms in LTE) while keeping the LTE frame structure unchanged. TTI could be extended to 2 or 5 ms for instance, while adding an interleaving scheme at Multicast CHannel (MCH) level, i.e., an intra-frame interleaving applied on coded blocks, in order to increase frequency/time diversity (noted π in the following figures).

In current 3GPP system, an MCH is mapped onto a single TTI of 1 ms. This MCH can be a collection of multiple codewords, independently coded as represented in Figure 9.6a. Now increasing the TTI to 2 ms while adding interleaving, data allocation within the frame can be carried out as described in Figure 9.6b.

In addition, time slicing inherited from DVB-H can also be applied in order to decrease power consumption on receiver side as depicted in the left side of Figure 9.7. This principle can be extended up to 5 ms TTI in a 10 ms frame as exemplified in the right side of Figure 9.7.

Based on a 10 ms frame length, a trade-off has to be found between reachable throughput, time interleaving depth, and power saving:

- If TTI is increased up to 5 ms, only 2 MCHs per frame can be allocated and time interleaving depth can reach 10 ms; time slicing ratio is then 50%.
- If TTI is set to 2 ms, 5 MCHs can be mapped; time interleaving depth is "only" 6 ms; time slicing can be quite efficient (ratio 20%), but reachable throughput is quite low.

The previous figures only present the concept of resource allocation (including time interleaving) in a unified system, derived from both E-MBMS (MCH and resource allocation concepts) and DVB-NGH (physical layer pipe [PLP]-like concept and time interleaving). Represented frames do not include yet any synchronization or signaling symbols.

To reach even larger time interleaving depths in order to address terrestrial or satellite coverage, the concept of inter-frame interleaving inherited from DVB-NGH could also be applied. In that case, data of a single MCH would not be mapped onto a single frame but could rather be mapped onto multiple frames, with or without frame hopping as described in Figure 9.8.

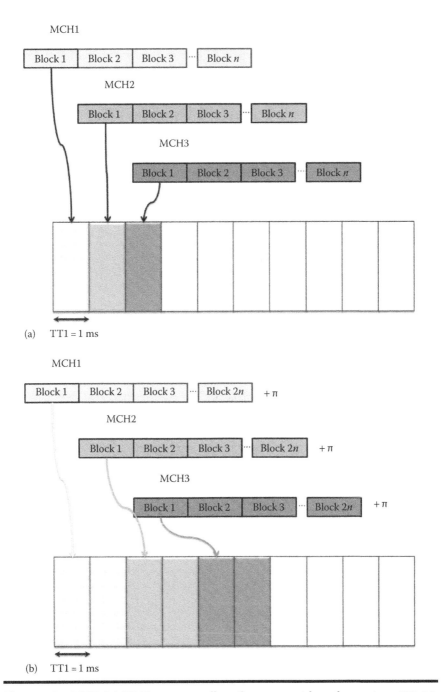

Figure 9.6 3GPP E-MBMS resource allocation concept based on a 1 ms TTI (a). Proposed resource allocation with a TTI enlarged to 2 ms and intra-frame interleaving stage (b).

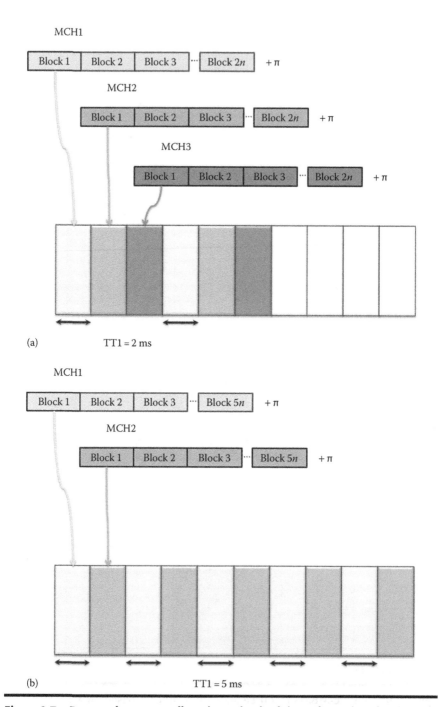

Figure 9.7 Proposed resource allocation using both intra-frame interleaving and time slicing. (a) Example for 2 ms TTI. (b) Extension to 5 ms TTI.

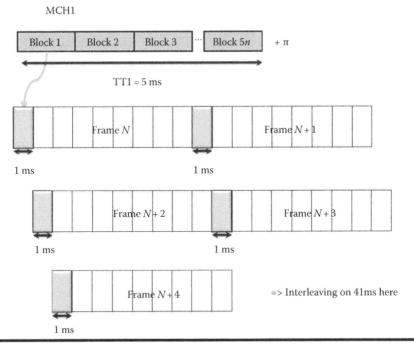

Figure 9.8 **Resource allocation based on a 5 ms TTI and using an inter-frame interleaving stage.**

9.3.4.3.2 Retransmission-Based Scheme

An alternative to TTI extension would be to reuse unicast HARQ mechanisms. Indeed, in LTE unicast transmission, time diversity is obtained by retransmissions. After initial transmission, the already-transmitted coded bits or new coded bits corresponding to the same information bits can be retransmitted in other subframes. The observations in the initial subframe and in retransmission subframes are combined in order to obtain improved decoding. When the same coded bits are transmitted, Chase combining is achieved. When new bits are transmitted, incremental redundancy is performed. Both types of retransmissions result in reduced code rate and increased diversity if the channel has sufficiently changed between two transmissions. In order to increase time diversity for a CBS based on E-MBMS while keeping the 1 ms TTI, automatic retransmissions could be used for MCH, since there is no ACK/NACK feedback available. For instance, in order to reach a code rate 1/3, three transmissions with code rate 1 could be used in three 1 ms subframes spread within a frame or located in different frames. For instance, in Figure 9.8, subframes carrying MCH1 data would represent five transmissions of the same information bits (e.g., via different coded bits) within a 41 ms duration. Each code block obtained after code block segmentation would be transmitted in each of these subframes. The delay between retransmissions would correspond to the interleaving depth. This method avoids the definition of an additional time interleaver, since the rate matching process already includes some interleavers. However, additional control signaling for MCH is probably needed in order to allow dynamic retransmissions (LTE redundancy versions used for unicast adaptive retransmissions could be used) or nonadaptive retransmissions (with a fixed number of retransmissions with a fixed delay). Furthermore, higher code rates beyond 1 should probably be defined, in order, for instance, to allow code rate 1/3 with five transmissions. In this case, initial transmission would have code rate 5/3, and the transport block size would be larger than the resource allocation, which is not currently supported in LTE.

9.3.4.4 Modulation

The modulation parameters, such as FFT size, subcarrier spacing (ΔF), and BW, are studied in this section in the perspective of a harmonized standard that shall be supported by both broadcasters and mobile operators. In particular, possible solutions are proposed to cover the missing BW cases for E-MBMS regarding DVB usual ones. Also subcarrier spacing reductions will be investigated to enable coverage extension taking into consideration Doppler effect limitations.

9.3.4.4.1 Bandwidth Harmonization

From Table 9.1, three native BW cases of the DVB-NGH standard, i.e., 6, 7, and 8 MHz, are not addressed* by current E-MBMS specifications. In what follows, we propose to deal with these three new BW cases using 3GPP principle that consists in addressing a given BW through the activation of the adequate number of subcarriers for a fixed ΔF and a given FFT size. Following 3GPP terminology, this translates into the allocation of a variable number of RBs per OFDM symbol.

The idea of the proposal is fairly simple and can be illustrated starting from the existing E-MBMS 10 MHz case with 1024 FFT size. To achieve a 10 MHz BW, a portion of subcarriers are modulated, numerically 600 subcarriers. From this basis, the aforementioned three missing BW cases can actually be covered by further reducing the number of modulated subcarriers. The principle is sketched in Figure 9.9.

Following this idea, we can calculate the number of subcarriers, or alternatively the number of RBs, being allocated according to the system BW. In Table 9.2, we layout some calculated parameters regarding the BW values under the 15 and 7.5 kHz legacy E-MBMS subcarrier spacing considerations.

The same approach can of course be followed starting from a 5 MHz BW with a 512-point FFT and then increasing the number of RBs. The choice between these

---

* 1.7 MHz is also a specific DVB BW but not addressed in this analysis.

| 10 MHz case | 8 MHz case |

Figure 9.9 **Way to address 8 MHz BW case starting from E-MBMS 10 MHz BW.**

Table 9.2 **E-MBMS Extended to UHF BWs Starting from the LTE 10 MHz BW**

BW (MHz)	ΔF (kHz)	FFT size	# of RBs	Effective BW (MHz)	Comments
10	**15/7.5**	**1024/2048**	**50**	**9**	**E-MBMS (Rel. 9)**
8	15/7.5	1024/2048	40	7.2	New BW
7	15/7.5	1024/2048	35	6.3	New BW
6	15/7.5	1024/2048	30	5.4	New BW

Table 9.3 **Complexity Efficiency Comparison of UHF BWs Using of 1024 or 512 FFTs**

FFT Size	BW (MHz)	Effective BW (MHz)	# of RBs	CE (%)
1024 (ΔF 15 kHz)	**10**	**9**	**50**	**58.58**
	8	7.2	40	46.88
	7	6.3	35	41.02
	6	5.4	30	35.16
512 (ΔF 15 kHz)	**5**	**4.5**	**25**	**93.75**
	6	5.4	30	82.03
	7	6.3	35	70.31
	8	7.2	40	58.59

two possibilities can actually be related to implementation strategies. To give more insight to this idea, it is interesting to introduce a so-called *complexity efficiency* (CE) indicator defined as the ratio between the modulated subcarriers and the FFT size. In Table 9.3, CE values obtained with FFT 1024 and 512 are compared for the different UHF BWs. For an 8 MHz BW, for instance, it turns out that the 512-point FFT is the most efficient solution.

Table 9.4 Comparison of "Mode 1 and 2" Allocation Solutions in Terms of BW Occupancy Ratio, for LTE and DVB Modes

Mode	BW (MHz)	ΔF (kHz)	FFT Size	# RBs	Effective BW (MHz)	BOR (%)
1 (LTE compatible)	8	7.5	2048	40	7.2	90
	7	7.5	2048	35	6.3	90
	6	7.5	2048	30	5.4	90
2 (DVB compatible)	8	7.5	2048	42	7.5	94.69
	7	7.5	2048	37	6.66	95.25
	6	7.5	2048	32[a]	5.76	96.13

[a] This mode should be used only for $\Delta F \leq 3.75$ kHz.

It is worth noting that the earlier proposal keeps approximately the same BW occupancy as in the legacy E-MBMS standard, in such a way that the ratio of effective BW over the total BW, denoted by BW occupancy ratio (BOR), is around 90%. This retains a fine regularity regarding 3GPP side, and this proposal could be considered as mode 1. On DVB side, an extended BW planning, i.e., higher BOR, is traditionally adopted hereby enabling increased system capacity. With this goal in mind, a second BW allocation mode can be considered with increased BW occupancy however compatible with DVB spectrum usage. The BOR figures for the two modes, namely, LTE compatible and DVB compatible, are reported in Table 9.4. As evident from the proposed BOR figures, the second mode provides about 5% higher occupancy. The related power spectral densities are reported in Figure 9.10 and compared with current DVB-T2 system to validate spectrum usage acceptability.

9.3.4.4.2 Coverage Area Improvement

One of the crucial issues for the enhanced E-MBMS standard is its short coverage area compared with DVB standards. A second step to the waveform parameter harmonization is to investigate to which extent the coverage area of the proposed CBS can be increased. Since the coverage size is directly proportional to the GI size, set to ¼ of the OFDM symbol duration in the current E-MBMS standard, one possible solution is to increase the OFDM symbol duration, which is equivalent to reduce the subcarrier spacing. Obviously, the weak point of doing this is that the resistance of Doppler spread is naturally affected. This issue has then to be analyzed as a trade-off between Doppler and echo resistance. To that end, Table 9.5 hereafter summarizes the Doppler and echo resistances of the current E-MBMS system for

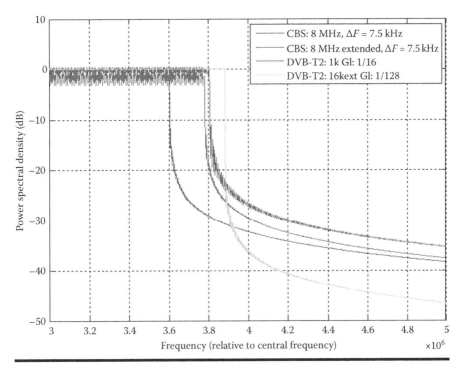

Figure 9.10 Comparison of spectral occupancy for CBS modes 1 and 2 with DVB-T2 system.

5 MHz BW case and under the practical assumption, which is widely used in wireless communications, that the supported Doppler value should not exceed 10% of the subcarrier spacing [14].

However, the maximum supported velocity depends on the carrier frequency. Assuming that the services are transmitted over UHF band (e.g., 600 MHz), these maximum Doppler values correspond to 2700 and 1350 km/h, respectively, which are way too far beyond a realistic situation and should give enough room to trade the Doppler resistance with the coverage area. Indeed, we introduce three new scenarios in Table 9.6.

Table 9.5 E-MBMS Doppler and Echoes Resistance for a 5 MHz BW

BW (MHz)	ΔF (kHz)	FFT Size	GI Duration (μs)	Coverage (km)	Max. Doppler (Hz)	# of OFDM Symbols per Subframe
5	15	512	16.67	5	1500	12
5	7.5	1024	33.33	10	750	6

Table 9.6   New CBS FFT/GI Proposed Parameters with Subcarrier Spacing Reduction

Scenario	BW (MHz)	ΔF (kHz)	FFT Size	GI Duration (μs)	Coverage (km)	Max. Doppler (Hz)	# of OFDM Symbols per Subframe
A	5	3.75	2048	66.67	20	375	3
B	5	2.5	3072	100	30	250	2
C	5	1.25	6144	200	60	125	1

The aforementioned three scenarios consist in dividing the reference subcarrier spacing (i.e., 7.5 kHz), by factors of 2, 3, and 6, respectively. Thus, the maximum coverage can be improved up to 60 km, leading in counterpart to a sacrifice of the Doppler resistance down to 125 Hz in the worst case. In UHF band however, this actually corresponds to a maximal supported velocity of 225 km/h, which is fairly acceptable. On the other hand, it has to be noticed that the related FFT sizes are not powers of 2 anymore but are still multiples of 2 and 3, which do not lead to complex implementations.

9.3.4.5 Sounding

So far, we have discoursed upon three possible scenarios where Doppler resistance has been traded for larger coverage area. In this section, the possible related pilot patterns (PPs) are introduced. It should be reminded that a good PP must meet several constraints: the channel estimation must support the Nyquist limit (SFN limit) and Doppler limit (see DVB-T2 implementation guidelines section 10.3.2.3.2), which can be numerically calculated as follows:

$$T_{Nyquist} = \frac{FFT}{F_s} \times \frac{X}{Y}. \tag{9.4}$$

$$D_{Nyquist} = \frac{F_S}{2 \cdot FFT \cdot (1+GI) \cdot X}, \tag{9.5}$$

where
   X denotes the spacing between two pilots in time domain
   Y denotes the spacing between two pilots in frequency domain

Table 9.7 Nyquist Limits Figures for E-MBMS Pilot Patterns

Scenario (kHz)	FFT Size	Fs (MHz)	X	Y	$D_{Nyquist}$ (Hz)	$T_{Nyquist}$ (µs)	GI (µs)	10% ΔF (Hz)
15	512	7.68	8	2	750	266.67	16.67	1500
7.5	1024	7.68	4	4	750	133.33	33.33	750

Complementary to this, DVB specifications adopt an additional margin defined as

$$T_{Nyquist} = \frac{4}{3} \times GI. \tag{9.6}$$

It is then interesting to confront the PPs as defined in the current E-MBMS standard with the Nyquist limits obtained from the previous equations. Table 9.7 gives the values computed for both subcarrier spacing 15 and 7.5 kHz. From these values, it appears that the 4/3 margin constraint is met by far for each scenario. However, one can note that the Doppler limit of the PP for 15 kHz is less than the conventionally used value based on 10% of subcarrier spacing. Nevertheless, the 750 Hz Doppler resistance is considered as being sufficient in practice.

In the sequel, the PP proposals suitable for the three new scenarios introduced in Table 9.7, namely, A, B, and C, are detailed. For each scenario, the proposed PPs are intuitively obtained following the same strategy as the one applied for the 7.5 and 15 kHz cases. Two different versions are however considered, denoted by v1 and v2, the latter corresponding to a denser PP that takes into account the 4/3 margin constraint. The PP proposals are depicted in Figure 9.11 for each scenario and each version and are shown only a single RB structure. The related Nyquist limits are listed in Table 9.8.

From these proposals, it turns out that the pilot overhead increases with the OFDM symbol length. In scenarios B and C for which the loss of spectral efficiency might become unacceptable, the overhead can be decreased by a factor 2–4 by considering a PP over several TTIs. With such a relaxed assumption, a more efficient PP can, for instance, be defined over two OFDM symbols in the 1.25 kHz configuration, as proposed in Figure 9.12: (in 1.25 kHz case, one TTI equals one OFDM symbol).

9.3.5 Common Broadcast Specification: Possible Follow-Up?

The study performed by the M[3] project showed the visibility of a physical layer unifying broadband and broadcast worlds by the way of a waveform providing a 100% broadcast downlink to a broadband wireless infrastructure.

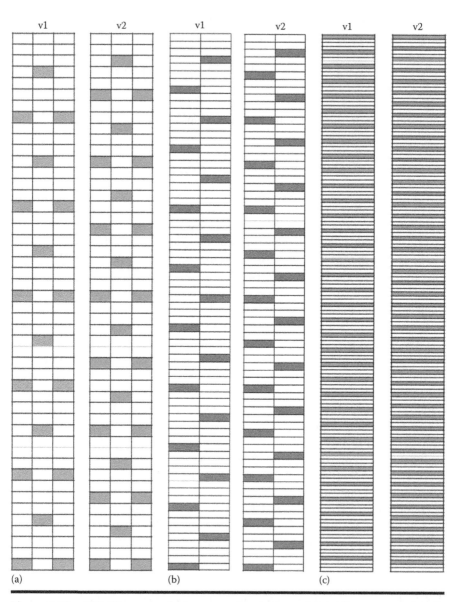

Figure 9.11 CBS PP proposals for scenarios A, B, and C of the proposed CBS. (a) $\Delta F = 3.75$ kHz, (b) $\Delta F = 2.5$ kHz, and (c) $\Delta F = 1.25$ kHz. For each scenario, version v2 is a denser pattern than version v1 to meet the 4/3 margin DVB constraint.

Table 9.8 Nyquist Limits Figures of the Pilot Patterns for the Three CBS Scenarios A, B, and C

Scenario	ΔF (kHz)	FFT Size	Fs (MHz)	X	Y	$D_{Nyquist}$ (Hz)	$T_{Nyquist}$ (μs)	GI (μs)	10% ΔF (Hz)	Feasibility
A (v1)	3.75	2048	7.68	2	8	750	66.67	66.67	375	OK, but critical for $T_{Nyquist}$
A (v2)	3.75	2048	7.68	2	6	750	88.33	66.67	375	OK
B (v1)	2.5	3072	7.68	2	8	500	100	100	250	OK, but critical for $T_{Nyquist}$
B (v2)	2.5	3072	7.68	2	6	500	133.33	100	250	OK
C (v1)	1.25	6144	7.68	1	4	500	200	200	125	OK, but critical for $T_{Nyquist}$
C (v2)	1.25	6144	7.68	1	3	500	266.67	200	125	OK

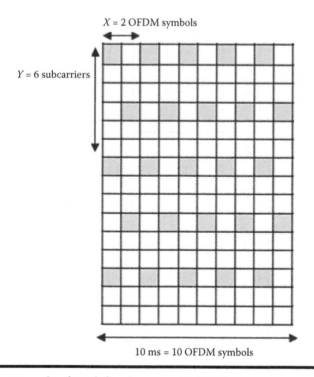

Figure 9.12 Example of PP defined over multiple TTI.

Mature technologies are on the shelf and the various trade-off to manage to meet the variety of frequency bands and network topologies and to optimize the covered areas are well-known and mastered.

The concepts delivered on the CBS are able to provide an efficient and elegant hybrid broadband/broadcast access infrastructure to the mobile multimedia data-hungry terminals. It is then the expectation of the M^3 project partners that they will be considered by broadcast and broadband standardization organizations.

9.4 Conclusions

The studies performed by the M^3 project have revealed the definitive benefits resulting from a one-to-many broadcast component added to the broadband wireless access facing a "data tsunami": the spectrum scarcity limits the expansion of BW that can be allocated to broadband wireless networks to a factor of 2–3, and this limitation will not allow to offset a data traffic growth reaching a magnitude of 30 or maybe more.

"Unfortunately," modern physical layers exhibit a spectral efficiency close to the Shannon limit; then in the state of current knowledge, neither revolutionary FEC nor ultraefficient waveform is expected to provide a viable solution to the "data tsunami" fear. M³ project partners have therefore sought to optimize the method of access to transmission resources by comparing the "one-to-one" and the "one-to-many" ones.

As far as network and terminal overheads and average cost per transmitted bit are considered, a clear advantage appears for the one-to-many method: only the "table of transmitted contents" (which allows service discovery) is required for the "one-to-many" terminal, while an extra signaling is required to manage accesses and to optimize wireless link of each terminal in a "one-to-one" broadband cell. The one-to-many delivery method maximizes the usage of the near-to-Shannon capacity for the multimedia content throughput.

The exploration of the business models relying on a hybrid wireless access, offering simultaneously a general purpose one-to-one broadband access and a spectrum-efficient one-to-many broadcast delivery, revealed the huge benefit such an NaaS behavior should provide in a B2B context: guaranteed QoS for a massive delivery of multimedia contents is noticeable for consumers and constitutes a competitive asset for the service providers, which will use the facility of a hybrid broadband/broadcast delivery proposed by network operators.

A key enabler for the deployment of multimedia services using the hybrid broadband/broadcast delivery is the availability of terminals implementing such a hybrid wireless access. Modern smart terminals already implement numerous wireless modems accessing IEEE (i.e., WiFi, BlueTooth) and/or 3GPP (i.e., 3G, LTE) networks, which led the M³ project partners to explore the possibilities to append a one-to-many "broadcast" component to the one-to-one "broadband" components of the IEEE and 3GPP standards. As LTE in its Release 10 implements an embryo of broadcast mode, the M³ project evaluated the possibility to extend its applicability to a 100% downlink component operated aside the broadband ones—this 100% broadcast downlink component being implemented on an independent "broadcast" infrastructure or sharing the broadcast infrastructure deployed to deliver traditional digital TV services.

Preliminary study of the CBS shows a very straightforward way to extend the set of E-MBMS parameters in order to make CBS suitable in various network topologies, including the usage of satellite coverage on top of the hot spot, the cellular and the broadcast ones.

Could a hybrid broadband/broadcast wireless access guaranteeing a high QoS to multimedia terminals on the move bring a differentiating advantage to mobile network operator?

The M³ project partners are definitively convinced!

Acknowledgments

As M³ (Mobile Multi-Media) project leader [6], I would like to thank all the project partners, namely, Alcatel Lucent Bell Labs France, Centre National d'Etudes

Spatiales, Parrot (formerly DiBcom), Institut National des Sciences Appliquées—Institut d'Electronique et de Télécommunications de Rennes, Mitsubishi Electric R&D Centre Europe, Institut Télécom, and TeamCast, for their valuable contributions to this chapter and each contributor one by one as well.

References

1. 3GPP TSG-SA S1-111210, Study item on converged mobile broadcast support for LTE, *WG1 Meeting #54*, Xi'an, China, May 2011.
2. CISCO white paper, Cisco visual networking index: Global mobile data traffic forecast update, 2011–2016, February 2012, http://www.cisco.com/en/US/solutions/collateral/ns341/ns525/ns537/ns705/ns827/white_paper_c11-520862.pdf
3. Ericsson, Traffic and market data report, November 2011, http://hugin.info/1061/R/1561267/483187.pdf
4. Byte Mobile, Mobile analytics report, February 2012.
5. Philip Hunter, Video-over-Wi-Fi taking off in Europe, Dec. 19, 2011, http://broadcastengineering.com/news/video-wifi-deutsche-telekom/
6. Christian Gallard, M^3 project website, Oct. 2010, https://m3.rd.francetelecom.com/espace-public
7. ETSI EN 302 755 v1.3.1, Digital video broadcasting (DVB); Frame structure channel coding and modulation for a second generation digital terrestrial television broadcasting system (DVB-T2), October 2011.
8. ETSI TS 102 831 v1.2.1, Digital video broadcasting (DVB); Implementation guidelines for a second generation digital terrestrial television broadcasting system (DVB-T2), June 2012.
9. 3GPP TS 36.104 v10.2.0, Base station radio transmission and reception (release 10), April 2011.
10. 3GPP TS 36.211 v9.1.0, Physical channels and modulation (release 9), March 2010.
11. 3GPP TS 36.212 v9.4.0, Multiplexing and channel coding (release 9), September 2011.
12. 3GPP TS 36.213 v9.0.1, Physical layer procedures (release 9), December 2009.
13. 3GPP TS 26.346 v10.2.0, MBMS protocols and codecs (release 10), November 2011.
14. European Celtic Wing TV Project, Services to wireless, integrated, nomadic, GPRS-UMTS & TV handheld terminals, http://projects.celtic-initiative.org/WING-TV/ (accessed, Oct. 2012).

Chapter 10

Overview of the HEVC Video Coding Standard

Benjamin Bross

Contents

This chapter provides an overview of the emerging high-efficiency video coding standard (HEVC). Its development was motivated by the need of increased coding efficiency compared to state-of-the-art hybrid video codecs like H.264/AVC. This increased coding efficiency is achieved by introducing new coding tools as well as by improving components already known from H.264/AVC. New tools introduced in HEVC are variable size block partitioning using quadtrees for the purpose of prediction and transformation and an additional in-loop filter, namely, sample adaptive offset (SAO). Improvements include more intra-prediction angles, advanced motion vector prediction (AMVP), a new block merging mode that enables neighboring blocks to share the same motion information, larger transform sizes, and a more efficient transform coefficient coding. HEVC incorporates only one entropy coder, which is basically CABAC from H.264/AVC. Since CABAC in H.264/AVC is more complex than the alternative CAVLC, a lot of efforts have been put into reducing complexity, memory requirements, and increasing the coding throughput thereof. While objective test results of the latest HEVC reference software show bit rate reductions up to 35% compared to H.264/AVC high profile (HP), preliminary subjective test results indicate that a 50% bit rate reduction for a comparable visual quality can be achieved.

10.1 Introduction

Around 10 years ago, the H.264/AVC video coding standard was developed to satisfy the need for higher coding efficiency, especially with regard to high-definition TV and transmission at channels having lower data rate like xDSL or UMTS. The state-of-the-art video coding standard these days, that is, the ITU-T Recommendation H.262, also known as ISO/IEC MPEG-2 Part 2, was approved in 1995 to extend the existing MPEG-1 standard by improving the efficiency for high bit rates and adding support for interlaced video. This resulted in MPEG-2 being used in TV broadcast such as DVB and storage on a DVD, for example. H.264/AVC successfully achieved an increase in the coding efficiency by 50% compared to MPEG-2 while being suitable for both low- and high-bit-rate coding as well as to accommodate the increasing diversification of transport layers and storage media.

Today, the resolutions of TVs, smartphones, and tablet devices are increasing, and this makes even small devices capable of showing high-definition video. At the same time, video application services like video on demand are becoming the majority of network traffic worldwide. These developments are asking for a more efficient coding of digital video. As a consequence, the premier video coding standardization organizations, namely, the ITU-T Video Coding Experts Group (VCEG) and the ISO/IEC Moving Pictures Expert Group (MPEG), have established a Joint Collaborative Team on Video Coding (JCT-VC) and have issued a joint call for proposals on video coding technology (CfP) [1]. One of the requirements is that the new technology should be capable of providing a bit rate reduction of approximately 50% at the same subjective quality compared to H.264/AVC [2].

The rest of the chapter is structured as follows. Section 10.2 describes the standardization process from the call for proposals to the second official milestone. Section 10.3 describes advances in coding tools, while Section 10.4 describes novelties in the high-level syntax including the abilities of HEVC with regard to parallel processing. Current objective performance results for HEVC, especially in comparison with H.264/AVC, are shown in Section 10.5. Finally, the chapter is concluded with Section 10.6.

An additional, detailed overview of HEVC and a performance as well as a complexity analysis of HEVC can be found in References 3–5.

10.2 Standardization Process

A video coding chain consists of an encoder that outputs the video as a bit stream, transmission of the bit stream, and a decoder that reconstructs the video from the bit stream. The output video can get displayed after an optional post-processing step. Only bit stream syntax/semantics and the decoding processes are in the scope of standardization. This allows for implementation specific optimization, for example, with regard to complexity, quality, and bit rate. Note that a video coding standard does not guarantee quality, but only interoperability.

In order to kick off the HEVC standardization process, all responses to the CfP have been evaluated in terms of both subjective and objective quality. During the first two meetings of the JCT-VC in April and July 2010, a test model under consideration (TMuC) including most promising tools from different responses to this call was created as a starting point. At the third meeting in October 2010, extensive tool experiments and evaluation of the results thereof led to the first HEVC test model reference software (HM1) and a first standard text specification, the working draft (WD1). HM software and WD text were further developed over the next four meetings, resulting in the HM5 software and WD5 text specification where the JCT-VC decided to not have context-based adaptive variable length coding (CAVLC) as a second entropy coder anymore and to use context-based adaptive binary arithmetic coding (CABAC) as the only entropy coding scheme. In February 2012, after the eighth meeting in San José, the first official milestone was achieved by issuing the sixth draft as committee draft (CD) for ISO/IEC MPEG two-month ballot period [6]. The second milestone, the draft international standard (DIS), was achieved at the end of July 2012 where DIS went out for five-month ISO/IEC MPEG ballot [7]. Finalization is planned for January 2013 with releasing the standard text as ISO/IEC final draft international standard (FDIS) and ITU-T consent.

10.3 Coding Tools

This section presents new coding tools introduced in HEVC. Like state-of-the-art video codecs, HEVC uses the hybrid video coding approach where the prediction of the video picture samples is followed by transform coding of the prediction

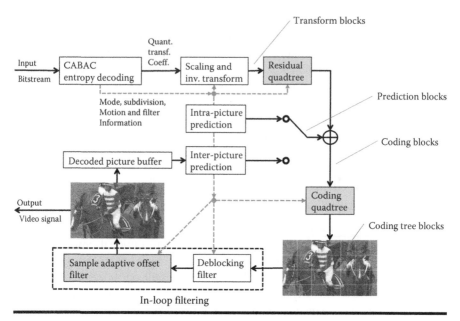

Figure 10.1   Block diagram of the HEVC MP decoder.

residual. Figure 10.1 shows a block diagram of the HEVC decoder including the basic processing blocks involved in decoding a video signal from an input bitstream. Completely new tools in video coding standardization are highlighted by gray-shaded blocks. For the purpose of prediction and transformation, the video picture is divided into blocks using the quadtree-based picture partitioning schemes described in Section 10.3.1. The new tools applied in intra-prediction using samples of the same picture are presented in Section 10.3.2, and the new tools applied in inter-prediction using samples of previously decoded pictures are presented in Section 10.3.3. The scaling of the quantized prediction residual transform coefficients and the new core transform itself are treated in Section 10.3.4. Section 10.3.5 is about changes in the CABAC entropy coding engine that is used to code prediction information and the quantized residual transform coefficients. The new in-loop filter operations that are applied prior to storing a decoded picture as a reference picture to be used for inter-prediction are described in Section 10.3.6.

10.3.1 Quadtree Picture Partitioning

In HEVC, a quadtree-based coding approach was introduced where a picture is divided into square coding tree blocks (CTBs) as shown in Figure 10.2. Each CTB is the root of a *coding tree*, which is used to further divide the CTB into

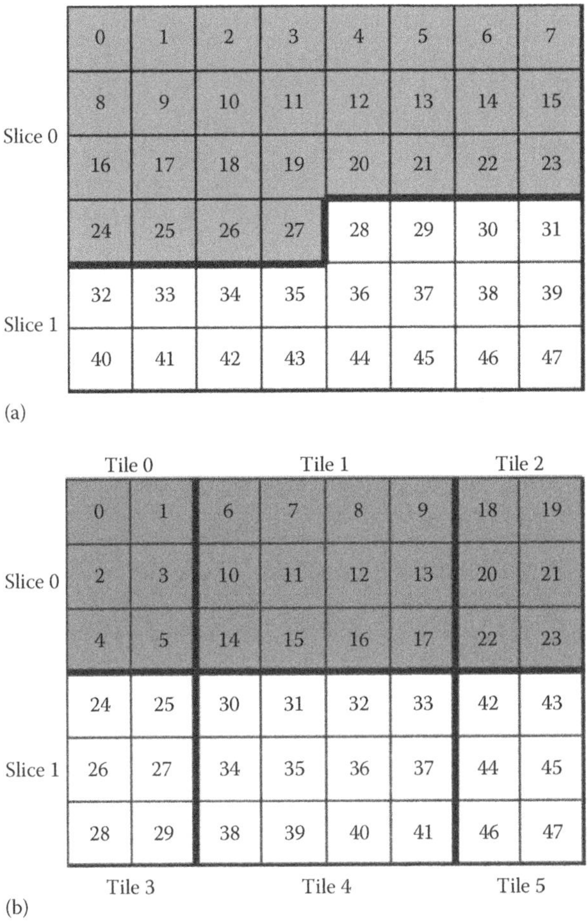

Figure 10.2 Subdivision of a picture into CTBs: (a) slices and (b) slices and tiles.

coding blocks (CBs). Their size can be adaptively chosen by using a quadtree-based partitioning with the leaves of the quadtree representing the CBs. Each CB is a root for a prediction and a transform tree. The prediction tree has only one level and describes how a CB can be further split into so-called prediction blocks (PBs), for each of which prediction parameters are specified. Figure 10.3 depicts all different ways allowed by current main profile (MP) to split a CB into inter-PBs. For transform coding of the prediction residual signal, each CB can also be split into smaller transform blocks (TBs) using another quadtree, namely, the *residual quadtree* (RQT). Figure 10.4a illustrates this quadtree partitioning, that is, a 64×64 luma CTB (solid bold line) being partitioned into CBs (white blocks), which, again, can be further divided into square TBs (gray blocks) of variable size.

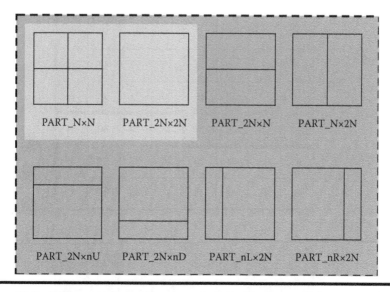

Figure 10.3 Different PB partition types in HEVC for intra- (light gray) and inter-prediction (dark gray). Note that PART_NxN is only allowed for CBs having minimum CB size for intra-prediction and for CBs having minimum CB size greater than 8 × 8 for inter-prediction.

The size of the CTBs is signaled for every video sequence and can range from 16×16 to 64×64 luma samples. Consequently, the maximum CB size is restricted by the CTB size. The minimum CB size is also signaled per sequence and shall range from 8×8 to the CTB size. Similarly, the minimum and maximum TB sizes are signaled as well as the maximum allowed RQT depth, whereas the maximum allowed TB size is 32×32 and the minimum allowed TB size is 4×4 luma samples. The HEVC default configuration uses a CTB size of 64×64, a minimum CB size of 8×8, and TB size ranging from 4×4 to 32×32. Figure 10.4a shows a possible partitioning using that configuration.

The TB sample arrays and associated syntax structures, for example, coded block flags or transform coefficient levels, are grouped together in a transform unit (TU). A prediction unit (PU) encapsulates everything that is related to prediction, that is, the PB sample arrays and associated syntax structures, for example, MVs or intra modes. The CB samples arrays, the associated syntax structures like the mode information whether intra- or inter-prediction is used, and the associated PUs and TUs are grouped together in a coding unit (CU). Consequently, the CTB sample arrays, associated coding tree syntax structures, and associated CUs are considered as a coding tree unit (CTU). Thus, it can be said that the CTU generalizes the concept of a macroblock as the basic processing unit in standardized video coding. A coding tree with a CTU as root (square), CU as leaves (triangle),

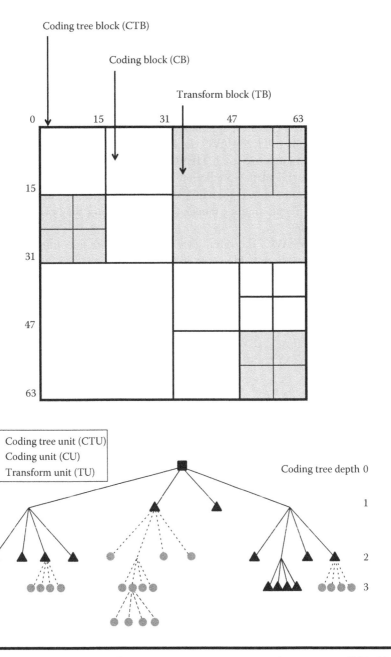

Figure 10.4 **HEVC coding and RQT structures. (a) A 64 × 64 luma CTB (solid bold) partitioned into CBs (white) and TBs (gray) of variable size. (b) Corresponding CTU (square) root with CU (triangle) leaves and nested RQT with CU root and transform unit (circle) leaves.**

and nested RQTs with TU leaves (circle) that correspond to the partitioning in Figure 10.4a is shown in Figure 10.4b.

10.3.2 Intra-Picture Prediction

The basic idea of predicting samples from already decoded neighboring samples of the same picture does not have significantly changed from the methods applied in H.264/AVC. While in H.264/AVC the DC, horizontal/vertical, directional, and plane prediction modes are used, HEVC specifies very similar DC, horizontal/vertical, angular, and planar prediction modes. The current HEVC MP defines 35 intra-prediction modes. Figure 10.5 shows all the modes from planar (0), over DC (1) to the different angular modes (2–34) including the horizontal (10) and vertical (26) mode.

Like in H.264/AVC, the intra-reference samples are also filtered (smoothed) prior to using as prediction. However, in H.264/AVC only 8 × 8 blocks are filtered; in HEVC, this smoothing is applied based on intra-mode-dependent thresholds.

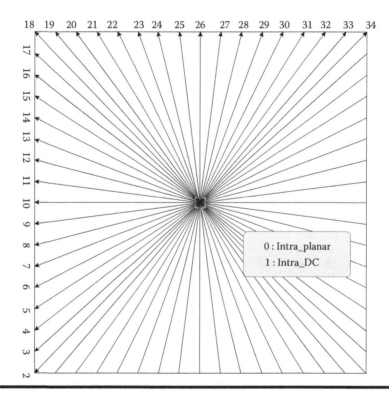

Figure 10.5 HEVC intra-prediction modes from 0 to 34. (From Bross, B. et al., High-efficiency video coding (HEVC) text specification draft 8, Document JCTVC-J1003 of JCT-VC, July 2012.)

For DC, horizontal, and vertical modes, the sample rows and/or columns of the current block are additionally filtered after being predicted to prevent "blockiness" at prediction boundaries.

As already mentioned in Section 10.3.1, more and larger block sizes are introduced in HEVC. Thus, the intra-prediction is performed on various and larger block sizes too. For every PB, an intra-mode is signaled while the prediction is applied to the TBs. Consequently, the maximum block size for intra-prediction is the maximum TB size, which is restricted to not exceed 32 × 32 luma samples. The larger blocks also allow an increased number of direction/angles when using angular prediction. One simplification over H.264/AVC is that the available prediction modes in HEVC do not depend on the PB size anymore.

When it comes to code the intra mode as part of PU syntax in HEVC, *most probable mode coding* can be used, which codes the mode as the difference from a predicted, most probable mode.

10.3.3 Inter-Picture Prediction

Inter-picture prediction in hybrid video coding uses already decoded blocks from reference pictures as a prediction for the current block. The way to describe the reference picture with a reference index and the displacement of the block in the reference picture with an MV is not changed in HEVC. Like in H.264/AVC, the MVs in HEVC represent displacements up to a quarter of a sample, and in B-slices, up to two predictions can be averaged (bi-prediction). The main changes are the way to code the motion information, that is, MVs and references indices, and the interpolation of the fractional sample values. A new MV prediction scheme and the new merge mode, used to code the motion information, and the interpolation filters used in HEVC are described in the following.

H.264/AVC only has one single motion vector predictor (MVP) to differentially code the MVs. It is computed as the median of three spatial neighboring MVs. HEVC improves the MV prediction by having a set of MVPs and signaling which one of these is actually used to predict the current MV. This technique, also known as MV prediction competition, was initially proposed in Reference 8 and further adapted to large block sizes with so-called *advanced motion vector prediction* (AMVP) in Reference 9. AMVP has two spatial MVPs and a temporal one (TMVP) competing for the prediction. They are selected among the positions shown in Figure 10.6: five spatial MVP candidate positions located on top of the current PB X (B0, B1, B2) and two on the left (A0, A1) as well as two TMVP candidate positions (T0, T1) in the collocated PB Y. When a candidate is not available, the zero MV is inserted in the candidate list. Considering the granularity of motion representation and that there are up to two vectors per PU for B slices, the memory size needed for storing motion data (including MVs, reference indices, and coding modes) could be significant. HEVC includes motion data storage reduction to reduce the size of the motion data buffer and the associated

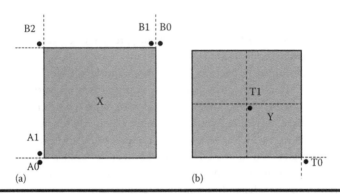

Figure 10.6 Positions of the MVPs. (a) Spatial MVP positions relative to current PB X. (b) TMVP positions in Y, which is the collocated block of X. (From Helle, P. et al., *IEEE Trans. Circ. Syst. Video Technol.*, 22(12), © 2012 IEEE.)

memory access bandwidth by storing all motion information of the collocated picture on a 16×16 block basis.

The *direct* and *skip* modes of H.264/AVC were introduced to avoid signaling MVs by deriving them from already decoded MVs. HEVC has a similar mode called *inter-PB merging*. The purpose of block merging is to compensate for the disadvantages of an initial quadtree-based subdivision by reducing the redundant sets of transmitted coding parameters. This is achieved by creating regions composed of neighboring PBs with associated PUs sharing identical motion information, which is signaled only once for each region. Thus, each region must contain at least one PB of a PU that is not merged, seeding new motion information. Figure 10.7 illustrates such a merge region where PB S corresponds to such a particular seeding block and PB X corresponds to the current PB to be coded. All the PBs along the block scanning pattern coded before X represent the set of causal PBs. Consequently, X can be merged only with one of the causal PBs. When X is merged with one of these blocks, the PU of X copies and uses the same motion information as the PU of this particular block. Out of this set of causal PBs, only a subset with neighboring PBs is used to build a list of merge candidates. A merge candidate in this list includes the motion information to be copied for X. In order to identify the candidate in the list, an index, called merge index, is signaled for X. It should be noticed that in current HEVC, the merge candidate seeding the prediction data is not necessarily the PU of a spatial neighboring PB like S in Figure 10.7. It can also be a candidate with prediction data from a temporally collocated PU or an additional candidate with prediction data derived otherwise. As a consequence, the merged region represented by the bold line in Figure 10.7 possibly extends to the temporal dimension and can be seen as a 3D region with similar motion data. Furthermore, block merging in HEVC features a mode that signals at the CU level that the prediction residual is

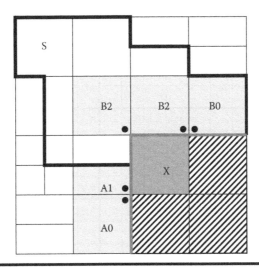

Figure 10.7 **Illustration of inter-PB merging in HEVC. Multiple PBs are merged into a region (bold line). PB S is not merged and codes the prediction data for that region. PB X is the current PB to be coded. The PBs in the striped area do not have associated prediction data yet as these PBs are successors to X in block scanning order. The dots indicate the sample positions directly adjacent to X, defining possible spatial merge candidates. (From Helle, P. et al.,** *IEEE Trans. Circ. Syst. Video Technol.,* **22(12), © 2012 IEEE.)**

zero and that the CU contains a single inter-predicted PU (PART_2N × 2N partition in Figure 10.3), thus being conceptually very similar to the H.264/AVC skip mode. Prior to any other information associated with a particular CB, this mode is signaled by a flag, called skip flag. Instead of the median-based approach in H.264/AVC, the motion information of the PU associated with a skip CU is inferred by the exact same procedure as used for block merging.

Although HEVC also performs motion compensation with an accuracy up to a quarter of a luma sample, the interpolation filter used to generate the fractional sample values is different. For luma fractional sample values, an eight-tap FIR filter is applied instead of a six-tap FIR filter for half-sample values and averaging for quarter-sample values as in H.264/AVC. Figure 10.8a shows a quarter-sample grid where the gray integer-sample positions are denoted with the uppercase letter $A_{i,j}$ and the quarter-sample positions are denoted with lowercase letters $a_{i,j}, ..., r_{i,j}$. The eight integer samples in a row, $(A_{-3,0}, ..., A_{4,0})^T$, are filtered with the coefficients $\mathbf{F}_Y$ in Equation 10.1 to calculate the fractional samples $(a_{0,0}, b_{0,0}, c_{0,0})^T$, while the eight integer samples in a column, $(A_{0,-3}, ..., A_{0,4})^T$, are multiplied with $\mathbf{F}_Y$ to calculate the fractional samples $(d_{0,0}, h_{0,0}, n_{0,0})^T$. The same coefficients $\mathbf{F}_Y$ are also used to calculate $(e_{0,0}, i_{0,0}, p_{0,0})^T$, $(f_{0,0}, j_{0,0}, q_{0,0})^T$, and $(g_{0,0}, k_{0,0}, r_{0,0})^T$ by performing the same filtering

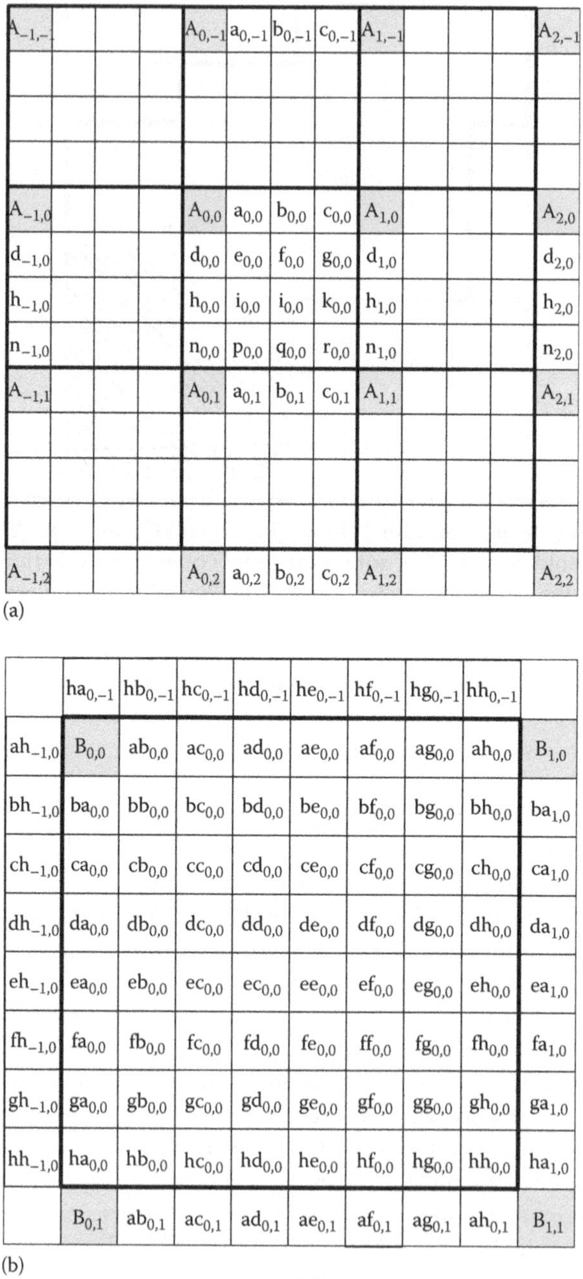

$A_{-1,-1}$				$A_{0,-1}$	$a_{0,-1}$	$b_{0,-1}$	$c_{0,-1}$	$A_{1,-1}$				$A_{2,-1}$
$A_{-1,0}$				$A_{0,0}$	$a_{0,0}$	$b_{0,0}$	$c_{0,0}$	$A_{1,0}$				$A_{2,0}$
$d_{-1,0}$				$d_{0,0}$	$e_{0,0}$	$f_{0,0}$	$g_{0,0}$	$d_{1,0}$				$d_{2,0}$
$h_{-1,0}$				$h_{0,0}$	$i_{0,0}$	$i_{0,0}$	$k_{0,0}$	$h_{1,0}$				$h_{2,0}$
$n_{-1,0}$				$n_{0,0}$	$p_{0,0}$	$q_{0,0}$	$r_{0,0}$	$n_{1,0}$				$n_{2,0}$
$A_{-1,1}$				$A_{0,1}$	$a_{0,1}$	$b_{0,1}$	$c_{0,1}$	$A_{1,1}$				$A_{2,1}$
$A_{-1,2}$				$A_{0,2}$	$a_{0,2}$	$b_{0,2}$	$c_{0,2}$	$A_{1,2}$				$A_{2,2}$

(a)

	$ha_{0,-1}$	$hb_{0,-1}$	$hc_{0,-1}$	$hd_{0,-1}$	$he_{0,-1}$	$hf_{0,-1}$	$hg_{0,-1}$	$hh_{0,-1}$	
$ah_{-1,0}$	$B_{0,0}$	$ab_{0,0}$	$ac_{0,0}$	$ad_{0,0}$	$ae_{0,0}$	$af_{0,0}$	$ag_{0,0}$	$ah_{0,0}$	$B_{1,0}$
$bh_{-1,0}$	$ba_{0,0}$	$bb_{0,0}$	$bc_{0,0}$	$bd_{0,0}$	$be_{0,0}$	$bf_{0,0}$	$bg_{0,0}$	$bh_{0,0}$	$ba_{1,0}$
$ch_{-1,0}$	$ca_{0,0}$	$cb_{0,0}$	$cc_{0,0}$	$cd_{0,0}$	$ce_{0,0}$	$cf_{0,0}$	$cg_{0,0}$	$ch_{0,0}$	$ca_{1,0}$
$dh_{-1,0}$	$da_{0,0}$	$db_{0,0}$	$dc_{0,0}$	$dd_{0,0}$	$de_{0,0}$	$df_{0,0}$	$dg_{0,0}$	$dh_{0,0}$	$da_{1,0}$
$eh_{-1,0}$	$ea_{0,0}$	$eb_{0,0}$	$ec_{0,0}$	$ec_{0,0}$	$ee_{0,0}$	$ef_{0,0}$	$eg_{0,0}$	$eh_{0,0}$	$ea_{1,0}$
$fh_{-1,0}$	$fa_{0,0}$	$fb_{0,0}$	$fc_{0,0}$	$fd_{0,0}$	$fe_{0,0}$	$ff_{0,0}$	$fg_{0,0}$	$fh_{0,0}$	$fa_{1,0}$
$gh_{-1,0}$	$ga_{0,0}$	$gb_{0,0}$	$gc_{0,0}$	$gd_{0,0}$	$ge_{0,0}$	$gf_{0,0}$	$gg_{0,0}$	$gh_{0,0}$	$ga_{1,0}$
$hh_{-1,0}$	$ha_{0,0}$	$hb_{0,0}$	$hc_{0,0}$	$hd_{0,0}$	$he_{0,0}$	$hf_{0,0}$	$hg_{0,0}$	$hh_{0,0}$	$ha_{1,0}$
	$B_{0,1}$	$ab_{0,1}$	$ac_{0,1}$	$ad_{0,1}$	$ae_{0,1}$	$af_{0,1}$	$ag_{0,1}$	$ah_{0,1}$	$B_{1,1}$

(b)

Figure 10.8 Fractional positions for quarter-sample interpolation: (a) showing luma quarter-sample grid and (b) showing chroma eighth-sample grid (4:2:0). (From Bross, B. et al., High-efficiency video coding (HEVC) text specification draft 8, Document JCTVC-J1003 of JCT-VC, July 2012.)

with previously obtained fractional samples $(a_{0,-3}, \ldots, a_{0,4})^T$, $(b_{0,-3}, \ldots, b_{0,4})^T$, and $(c_{0,-3}, \ldots, c_{0,4})^T$:

$$
\mathbf{F}_Y = \begin{pmatrix} -1 & 4 & -10 & 58 & 17 & -5 & 1 & 0 \\ -1 & 4 & -11 & 40 & 40 & -11 & 4 & -1 \\ 0 & 1 & -5 & 17 & 58 & -10 & 4 & -1 \end{pmatrix}
\tag{10.1}
$$

The chroma interpolation in HEVC is also more complex. A four-tap FIR filter is used to calculate the fractional sample values instead of the bilinear interpolation used in H.264/AVC. As in H.264/AVC, given 4:2:0 chroma subsampling, quarter-sample accurate luma MVs correspond to an eight-sample chroma accuracy. Figure 10.8b illustrates these fractional positions between integer chroma samples, denoted by two lowercase letters, where the first letter represents the rows a–h and the second letter represents the columns a–h. The coefficients $\mathbf{F}_C$ of the four-tap chroma FIR filter are given in Equation 10.2. Like in luma interpolation, the fractional samples $(ab_{0,0}, \ldots, ah_{0,0})^T$, with one horizontal integer component, and $(ba_{0,0}, \ldots, ha_{0,0})^T$, with one vertical integer component, are derived first by multiplying $\mathbf{F}_C$ with the four integer samples $(B_{-1,0}, \ldots, B_{2,0})^T$ in the same row and the integer samples $(B_{0,-1}, \ldots, B_{0,2})^T$ in the same column, respectively. Then, these fractional values are again used to calculate the other fractional sample values in the same row; for example, $(bb_{0,0}, \ldots, bh_{0,0})^T$ are obtained by multiplying $\mathbf{F}_C$ with $(ba_{-1,0}, \ldots, ba_{2,0})^T$:

$$
\mathbf{F}_C = \begin{pmatrix} -2 & 58 & 10 & -2 \\ -4 & 54 & 16 & -2 \\ -6 & 46 & 28 & -4 \\ -4 & 36 & 36 & -4 \\ -4 & 28 & 46 & -6 \\ -2 & 16 & 54 & -4 \\ -2 & 10 & 58 & -2 \end{pmatrix}
\tag{10.2}
$$

10.3.4 Transform, Scaling, and Quantization

The current design of the *core transform*, as introduced in Reference 10, is based on matrix multiplications. For the inverse transform as defined in HEVC, the columns of a TB are processed before the rows. Let $\mathbf{C}$ be an $N \times N$ block of transform coefficients and $\mathbf{T}_{ver}$, $\mathbf{T}_{hor}$ be the $N \times N$ transform matrices. Then, the columns of $\mathbf{C}$ are transformed as in Equation 10.3, and afterward, the columns of the output $\mathbf{Y}$ are transformed as in Equation 10.4, resulting in an $N \times N$ residual TB $\mathbf{R}$. It should be noted that the coefficients have 16-bit precision while the transform matrix values can be represented using 8-bit. Consequently, after multiplication, intermediate rounding and clipping to 16 bits is needed whereas the final rounding and clipping depend on the bit depth.

$$\mathbf{Y} = \mathbf{C}^T \cdot \mathbf{T}_{ver} \tag{10.3}$$

$$\mathbf{R} = \mathbf{Y}^T \cdot \mathbf{T}_{hor} \tag{10.4}$$

In general, both transform matrices are identical and an approximation of the DCT. Since HEVC allows TB sizes from 4×4 to 32×32, a more flexible transform matrix design than the one for the 4×4 and 8×8 transforms in H.264/AVC is needed. Note that the key property of the H.264/AVC integer transforms, to be reversible, is kept. The DCT-based HEVC core transform matrices $\mathbf{T}_{DCT,N \times N}$ can all be represented by the same 32×32 matrix, that is, $\mathbf{T}_{DCT,32 \times 32}$. In order to get the smaller matrices, columns i and rows j ($1 \ll (5 - Log\,2(N))$) are taken from $\mathbf{T}_{DCT,32 \times 32}$, where $i, j = 0, \ldots, N - 1$. As an example, $\mathbf{T}_{DCT,4 \times 4}$, which consists of columns $0, \ldots, 3$ and rows $0, 6, 12, 18$ of $\mathbf{T}_{DCT,32 \times 32}$, is given in Equation 10.5. The complete 32×32 matrix can be found in Reference 7.

$$\mathbf{T}_{DCT,4 \times 4} = \begin{pmatrix} 64 & 64 & 64 & 64 \\ 83 & 36 & -36 & -83 \\ 64 & -64 & -64 & 64 \\ 36 & -83 & 83 & -36 \end{pmatrix} \tag{10.5}$$

For intra-coded 4×4 luma TBs, an alternative approximation of a four-point DST as in Equation 10.6 is defined. Consequently, both horizontal and vertical matrices are DST based:

$$\mathbf{T}_{hor} = \mathbf{T}_{ver} = \mathbf{T}_{DST}$$

$$\mathbf{T}_{DST} = \begin{pmatrix} 29 & 55 & 74 & 84 \\ 74 & 74 & 0 & -74 \\ 84 & -29 & -74 & 55 \\ 55 & -84 & 74 & -29 \end{pmatrix} \tag{10.6}$$

10.3.5 Entropy Coding

In general, entropy coding is the last stage in video encoding or the first stage in video decoding. It converts all syntax elements to a sequence of bits and vice versa. The basic entropy coding engine in HEVC is CABAC from H.264/AVC [11] with some modifications. CABAC performs entropy coding in three steps: *binarization, context modeling*, and *binary arithmetic coding*. Binarization maps a syntax element, for example, a transform coefficient level, to a string of binary-valued symbols, that is, the bin string. For bins of this bin string, context modeling can derive a context, which relates to a probability used in the binary arithmetic coding engine when coding the respective bin. While the arithmetic coding engine, that is, the modulo or M-coder, is not changed in HEVC, modifications made in binarization and context modeling are described in the following.

There are two ways to code one bin from a bin string: the so-called *regular* mode and the *bypass* mode. The regular mode incorporates context modeling whereas the bypass mode, as the name implies, bypasses this stage and codes the bin with 1 bit. Using an adequate context modeling results in one bin being represented by less than one bit, that is, a bin-to-bit ratio greater than 1. Bypass coding operates with a bin-to-bit ratio equal to 1. Although being very efficient, context modeling adds a lot of complexity and may add dependencies on previously coded bins. For syntax elements that might become quite large, resulting in long bin strings, the first n bins are considered as prefix and the remaining bins represent the suffix. The prefix bins are coded in regular mode and the suffix uses bypass mode coding. This division assures that, how large one syntax element value might be, only a maximum of n bins are coded using the more complex regular mode. In H.264/AVC, as well as in HEVC, the MV differences and the transform coefficient levels are binarized that way. In order to further reduce the number of regular coded bins, *Golomb–Rice codes* have been included in the binarization suffix of the transform coefficient levels [12]. This allows to reduce the number n of prefix bins from 15 to 3 for the transform coefficient levels without sacrificing coding efficiency.

Transform coding, or more precise *transform coefficient level coding*, accounts for most of the coded data. This portion increases significantly when coding at high bit rates. Consequently, improving the coding efficiency of the transform coding directly leads to an improved overall coding efficiency. As HEVC introduces TB sizes greater than 8×8, it seems reasonable to adapt the H.264/AVC transform coding, which was designed for 4×4 and 8×8 blocks. A straightforward extension of the H.264/AVC techniques has been shown to be suboptimal for large TB sizes [13]. Consequently, the techniques proposed in Reference 13 have been integrated in HEVC and further improved w.r.t. coding efficiency, parsing throughput, and parallel context derivation. In general, the transform coefficient levels of a TB are parsed as follows. The horizontal (x) and vertical (y) offset of the last nonzero, that is, significant, coefficient in the TB in scan order is coded first. This replaces the interleaved coding of the first coefficient level bin, that is, the significant coefficient flag, and a flag signaling whether the current coefficient level is the last one. Removing this dependency solves the problem of deriving the contexts of both flags in parallel.

To actually parse the coefficient levels, the TB is divided into 4×4 *subblocks*. The scan order, in which the subblocks are processed, is an upright *diagonal scan*

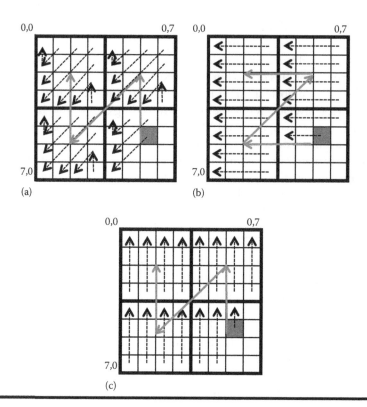

Figure 10.9 Different transform coefficient scan pattern for an 8×8 TB: (a) subblock upright diagonal scan, (b) horizontal scan for luma TBs with intra-mode 22–30, and (c) vertical scan for luma TBs with intra-mode 6–14.

going from the subblock containing the last significant position to first top left subblock. In the same scan order, the 16-coefficient levels in a 4×4 subblock are parsed. Figure 10.9a shows an example of how coefficients in an 8×8 TB are parsed, or scanned, using the upright diagonal scan. Note that the scanning starts at the last significant position, which is represented by the gray position (6,5) in Figure 10.9. Another functionality that was incorporated on top of the subblock processing is a so-called coefficient group flag [14]. This flag, which is coded for the subblocks between the first and the last one, signals whether all coefficient levels are equal to zero in the respective subblock. As a special case, for some 4×4 and 8×8 intra-predicted TBs, a *vertical or horizontal scan* pattern is used as illustrated in Figure 10.9b and c.

At high bit rates, the context modeling for the first three bins might still represent a parsing bottleneck. Therefore, scheme II from Reference 14 was incorporated in HEVC to even further reduce the number of regular coded bins. Whether this is done or not is decided adaptively based on the previously coded coefficient levels in a subblock. The third bin of the first coefficient level greater than one is regular coded whereas the third bin for all following coefficients greater than one is bypass coded. After eight nonzero coefficients, the second bin is also coded in bypass mode. When we assume the worst case, that is, all 16 coefficients in a subblock are greater than 1, the first coefficient has three bins regular coded, the following seven have two bins regular coded, and the remaining eight just have the first bin regular coded. This reduces the average regular coded bins for a subblock from 3 to about 1.6.

Besides throughput and parallel decoding issues, the memory required to store and derive contexts has been addressed in HEVC as well. On one hand, the two 8-bit values for the initialization variables m and n in H.264/AVC CABAC have been replaced by two 4-bit values resulting in *8-bit context initialization values*. On the other hand, a lot of efforts have been put into reducing the number of contexts. Furthermore, context derivation, which uses previously coded data from the neighboring block above or to the left, was avoided to reduce the number of line buffers used to store the neighboring data. Only the contexts for flags signaling a coding tree split and the skip mode are still derived by evaluating the syntax element to the left and above.

10.3.6 In-Loop Filtering

The block-based hybrid video coding approach might introduce large differences between two neighboring sample values at a block boundary. These artifacts lead to a certain "blockiness" that can be perceived in the picture. In order to reduce this blockiness, an edge filtering process is applied to the reconstructed picture in H.264/AVC. It is designed in a way that the block artifacts are removed or softened while sharp image content is preserved. The filtered pictures can be fed again into the coding loop, that is, as reference pictures. Therefore, in general, filters applied at this place are called *in-loop filters*.

In principle, the *deblocking filter* from H.264/AVC is kept in HEVC [15]. While the H.264/AVC deblocking filter is applied on edges of a minimum 4×4 luma sample block raster, HEVC performs the deblocking filter operation based on minimum 8×8 luma sample blocks. Another change was introduced in order to simplify parallel deblocking. Instead of filtering vertical and horizontal edges block by block, the vertical edges of the whole picture are filtered prior to the horizontal one.

After deblocking, an additional SAO can add specific offset values to the picture samples [16]. The offset values are restricted to be represented by 4 bits less than the current bit depth. Given 8- bits, the offset can range from −8 to 7. SAO is applied and controlled on a CTB basis. If applied, individual parameters may be specified for each CTB. One of these parameters is the offset type: edge offset or band offset. Edge offset is controlled by classifying a sample based on the differences between this sample and two of its neighboring values. The band offset performs a classification based on the sample value itself by dividing the range of values into four classes. For each class, an offset value is signaled, which is added to the respective sample.

10.4 High-Level Syntax

The basic concept of representing a video codec by a network abstraction layer (NAL) and a video coding layer (VCL) as introduced in H.264/AVC was kept and extended in HEVC. An overview of the NAL/VCL structure can be found in Reference 17. One of the NAL features is the usage of parameter sets to include parameters used for coding video sequences in a sequence parameter set (SPS) and pictures therein in a picture parameter set (PPS). The VCL incorporates a coded representation of the video content in a slice. A slice consists of a slice header and slice data, and the partitioning of a picture into slices is described in Section 10.4.3. Among other data, the parameter sets and the slice header, which are formatting the coded video data, are considered as *high-level syntax*.

Section 10.4.1 shortly describes how the parameters sets known from H.264/AVC are used in HEVC. The new high-level syntax used to signal how reference pictures are stored and how the reference picture list is built is presented in Section 10.4.2. An overview of high-level syntax, which allows for parallel processing beyond slices, is given in Section 10.4.3 with focus on a partitioning called tiles.

10.4.1 Parameter Sets

The H.264/AVC parameter sets were introduced to decouple parameters, which are expected to rarely change, from the coded picture sample values. This decoupling allows to transmit the parameters either "in-band," that is, within the same channel as the coded sample values, or "out-of-band." When coding the parameters "out-of-band," more reliable, error-resilient transmission channels can be used, for example, for videoconferencing when the call is initiated. As in H.264/AVC, the SPS includes, among other parameters, profile/level information, the bit depth of the sample values, the picture size, cropping, and video usability information. In HEVC, syntax describing the new quadtree-based coding structures, presented in Section 10.3.1, is added to the SPS. Syntax for tiles and the other tools enabling high-level parallelism, presented in Section 10.4.3, is present in the PPS. Quantization syntax is still signaled in SPS and PPS while the information whether weighted prediction is applied remains in PPS.

10.4.2 Reference Picture Buffering and List Construction

For inter-prediction, which uses previously decoded pictures as a reference for prediction, these previously decoded pictures are stored in the decoded picture buffer

(DPB). To indicate which picture is used to predict the block of samples, an index to a list of reference pictures is signaled for every inter-predicted block. Thus, the reference picture list maps the reference index to a picture in the DPB. In H.264/AVC, the reference picture buffering in the DPB is controlled by so-called memory management control operation (MMCO) syntax, and the reference picture list construction is described by means of reference picture list modification syntax. While all pictures in a video sequence are addressed using a unique, increasing, picture order count (POC), pictures that are used as a reference are assigned another identifier, the frame number, which increases from reference picture to reference picture.

In HEVC, the buffer management is simplified by describing the buffer explicitly with *reference picture sets* (RPSs) instead of complicated MMCO syntax with several exceptions and rules [18]. An RPS consists of deltas to the current POC, for example, when current picture with POC 7 uses pictures with POC 6 and 8 as reference, the delta POCs -1 and 1 are included in the RPS. Several RPS can be signaled in the SPS and either indexed or overwritten by new RPS in the slice header. A prediction scheme, which predicts current RPS delta POCs from previous RPSs, reduces the signalization overhead of the RPSs. The whole reference picture handling for a slice can be described by the following steps. In the slice header, parse the RPS and mark all pictures in the DPB that are not in the RPS as "unused." Based on the marking, the reference picture list is initialized. After decoding the slice data, when the current picture is further used as a reference, it is inserted in the DPB and marked as "used." This explicit reference picture list modification gets rid of the frame numbers, gaps in frame numbers, and nonexisting picture exceptions in H.264/AVC. Furthermore, the explicit marking of RPS pictures in the DPB allows to detect missing reference pictures before decoding the slice data. This facilitates error concealment when a picture is lost.

10.4.3 Parallelism

In H.264/AVC, slices allow parallel coding of pictures. In times of multicore hardware architectures, several new tools have been introduced to HEVC to improve parallel processing, namely, tiles, dependent slices, and wavefront parallel processing [19].

In HEVC, a picture is still divided into slices, but instead of macroblocks, each slice contains an integer number of CTBs. CTBs are further described in Section 10.3.1. As an example, Figure 10.2a shows how a picture can be divided into CTBs included in two slices. The numbering of the CTBs corresponds to the raster scan processing order of these. In addition to slices, *tiles* allow a picture to be divided into columns and rows. A possible division into six tiles, residing in three columns and two rows, is illustrated in Figure 10.2b. It can be seen that the CTBs in a tile are also processed in raster scan order. Note that, when more than one column is used, CTBs are not necessarily processed in raster scan order in a picture or slice.

The combination of slices and tiles is restricted by the condition that either all slices within a tile shall be complete or all tiles within a slice shall be complete. Tiles are independent, and they can be processed in parallel while not carrying the signalization overhead of the slice header.

Dependent slices are basically slices without the parameter overhead signaled in the slice header and without resetting the CABAC engine at the beginning of each slice.

Wavefront parallel processing is a mechanism that allows to store the states of the CABAC engine after the second CTU of each CTU row has been decoded. These states can be used to initialize the CABAC engine when starting the decoding of the first CTU in the next CTU row. Thus, with a delay of two CTUs, the decoding of a subsequent CTU row can be started in parallel without sacrificing coding efficiency compared to not inheriting the CABAC states.

10.5 Performance

A first study of objective HEVC coding efficiency performance with regard to H.264/AVC was conducted in Reference 20. The software version used to be compared with H.264/AVC was the latest TMuC, TMuC 0.9 [21]. For the most recent versions HM4, HM5, and HM6 of the HEVC test model reference software, [22–24] provide objective performance analyses.

Three test conditions, namely, *Intra*, *Random Access*, and *Low Delay B*, are chosen to represent common use cases of the codec. Intra, as the name implies, is restricted to code every picture using intra-prediction. This targets applications that require the decoding of the pictures to be independent from previously coded pictures, for example, video editing. The Random Access coding structure allows more advanced coding structures to further increase the coding efficiency. For example, averaging predictions from previous and following pictures lead to a higher coding efficiency. To guarantee causality, the subsequent pictures used as a reference in prediction have to be coded before the picture that references these. Therefore, the coding order does not equal the display order anymore, and the encoder delay is increased. One typical application for Random Access would be storage of video, for example, on an optical disk or on a video on demand server, where all the material is already coded and compression efficiency is more important than encoding time or delay. In order to be able to start decoding the video somewhere between the beginning and the end, for example, forward/rewind operations in a video player, intra-coded pictures are inserted after a predefined number of inter-predicted pictures. This periodical breakup of inter-picture dependencies allows for such a random access. Applications that require a low encoder delay, for example, video conferencing, are likely to use coding structures similar to the Low Delay B one. The inter-prediction references only preceding pictures. Consequently, the

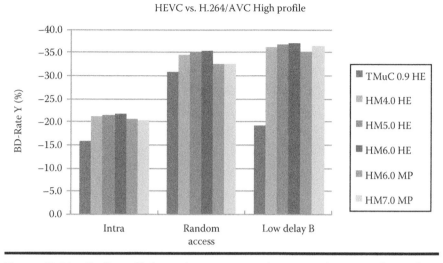

Figure 10.10 Coding efficiency improvements in terms of BD rate for various HEVC test model versions compared to an H.264/AVC HP anchor (0%).

coding order of the pictures corresponds to the display order. Inter-predicted pictures can use up to two predictions from reference pictures that are averaged. These pictures are referred to as bi-predictive (B) pictures.

All the aforementioned preliminary evaluations and data from the most recent test model at the time of writing, HM7.0, are summarized in Figure 10.10. The results are given in terms of bit rate savings expressed by the Bjøntegaard delta (BD) rate, where negative numbers represent an average bit rate reduction [25]. In order to allow for a fair comparison, the optimized H.264/AVC HP anchor from Reference 24 was used as the reference for all HEVC versions. This reference anchor was generated using the most recent JM version at the time, that is, JM18.3 [26]. The HM high-efficiency (HE) configurations used for the different HEVC versions are described in References 27–30. For all HE configurations, the bit depth was set to eight to allow fair comparison with the 8-bit H.264/AVC HP. Since HM6.0, common test conditions additionally include a configuration that reflects the MP established in CD [30, 31]. It can be seen that from the early 0.9 version to the most recent versions, the savings are increasing, while from HM4.0 to HM7.0, the coding efficiency is more or less stable. This reflects recent efforts in HEVC development that have been aimed to stabilize the overall design, for example, to resolve parsing throughput issues. Figure 10.10 also shows slightly decreasing coding efficiency when using MP instead of HE configuration. The motivation of having a slightly less efficient MP is illustrated in Figure 10.11. As a rough indicator of complexity, encoder and decoder runtimes of HM6.0 HE and MP configurations are given. Average encoder/decoder runtime reductions of 15%/10% come

HEVC High efficiency vs. main profile

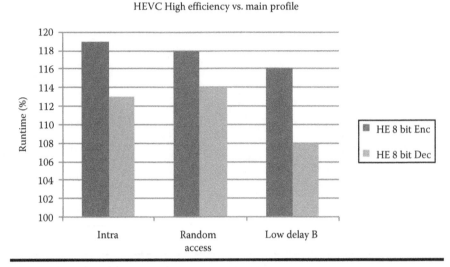

Figure 10.11 Relative runtimes for the HEVC CD test model HM6.0 HE 8-bit configuration compared to the MP anchor (100%).

along with a maximum BD-rate decrease of 3%. This was considered as a fair complexity–performance trade-off for the MP. Some rate-distortion curves, where the PSNR of the Y component is plotted over the bit rate, can be found in Figure 10.12 for Intra, in Figure 10.13 for Random Access, and in Figure 10.14 for Low Delay B. In each figure, curves for H.264/AVC HP and HM7.0 MP are shown for two sequences. The first one, RaceHorses, represents the low-bit-rate case with rates ranging from 200 kbps to 5.5 Mbps. The second sequence, BQTerrace, represents the high-bit-rate case with rates from 4 to 200 Mbps. It can be seen that the PSNR for HM7.0 MP is always at least 1 dB higher than the one for H.264/AVC HP for the same bit rate.

During the standardization process, objective performance is constantly monitored using PSNR distortion measurement. However, increased visual quality does not always manifest in a high PSNR value as well as artifacts may also be not taken into account by means of PSNR. Therefore, the JCT-VC decided to initiate a preliminary subjective evaluation of HEVC compared to H.264/AVC using the HM and JM software. The results of this evaluation are reported in Reference 32. An overall bit rate reduction of 58% on average is reported for a given quality in terms of mean opinion score (MOS).

Note that all these results are just indicators and represent only the respective snapshot in HEVC development. All results and a detailed description of the H.264/AVC references and test configurations can be found in References 22–24.

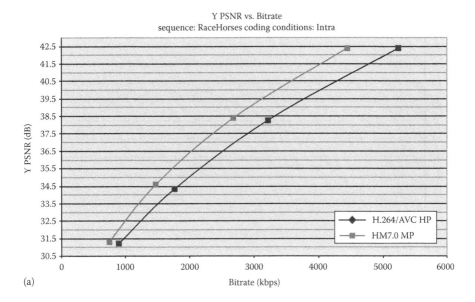

(a)

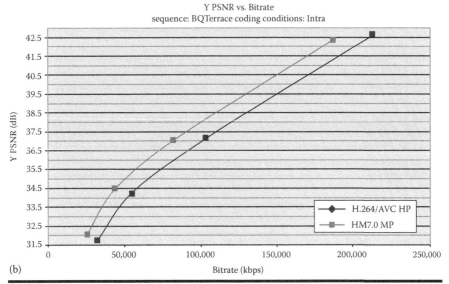

(b)

Figure 10.12   Rate-distortion curves for H.264/AVC HP and HM7.0 MP Intra. (a) RaceHorses 416 × 240 sequence. (b) BQTerrace 1920 × 1080 sequence.

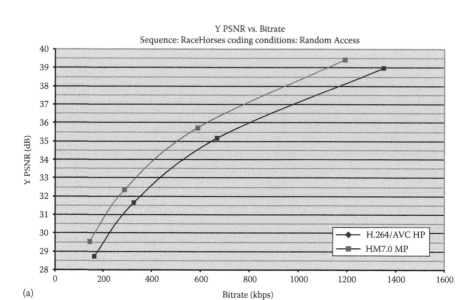

(a)

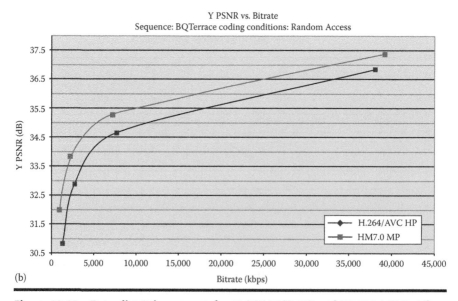

(b)

Figure 10.13 Rate-distortion curves for H.264/AVC HP and HM7.0 MP Random Access. (a) RaceHorses 416 × 240 sequence. (b) BQTerrace 1920 × 1080 sequence.

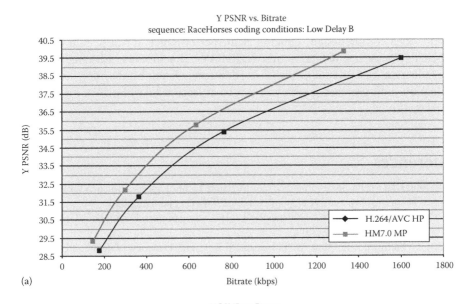

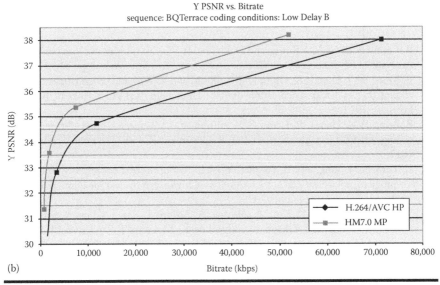

Figure 10.14   Rate-distortion curves for H.264/AVC HP and HM7.0 MP Low Delay B. (a) RaceHorses 416 × 240 sequence. (b) BQTerrace 1920 × 1080 sequence.

10.6 Conclusions

With issuing a DIS of the HEVC video coding standard, the penultimate milestone was achieved. Compared to previous, state-of-the-art video codecs, HEVC provides improved coding performance due to advances in tools all over the hybrid video coding design. Looking at the performance, it can be said that the targeted 50% bit rate reduction over H.264/AVC is achievable. Although the bit rate reduction for a given objective distortion seems to not hit the 50%, subjective evaluation indicates that the bit rate can be at least reduced by 50% for a given visual quality.

References

1. Joint Call for Proposals on Video Compression Technology. Document VCEG-AM91 of ITU-T Q6/16 and N1113 of JTC1/SC29/WG11, January 2010.
2. Draft Requirements for next-generation video coding project. Document VCEG-AL96 of ITU-T Q6/16, July 2009.
3. G. J. Sullivan, J.-R. Ohm, W.-J.Han, and T. Wiegand. Overview of the High Efficiency Video Coding (HEVC) Standard. *IEEE Transactions on Circuits and Systems for Video Technology*, 22(12), 2012.
4. J.-R. Ohm, G. J. Sullivan, H. Schwarz, T. K. Tan, and T. Wiegand. Comparison of the Coding Efficiency of Video Coding Standards – Including High Efficiency Video Coding (HEVC). *IEEE Transactions on Circuits and Systems for Video Technology*, 22(12), 2012.
5. F. Bossen, B. Bross, K. Sühring, and D. Flynn. HEVC Complexity and Implementation Analysis. *IEEE Transactions on Circuits and Systems for Video Technology*, 22(12), 2012.
6. B. Bross, W.-J. Han, J.-R. Ohm, G. J. Sullivan, and T. Wiegand. High efficiency video coding (HEVC) text specification draft 6. Document JCTVC-H1003 of JCT-VC, February 2012.
7. B. Bross, W.-J. Han, J.-R. Ohm, G. J. Sullivan, and T. Wiegand. High efficiency video coding (HEVC) text specification draft 8. Document JCTVC-J1003 of JCT-VC, July 2012.
8. G. Laroche, J. Jung, and B. Pesquet-Popescu. RD optimized coding for motion vector predictor selection. *IEEE Transactions on Circuits and Systems for Video Technology*, 18(9): 1247–1257, September 2008.
9. W.-J. Han, J. Min, I.-K. Kim, E. Alshina, A. Alshin, T. Lee, J. Chen, V. Seregin, S. Lee, Y. M. Hong, M.-S. Cheon, N. Shlyakhov, K. McCann, T. Davies, and J.-H.Park. Improved Video Compression Efficiency Through Flexible Unit Representation and Corresponding Extension of Coding Tools. *IEEE Transactions on Circuits and Systems for Video Technology*, 20(12): 1709–1720, 2010.
10. A. Fuldseth, G. Bjøntegaard, M. Budagavi, and V. Sze. CE10: Core transform design for HEVC. Document JCTVC-G495 of JCT-VC, November 2011.
11. D. Marpe, H. Schwarz, and T. Wiegand. Context-based adaptive binary arithmetic coding in the H.264/AVC video compression standard. *IEEE Transactions on Circuits and Systems for Video Technology*, 13(7): 620–636, July 2003.

12. T. Nguyen, D. Marpe, H. Schwarz, and T. Wiegand. Reduced-complexity entropy coding of transform coefficient levels using truncated golomb-rice codes in video compression. In *IEEE International Conference on Image Processing 2011*, pp. 753–756, Brussels, Belgium. IEEE, September 2011.

13. T. Nguyen, H. Schwarz, H. Kirchhoffer, D. Marpe, and T. Wiegand. Improved context modeling for coding quantized transform coefficients in video compression. In *Picture Coding Symposium*, pp. 378–381, Nagoya, Japan, 2010.

14. J. Sole, R. Joshi, N. Nguyen, T. Ji, M. Karczewicz, G. Clare, F. Henry, and A. Duenas. Transform Coefficient Coding in HEVC. *IEEE Transactions on Circuits and Systems for Video Technology*, 22(12), 2012.

15. A. Norkin, G. Bjøntegaard, A. Fuldseth, M. Narroschke, M. Ikeda, K. Andersson, M. Zhou, and G. Van der Auwera. HEVC Deblocking Filter. *IEEE Transactions on Circuits and Systems for Video Technology*, 22(12), 2012.

16. C.-M. Fu, E. Alshina, A. Alshin, Y.-W. Huang, C.-Y. Chen, C.-Y. Tsai, C.-W. Hsu, S.-M.Lei, J. H. Park, and W.-J. Han. Sample Adaptive Offset in the HEVC Standard. *IEEE Transactions on Circuits and Systems for Video Technology*, 22(12), 2012.

17. T. Wiegand, G. J. Sullivan, G. Bjøntegaard, and A. Luthra. Overview of the H.264/AVC video coding standard. *IEEE Transactions on Circuits and Systems for Video Technology*, 13(7): 560–576, 2003.

18. R. Sjöberg, Y. Chen, A. Fujibayashi, M. M. Hannuksela, J. Samuelsson, T. K. Tan, Y.-K. Wang, and S. Wenger. Overview of HEVC high-level syntax and reference picture management. *IEEE Transactions on Circuits and Systems for Video Technology*, 22(12), 2012.

19. C. C. Chi, M. Alvarez-Mesa, B. Juurlink, G. Clare, F. Henry, S. Pateux, and T. Schierl. Parallel Scalability and Efficiency of HEVC Parallelization Approaches. *IEEE Transactions on Circuits and Systems for Video Technology*, 22(12), 2012.

20. S. Park, J. Park, and B. Jeon. Report on the evaluation of HM versus JM. Document JCTVC-D181 of JCT-VC, January 2011.

21. JCT-VC. Subversion repository for the HEVC reference software. https://hevc.hhi.fraunhofer.de/svn/svn_HEVCSoftware/tags/, 2012.

22. B. Li, G. J. Sullivan, and J. Xu. Comparison of compression performance of HEVC working draft 4 with AVC. Document JCTVC-G399 of JCT-VC, November 2011.

23. B. Li, G. J. Sullivan, and J. Xu. Comparison of compression performance of HEVC working draft 5 with AVC high profile. Document JCTVC-H0360 of JCT-VC, February 2012.

24. B. Li, G. J. Sullivan, and J. Xu. Comparison of compression performance of HEVC draft 6 with AVC high profile. Document JCTVC-I0409 of JCT-VC, April 2012.

25. G. Bjøntegaard. Calculation of average PSNR differences between RD curves. Document VCEG-M33 of ITU-T Q6/16, April 2001.

26. JVT. H.264/AVC reference software. http://iphome.hhi.de/suehring/tml/, 2012.

27. F. Bossen. Common test conditions and software reference configurations. Document JCTVC-C500 of JCT-VC, October 2010.

28. F. Bossen. Common test conditions and software reference configurations. Document JCTVC-F900 of JCT-VC, July 2011.

29. F. Bossen. Common test conditions and software reference configurations. Document JCTVC-G1200 of JCT-VC, November 2011.

30. F. Bossen. Common test conditions and software reference configurations. Document JCTVC-H1100 of JCT-VC, February 2012.

31. F. Bossen. Common test conditions and software reference configurations. Document JCTVC-I1100 of JCT-VC, April 2012.

32. G. J. Sullivan and J.-R. Ohm. Report on preliminary subjective testing of HEVC compression capability. Document JCTVC-H1004 of JCT-VC, February 2012.

33. P. Helle, S. Oudin, B. Bross, D. Marpe, M. O. Bici, K. Ugur, J. Jung, G. Clare, and T. Wiegand. Block merging for quadtree-based partitioning in HEVC. *IEEE Transactions on Circuits and Systems for Video Technology*, 22(12), 2012.

NEXT GENERATION HANDHELD DVB TECHNOLOGY

Basic Sheer Terrestrial Profile

Chapter 11

Bit-Interleaved Coded Modulation in Next-Generation Mobile Broadcast Standard DVB-NGH

Catherine Douillard and Charbel Abdel Nour

Contents

11.1 Introduction

Bit-interleaved coded modulation (BICM) is the state-of-the-art pragmatic approach for combining channel coding with digital modulations in fading transmission channels [1]. The modulation constellation can thus be chosen independently of the coding rate. The core of the BICM encoder consists of the serial concatenation of a forward error correcting (FEC) code, a bit interleaver, and a binary labeling that maps blocks of bits to the constellation symbols.

FEC coding in the first generation of DVB standards was based on an association of convolutional and Reed–Solomon codes. In the second generation, it calls for the serial concatenation of a Bose–Chaudhuri–Hocquenghem (BCH) code and a low-density parity-check (LDPC) code. This structure ensures a better protection, allowing more data to be transported through a given channel. The gain achieved by the second-generation code in stationary transmission channels ranges from 3 dB for low rate services to more than 5 dB for high rate services at the *quasi error free** (QEF) point. For an additive white Gaussian noise (AWGN) channel and this coding structure, the gap to capacity varies from 1 to 2.5 dB† at QEF.

The BICM scheme can also be complemented with a set of interleavers that provide further protection against impulsive noise and time-selective fading.

In 2009, when the DVB-NGH Call for Technologies [2] was issued, two technical state-of-the-art DVB standards could have been used as a starting point

---

* QEF is defined as "less than one uncorrected error event per hour of transmission at the throughput of a 5 Mbps single TV service decoder" and approximately corresponds to a transport stream frame error ratio FER < 10^{-7}.

† These figures take the capacity loss due to the P1 preamble and the pilot patterns into account.

for DVB-NGH: DVB-SH [3] and DVB-T2 [4]. Both standards include state-of-the-art BICM technologies. In particular, they both use a capacity approaching coding scheme: a turbo code [5] is used in DVB-SH, while the aforementioned concatenation of BCH and LDPC codes was adopted in DVB-T2. Moreover, the DVB-NGH Commercial Requirements [6] mention the possibility to combine DVB-NGH and DVB-T2 signals in one radio frequency (RF) channel. The natural way for this combination calls for the use of the so-called future extension frames (FEFs) of DVB-T2. Although a DVB-T2 FEF can contain BICM components totally different from the DVB-T2 BICM module, the existence of combined DVB-T2/NGH receivers finally pushed the elaboration of a DVB-NGH physical layer strongly inspired by DVB-T2.

Therefore, DVB-NGH was designed to provide an extension of the DVB-T2 system capabilities, to ease the introduction of TV services to mobile terminals within an existing terrestrial digital TV platform. In particular, the objective to keep reasonable receiver complexity and power consumption while increasing the receiver robustness for mobile reception has guided the choice for the BICM components. Following the approach initiated with the definition of T2-Lite, the BICM module in the DVB-NGH standard is mainly derived from a subset of DVB-T2 BICM components, with a set of additional features allowing for higher robustness and coverage.

The complexity reduction requirements led to the following restrictions in the BICM module: only the short FEC block size ($N_{ldpc} = 16,200$ bits) is allowed, the size of the time interleaver (TI) memory is halved compared to DVB-T2, the use of rotated constellations is prohibited in 256-QAM, and the maximum coded data bit rate is limited to 12 Mbps at the LDPC decoder input. On the other hand, higher robustness and coverage are achieved through some improvements at the FEC code and constellation levels and in the definition of the interleavers. In short,

- Lower code rates (1/5, 4/15, 1/3, and 2/5) were introduced
- Nonuniform 64- and 256-QAM constellations and four-dimensional (4D) rotated QPSK constellations came on top of uniform QAM and two-dimensional (2D) rotated constellations
- The interleaver chain was improved with the use of a parity bit interleaver for all constellation sizes, enhanced I/Q component interleaving, convolutional inter-frame time interleaving, and adaptive cell quantization in the TI

The rest of the chapter is structured as follows. Section 11.2 describes the structure of the BICM module in DVB-T2. Then, Section 11.3 presents the modifications and the new elements that have been introduced in DVB-NGH. Section 11.4 provides some performance results that are compared with a DVB-H system. Finally, Section 11.5 concludes the chapter.

11.2 Bit-Interleaved Coded Modulation in DVB-T2

In DVB-T2, the input to the BICM module consists of one or more logical data streams. Each stream is carried by one physical layer pipe (PLP). Service-specific robustness is then made possible since each PLP can have its own constellation, FEC code rate, and time interleaving depth. At the transmitter, the structure of the DVB-T2 BICM module for data PLPs is described in Figure 11.1. It is followed by the frame builder whose role is to collect PLP cells before the OFDM waveform generation.

11.2.1 Forward Error Correction

The LDPC codes adopted in the second generation of DVB standards (DVB-x2) [7] ensure low-complexity encoding due to their irregular-repeat accumulate structure [8] and an efficient description of the parity-check matrix. The shape of the parity-check matrix is described in Figure 11.2. It is divided into two submatrices: the first one, on the right, is a staircase lower triangular $(N_{ldpc} - K_{ldpc})$-by-$(N_{ldpc} - K_{ldpc})$ matrix, where N_{ldpc} and K_{ldpc}, respectively, denote the sizes of the coded and uncoded LDPC blocks. This staircase lower triangular part makes the encoding easy since the parity bits can directly be computed by the successive accumulation of the ones contained in the left-hand submatrix. The second submatrix, on the left, is constructed using a template matrix, which is a (K_{ldpc}/M)-by-$(N_{ldpc} - K_{ldpc})/M$ binary matrix where the nonzero entries are replaced by an M-by-M identity matrix whose rows have been cyclically shifted. Parameter M is the periodicity factor of the code. It is equal to 360 in DVB-x2 standards. Finally, cyclic shifts yield simple interleavers, implemented with shift registers, between check node processors and variable node processors.

LDPC codes are decoded with different variants of the *belief propagation* algorithm [9], which is an iterative algorithm using the bipartite graph representation of the parity-check matrix of the code [10]: coded bits are represented by *variable nodes* that are connected to *check nodes* according to the parity relations of the code. At each iteration, extrinsic information is passed from variable nodes to check nodes and from check nodes back to variable nodes. The messages sent from variable nodes to check nodes are computed based on the observed value of the variable node and some of the messages stemming from the neighboring check nodes. The partitioning of the DVB-x2 parity-check matrix into groups of M variables and checks allows an intrinsic parallelism as well as reduced storage requirements for decoding: the variables can be saved into M blocks of memory and thus M memory accesses can be done simultaneously.

In order to reach the QEF target without any change in the slope of the error rate curves, an outer t-error-correcting BCH code with t = 10 or 12 has been added that allows residual errors to be removed.

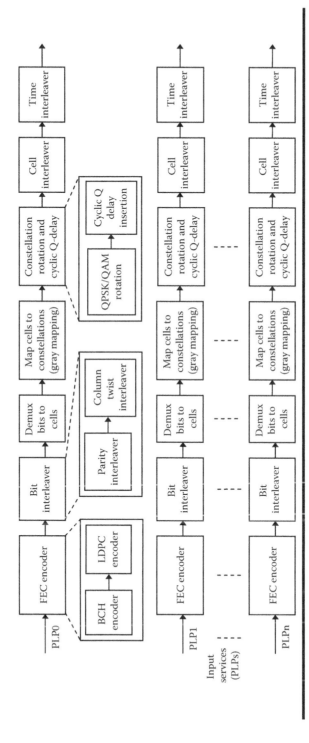

Figure 11.1   DVB-T2 BICM module.

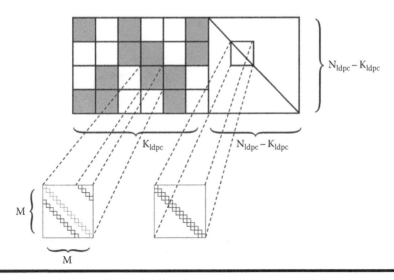

Figure 11.2   Shape of the parity-check matrix for DVB-x2 LDPC codes.

The set of coding rates and blocks sizes specified in DVB-T2 are summarized in Table 11.1. This table gives the FEC coding parameters for the normal FEC frame length ($N_{ldpc} = 64,800$ bits) and for the short FEC frame length ($N_{ldpc} = 16,200$ bits). Short FEC blocks allow a finer bit-rate granularity but incur a greater overhead and slightly worse performance than long blocks: a typical loss of 0.2–0.3 dB is observed in the AWGN channel. This penalty can reach 0.5 dB in memoryless fading channels.

11.2.2 Bit Interleaver and Bit-to-Cell Demultiplexer

LDPC codes can be very sensitive to deep fading or erasure events: when several bits connected to a given check node are deeply faded or erased, the check node is not able to feed the updated information back to the corresponding variable nodes and does not contribute to the iterative process in an efficient way any longer, thus entailing a performance loss. When high-order modulations are used, fading and erasures act at the cell level, thus affecting all the bits contained in a cell identically. Therefore, the role of the bit interleaver is to prevent bits connected to the same check node from being transmitted in the same constellation cell. This avoids that one faded cell may result in the loss of multiple bits connected to the same check node.

The DVB-T2 bit interleaver is a block interleaver applied at the LDPC codeword level, consisting of parity interleaving followed by column-twist interleaving. It is used in all data transmission configurations except when QPSK constellation is used. If basic block interleaving—column-wise writing and row-wise reading—were applied directly to the LDPC codewords, many constellation symbols would

Table 11.1 Data Coding Parameters for DVB-T2

a. $N_{ldpc} = 64{,}800$ bits

LDPC Code Rate	BCH Uncoded Block Size K_{bch}	BCH Coded Block Size N_{bch} (or LDPC Uncoded Block Size K_{ldpc})	BCH t-Error Correction	$N_{bch} - K_{bch}$
1/2	32,208	32,400	12	192
3/5	38,688	38,880	12	192
2/3	43,040	43,200	10	160
3/4	48,408	48,600	12	192
4/5	51,648	51,840	12	192
5/6	53,840	54,000	10	160

b. $N_{ldpc} = 16{,}200$ bits

LDPC Code Identifier	BCH Uncoded Block Size K_{bch}	BCH Coded Block Size N_{bch} (or LDPC Uncoded Block Size K_{ldpc})	BCH t-Error Correction	$N_{bch} - K_{bch}$	Effective LDPC Code Rate
1/2	7,032	7,200	12	168	4/9
3/5	9,552	9,720	12	168	3/5
2/3	10,632	10,800	12	168	2/3
3/4	11,712	11,880	12	168	11/15
4/5	12,432	12,600	12	168	7/9
5/6	13,152	13,320	12	168	37/45

Table 11.2 Bit Interleaver Structure and Number of Substreams at the Output of the Demultiplexer for DVB-T2

| Modulation | Rows N_r | | Columns N_c | Number of Substreams |
	$N_{ldpc} = 64,800$	$N_{ldpc} = 16,200$		
QPSK	—	—	—	2
16-QAM	8,100	2,025	8	8
64-QAM	5,400	1,350	12	12
256-QAM	4,050	—	16	16
	—	2,025	8	8

contain multiple coded bits connected to the same check node. Therefore, the parity interleaver is designed in such a way that the parity part of the parity-check matrix has the same structure as the information part. Then, the information bits and the parity interleaved bits are serially written into the so-called column-twist interleaver column-wise and serially read out row-wise. The write start position of each column is twisted by an integer value t_c, depending on the code size, the constellation order, and the column number.

The configuration of the overall bit interleaver for each modulation format is specified in [4].

An additional bit-to-cell demultiplexer is inserted between the bit interleaver and the constellation mapper in order to allow a finer optimization of the correspondence between the LDPC code and the modulation. It divides the bit stream at the output of the bit interleaver into a number of substreams, which is a multiple of the number of bits per constellation cell, as also reported in Table 11.2. The bit-to-cell demultiplexing mapping is a function of the code size, the code rate, and the constellation order.

11.2.3 Modulation Constellations

11.2.3.1 QAM Modulation

The high error-correcting performance of the DVB-T2 FEC allows for the application of high-order constellation schemes. Besides the use of QPSK, 16-QAM, and 64-QAM as in DVB-T, DVB-T2 advocates 256-QAM, increasing the gross data rate to 8 bits per OFDM cell and yielding a 33% increase in spectral efficiency for a given code rate compared to DVB-T. Constellation symbols are labeled following a Gray mapping, in order to minimize the bit error probability at the output of the constellation demapper.

11.2.3.2 Rotated and Q-Delayed Constellations

The DVB-T2 constellations can be implemented according to two different modes: conventional nonrotated or rotated constellations. When using conventional QAM constellations, each signal component, in-phase I (real) or quadrature Q (imaginary), carries half of the binary information held in the signal. Consequently, when a constellation signal is subject to a fading event, I and Q components fade identically. In case of severe fading, the information transmitted on I and Q components suffers an irreversible loss.

When a rotation is applied to the constellation, components I and Q both carry the whole binary content of the signal, as every point in the constellation now has its own projections over the I and Q axes. This is illustrated in Figure 11.3a for a 16-QAM constellation. The rotation is performed by multiplying each I/Q component vector by a 2×2 orthogonal matrix:

$$\begin{bmatrix} y_I \\ y_Q \end{bmatrix} = \begin{bmatrix} \cos\Phi & -\sin\Phi \\ \sin\Phi & \cos\Phi \end{bmatrix} \begin{bmatrix} x_I \\ x_Q \end{bmatrix} \tag{11.1}$$

Next, the Q component of the resulting vector is cyclically delayed by one cell within the FEC block. Consequently, due to the subsequent effect of the cell, time, and frequency interleavers, the two copies or projections of the signal are sent separately in two different time periods and/or two different OFDM subcarriers, in order to benefit from time or frequency diversity, respectively. With this technique,

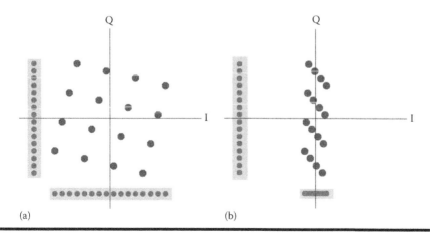

Figure 11.3 Effect of rotated constellations on transmissions with deep fading events. (a) Each component I/Q carries the whole binary information, and both components are transmitted separately. (b) After a deep fading event on I component, the whole binary content can be retrieved from Q.

Table 11.3 Rotation Angle for Each Constellation in DVB-T2

Constellation	QPSK	16-QAM	64-QAM	256-QAM
Φ (°)	29.0	16.8	8.6	3.6

the diversity order of BICM is doubled compared to the case of nonrotated constellation. Thus, when a transmitted component I or Q of a rotated QAM symbol is deeply faded or erased, the binary information carried by the whole QAM symbol can nevertheless be retrieved thanks to the remaining corresponding component, as illustrated in Figure 11.3b.

The choice of the rotation angle results from a compromise between minimizing the pairwise error probability between two different transmitted sequences in fading channels at high signal-to-noise ratios (SNRs) and minimizing the bit error probability at the demapper output when one of the components I or Q is erased [11]. The rotation angle values Φ depend on the constellation and are given in Table 11.3.

From the receiver implementation point of view, the main effect is a complexity increase in the demapper, as I and Q components have to be demapped jointly. For an order M conventional constellation, the I and Q components can be demapped independently by computing $\sqrt{M}$ one-dimensional (1D) Euclidean distances whereas for an order M rotated constellation, the I and Q components have to be demapped jointly by computing M 2D Euclidean distances. However, some simplified rotated constellation demappers, requiring a reduced number of distance computations, can be implemented without visible loss of performance [12].

Rotated constellations have no impact on performance in the AWGN channel. In fading channels, they give maximum gain when used with low constellation sizes (such as QPSK) and higher code rates (such as 4/5 and 5/6). Conversely, the gain is much reduced for high constellation sizes (such as 256-QAM) and lower code rates. The robustness of rotated constellations against erasure events makes it of particular interest when time-frequency slicing (TFS) is enabled (see Chapter 13) in the case where one or several RF channels are strongly attenuated with respect to the others. Adopting rotated constellations can prevent the system from collapsing when the erasure percentage approaches the transmission limit given by the redundancy ratio.

11.2.4 Cell Interleaver

The cell interleaver and the subsequent TI are essentially intended to provide protection against impulsive noise and time-selective fading.

The cell interleaver applies a pseudo-random permutation in order to uniformly spread the cells in the FEC codeword. It aims at ensuring an uncorrelated distribution of channel distortions and interference along the FEC codewords in the receiver and at increasing the cell separation when rotated constellations are used. This pseudo-random permutation varies from one FEC block to another.

11.2.5 Time Interleaver

The TI provides the core of the interleaving in the time domain, spreading the cells of each FEC block over several symbols and possibly over several T2 frames. It operates at the PLP level.

The smallest unit of the frame structure used for signaling and mapping interleaved FEC blocks is called the interleaving frame (IF). Since the data rate of each PLP can vary, each IF can contain a variable number of FEC blocks, which is signaled by the modulator.

The TI is a row–column block interleaver that operates on TI blocks. The memory dedicated to a TI block is limited to the fixed amount of $M_{TI} = 2^{19} + 2^{15}$ cells, thus allowing the receiver to store and process one data PLP and its associated common PLP, if any.

The relation between the maximum interleaving duration T_{intmax} and the PLP data rate R_{PLP} is given by [13]

$$T_{intmax} = \frac{\left\lfloor \dfrac{M_{TI} \times \eta_{MOD}}{N_{LDPC}} \right\rfloor \times \left(K_{bch} - 80\right)}{R_{PLP}} \tag{11.2}$$

where η_{MOD} denotes the number of bits per constellation cell.

In the simplest case, each IF consists of one TI block and is mapped directly to one T2 frame as shown in Figure 11.4a. Because of the limit on the number of cells contained in a TI block, resulting data rate for a PLP can then be insufficient. In this case, each IF can consist of more than one TI block in order to increase the maximum data rate for a PLP, as illustrated in Figure 11.4b. On the other hand, increasing the number of TI blocks per IF for a given frame duration reduces the interleaving time, so reducing the time diversity and therefore the system's resistance to impulsive interference and fast time-varying channels.

Conversely, greater time diversity can be achieved through *multi-frame interleaving*, by mapping each IF across more than one T2 frame. This option is typically used to provide greater time diversity for low data-rate services since, in this case, only one TI block is allowed per IF.

Another way of increasing the time diversity of each PLP, which can be combined with the previous one, involves using type 2 data PLPs that are carried in multiple *subslices* per T2 frame. Increasing the number of subslices leads to a more uniform distribution of information over time and increases time diversity but reduces the power-saving opportunities in receivers. An additional mechanism, called *frame skipping*, allows the PLPs to be transmitted only in a subset of T2

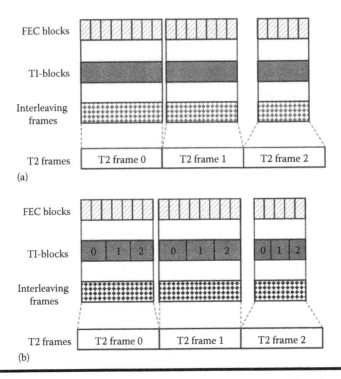

Figure 11.4 Examples of time interleaving configurations for data PLPs in DVB-T2. (a) Basic configuration: each TI block is mapped directly to one T2 frame. (b) Increasing data rate: each IF consists of three TI blocks.

frames regularly spaced over time, which enables better power saving at the expense of worsened time diversity.

A detailed discussion of the possible trade-offs related to time diversity, data rate, latency, and power saving in DVB-T2 can be found in the implementation guidelines [13], in [14], and in Chapter 12.

11.2.6 First Step toward DVB-NGH: The T2-Lite Profile

A new profile of DVB-T2, called T2-Lite, was defined during the course of the DVB-NGH standardization process and was added to the DVB-T2 specifications in version 1.3.1 [4], Annex I in October 2011. The T2-Lite profile allows simpler receiver implementations for low-capacity applications such as mobile broadcasting. It is based on a limited subset of the modes available in DVB-T2. The modes that require the most complexity and memory have been removed in order to reduce the complexity of T2-Lite-only receivers so as to minimize the cost and power consumption of handheld devices. In particular, in T2-Lite, the sum of the maximum coded data rates for any data PLP and its associated common PLP is limited to 12 Mbps.

On the other hand, a few new elements have been added, such as two new low coding rates in order to increase the robustness of the T2-Lite transmissions. The restrictions and additions related to the T2-Lite BICM module are shortly reviewed in the remainder of this section. For more details on the mobile T2-Lite profile of DVB-T2, see Chapter 6.

11.2.6.1 Forward Error Correction and Modulation Constellations

In order to allow for simpler receivers with lower latency, the normal FEC frame length of DVB-T2 has been removed from the T2-Lite profile: only the short 16,200-bit LDPC codes need to be implemented in T2-Lite receivers. Furthermore, since T2-Lite aims at providing high robustness for portable and mobile reception, the highest values of coding rates, 4/5 and 5/6, are not included in this profile, nor the combination of coding rates 2/3 and 3/4 with 256-QAM constellation. Conversely, coding rates 1/3 and 2/5 have been added, together with their respective bit-to-cell demultiplexers. The addition of these low rates allows the coverage of T2-Lite services to be extended at the expense of lower capacity. The utilization of rotated constellations together with the 256-QAM constellation is also not included in the profile.

11.2.6.2 Bit Interleaver

For the two new coding rates of T2-Lite, 1/3 and 2/5, the parity interleaver specified in DVB-T2 for higher-order constellations is also used for QPSK. It was shown that, for these low coding rates, the presence of a parity interleaver is beneficial in fading channels. As an example, Figure 11.5 shows a performance gain of 0.5 dB at FER 10^{-3} for coding rate 1/3 in a Rayleigh fading channel when the parity interleaver is enabled.

11.2.6.3 Time Interleaver

In order to further reduce the implementation complexity of T2-Lite-only receivers, T2-Lite has reduced the time interleaving memory from $2^{19} + 2^{15}$ cells down to 2^{18} cells. The number of cells that can be stored in the time deinterleaver memory has to be shared between one data PLP and its associated common PLP, if any. Halving the time interleaving memory capacity is acceptable in the context of mobile services, as they are transmitted at rather low data rates and so the corresponding duration remains adequate.

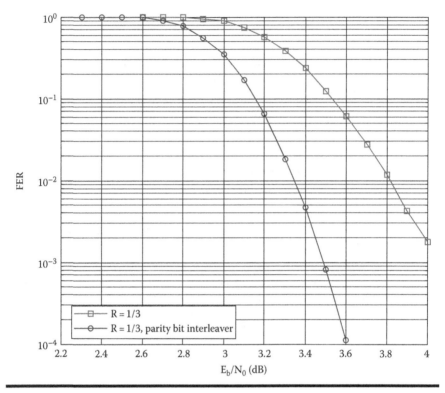

Figure 11.5   Effect of the parity bit interleaver with QPSK in a Rayleigh memory-less channel for coding rate 1/3.

11.2.6.4 Summary

The different blocks of the DVB-T2 BICM module modified in the T2-Lite profile are identified in Figure 11.6.

11.3 Bit-Interleaved Coded Modulation in DVB-NGH

Following the approach initiated with the definition of T2-Lite, DVB-NGH was designed to provide an extension of the DVB-T2 system capabilities and to ease the introduction of TV services to mobile terminals within an existing terrestrial digital TV platform. In particular, the objectives to keep reasonable receiver complexity and power consumption and to increase the receiver robustness and system coverage for mobile reception have guided the choice for the BICM components.

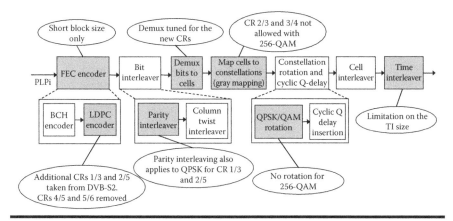

Figure 11.6   Modifications made to the DVB-T2 BICM module in the T2-Lite profile (shaded blocks).

This section addresses only the BICM module related to the data PLPs and mainly focuses on the terrestrial path in DVB-NGH. The L1 signaling path calls for slightly different coding, interleaving, and constellation parameters that are detailed in Chapter 15. Moreover, in a hybrid system, the BICM module in the satellite path also has specific features presented in Chapter 22.

The DVB-NGH BICM module structure at the transmitter for data PLPs is similar to the one of DVB-T2 (see Figure 11.1), except that the cell interleaver position has been moved in front of the constellation mapper, as explained in Section 11.3.4.

11.3.1 Forward Error Correction

As in T2-Lite, only the short 16,200-bit LDPC codes have been implemented in order to reduce receiver complexity. Furthermore, the code rate values were chosen to uniformly cover the range 5/15 (1/3) to 11/15, thus providing equidistant performance curves with respect to SNR. The set of coding rates and blocks sizes are summarized in Table 11.4.

Coding rates 1/5, 1/3, 2/5, 3/5, 2/3, and 11/15 are directly taken from T2-Lite and DVB-T2. On the contrary, code rates 4/15, 7/15, and 8/15 call for new codes specific to DVB-NGH. The BCH code is identical to the one used in DVB-T2 for the short block size.

The use of rates 1/5 and 4/15 is restricted to constellations QPSK and 16-QAM as in the hybrid profile. They are intended to enable the combination of a terrestrial and a satellite signal in a hybrid single-frequency network.

Figure 11.7 shows the error correction performance of the resulting code in a Rayleigh memoryless channel.

Table 11.4 Data Coding Parameters for DVB-NGH

LDPC Code Rate	BCH Uncoded Block Size K_{bch}	BCH Coded Block Size N_{bch} (or LDPC Uncoded Block size K_{ldpc})	BCH t-Error Correction	$N_{bch} - K_{bch}$
3/15 (1/5)	3,072	3,240		
4/15	4,152	4,320		
5/15 (1/3)	5,232	5,400		
6/15 (2/5)	6,312	6,480		
7/15	7,392	7,560	12	168
8/15	8,472	8,640		
9/15 (3/5)	9,552	9,720		
10/15 (2/3)	10,632	10,800		
11/15	11,712	11,880		

11.3.2 Bit Interleaver and Bit-to-Cell Demultiplexer

DVB-NGH inherited the bit interleaver structure from DVB-T2, consisting of parity interleaving followed by column-twist interleaving. Parity interleaving is applied to all constellations, including QPSK and for all coding rates, since it was shown to improve performance in fading channels. Column-twist interleaving is used for all constellations but QPSK. The configuration of the column-twist interleaver for each modulation format is specified in Table 11.5. The write start position of each column is twisted by t_c, where the value of t_c depends on the column number.

As in DVB-T2, an additional bit-to-cell demultiplexer is inserted between the bit interleaver and the constellation mapper. It divides the bit stream at the output of the bit interleaver into a number of substreams, which is a multiple of the number of bits per constellation cell, as shown in Table 11.5. In DVB-NGH, the bit-to-cell demultiplexing parameters have been specifically tuned in order to allow a finer optimization for each constellation size and code rate.

11.3.3 Modulation Constellations

DVB-NGH has inherited the four constellations of DVB-T2: QPSK, 16-QAM, 64-QAM, and 256-QAM. For 64-QAM and 256-QAM, in addition to conventional uniform constellations, constellations with quasi-Gaussian distributions

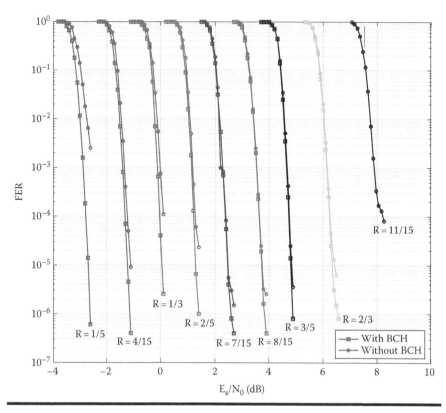

Figure 11.7 Frame error rate curves of the DVB-NGH FEC code in Rayleigh memoryless channel with QPSK constellation and 50 LDPC decoding iterations.

Table 11.5 Column-Twist Interleaver Sizes and Number of Substreams at the Output of the Demultiplexer for DVB-NGH

Modulation	Rows N_r	Columns N_c	Number of Substreams
QPSK	–	–	2
16-QAM	2025	8	8
64-QAM	1350	12	12
256-QAM	2025	8	8

have been added. Except for the 256-QAM, which is always nonrotated, the uniform and nonuniform constellations can be implemented according to two different modes: nonrotated or rotated constellations. Furthermore, the rotated constellation principle was extended to four dimensions for QPSK and high coding rates.

11.3.3.1 Nonuniform QAM Constellations

Nonuniform constellations have been introduced to bridge the observed gap between capacity curves of uniform constellations and the Shannon limit. In fact, when the received signal is perturbed by Gaussian-distributed noise, the mutual information expression is maximized for a Gaussian distribution of the transmitted signal. Applying this assumption leads to the famous Shannon capacity formula. However, the distribution of conventional QAM constellations is far from Gaussian: it is both discrete, as only a limited number of signal values are transmitted, and uniform, since the constellation points are regularly spaced and transmitted with equal probabilities.

Nonuniform constellations try to make the transmitted constellation distribution appear "more" Gaussian. Called *shaping gain*, the corresponding improvement adds up to the coding gain of coded modulation schemes. It has been shown that the shaping gain of discrete constellations in the AWGN channel cannot exceed $10 \log(\pi e/6) \approx 1.53$ dB, where e represents Euler's number [15]. Nevertheless, this is an asymptotic value that can be approached only for very high-order constellations and high SNR values.

Two main shaping techniques have been investigated so far: using a classical constellation with a regular distribution of the signal points and transmitting the signal points with different probabilities or using a constellation whose signal points are nonuniformly spaced and transmitting all the signal points with the same probability. The nonuniform constellations proposed in DVB-NGH belong to the second category.

Constellation point coordinates are chosen to maximize the BICM capacity of the underlying QAM. Let us detail the approach in the example of 16-QAM since the optimization principle is simpler to explain in this case, although nonuniform 16-QAM has not been adopted in DVB-NGH. If we consider that uniform 16-QAM uses positions $\{-3, -1, +1, +3\}$ on each axis, then we can make a nonuniform version having positions $\{-\gamma, -1, +1, +\gamma\}$, using only one parameter γ. For any particular SNR, we can plot the BICM capacity as a function of γ. For example, Figure 11.8 shows the BICM capacity of the nonuniform 16-QAM at an SNR of 10 dB. γ equal to 3 corresponds to the uniform case, while the maximum capacity is obtained for a value of γ between 3.35 and 3.4. Selecting the values of γ yielding the maximum capacity for a large range of SNRs can provide the basis for the construction of an adaptive nonuniform 16-QAM.

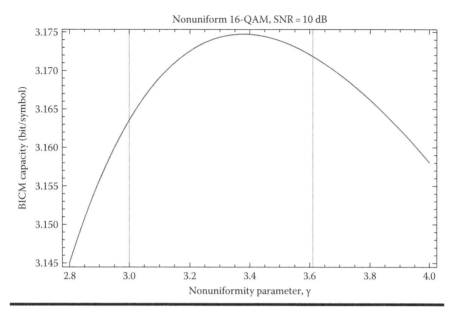

Figure 11.8   BICM capacity curve as a function of nonuniformity parameter γ for 16-QAM in AWGN at 10 dB SNR. (From a report produced by Jonathan Stott Consulting under contract to BBC R&D.)

When considering higher-order constellations, where larger gains are expected, the capacity maximization involves more than one nonuniformity parameter: three parameters are required for nonuniform 64-QAM whose coordinates on I and Q axes are $\{-\gamma, -\beta, -\alpha, -1, +1, +\alpha, +\beta, +\gamma\}$ and seven parameters for nonuniform 256-QAM whose coordinates on I and Q axes are $\{-\eta, -\zeta, -\varepsilon, -\delta, -\gamma, -\beta, -\alpha, -1, +1, +\alpha, +\beta, +\gamma, +\delta, +\varepsilon, +\zeta, +\eta\}$.

A solution to this problem was provided numerically for a large range of SNRs. As a consequence of the dependence of the nonuniform constellation points on the SNR, a given nonuniform constellation cannot provide the maximum coding gain for any operation point and accordingly for any code rate. Therefore, a specific nonuniform constellation has been defined for each code rate. The corresponding constellation mappings are given in Tables 11.6 and 11.7. Note that the I/Q coordinates do not have the aforementioned form since a normalization operation was performed in order to keep the same transmit power as for the uniform constellations. An example of nonuniform 256-QAM constellation is shown is Figure 11.9.

Figure 11.10 shows the performance gain of the nonuniform 256-QAM in the AWGN channel compared with the classical constellation. The observed gain ranges from 0.4 to 0.9 dB, depending on the coding rate. For nonuniform 64-QAM, the maximum gain amounts to 0.3 dB. The actual gain in mobile channels has still to be assessed.

Table 11.6   Constellation Mapping of the I and Q Components for the Uniform and Nonuniform 64-QAM

I/Q Values		Binary Mapping							
		100	*101*	*111*	*110*	*010*	*011*	*001*	*000*
Uniform		−7	−5	−3	−1	1	3	5	7
Nonuniform	1/3	−7.2	−5.2	−1.9	−1.4	1.4	1.9	5.2	7.2
	2/5	−7.4	−4.9	−2.0	−1.3	1.3	2.0	4.9	7.4
	7/15	−7.5	−4.6	−2.3	−1.0	1.0	2.3	4.6	7.5
Coding rate	8/15	−7.5	−4.6	−2.4	−0.9	0.9	2.4	4.6	7.5
	3/5	−7.5	−4.6	−2.5	−0.9	0.9	2.5	4.6	7.5
	2/3	−7.4	−4.7	−2.6	−0.9	0.9	2.6	4.7	7.4
	11/15	−7.3	−4.7	−2.7	−0.9	0.9	2.7	4.7	7.3

11.3.3.2 Rotated Constellations

Like in the T2-Lite profile, the utilization of rotated constellations together with the 256-QAM constellation is not allowed in DVB-NGH, to limit the demapper complexity. On the other hand, nonuniform 64-QAM can be used in its rotated form, but with a rotation angle different from the uniform case. Furthermore, for QPSK and high coding rates, the concept of rotated constellations is extended to a higher diversity scheme, called *4D rotated constellations*. The cyclic shift delay applied to the quadrature Q component in DVB-T2 has been replaced in DVB-NGH by a more sophisticated I/Q component interleaver providing better time separation and channel diversity for both diversity schemes.

11.3.3.2.1   Rotated 16-QAM and Uniform 64-QAM

In DVB-T2, the rotation angle values chosen for 16-QAM and 64-QAM were the result of a compromise between degrees of selectivity in various types of fading channels [11]. The same angles were adopted in DVB-NGH.

11.3.3.2.2   Rotated Nonuniform 64-QAM

For nonuniform 64-QAM, a different approach was adopted: for each case corresponding to the different coding rate values, the angle giving the best performance for channels with a high amount of erasures was chosen. These are the scenarios

Table 11.7 Constellation Mapping of the I and Q Components for the Uniform and Nonuniform 256-QAM

I/Q Values		*Binary Mapping*															
		0000	0001	0011	0010	0110	0111	0101	0100	1100	1101	1111	1110	1010	1011	1001	1000
Uniform		15	13	11	9	7	5	3	1	−1	−3	−5	−7	−9	−11	−13	−15
Nonuniform	Coding rate																
	1/3	17.2	12.6	9.7	9.3	3.8	4.1	2.5	2.4	−2.4	−2.5	−4.1	−3.8	−9.3	−9.7	−12.6	−17.2
	2/5	17.3	13.1	9.4	8.8	4.2	4.3	2.1	2.1	−2.1	−2.1	−4.3	−4.2	−8.8	−9.4	−13.1	−17.3
	7/15	17.5	13.1	9.2	8.2	4.7	4.6	1.6	1.7	−1.7	−1.6	−4.6	−4.7	−8.2	−9.2	−13.1	−17.5
	8/15	17.5	13.0	9.3	8.1	5.0	4.6	1.6	1.5	−1.5	−1.6	−4.6	−5.0	−8.1	−9.3	−13.0	−17.5
	3/5	16.7	13.1	10.3	8.0	5.9	4.2	2.3	0.9	−0.9	−2.3	−4.2	−5.9	−8.0	−10.3	−13.1	−16.7
	2/3	16.7	13.1	10.3	8.0	5.9	4.2	2.3	0.9	−0.9	−2.3	−4.2	−5.9	−8.0	−10.3	−13.1	−16.7
	11/15	16.6	13.1	10.3	8.0	6.0	4.2	2.4	0.9	−0.9	−2.4	−4.2	−6.0	−8.0	−10.3	−13.1	−16.6

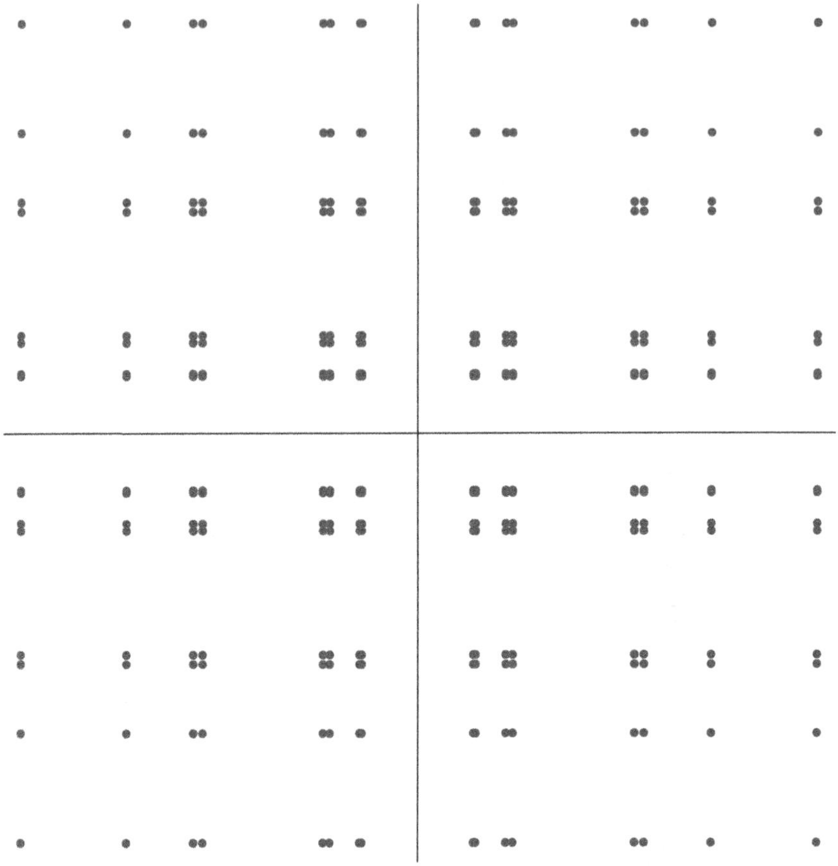

Figure 11.9 Nonuniform 256-QAM constellations adopted in DVB-NGH for coding rate 1/3.

where significant gains can be observed for 64-QAM. They are of particular interest when TFS is enabled since, in the case where one or several RF channels are strongly attenuated, a high amount of cells can be erased.

11.3.3.2.3 4D Rotated Constellations

Conventional 2D rotated constellations have been replaced by 4D rotated constellations when QPSK constellation is combined with high coding rates: 8/15, 3/5, 2/3, and 11/15. The 4D rotation is performed by multiplying two vectors consisting of the I/Q components of two adjacent input cells by a 4×4 orthogonal matrix:

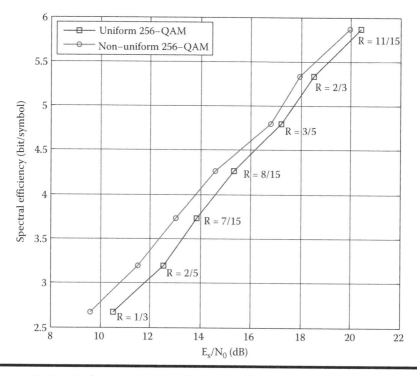

Figure 11.10 **Performance comparison of uniform and nonuniform 256-QAM over the AWGN channel. Both curves display the required SNR to achieve an FER = 10⁻⁴ after LDPC decoding.**

$$\begin{bmatrix} y_{0I} \\ y_{0Q} \\ y_{1I} \\ y_{1Q} \end{bmatrix} = \begin{bmatrix} +a & -b & -b & -b \\ +b & +a & -b & +b \\ +b & +b & +a & -b \\ +b & -b & +b & +a \end{bmatrix} \begin{bmatrix} x_{0I} \\ x_{0Q} \\ x_{1I} \\ x_{1Q} \end{bmatrix}. \tag{11.3}$$

The 4D rotation matrix is characterized by a single parameter r taking values in the range [0,1], referred to as the *rotation factor*, which is defined as

$$r = \frac{3b^2}{a^2}. \tag{11.4}$$

Since the rotation matrix is by definition orthogonal, $a^2 + 3b^2 = 1$. Thus, a and b are derived from r as

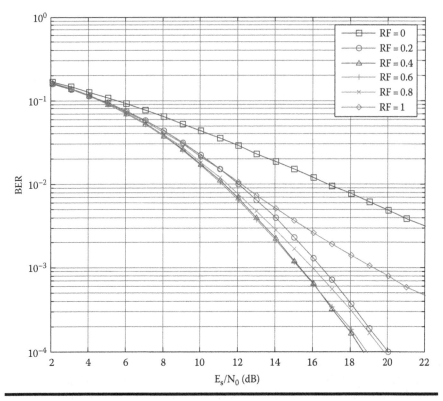

Figure 11.11 **Bit error rate performance of 4D-rotated QPSK for several values of the rotation factor in Rayleigh memoryless channel.**

$$a = \sqrt{\frac{1}{(1+r)}}, \quad b = \sqrt{\frac{r}{3(1+r)}}. \tag{11.5}$$

The optimal value for r was actually chosen to minimize the bit error rate at the demapper output through simulations in Rayleigh fading channels. Figure 11.11 shows the bit error rate performance of a 4D rotated QPSK for different values of the rotation factor.

With 4D rotated constellations, the diversity order of the BICM is doubled compared to the case of DVB-T2-like rotated constellations, now called 2D rotated constellations and quadrupled in comparison with nonrotated constellations. Consequently, 4D rotated constellations behavior trends are similar to 2D rotated constellations but with more pronounced performance: in fading channels, they provide gain only when used with very low constellation sizes and high code rates and they show high robustness in the face of large amounts of deep fades or erasures. From a complexity point of view, at the receiver side, M^2 4D Euclidean distances have to be computed by the demapper if M is the constellation order. Consequently, only 4D rotated QPSK have been considered for DVB-NGH.

11.3.3.2.4 Synthesis

Table 11.8 summarizes the rotated constellations modes and parameters adopted in the DVB-NGH standard.

Figure 11.12 shows the impact of rotated QPSK in Rayleigh fading channels as a function of the coding rate. As already mentioned, the gain of rotated constellations increases as the coding rate rises. In the absence of erasure events (a), rotated QPSK turns out to be of particular interest for coding rates greater than 7/15. In presence of erasures (b), the related gain gets larger and reaches several decibels for the high coding rates.

Applying iterative demapping (ID) at the receiver side guarantees that rotated constellation brings a performance gain in all transmission conditions. For QPSK, this additional gain is limited, but it increases for higher constellation orders.

11.3.3.3 I/Q Component Interleaver

In DVB-T2, the Q component of each rotated constellation signal is cyclically delayed by one cell within the FEC block. In DVB-NGH, the component separation for 2D and 4D rotated constellations has been reconsidered in order to provide maximum time diversity according to the new structure of the TI. Special attention has been paid to guarantee a minimum time and channel separation between

Table 11.8   Summary of the Rotated Constellation Modes in DVB-NGH

Modulation	Code Rate						
	1/3	2/5	7/15	8/15	3/5	2/3	11/15
QPSK	2D ($\Phi = 29.0°$)			4D ($r = 0.4285$)			
16QAM	2D ($\Phi = 16.8°$)						
Uniform 64QAM	2D ($\Phi = 8.6°$)						
Nonuniform 64-QAM	2D ($\Phi = 11.8°$)	2D ($\Phi = 8.5°$)	2D ($\Phi = 6.8°$)	2D ($\Phi = 6.3°$)	2D ($\Phi = 5.5°$)	2D ($\Phi = 5.8°$)	2D ($\Phi = 6.2°$)
256QAM	N/A						

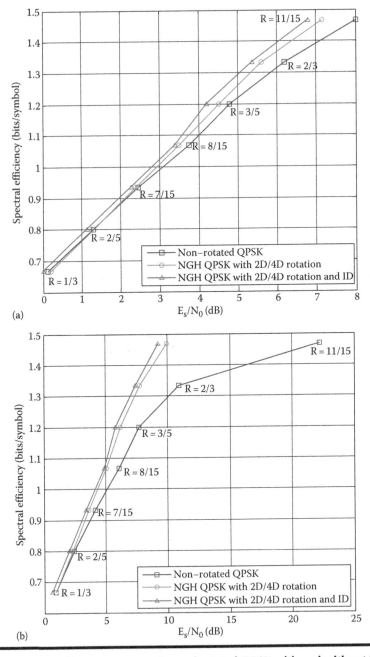

(a)

(b)

Figure 11.12   Performance comparison of rotated QPSK with and without ID and classical QPSK (a) in a Rayleigh memoryless channel and (b) in a Rayleigh memoryless channel with 20% erasure. All curves display the required SNR to achieve an FER = 10^{-4} after LDPC decoding.

the I and Q pairs of the original non-interleaved cells, especially when multi-frame time interleaving (see Chapter 12) or TFS (see Chapter 13) is enabled.

I/Q component interleaving is applied after the 2D or 4D rotation and is performed on each FEC block independently according to the following three steps:

1. The I and Q components of the cells belonging to an FEC block are separately written column-wise into two matrices of the same size.
2. A cyclic shift is applied to each column of the Q-component matrix.
3. The two matrices are read out synchronously row-wise, and complex cells are formed by each read pair of a real (I) and an imaginary (Q) component.

The number of rows N_r in the matrices and the values of the cyclic shifts depend on whether TFS is enabled or not.

When TFS is off, the I/Q component interleaver distributes the D = 2 or 4 dimensions of each constellation evenly over the FEC block, the resulting distance between the D components of each constellation signal being (1/D)th of the FEC length. In this case, N_r is equal to D, and the cyclic shifts of all columns are equal to D/2.

When TFS is on, parameter N_r is a function of the number of RF channels N_{RF}, and the cyclic shift can take $N_{RF}-1$ different values. The values of these parameters are chosen so as to ensure that the D dimensions of each constellation signal are transmitted over all possible combinations of RF channels.

11.3.4 Cell Interleaver

The cell interleaver of DVB-NGH is the same as in DVB-T2. It uniformly spreads the cells in the FEC codewords to ensure in the receiver an uncorrelated distribution of channel distortions and interference along the FEC codewords. This pseudo-random permutation varies from one FEC block to another. However, in contrast to DVB-T2, it is placed *before* the constellation rotation in order not to interfere with the I/Q component interleaver.

When 4D rotated QPSK are used, cell interleaving is applied pair-wise. Taking this constraint into account, applying cell interleaving before or after the constellation rotation—but necessarily before the I/Q component interleaver— is equivalent for 2D and 4D rotations. This allows an ID process to be easily implemented between the LDPC decoder and the rotated constellation demapper at the receiving side, since it does not involve the cell deinterleaver, as shown in Figure 11.13.

11.3.5 Time Interleaver

The TI is placed at the output of the I/Q component interleaver or at the output of the cell interleaver, depending on whether rotated constellations are used or not.

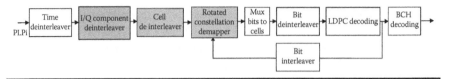

Figure 11.13   Principle of ID implementation at the receiver.

Its structure is inspired from DVB-T2, but it additionally offers the possibility to combine a convolutional interleaver on top of the core block interleaver when interframe interleaving is enabled. It operates at the PLP level, and the TI parameters can vary from a PLP to another.

As in T2-Lite profile, the maximum time interleaving size has been reduced compared to DVB-T2: for the terrestrial link, the memory dedicated to a TI block cannot exceed 2^{18} *memory units* (MU) to store all the PLPs associated with a given service. However, the number of cells contained in an MU is not constant. Actually, low-order constellations such as QPSK and 16-QAM are less sensitive to quantization noise than 64-QAM and 256-QAM. Thus, QPSK and 16-QAM constellations can afford coarser cell quantization than 64-QAM and 256-QAM. For instance, if the quantization of the I and Q components is based on the worst case 256-QAM, 24 bits per cell are typically required (10 bits for I and Q and 4 bits for the channel state information). Some simulations with 16-QAM have shown that the use of 12 bits per cell (4 bits for I and Q and 4 bits for the channel state information) only entails a negligible performance while saving half of the memory. Consequently, *adaptive quantization* has been implemented, which makes the amount of memory needed to store a cell depend on the constellation. In practice, an MU can contain one 64-QAM or 256-QAM constellation cell or two consecutive QPSK or 16-QAM constellation cells. This latter case allows higher time diversity for the most robust modes, since the TI block memory can then store up to 2^{19} cells. It is referred to as *pair-wise interleaving* since both cells in an MU are written and read together.

The core element remains a block row–column interleaver. The cells collected from one PLP constitute an IF. In the simplest case, each IF is implemented as a single TI block interleaver. However, this configuration limits the maximum data rate because of the aforementioned size limitation. To increase the data rate, it is therefore possible to divide the IF into several TI block interleavers (recall that the maximum bit rate including source and parity data is 12 Mbps). Conversely, for low data rate services, longer time interleaving and hence higher time diversity can be achieved by spreading the IF over several NGH logical frames. In this case, the overall TI is implemented as a combination of a convolutional interleaver with a block interleaver. Figure 11.14 illustrates this combined structure.

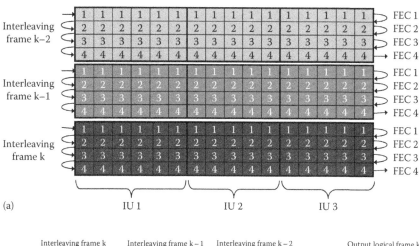

(a)

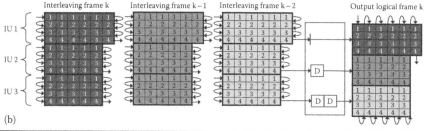

(b)

Figure 11.14 Time interleaving for $N_{IU} = 3$ in the hypothetical case where each FEC codeword length contains 16 cells and each IF contains 4 FEC blocks. (a) Writing process. (b) Delaying and reading process for convolutional interleaving.

The cells to be interleaved are written row-wise into the TI memory, FEC block by FEC block as illustrated in Figure 11.14a. The IF is then partitioned into N_{IU} interleaver units (IU) of nearly the same size, N_{IU} being equal to the number of NGH logical frames mapped by each TI block. Each IU is passed in one of the delay lines of the convolutional interleaver and the cells are afterward read column-wise, as shown in Figure 11.14b. Each input IF is therefore spread over N_{IU} NGH logical frames. This combined block/convolutional TI structure allows for time interleaving depths greater than 1 sec on the terrestrial segment. The depth can be increased to up to 10 s for the satellite link, since the TI memory limitation is then 2^{21} MUs (see Chapter 22). For the terrestrial segment, the IUs are uniformly spread over the TI length, whereas for the satellite link, a *uniform-late* profile can be adopted, making it possible to transmit fewer cells of an FEC block in the first logical frames than in the last logical frames and therefore to decrease the zapping time (see Chapter 12).

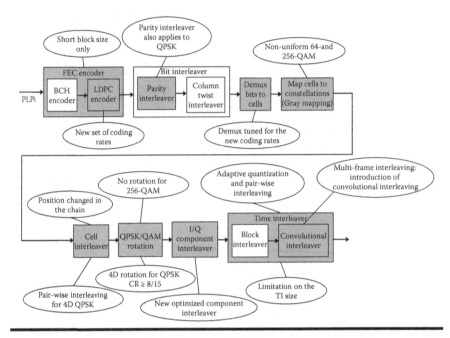

Figure 11.15 Structure of the DVB-NGH BICM module and its differences with DVB-T2.

11.3.6 Summary

The main differences between the DVB-T2 and DVB-NGH BICM are summarized in Figure 11.15.

11.4 Performance Evaluation

The performance of the DVB-NGH BICM module has been evaluated in several transmission channels. Figures 11.16 and 11.17 show simulation results of the DVB-NGH BICM in AWGN and Rayleigh channels for modulation and coding rate combinations having best performance. They also include the unconstrained Shannon capacity and the DVB-H coded modulation scheme for comparison purposes. The curves display the required SNR to achieve an FER = 10^{-4} after LDPC decoding for the DVB-NGH case and the required SNR to achieve QEF performance for DVB-H. In the AWGN channel, DVB-NGH outperforms the first-generation system by around 2.0–2.5 dB. In Rayleigh fading channels, the gain ranges from 3.0 to 7.0 dB. One worthwhile observation is the behavior of the curves for high coding rates. Indeed, DVB-NGH shows performance curves quasi-parallel to the Shannon capacity curve when coding rate increases, while DVB-H curves

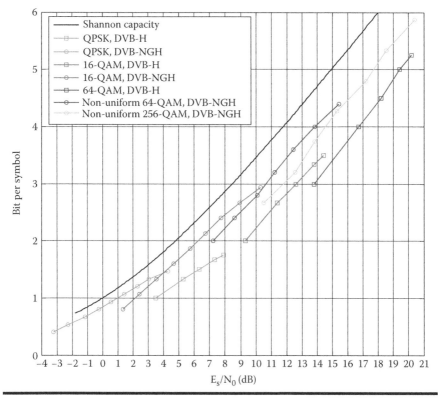

Figure 11.16   Required SNR to achieve an FER = 10⁻⁴ after LDPC decoding over the AWGN channel. Comparison with the Shannon limit and DVB-H. LDPC code rates for DVB-NGH: 1/5 and 4/15 (QPSK and 16-QAM only), 1/3, 2/5, 7/15, 8/15, 3/5, 2/3, 11/15. Convolution code rates for DVB-H: 1/2, 2/3, 3/4, 5/6, and 7/8.

move away from the capacity at high coding rates, and gaps reaching 10 decibels can be observed. This behavior shows that DVB-H suffers from a weaker FEC code and a lack of diversity techniques.

11.5 Conclusions

This chapter provides an overview of the BICM modules of DVB-T2 and DVB-NGH. Important similarities can be seen between both modules. Thanks to state-of-the-art coded modulation technologies, the DVB-T2 BICM proved to exhibit excellent performance, close to capacity. In addition to performance, the strategy of "family of standards" has favored the reuse of technologies adopted in other second-generation DVB standards and has greatly impacted the definition of the BICM module of DVB-NGH. However, some modifications have been introduced

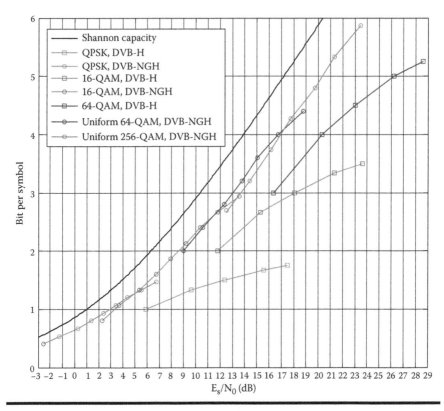

Figure 11.17 Required SNR to achieve an FER = 10⁻⁴ after LDPC decoding over Rayleigh fading channel. Comparison with the Shannon limit and DVB-H. LDPC code rates for DVB-NGH: 1/5 and 4/15 (QPSK and 16-QAM only), 1/3, 2/5, 7/15, 8/15, 3/5, 2/3, 11/15. Convolution code rates for DVB-H: 1/2, 2/3, 3/4, 5/6, and 7/8.

in order to improve robustness, increase spectral efficiency, and optimize existing blocks while limiting the receiver complexity. Indeed, the operating SNR range was extended to include low values via the introduction of low FEC coding rates, essential for mobile operation. Rotated constellations were extended to four dimensions for QPSK in order to increase robustness to deep fade events; nonuniform high-order constellations were introduced in order to achieve a shaping gain; an improved I/Q component interleaver was introduced for rotated constellations in order to better average channel impairments; a convolutional interleaver was added on top of the existing block DVB-T2 TI in order to increase the interleaving depth. Thanks to these techniques, the BICM module of DVB-NGH is at the cutting edge of coded modulation technologies: it spans a SNR range of more than 20 dB and supports spectral efficiencies ranging from 0.67 to 5.87 bits per constellation symbol for the single antenna mode with a gap to capacity ranging from 2 to 3 dB at the QEF point in fading static channels.

Acknowledgments

The authors would first like to thank all their colleagues from the DVB TM-H working group that have contributed to the development of advanced techniques in the BICM module of DVB-NGH.

Special thanks to Jonathan Stott from Jonathan Stott Consulting, Peter Moss from BBC, Mihail Petrov from Panasonic, and Clément Rousseau from Telecom Bretagne for their valuable contribution to this chapter: results, figures, and corrections.

Part of the work dedicated to the BICM module of DVB-NGH was funded by the Eurêka/Celtic-plus CP07-005 ENGINES project.

References

1. A. Guillén i Fàbregas, A. Martinez, and G. Caire, Bit-interleaved coded modulation, *Foundations and Trends® in Communications and Information Theory*, 5(1–2), 1–153, 2008.
2. DVB Technical Module sub-group on Next Generation Handheld, DVB TM-H NGH Call for technologies (CfT), DVB TM-NGH 019r6, November 2009.
3. ETSI EN 302 583, v 1.2.1, Digital video broadcasting (DVB), framing structure, channel coding and modulation for satellite services to handheld devices (SH) below 3 GHz, December 2011.
4. ETSI EN 302 755 v1.3.1, Digital video broadcasting (DVB), frame structure channel coding and modulation for a second generation digital terrestrial television broadcasting system (DVB-T2), October 2011.
5. C. Berrou and A. Glavieux, Near optimum error correcting coding and decoding: Turbo-codes, *IEEE Transactions on Communications*, 44(10), 1261–1271, October 1996.
6. DVB Commercial Module sub-group on Next Generation Handheld, Commercial requirements for DVB-NGH, CM-1062R2, June 2009.
7. M. Eroz, F.-W. Sun, and L.-N. Lee, An innovative low-density parity-check code design with near-Shannon-limit performance and simple implementation, *IEEE Transactions on Communications*, 54(1), 13–17, January 2006.
8. H. Jin, A. Khandekar, and R. J. McEliece, Irregular repeat–accumulate codes, *Proceedings of the International Symposium on Turbo Codes and Related Topics*, Brest, France, 2000.
9. R. G. Gallager, Low-density parity-check codes, *IRE Transactions on Information Theory*, 8, 21–28, January 1962.
10. F. R. Kschischang, B. Frey, and H.-A. Loeliger, Factor graphs and the sum-product algorithm, *IEEE Transactions on Information Theory*, 47(2), 498–519, February 2001.

11. C. Abdel Nour and C. Douillard, Improving BICM performance of QAM constellations for broadcasting applications, *Proceedings of the International Symposium on Turbo Code sand Related Topics*, Lausanne, Switzerland, 2008.

12. M. Li, C. Abdel Nour, C. Jego, and C. Douillard, Design of rotated QAM Mapper/Demapper for the DVB-T2 standard, *Proceedings of the IEEE Workshop on Signal Processing Systems (SiPS)*, Tampere, Finland, 2009.

13. ETSI TS 102 831 v1.2.1, Digital video broadcasting (DVB), implementation guidelines for a second generation digital terrestrial television broadcasting system (DVB-T2), February 2012.

14. D. Gozálvez, D. Gómez-Barquero, D. Vargas, and N. Cardona, Time diversity in mobile DVB-T2 systems, *IEEE Transactions on Broadcasting*, 57(3), 617–628, March 2011.

15. G. D. Forney Jr. and L.-F. Wei, Multidimensional constellations—Part I: Introduction, figures of merit and generalized cross constellations, *IEEE Journal on Selected Areas in Communications*, 1(6), 877–892, August 1989.

Chapter 12

Time Interleaving in DVB-NGH

Pedro F. Gómez, David Gómez-Barquero,
David Gozálvez, Amaia Añorga, and Marco Breiling

Contents

12.1 Introduction

Although Digital Video Broadcasting—Terrestrial Second Generation (DVB-T2) primarily targets static and portable reception, one of the main novelties compared to DVB-T (Terrestrial) is that it incorporates a time interleaver at the physical layer in order to cope with impulse noise and benefit from time diversity in mobile scenarios. The first-generation mobile broadcasting standard DVB-H (Handheld) added an optional forward error correction (FEC) scheme at the link layer known as Multiprotocol Encapsulation—Forward Error Correction (MPE-FEC) to improve the mobile performance of DVB-T. MPE-FEC provides time interleaving durations in the order of 100–400 ms (one time-sliced burst duration), and it compensates the performance degradation of DVB-T under mobility conditions such that the required signal strength in low-speed reception is reduced [1]. However, the protection offered by MPE-FEC in static reception conditions is not useful, while it requires some redundancy overhead. In this sense, DVB-T2 clearly outperforms both DVB-T and DVB-H in mobile channels because it has a single FEC with TI at the physical layer [2].

The TI in DVB-T2 can be configured on a Physical Layer Pipe (PLP) basis and it is very flexible. It allows different trade-offs in terms of time diversity, latency, and power saving. The time interleaver (TI) is a block interleaver (BI) that operates with cells (constellation symbols). The size of the time de-interleaving (TDI) memory in the receivers specified by the standard is approximately 2^{19} cells. The available TDI memory in the receivers limits the maximum TI duration, but it also depends on the data rate, FEC code rate, and signal constellation of the PLP. Although the physical layer of DVB-T2 can provide interleaving durations of up to several seconds for typical mobile TV data rates, the utilization of long TI is limited by the zapping time. The use of TI increases the end-to-end latency and the zapping time, where the latter is a crucial parameter for TV usability. Generally, it is considered that zapping times longer than two seconds are felt as annoying [3]. The average zapping time in DVB-T2 is approximately 1.5 times the TI duration. The reason is that terminals must wait to receive one entire TI-block before they are able to de-interleave it.

The mobile profile of DVB-T2, known as T2-Lite, reuses exactly the same TI as DVB-T2. However, in order to reduce the complexity of stand-alone T2-Lite chipsets, the size of the TDI memory in the receivers has been halved from 2^{19} down to 2^{18} cells. For more details on T2-Lite, see Chapter 6.

The next-generation mobile broadcasting standard DVB-NGH (Next Generation Handheld) has adopted the same TDI memory size as T2-Lite (i.e., 2^{18} cells) for the terrestrial single-input single-output (SISO) and multiple-input multiple-output (MIMO) profiles.* However, DVB-NGH makes a much more efficient use of the available memory, allowing longer TI durations (or larger service data rates for a given TI duration). Furthermore, for a given TI duration, it introduces a shorter end-to-end latency and zapping time. All these benefits are obtained thanks to two new techniques that have been incorporated with respect to DVB-T2.

DVB-NGH has adopted a *convolutional interleaver* (CI) for inter-frame interleaving (i.e., across multiple frames), keeping a BI like in DVB-T2 only for intra-frame interleaving (i.e., within the frame). The CI is introduced in such a way that it is possible to combine both de-interleavers into a single one at the receivers, such that each interleaver does not need dedicated memory. The benefits of the CI are twofold. On the one hand, the CI requires around half the memory than a BI to provide the same time interleaver duration. On the other hand, the zapping time is approximately equal to the TI duration instead of the factor 1.5 like that for the BI. It should be noted that these benefits apply only for inter-frame interleaving, because when only intra-frame interleaving is used, the CI is not employed.

DVB-NGH has also adopted a technique known as *adaptive cell quantization*. This technique allows a higher number of cells to be stored within a given physical TDI memory for robust transmission modes that tolerate a higher quantization noise. DVB-NGH has the same TDI memory size as T2-Lite for 64-QAM and 256-QAM (i.e., 2^{18} cells), but for QPSK and 16-QAM it is possible to interleave up to 2^{19} cells in time. This optimization of the TDI memory allows longer TI durations for a given service data rate or higher service data rates for a given TI duration with both intra-frame and inter-frame interleaving.

This chapter provides an overview of the TI mechanism adopted in the next-generation mobile broadcasting standard DVB-NGH for the terrestrial profile and discusses the different trade-offs related to the use of TI: latency, time diversity, and power consumption. The rest of the chapter is structured as follows. Section 12.2 reviews the TI mechanism of DVB-T2. Section 12.3 is devoted to describe in detail the TI of DVB-NGH, consisting in the combination of intra-frame block interleaving and inter-frame convolutional interleaving. The technique known as adaptive cell quantization is also presented. Section 12.4 addresses implementation

---

* For the MIMO profile, each transmit antenna can make use of only half of the TDI memory.

aspects of the time interleaver at the transmitter side, and the time de-interleaver at the receiver side. Section 12.5 presents illustrative results. Finally, the chapter is concluded with Section 12.6.

12.2 Time Interleaving in DVB-T2

12.2.1 DVB-T2 Time Interleaver

The time interleaver of DVB-T2 consists of a BI that operates on the cells of a PLP. Interleaving cells is more efficient for high-order constellations, such as 64QAM and 256QAM, whereas interleaving bits is more efficient for low-order constellations, such as QPSK and 16QAM. The reason is apparent in the receiver: for cell interleaving, the demapper is located after the TDI (see Figure 12.1a), and the receivers need to store three values for each cell regardless of the constellation order: the components in-phase (I) and quadrature (Q), and the channel state information (CSI), whereas for bit interleaving, the demapper is located before the TDI (see Figure 12.1b) and the receivers need to store only the log-likelihood ratio (LLR) of each bit.

Typically, five quantization bits are required per LLR for soft decoding of terrestrial transmissions (for satellite transmissions, only four bits per LLR are required). For cell interleaving, the DVB-T2 implementation guidelines recommend to store 10 bits for both I and Q components, plus some bits for the CSI [4]. The size of the TDI memory has been set by the standard to $2^{15} + 2^{19}$ cells (recall that T2-Lite has a reduced TDI memory size of 2^{18} cells). At the receivers, there is one single block of on-chip TDI memory, which is shared by the data PLP and its associated common PLP, if any. The memory size is constant for all modulation and code rates combinations (MODCODs).

At the transmitter side, the TI is the last block of the so-called Bit-Interleaved Coding and Modulation (BICM) module. It is located between the cell interleaver and the frame builder, which is responsible to assemble the data and signaling path and construct the T2 frame (see Figure 12.2). For an overview of the BICM modules of DVB-T2 and DVB-NGH, see Chapter 11.

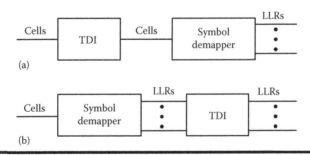

Figure 12.1 **Comparison between cell de-interleaving (a) and bit de-interleaving (b).**

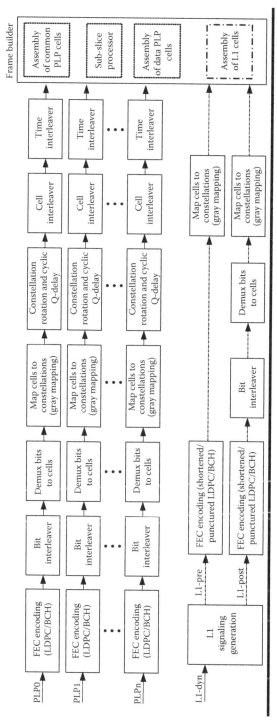

Figure 12.2 DVB-T2 BICM module in the transmitter chain.

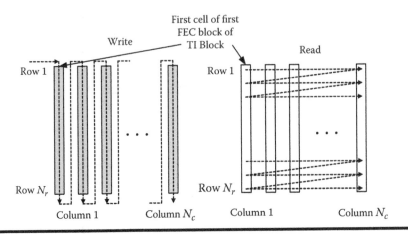

Figure 12.3 DVB-T2 time interleaver based on a BI.

Once an FEC block (LDPC code word) has been processed by the cell inter-leaver, it is fed to the BI. The BI operates on units referred to as TI-blocks, each of them containing an entire number of FEC blocks. The BI consists of a rectangular array of N_r rows and N_c columns. The FEC blocks that form the TI-block are fed into the array column by column and read out row by row (see Figure 12.3). With this TI structure, consecutive symbols before interleaving appear after the BI $N_r - 1$ symbols apart. The received sequence is restored by an inverse operation using an N_r-by-N_c array as de-interleaver. Each FEC block is divided into five columns, such that the number of rows, N_r, is fixed and equal to the size of the FEC block in cells divided by five. The number of columns is variable, and it is five times the maximum number of FEC blocks in the TI-block. Since the data rate of the PLP can vary over time (variable bit rate—VBR), more or less columns of the array are filled depending on the instantaneous PLP data rate, that is, the current number of FEC blocks in the TI-block.

DVB-T2 receivers have to wait until the complete reception of one entire TI-block before they can de-interleave it and process the FEC blocks. Consequently, the zapping time is proportional to the TI duration.

12.2.2 Mapping Options in DVB-T2

There are two basic TI configurations in DVB-T2, known as *inter-frame* and *intra-frame interleaving* [2]. Inter-frame interleaving is achieved when one TI-block is transmitted across multiple T2 frames. If the TI-block is transmitted in a single T2 frame, the TI configuration is known as intra-frame interleaving. By means of inter-frame interleaving, it is possible to extend the interleaving duration beyond

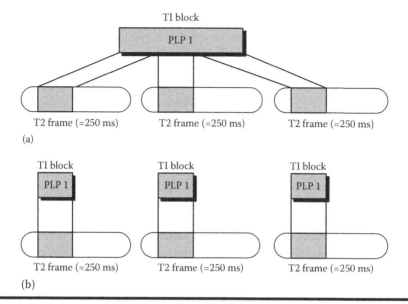

Figure 12.4   Mapping options in DVB-T2: inter-frame (a) and intra-frame (b) interleaving.

one T2 frame (whose maximum duration is 250 ms). An example of both inter-frame and intra-frame interleaving is illustrated in Figure 12.4, where one TI-block is mapped to three (Figure 12.4a) or one single T2 Frame (Figure 12.4b).

12.2.3 Additional Interleaving Configuration Options

DVB-T2 allows additional interleaving configuration options when mapping the TI-blocks to the T2 frames: sub-slicing and frame jumping. Sub-slicing consists of dividing the set of cells of one PLP into smaller units defined as sub-slices that are then spread across the T2 frame. There are two types of data PLPs, which are intended to carry the actual T2 payload services*: data PLPs of type 1 and data PLPs of type 2 [4]. The difference between the two types is the possibility of performing sub-slicing. Data PLPs of type 1 are transmitted in a single burst (slice) within the T2 frames, which maximizes the power saving at the receivers.

---

* A third type of PLP is the common PLP, which is intended to carry data shared between several services, such as signaling (e.g., PSI/SI tables or conditional access information), electronic program guides, or shared service components. This way, there is no need to duplicate this common information for each PLP. Several PLPs can form a group of PLPs, which share one common PLP. The common PLPs are similar to data PLPs of type 1, in the sense that they are transmitted in a single slice after the L1 signaling information.

Data PLPs of type 2 are transmitted in at least two sub-slices within each T2 frame, which increases the time diversity at the expense of higher power consumption. If the number of sub-slices is high enough, it is possible to achieve a quasi-continuous transmission of each type 2 PLP, which maximizes the time diversity but does not save power at the receivers. The number of sub-slices per T2 frame is dynamically configurable, but it is the same for all data PLPs of type 2 carried in the same T2 frame. Sub-slicing can be applied to both intra-frame and inter-frame interleaving. But for all sub-slices of a PLP to have the same size, only certain values of sub-slicing and combinations with inter-frame interleaving are allowed. It should be noted that each PLP can have a specific TI configuration, in addition to a specific MODCOD. However, if sub-slicing is used, the number of sub-slices in one T2 frame is common for all data PLPs employing sub-slicing.

In addition to sub-slicing, DVB-T2 allows *frame jumping* when mapping the interleaving frames to the T2 frames. Frame jumping is also applied on a PLP basis and can be combined with inter-frame interleaving and sub-slicing. It consists in skipping some T2 frames between two T2 frames carrying a PLP, such that such PLPs are not transmitted in every T2 frame but only in a subset of frames regularly distributed over time. This discontinuous transmission mode is also known as time-slicing. The separation between frames carrying information from one PLP is referred to as the *jumping interval* (I_JUMP). When frame jumping is used, the information from the PLPs must be buffered during one jumping interval before it can be transmitted over the air. This increases the amount of information to be transmitted in one T2 frame, which results in a higher TDI memory utilization. The usage of frame jumping reduces the power consumption at the receivers, since they can switch off their RF front ends and OFDM demodulators during the off-times when the PLP is not transmitted. A jumping interval of two T2 frames is represented in Figure 12.5b.

12.2.4 Time Interleaving Trade-Offs in DVB-T2

The TI of DVB-T2 is very flexible, and it allows different trade-offs in terms of robustness (time diversity), power consumption, latency (zapping time), and service data rate. Inter-frame interleaving can be used to extend the interleaving beyond the duration of one T2 frame. By means of inter-frame interleaving, it is possible to achieve interleaving durations up to several seconds at the expense of an increased zapping time and a restriction in the maximum data rate. Frame jumping, on the other hand, reduces the power consumption of the terminals but increases the latency and zapping time because nonconsecutive frames are used. This may reduce the time diversity achieved with intra-frame interleaving due to the nonuniform distribution of information over time. Sub-slicing also improves the time diversity within the frame but increases the power consumption. Next, we review the different trade-offs involved in the TI configuration.

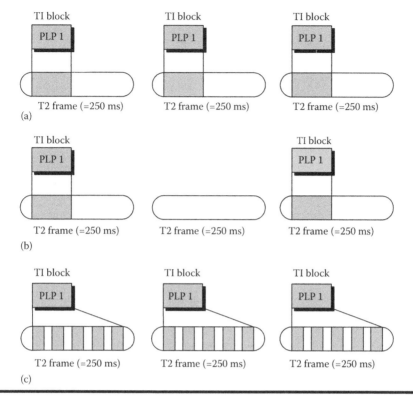

Figure 12.5 Additional mapping options in DVB-T2: $I_{JUMP} = 1$, no sub-slicing (a), $I_{JUMP} = 2$, no sub-slicing (b), $I_{JUMP} = 1$, sub-slicing (c).

12.2.4.1 Time Diversity Trade-Off

Time diversity in DVB-T2 is conditioned by the utilization of inter-frame interleaving, sub-slicing, and frame jumping. Naturally, the time diversity increases as a function of the number of frames interleaved. For a given number of interleaved frames, the maximum time diversity is achieved without frame jumping and with maximum sub-slicing. That is, the maximum time diversity is achieved when the services are transmitted continuously over time (in all T2 frames and in all OFDM symbols).

12.2.4.2 Time Interleaving Duration Trade-Off

The maximum TI duration, $T_{int,max}$, depends on the amount of TDI memory available in receivers and the data rate, the FEC code rate, and the constellation order of the PLP. It can be computed as follows:

$$TI_{int,max} = \frac{TDI \times CR \times \log_2(\mu)}{R_b},$$ (12.1)

where
CR is the code rate
μ is the number of points in the constellation (e.g., 4 for QPSK)
R_b is the PLP data rate (in bps)
TDI is the available TDI memory ($2^{15} + 2^{19}$ cells)

12.2.4.3 Zapping Time Trade-Off

The latency mainly depends on the inter-frame interleaving and the frame jumping configuration. The average zapping time that a receiver must wait until the complete reception of the first interleaving frame, T_{change}, can be computed as follows:

$$T_{change} = T_{wait} + T_{int},$$ (12.2)

where
T_{int} is the time interleaving and de-interleaving duration
T_{wait} is the average period of time until the arrival of a T2 frame carrying the beginning of an interleaving frame

Assuming that PLPs are distributed across the total length of the T2 frames by means of sub-slicing, the value of T_{int} as a function of the inter-frame interleaving and frame jumping settings can be computed as follows:

$$T_{int} = T_{frame} \times \left(I_{JUMP} \cdot (N-1) + 1\right),$$ (12.3)

where
T_{frame} is the duration of the T2 frame
I_{JUMP} is the jumping interval
N is the number of interleaved frames

In DVB-T2, the average value of T_{wait} can be approximately computed as $T_{int}/2$, which is the mean between the best case (an interleaving frame starts immediately after the zapping event) and the worst case (an interleaving frame started shortly before the zapping event, so this interleaving frame is lost). Hence, the average zapping time is approximately one-and-a-half-times the TI duration. If sub-slicing is not used, the values presented here may differ in one T2 frame more or less, depending on the position of the PLPs within the frame.

12.2.4.4 Power Consumption Trade-Off

The receiver power consumption depends only on the frame jumping and sub-slicing settings. The power consumption decreases with a lower number of sub-slices and a higher jumping interval. If the number of sub-slices is rather high, receivers cannot save power because they must have their RF components switched on during the entire T2 frame. With frame jumping, it is possible to save power by skipping the reception of entire T2 frames, even when the PLPs are continuously transmitted within the T2 frames with sub-slicing. The power saving at the receivers, P_{saving}, can be computed as follows:

$$P_{saving}(\%) = 100 \times \left(1 - \frac{\min\{T_{frame} + T_{synch}, N_{slices} \times (T_{synch} + T_{slice})\}}{I_{JUMP} \times T_{frame}} \right), \quad (12.4)$$

where

N_{slices} is the number of slices in the T2 frame

T_{synch} is the time required for re-synchronization after awakening the RF front end from sleep mode

T_{slice} is the duration of one sub-slice, I_{JUMP} is the jumping interval

T_{frame} is the duration of a T2 frame

12.3 Time Interleaving in DVB-NGH

DVB-NGH introduces two main modifications in the TI of DVB-T2: convolutional interleaving for inter-frame interleaving and adaptive cell quantization for QPSK and 16-QAM. Both modifications improve the memory efficiency of the interleaving process, being able to provide longer TI durations for a given service data rate, or higher service data rates for a given TI duration. The CI also reduces the end-to-end delay and the zapping time compared to the sheer BI adopted in DVB-T2. In addition, a new component interleaver has been adopted in order to optimize the performance when using rotated constellations together with Time-Frequency Slicing (TFS) and CI.

12.3.1 Inter-Frame Convolutional Interleaving

DVB-NGH has adopted a CI for inter-frame interleaving, keeping a BI only for intra-frame interleaving like in DVB-T2. Although the time interleaver contains separate inter-frame CI and intra-frame BI functionalities, its structure allows combining both interleavers into one single interleaver at the receivers, such that each interleaver does not need dedicated memory. Figure 12.6 depicts the modifications of the BICM module of DVB-NGH regarding TI. In the figure, we can see the new component interleaver placed after the constellation rotation and before the TI.

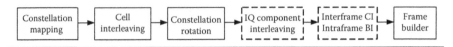

Figure 12.6   Modifications applied at the DVB-T2 BICM module in DVB-NGH that affect the TI.

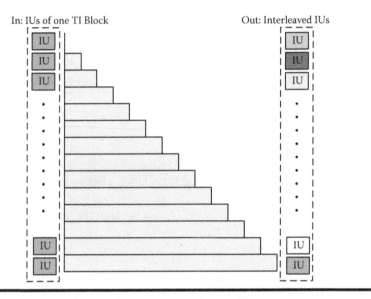

Figure 12.7   Convolutional uniform interleaving profile.

Convolutional interleaving allows different interleaving profiles with different trade-offs regarding time diversity, fast zapping, and memory requirements. Figure 12.7 shows the uniform profile adopted in DVB-NGH for terrestrial receivers. The uniform profile maximizes the time diversity (and hence robustness) although it leads to an increased zapping time. For the hybrid terrestrial–satellite profiles, it is also possible to employ the uniform–late profile that provides long TI with lower zapping times. For more information regarding the uniform–late profile adopted in DVB-NGH, see Chapter 22.

Compared to sheer block interleaving, convolutional interleaving requires less memory and introduces a shorter end-to-end latency and zapping time for the same TI duration. For a given time interleaver duration of N frames, the uniform CI profile requires about half the memory and introduces half the end-to-end delay of the corresponding BI (all interleaving units [IUs] experience a total delay of N frames whereas it is $2N$ with a BI). The average zapping time is reduced by a factor of at least 1/3.

The CI adopted in DVB-NGH operates with sets of cells referred to as *interleaving units* on an FEC block basis. For the terrestrial profiles, only the uniform

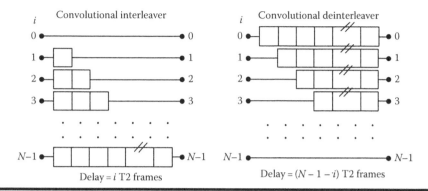

Figure 12.8   Convolutional interleaving.

profile has been adopted, and the maximum interleaving length has been limited to 16 frames. Each FEC block is divided into N IUs, where N is the number of interleaved frames. At the transmitter side, IUs are delayed from 0 to $N - 1$ frames. At the receiver, an inverse operation is performed in order to recover the original data. That is, IUs delayed by i frames at the transmitter are delayed $(N - 1 - i)$ frames at the receiver (Figure 12.8).

12.3.2 Intra-Frame Block Interleaving

Intra-frame BI ensures good interleaving within each NGH frame. The BI of DVB-NGH is similar to the one adopted in DVB-T2 but with some differences. The time interleaver cuts each FEC block into N IUs. Each IU is fed into one column of a BI, and the cells are later read out row by row (see Figure 12.9). Between the writing and the reading, the convolutional interleaving takes place. In contrast to T2 BI, where there is a single BI and where each FEC block is split into five columns, in NGH, there are N small BIs per frame for each PLP, and the number of rows in each BI is identical to or one row less than that for the largest IU in the FEC blocks. The number of columns is variable and equals the number of FEC blocks per NGH frame.

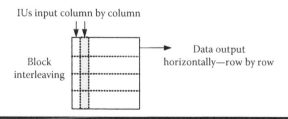

Figure 12.9   Intra-frame block interleaving for one BI.

12.3.3 Adaptive Cell Quantization

DVB-T2 specifies the size of the TDI memory of the receivers in number of cells (constellation symbols). The memory is shared by the data PLP and its common PLP (in case it exists). The size is $2^{19} + 2^{15}$ cells for DVB-T2 and 2^{18} cells for T2-Lite. The memory size is constant for all MODCODs. Using a constant number of cells irrespective of the MODCOD leads to an important reduction in the maximum PLP data rate and maximum TI duration for MODCODs with low spectral efficiencies. Hence, the maximum TI duration for robust transmission modes with low spectral efficiencies is lower than for high-throughput transmission modes.

In contrast to the fixed quantization used in DVB-T2, DVB-NGH has adopted an adaptive scheme. The TDI size of the receivers is 2^{18} cells for 64-QAM and 256-QAM, and 2^{19} cells for QPSK and 16-QAM. This optimization of the TDI memory allows longer TI durations or higher service data rates with QPSK and 16QAM, for both intra-frame and inter-frame interleaving.

The reason for the increased number of cells is that those modulations tolerate a higher quantization noise. The DVB-T2 implementation guidelines recommend storing 10 bits for the I and Q components, plus some bits for the CSI. It is clear that the fixed quantization mentioned earlier was specified to cope with the worst case of 256-QAM and very high spectral efficiencies. Quantization was chosen fine enough such that the quantization noise density, Q_0, does not affect the signal-to-noise ratio (*SNR*) significantly.

$$SNR = \frac{E_S}{N_0 + Q_0}. \tag{12.5}$$

For a given spectral efficiency, the quantization noise density Q_0 must be considerably lower than the background noise density N_0. For 256-QAM, which requires high SNR values, a very high E_S/Q_0 is required. However, for QPSK, which allows for even negative SNRs, the quantization noise can be quite large without degrading performance. For a given TDI memory size in bits, the number of available cells for TI can be larger than that in DVB-T2 for low spectral efficiencies (coarser quantization) and the same number of cells as currently for high spectral efficiencies. Simulation of different degrees of quantization has shown that about half the number of memory bits suffices to store QPSK and 16-QAM cells than for 64- and 256-QAM cells for an acceptable performance loss of less than 0.25 dB. Therefore, the number of available TI cells for QPSK and 16-QAM can be made double that of 64- and 256-QAM.

12.3.4 Component Interleaver

When constellation rotation is used in T2, the normalized constellation symbols of each FEC block, coming from the constellation mapper, are rotated in the complex plane, and the imaginary part is cyclically delayed by one cell within an FEC block. This way, I and Q components are separated by the subsequent interleaving process: At first, the cell interleaver performs a random permutation within the FEC block, and afterwards the time interleaver disperses the cells of an FEC block over time and mixes them with cells from other FEC blocks. Thanks to this scheme, in general, the I and Q components of a constellation symbol travel on different frequencies and at different times. If the channel destroys one of the components, the other component can be used to recover the symbol.

However, this structure is suboptimum in fading channels because often both I and Q of a symbol go to cells, which are close to each other in time. Hence, both I and Q are likely to fade together, that is, both components are often lost simultaneously. This drawback is even more pronounced with TFS as used in NGH. In this case, both I and Q of a symbol would often go to the same time-frequency slice (same time slice and same RF channel). This involves that if this time-frequency slice is lost (e.g., an RF channel is suffering deep fading), the symbol is completely lost, too. For that reason, a new structure defined as component interleaver is proposed to replace the cyclic Q delay block, while the cell interleaver block of T2 (Figure 12.10) is moved to before the constellation rotation block. Figure 12.10 shows the new order of the three blocks within the transmitter chain. Note that the cyclic Q delay of T2 is basically a special type of IQ component interleaver. The IQ component interleaver of NGH is another special type of component interleaving that is optimized for transmission over fading channels and TFS. The reason for moving the cell interleaver will be presented further below.

The new component interleaver improves the T2 concept by moving the Q components within the FEC block more systematically than the combination of cyclic Q delay and cell interleaver. Basically, two modes can be distinguished: one mode specialized to be used with TFS, and another mode when TFS is not used.

Let us first consider the non-TFS mode, which is simpler. When TI over a duration T_{int} is used over a fading channel with coherence time T_{coh} (for the sake

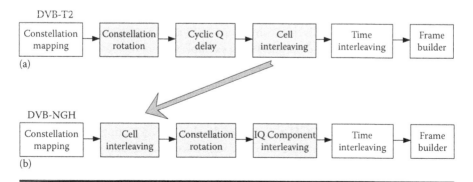

Figure 12.10 **Order of blocks in T2 transmitter (a) and NGH transmitter (b).**

of simplicity, let us assume flat fading—observe that we consider slow-fading here, i.e., shadowing), we have basically three different cases:

- $T_{int} \ll T_{coh}$, that is, the fades last in general much longer than the time interleaver: in this case, the constellation rotation is not really helpful anyway, so the type of component interleaver is irrelevant.
- $T_{int} \approx T_{coh}$, that is, the fade duration is in the same order as the time interleaver duration. Here, a maximum de-correlation of the fading affecting the I and Q components is achieved by separating the two components as far as possible from each other by the component interleaver.
- $T_{int} \gg T_{coh}$, that is, there are many independent fades within the time interleaver duration. The time-interleaved channel has a high diversity, such that the type of component interleaving is quite uncritical.

We find therefore that the second case is most important for defining the type of component interleaving. The component interleaving selected by NGH starts by applying a conventional row–column BI to its input, where the rotated cells are written column-wise and read row-wise. For two-dimensional constellation rotation, this BI has only two rows. The aforementioned maximum separation of the I and Q components is achieved by cyclically shifting all Q components of an FEC block by half an FEC block with respect to their I component counterparts in the next step, that is, after the row–column interleaving. These operations are displayed in Figure 12.11. Note that the NGH standard document describes this cyclic shift of the Q components indirectly by cyclically shifting the columns in the row–column interleaver for the Q component, that is, swapping the two rows. However, this is equivalent to the displayed method. For the four-dimensional constellation rotation that has been introduced in NGH for QPSK constellations, the row–column interleaver contains four rows, and the Q components are cyclically shifted by half an FEC block, too.

Note that in contrast to the cyclic Q delay in T2, the NGH component interleaver changes the position not only of the Q components but also of the I components because of the row–column interleaving. After applying the cyclic shift to the Q components, the new cells made up by a pair of I and Q components (I above Q in Figure 12.11) are read and output. In the example shown, the output cells would be (0, 1), (2, 3), (4, 5), until (9, 8), where the first value is the index of the rotated symbol before the component interleaving, whose I component is taken, and the second value is the respective index for the Q component.

Let us next consider the mode, which applies to the component interleaver, when TFS is used in the NGH transmitter. In this case, we can assume that the different RF channels experience (more or less) uncorrelated fading. However, by contrast to the third case ($T_{int} \gg T_{coh}$) discussed earlier for non-TFS, we benefit only from a limited diversity here, as the number of RF channels is in the order of 2 to at most 7. Therefore, the type of component interleaving does indeed play

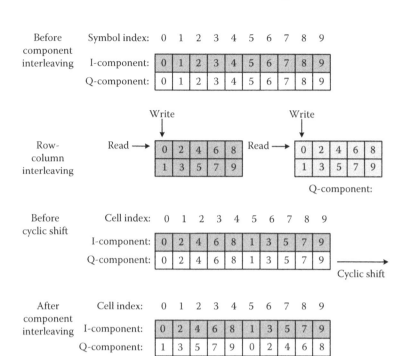

Figure 12.11 **Example of an FEC block of 10 symbols/cells (indexed 0–9), whose I and Q components are shown before and after the component interleaving. A value of 7 means, for example, that the respective component belongs to the rotated symbol with index 7 prior to the component interleaving. Component interleaving comprises the two steps of row–column interleaving (for I and Q) and cyclic shift (only for Q).**

an important role here. The optimum component interleaver should satisfy two conditions:

1. Similar to the non-TFS case described earlier, the separation in time should be maximum, as the same arguments for the temporal fading apply. Once again, this can be achieved by applying a cyclic shift of the Q components by half an FEC block with respect to their I counterparts.
2. Moreover, the component interleaver must not place both the I and Q components of a rotated symbol on the same RF channel.

These requirements are resolved by a row–column interleaver as described for the non-TFS case, where the rotated symbols are column-wise written and row-wise read. However, in the TFS case, each column is cyclically shifted by an individual number of rows before reading the rows. The number of rows to shift takes consecutively one value after the other from a set of $N_{RF} - 1$ "favorable" shift values.

These values ensure that I and Q of a rotated symbol always end up on different RF channels and that any Q component is cyclically shifted by approximately half an FEC block. This will be illustrated by the following example:

Assume that the NGH transmitter uses TFS over $N_{RF} = 3$ RF channels and with $N_K = 2$ TFS cycles, over which the FEC block is dispersed, that is, the FEC block is subsequently transmitted over RF channels 0, 1, and 2 (during TFS cycle 0), and again over 0, 1, and 2 (TFS cycle 1). In this case, the row–column interleaver has $N_R = N_{RF} \times N_K = 6$ rows. Only $N_{RF} - 1 = 2$ "favorable" shift values will be applied for cyclically shifting the columns, which are 2 and 4 in this case. In our example, an FEC block contains 30 cells (see Figure 12.12). After column-wise writing the I and Q components into their respective BIs, the first column of the Q BI is cyclically shifted downward by two rows, the second column by four rows, the third by two rows, the fourth by four rows, and finally the fifth by two rows again. Then both BIs are read out row-wise and the new I–Q pairs form the output cells.

Observe that all symbols, whose I components are transmitted on RF channel 0, have their Q components on either RF channel 1 or 2. Note moreover that the shift values 2 and 4 are close to 3, which would correspond to cyclically shifting the Q component by *exactly* half an FEC block. However, a cyclic shift of three rows would put each Q component on the same RF channel as its I component; therefore, this shift value is not admissible. Using 2 and 4 instead corresponds to shifting the Q component *approximately* half an FEC block with respect to the corresponding I component. Using other shift values (1 and 5) would result in a smaller and therefore suboptimum separation of the I and Q components on average.

In DVB-T2, component interleaving is represented not only by the cyclic Q delay block. The cell interleaver block is also a necessary part of the component interleaving, as the cyclic Q delay alone is too simplistic and does hardly separate the I and Q components. Therefore, during the NGH standardization, it was at first assumed that the cell interleaver block could be dropped after the introduction of the new component interleaver. However, it proved in simulations that the performance without the cell interleaver was considerably degraded with respect to the performance with the cell interleaver. The assumed reason is that the bit interleaver inside the BICM chain is not efficient enough in arbitrarily mixing up the code bits (data and parity bits) of an FEC block. The bit interleaver structure seems to retain too much the vicinity of code bits, which were proximate before the bit interleaving. Hence, in T2, the cell interleaver acts as a kind of an additional "bit interleaver" (although it does not operate on code bits, but on cells representing groups of bits). It was therefore decided to keep the cell interleaver in NGH. As the cell interleaver block performs random interleaving, it must be placed before the component interleaver. Otherwise, it would destroy the carefully optimized order of the I and Q components. Moreover, since the new task of the cell interleaver is related to bit interleaving, it was moved to immediately after the BICM chain, whose last block is the constellation mapper.

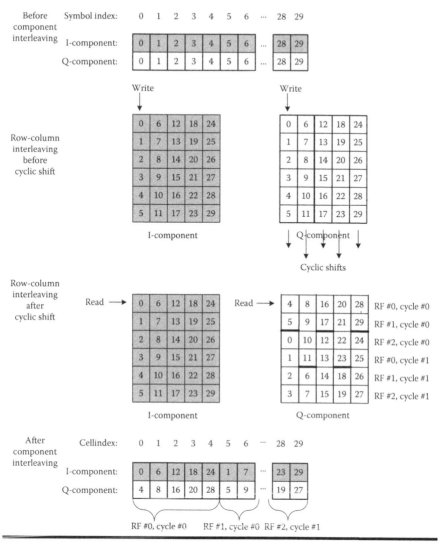

Figure 12.12 Component interleaving for the TFS case with three RF channels and two TFS cycles. At first, the row–column interleavers are written column-wise, then the columns of the BI of the Q component are cyclically shifted by the favorable shift values, and finally both BIs are read row-wise, shaping the new I–Q cells.

12.3.5 Time Interleaving Trade-Offs in DVB-NGH

12.3.5.1 Time Interleaving Duration Trade-Off

As in DVB-T2, the maximum TI duration depends on the amount of TDI memory available in receivers and the data rate, the FEC code rate, and the constellation order of the PLP. The differences in DVB-NGH are that the amount of TDI memory available in receivers depends on the modulation due to the use of adaptive cell quantization and that the CI is used for inter-frame interleaving.

If intra-frame interleaving is applied, the same formula (Equation 12.1) as for DVB-T2 applies because CI is not used. For inter-frame interleaving, the memory saving compared to DVB-T2 increases with the number of interleaved frames, tending to 50%. In particular, the CI requires a memory factor of $(N+1)/2$ instead of N like the BI of DVB-T2, where N is the number of interleaved frames. For the CI uniform profile, the maximum TI duration, $T_{int,max}$, can be computed (assuming no frame jumping) as follows:

$$T_{int,max} = \frac{TDI \cdot CR \cdot \log_2(\mu)}{R_b} \times \frac{2N}{N+1}. \tag{12.6}$$

12.3.5.2 Latency Trade-Off

The latency mainly depends on the inter-frame interleaving and the frame jumping configuration. The CI introduces a shorter end-to-end latency and zapping time in comparison with BI. Thanks to the CI, receivers do not need to wait until the arrival of the next T2 frame carrying the beginning of an interleaving frame, but to the next NGH frame carrying cells from this PLP. As a result, the average zapping time, T_{change}, can be computed in DVB-NGH as follows:

$$T_{change} = T_{wait} + T_{int} = \frac{T_{frame} \times I_{JUMP}}{2} + T_{frame} \times \left(I_{JUMP} \times (N-1) + 1\right). \tag{12.7}$$

12.3.5.3 Power Consumption Trade-Off

The receiver power consumption in DVB-NGH is exactly the same as in DVB-T2, because it depends only on sub-slicing and frame jumping. For more details on these concepts, see Section 12.2.4.4.

12.4 Time Interleaving Implementation in DVB-NGH

12.4.1 Transmitter Side

The time interleaver used in DVB-NGH has two main characteristics that allow it having higher memory efficiency, without degrading the interleaving depth: convolutional interleaving and adaptive cell quantization. In this section, the most important concepts for the implementation of a time interleaver including those two functionalities will be described.

The TI is done at the PLP level. It means that the interleaver configuration (intra-frame or inter-frame interleaving, interleaving length, etc.) might be different for each PLP. Therefore, some PLP-level configuration parameters have to be defined for the implementation of the time interleaver. These parameters are first presented, followed by a detailed description of the time interleaver implementation.

12.4.1.1 Configuration Parameters

Some of the required configuration parameters for the time interleaver are included in the configurable part of the L1 signaling. These parameters are described hereafter, as they are going to be referred to along the whole section (Table 12.1).

Apart from the parameters described in Table 12.1, there is another important parameter to be configured, that is, the number of FEC blocks in the current interleaving frame. As it is a parameter that can vary from one interleaving frame to another, it must be indicated in the dynamic part of the L1 signaling (Table 12.2).

For a better understanding of the parameters presented earlier, some examples are presented in the following lines, considering the three basic TI cases. The first case analyzed (the left-most in Figure 12.13) is the one in which an interleaving frame is divided into N_{TI} TI-blocks. This case does not allow inter-frame interleaving; in other words, each interleaving frame is mapped to exactly one logical frame.

Table 12.1   Time Interleaving Configuration Parameters in the Configurable L1 Signaling

Parameter	Range	Description
TIME_IL_TYPE	0 or 1	**if 0:** each interleaving frame might be divided into several TI-blocks. Only intra-frame interleaving is allowed.
		if 1: each interleaving frame is composed of exactly one TI-block. Intra- and inter-frame interleaving are allowed.
TIME_ILV_ DEPTH	0–16	**if TIME_ILV_TYPE = 0:** number of TI-blocks per interleaving frame.
		if TIME_ILV_TYPE = 1: number of NGH frames which each interleaving frame is mapped to.
PLP_NUM_ BLOCKS_MAX	0–1023	Maximum number of FEC blocks per interleaving frame.
PLP_LF_ INTERVAL	1–16	Distance between two consecutive logical frames carrying information of an interleaving frame.

Table 12.2 Time Interleaving Configuration Parameters in the Dynamic L1 Signaling

Parameter	Range	Description
PLP_NUM_BLOCKS	0 to PLP_NUM_BLOCKS_MAX	Number of FEC blocks in the current interleaving frame

In the other two cases, each interleaving frame is composed of one TI-block. The difference between the two is that in one case, each interleaving frame is mapped to exactly one logical frame (case 2, in the middle of Figure 12.13), while in the other case, the information of one interleaving frame is spread over P_I logical frames (case 3, the right-most in Figure 12.13).

Apart from the configuration parameters already mentioned, a new parameter called $N_{FEC\_TI}(n,s)$ is indicated in Figure 12.13. It represents the number of FEC blocks for the TI-block s of the interleaving frame n, and it does not require a specific signaling parameter, as it can be calculated from the parameters already available. For cases 2 and 3 where an interleaving frame contains only one TI-block (with index $s = 0$), the number of FEC blocks in this unique TI-block must be equal to the number of FEC blocks in the current interleaving frame n ($N_{BLOCKS\_IF}(n)$).

For case 1, the FEC blocks in the interleaving frame n are divided into N_{TI} TI-blocks in the most equitable way. If $N_{BLOCKS\_IF}(n)$ is an entire multiple of N_{TI}, all TI-blocks will have the same number of FEC blocks. Otherwise, all TI-blocks within an interleaving frame will differ by at most one FEC block, and the TI-blocks with the lowest number of FEC blocks will come first.

12.4.1.2 Writing TI-Blocks into Interleaver Memory

The TI requires a whole TI-block to be written into the interleaver memory to start with its operation. For this, each TI-block must be divided into IUs, which are composed of several cells. Each FEC block inside a TI-block is divided into N_{IU} IUs, which must coincide with the number of logical frames to which the current interleaving frame is mapped:

$$N_{IU} = P_I. \tag{12.8}$$

The number of cells of each FEC block (N_{cells_FEC}) must be divided into N_{IU} IUs in the most equitable way. If N_{cells_FEC} is an entire multiple of N_{IU}, all IUs will contain the same number of cells. Otherwise, the IUs of the same FEC block will differ by at most one cell.

The cells within an IU are grouped into memory units (MUs). This is the unit used for writing data into and reading from the interleaving memory. Thanks to

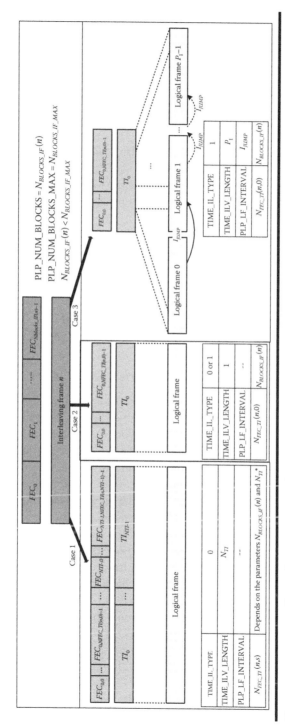

Figure 12.13 TI configuration parameters for the three interleaving cases.

the adaptive cell quantization mechanism adopted in DVB-NGH, the number of cells per MU depends on the modulation used. For the case of QPSK and 16-QAM modulation, an MU is composed of two consecutive cells. If the number of cells within an FEC block is odd, the last MU contains just one cell and the second half of the MU is left empty. For 64-QAM and 256-QAM, each MU contains just one cell. The write operation of the MUs into the interleaver memory is depicted in Figure 12.14 and described as follows:

■ The interleaving is done using $N_{IU} = P_I$ BIs. Each IU within an FEC block is written in one of those interleavers.
■ The number of rows of each BI must be equal to the number of MUs within an IU. This way, each FEC block will occupy exactly one column inside each interleaver.
■ Taking into account that a whole TI-block must be written into the interleaver memory before the read operation starts and that each FEC block occupies one column inside each BI, the number of columns of each BI must be equal to the maximum number of FEC blocks inside a TI-block ($N_{FEC\_TI\_MAX}$). For TI-blocks with less than $N_{FEC\_TI\_MAX}$ FEC blocks, the columns not required are just left empty.
■ The MUs of each TI-block are written column-wise into the BIs.

12.4.1.3 Reading TI-Blocks from Interleaver Memory

The MUs written into the interleaver memory must be read in a predefined way so that the mapping from interleaving frames to NGH frames is correctly done.

As mentioned at the beginning of this section, there are some interleaver configuration cases for which only inter-frame interleaving is performed ($P_I = 1$). For these cases, the time interleaver is composed of only $P_I = 1$ BI and the TI-block that has been written column-wise into it is read out row-wise. If the interleaving frame is composed of several TI-blocks, then the same BI is used in the same way sequentially for each TI-block. The reading operation for this configuration case is shown in Figure 12.15.

It has been already explained that it is also possible to configure the time interleaver in such a way that each interleaving frame is mapped to more than one NGH frame ($P_I > 1$). Also in this case, the IUs that have been written column-wise into the interleaver memory are read out row-wise, but before that, a different delay is applied to each IU. The following formula shows how this delay is calculated for each IU:

$$D(k) = k \cdot I_{JUMP} \quad k: \text{IU Index, from 0 to } P_I - 1. \tag{12.9}$$

From this formula, it can be deduced that the NGH frame with index n will contain the first IU from interleaving frame n, the second one from interleaving

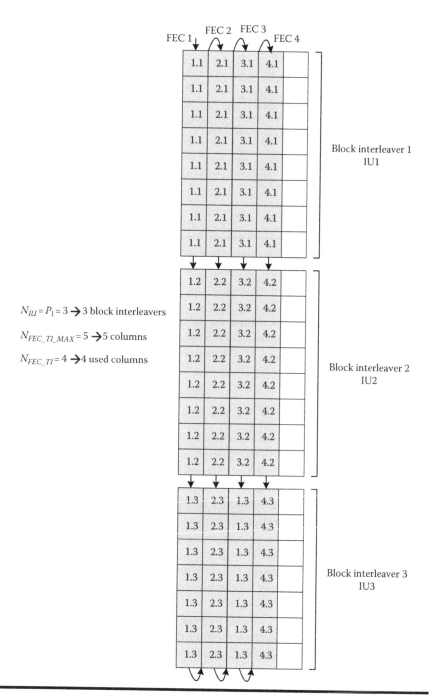

Figure 12.14 Writing a TI-block into the interleaver memory. Each small box represents one MU.

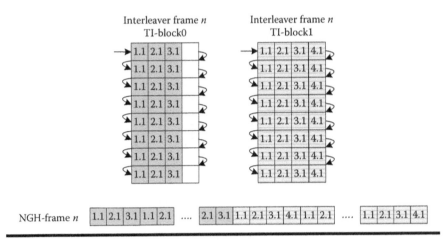

Figure 12.15   Read operation for the case $P_I = N_{IU} = 1$ and $N_{TI} = 2$.

frame $n - I_{JUMP}$, the third one from interleaving frame $n - 2 \times I_{JUMP}$, and so on. This idea is represented in Figure 12.16 for the case $I_{JUMP} = 1$.

As mentioned before, also for this case, the read operation is done row-wise. At this point, it is important to notice that each row with index r is first read for all IUs before jumping to the next row $r + 1$. This read operation is represented in Figure 12.17, for the NGH frame n of Figure 12.16.

12.4.2 Receiver Side

The implementation of the time de-interleaver in the receiver side basically consists of doing the inverse operation to that carried out by the time interleaver presented in the previous section. Therefore, the parameters required are exactly the same as those used for the time interleaver at the transmitter side.

The time de-interleaver is composed of $P_I = N_{IU}$ block de-interleavers, each aimed at saving the MUs corresponding to one IU. The number of columns in each BI has to be equal to the maximum number of FEC blocks per TI-block ($N_{FEC\_TI\_MAX}$), while the number of rows must coincide with the number of MUs per IU. This memory arrangement correctly done together with the information about the number of FEC blocks for the current TI-block (N_{FEC_TI}) is sufficient to start with the de-interleaving operation.

The MUs must be written row-wise into the BIs, as shown in Figure 12.18.

The next step consists of compensating the delays introduced at the IU level in the time interleaver. For that aim, the IUs with the highest delay at transmitter side will be the less delayed at the receiver and vice versa. The formula for the calculation of the new delays is shown here:

$$D(k) = ((P_I - 1) - k) \cdot I_{JUMP} \quad k \text{: IU Index, from 0 to } P_I - 1. \qquad (12.10)$$

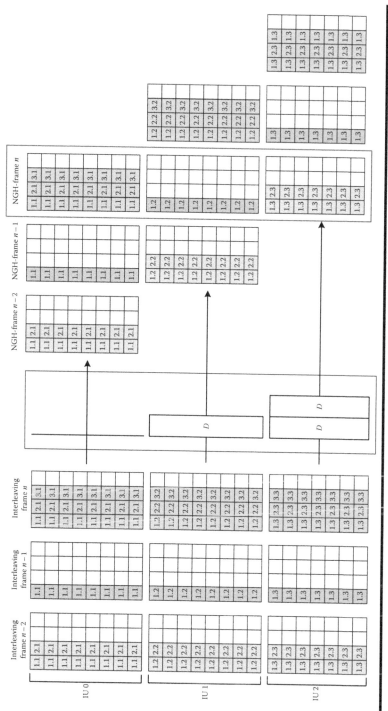

Figure 12.16 Delay operation for the case $P_I = N_{IU} = 3$ and $I_{JUMP} = 1$.

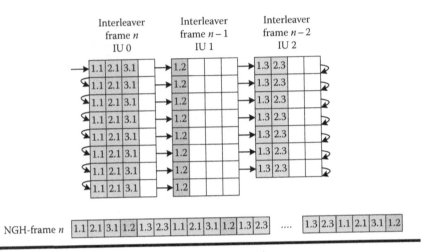

Figure 12.17 Read operation of NGH frame *n* in Figure 12.13.

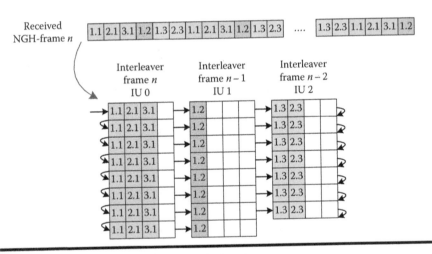

Figure 12.18 Writing MUs into time de-interleaver.

The delay operation is shown in Figure 12.19 for the NGH frame *n* of the example in Figure 12.18.

Once the delays have been compensated, the MUs can be read column-wise from the block de-interleavers. This is the last step within the de-interleaving operation. The read operation for the interleaving frame *n* of Figure 12.19 is depicted in Figure 12.20.

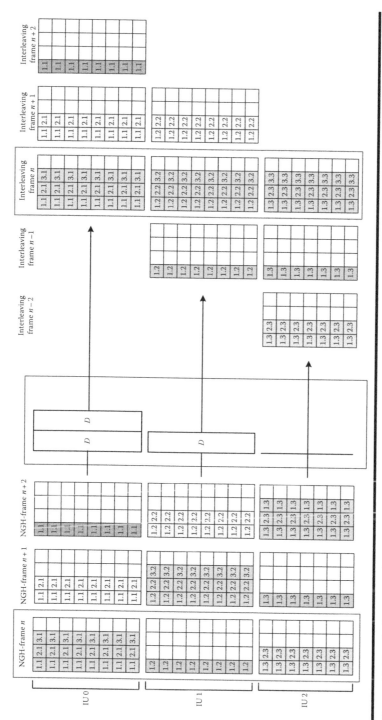

Figure 12.19 Delay operation in the time de-interleaver.

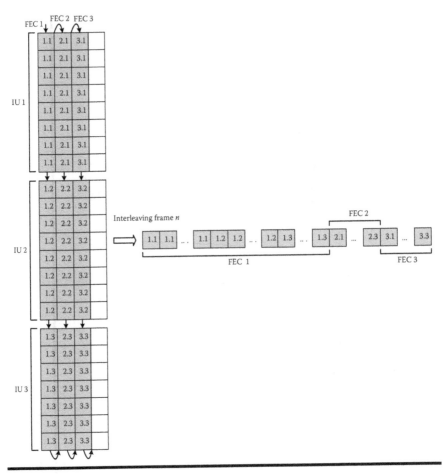

Figure 12.20 Read operation of the time de-interleaver.

12.5 Performance Evaluation Results and Discussions

12.5.1 Time Interleaving Trade-Offs in DVB-T2 and DVB-NGH

12.5.1.1 Power Consumption Trade-Off

Figure 12.21 illustrates the power saving in DVB-T2 and DVB-NGH for different combinations of sub-slicing and jumping interval. The results are given for 8K FFT, guard interval 1/4, PP1, and frame length 250 ms. It has been assumed that that a receiver needs 15 ms to wake up and start receiving each time slice. A total of 2^{18} cells of data from a PLP are transmitted in each frame. In the figure, we can see that the power saving diminishes with the number of sub-slices per frame and that values

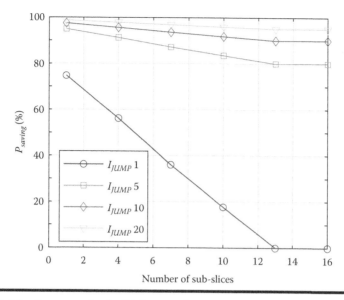

Figure 12.21 Power saving in DVB-T2 and DVB-NGH.

higher than 90% cannot be achieved without the use of a jumping interval. It should be pointed out that a faster wake-up time and a lower number of cells transmitted per PLP (e.g., due to lower data rates or the utilization multi-frame interleaving) will result in better power-saving values than those shown in Figure 12.21 [5].

12.5.1.2 Time Interleaving Duration Trade-Off

In Figure 12.22, we compare the maximum interleaving duration of DVB-T2, T2-Lite, and DVB-NGH. The results have been obtained for QPSK, as this is the most limiting constellation. It should be noted that the combination of convolutional interleaving and adaptive cell quantization allows terrestrial DVB-NGH receivers to support interleaving durations four times longer than T2-Lite receivers and two times longer than DVB-T2 [5]. According to Figure 12.22, the highest interleaving duration in DVB-T2 and T2-Lite receivers is 1 and 2 s respectively, whereas interleaving durations up to more than 4 s can be achieved in terrestrial DVB-NGH receivers.

12.5.1.3 Zapping Time Trade-Off

Figure 12.23 compares the average zapping time of DVB-T2 and DVB-NGH without fast access decoding. In both cases, the zapping time increases with the interleaving duration and the jumping interval (I_{JUMP}). We can see that the zapping time of DVB-NGH is approximately 33% lower than in DVB-T2 due to the utilization of a CI instead of a BI for inter-frame interleaving. In spite of this,

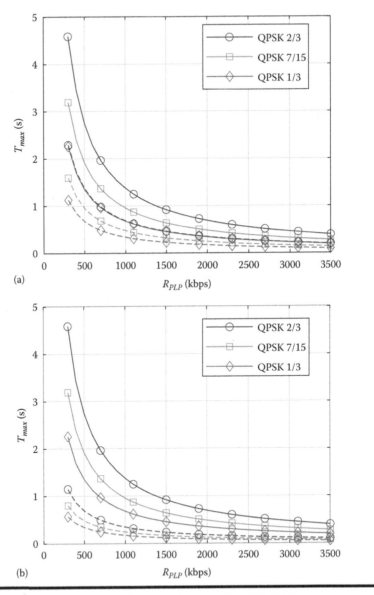

Figure 12.22 Maximum interleaving duration in DVB-NGH according to the PLP data rate. Dashed lines correspond to DVB-T2 (a) and to T2-Lite (b).

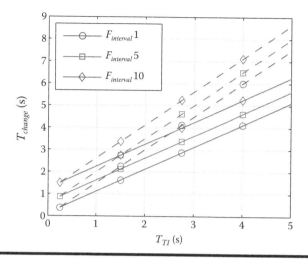

Figure 12.23 **Average zapping time in DVB-NGH according to the interleaving duration. Dashed lines correspond to DVB-T2.**

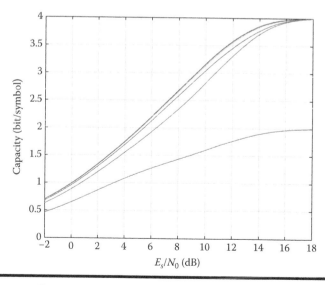

Figure 12.24 **Cell quantization of 16-QAM with 1 bit (bottom) to 7 bits (top) per I or Q component.**

interleaving durations of several seconds are not possible with tolerable zapping times (i.e., below 2 s) without fast access decoding [5].

12.5.2 Adaptive Cell Quantization Results

In Figure 12.24, we compare the capacity of nonrotated 16-QAM according to the number of bits used for quantizing the cells' I and Q components. The results show that, in this case, 4 bits for quantized I/Q plus 4 bits for quantized CSI are sufficient to achieve a performance very close to the case with nonquantized inputs. In particular, the degradation caused by the cell quantization is lower than 0.1 dB. The utilization of 4 bits for the I/Q components plus 4 bits for CSI results in 12 bits/cell instead of 24 bits/cell as currently recommended in the implementation guidelines of DVB-T2.

12.6 Conclusions

This chapter has focused on the TI scheme adopted in DVB-NGH. In particular, DVB-NGH has included a BI for intra-frame interleaving in the order of hundreds of milliseconds together with a CI for inter-frame interleaving up to several seconds. The combination of block and convolutional interleaving in DVB-NGH provides an advantage in terms of maximum interleaving duration and average zapping time. On one hand, the utilization of convolutional interleaving and adaptive cell quantization multiplies by four the maximum interleaving duration compared to T2-Lite. On the other hand, the CI reduces the average zapping time by about 33% for inter-frame interleaving compared to the BI in DVB-T2. The new component interleaver has been optimized for the combination of rotated constellations with inter-frame interleaving and TFS in such a manner that the components of each rotated symbol end up being transmitted with the maximum possible separation in the time domain and in different RF channels.

References

1. G. Faria, J. A. Henriksson, E. Stare, and P. Talmola, DVB-H: Digital broadcast services to handheld devices, *Proceedings of the IEEE*, 94(1), 194–209, 2006.
2. D. Gozálvez, D. Gómez-Barquero, D. Vargas, and N. Cardona, Time diversity in mobile DVB-T2 systems, *IEEE Transactions on Broadcasting*, 57(3), 617–628, 2011.
3. H. Fuchs and N. Färber, Optimizing channel change time in IPTV applications, *Proceedings of the IEEE International Symposium on Broadband Multimedia Systems and Broadcasting (BMSB)*, Las Vegas, NV, 2008.
4. ETSI TR 102 831 v1.2.1, Implementation guidelines for a second generation digital terrestrial television broadcasting system (DVB-T2), 2012.
5. D. Gozálvez, Combined time, frequency and space diversity in multimedia mobile broadcasting systems, PhD dissertation, Universitat Politècnica de València, Valencia, Spain, June 2012.

Chapter 13

Time-Frequency Slicing for DVB-NGH

Jordi Joan Giménez, David Gómez-Barquero, Staffan Bergsmark, and Erik Stare

Contents

13.1 Introduction

Time-frequency slicing (TFS) is a novel transmission technique that consists in transmitting the digital TV services across several radio frequency (RF) channels by means of frequency hopping and time-slicing (i.e., discontinuous transmission of each service). TFS breaks the existing paradigm of transmitting TV services in a single RF channel (multiplex), where the reception of a particular service is simply performed by tuning the RF channel that carries the desired service. With TFS, services are sequentially and discontinuously transmitted over a set of several RF channels. Figure 13.1 depicts the differences between the transmission of three services in the traditional way (each multiplex in a different RF channel) and using TFS over three RF channels. With TFS, the reception of a particular service is performed following dynamically over time the frequency hops among different RF channels.

The advantages of using TFS for the transmission of TV services can be addressed from two different points of view: a coverage gain due to improved frequency diversity and a capacity gain due to improved statistical multiplexing (StatMux) for variable bit rate (VBR) services. The frequency diversity provided by TFS can be very significant, since services can be potentially spread over the whole

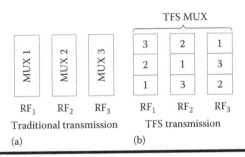

Figure 13.1 Traditional digital TV transmission (a) and TFS (b) over three RF channels.

RF frequency band. Furthermore, the combination of multiple RF channels into a single TFS multiplex allows for an almost ideal StatMux gain [1].

TFS was originally proposed within the standardization process of DVB-T2 (Digital Video Broadcasting—Second Generation Terrestrial) [2]. Although TFS was fully specified in the DVB-T2 standard, the TFS part was labeled only as *informative* due to the need of implementing two tuners (front ends) at the receivers, which would have caused an important increase in receiver complexity and implementation cost. The next-generation mobile broadcasting standard DVB-NGH (Next Generation Handheld) has adopted TFS because it can be operated with a single tuner without adding excessive complexity at the receivers.

13.1.1 TFS in DVB-T2

The main driving force for DVB-T2 was to increase the spectral efficiency of digital terrestrial TV (DTT) networks for the transmission of HDTV services. TFS was originally proposed by the Swedish broadcast network operator Teracom in order to increase the efficiency of the StatMux of high-quality TV services and to improve coverage due to increased frequency diversity. Although DVB-T2 provides a significant capacity increase over its predecessor DVB-T (Terrestrial), in the order of 50% up to 70% [3], the number of HDTV services that can be allocated per multiplex is limited. Hence, it is not possible to exploit the maximum StatMux gain. The main idea behind TFS was to combine the capacity of several RF channels (up to six) to create a high-capacity multiplex that would allow for an almost ideal StatMux gain [4]. But TFS was also proved to provide an important gain in terms of coverage due to enhanced frequency diversity. Large-scale field measurements performed during the T2 standardization process by Teracom reported gains in the order of 4–5 dB with four RF channels for both fixed roof-top and portable reception [1]. The TFS coverage gain was expressed as the difference between the average signal strength and the minimum signal strength calculated over all RF frequencies in each location. It should be noted that this coverage gain can be turned into a capacity gain, since it would be possible to use a transmission mode with a higher spectral efficiency keeping the same coverage. Furthermore, this gain is independent of the gain due to improved StatMux, so the two gains will add (i.e., gain factors can be multiplied).

Although not evaluated in the T2 standardization process, it was acknowledged that TFS can also improve the transmission robustness against channel time variations and the tolerance against static and time-varying interferences, since the interference level usually varies with frequency. Thanks to the frequency interleaving across several RF channels, depending on the error correction capability of the Forward Error Correction (FEC) scheme, one, or even several, of the RF signals could be completely lost, provided that the other RF signals are good enough.

The major disadvantage of TFS in DVB-T2 is the requirement of two tuners at the receivers to receive a single service, which makes the receivers more complex and expensive. It was not possible to always guarantee a time interval between successive frames of the same service long enough for frequency hopping among RF channels with a single tuner. However, internally in a frame, a single tuner would have been sufficient. Hence, the technique was finally moved from the normative part of the T2 specification and was left as an informative annex, although specified in detail, for future implementations [2]. Full support for TFS can be also found in surrounding specifications such as PSI/SI (e.g., the T2 delivery system descriptor includes a field for TFS) [5], and the interface between the gateway and the modulators known as T2-MI [6].

13.1.2 TFS in DVB-NGH

In a similar way as for DVB-T2, TFS could also be very beneficial for DVB-NGH, offering both a gain in capacity due to more efficient StatMux and a gain in coverage due to increased frequency diversity. However, the point of view from which TFS has been addressed in DVB-NGH is slightly different from that of DVB-T2. For DVB-NGH, it is considered as a likely scenario that NGH services are introduced within one or more existing T2 multiplexes by the future extension frame (FEF) mechanism [7]. On each RF channel, there would therefore be a time division between the T2 signal and the NGH signal. For an NGH receiver, the T2 frame (which does not have to be received) would provide a natural guard period, allowing the receiver to perform frequency hopping at frame boundaries. Furthermore, in contrast to DVB-T2, where the focus was on HDTV services with very high bit rate, in DVB-NGH, the maximum coded service data rate (i.e., including source and parity data) has been limited to 12 Mbps to reduce the complexity of the receivers. Such limitation relaxes the time constraints to correctly follow the frequency variations of the transmitted TFS signal, and allows, together with the aforementioned guard period between NGH frames, the use of receivers with a single tuner.

The main issues that prevail in mobile broadcasting are related to improvements in coverage to cope with the more severe propagation conditions and reduce the network infrastructure investments, and improvements in the power consumption of the receivers to increase the battery life. Hence, whereas the main driver for DVB-T2 was increased capacity, the main goal of DVB-NGH is improved coverage, especially for pedestrian indoor reception [7].

The benefits of TFS in terms of coverage in DVB-NGH go beyond those identified in the DVB-T2 standardization, where fixed reception was the most important reception case. TFS cannot only provide a significant coverage gain, but also improve the robustness of the transmitted signal. The additional frequency diversity is especially important for pedestrian reception conditions, where the time diversity is very little or inexistent. Moreover, for mobile reception, the increased frequency diversity can reduce the requirements for time interleaving, reducing the end-to-end latency and zapping time. This applies to fixed as well as portable and mobile reception.

Another important difference in TFS operation is that it is not feasible to implement TFS frequency hopping within a frame (i.e., intra-frame TFS) when using fairly short FEFs to transmit NGH services. The frequency hopping is then instead done across frames (i.e., inter-frame TFS).* Inter-frame TFS does not allow to exploit StatMux in the same way as intra-frame TFS, because the total multiplex capacity is limited to that of one RF channel (although hopping), whereas with intra-frame TFS, several RF channels are used simultaneously for TFS. However, it maintains the coverage gain due to combined frequency and time diversity.

Another important benefit of TFS is the possibility to find spectrum for DVB-NGH services more easily and in a more flexible way, since it is possible to combine several RF channels with different percentages of utilization allocated to DVB-NGH, thanks to the new FEF bundling mechanism defined in DVB-NGH. For more details on FEF bundling, see next Chapter 14.

TFS has been included in the sheer terrestrial single-input single-output/multiple-input single-output (SISO/MISO) base profile as well as in the multiple-input multiple-output (MIMO) profile of DVB-NGH. The maximum MIMO performance is achieved when both horizontal (H) and vertical (V) polarization signals are received with the same power. However, for a given RF channel, the frequency and directional variations of the H and V antenna diagrams can be highly different, and there can therefore be large variations already in the transmitted signal. TFS can smooth the negative effects of such variations, since the difference varies with the frequency. For more details on the MIMO terrestrial profile of DVB-NGH, see Chapter 19.

This chapter provides an overview of the transmission technique TFS, describes its implementation in DVB-NGH, and provides illustrative performance evaluation results about the coverage gain with field measurements and physical layer simulations. The rest of the chapter is structured as follows. Section 13.2 provides some background information for TFS. It describes the two transmission modes possible: inter-frame TFS[†] and intra-frame TFS,[‡] and elaborates the TFS gains in terms of capacity, coverage, and against interferences. Section 13.3 describes implementation aspects of TFS in DVB-NGH at the gateway, modulators, and receivers. Section 13.4 presents field measurement results about the TFS coverage gain, and Section 13.5 physical layer simulations results. Finally, the chapter is concluded with Section 13.6.

---

* DVB-NGH has defined a new logical frame structure especially suited to the transmission using FEFs. Logical frames do not need to be synchronized to physical frames and may span several physical frames. In this chapter, unless otherwise stated, the term frame is used for physical layer frames. For a detailed description of the logical frame structure of DVB-NGH, see Chapter 14.

[†] Using logical channels of Type D.

[‡] Using logical channels of Type C.

13.2 Time-Frequency Slicing Background

TFS can be implemented within the same frame (intra-frame TFS), by means of frequency hopping between the sub-slices of one frame that are transmitted in different RF channels, and frame-by-frame (inter-frame TFS), using frequency hopping only at frame boundaries. The implementation of one or the other depends on the availability of capacity for DVB-NGH on the different RF channels, the type of the Physical Layer Pipe (PLP),* and the frame length.

13.2.1 TFS Operation Modes

For intra-frame TFS, frequency hopping is performed within the frames between the sub-slices of the PLPs, as depicted in Figure 13.2. The sub-slices of one PLP are uniformly distributed among the set of RF channels of the TFS multiplex. The number of sub-slices must be an entire multiple of the number of RF channels. It should be noted that the different RF channels must be synchronized such that the receivers can perform frequency hopping between sub-slices. This mode of operation allows for intra-frame and inter-frame time interleaving. With intra-frame time interleaving, all sub-slices of one PLP within one frame are jointly interleaved in frequency and time and, thus, increasing time diversity. When inter-frame interleaving is used, the FEC blocks of each PLP are convolutionally interleaved over several frames prior to this. Intra-frame TFS allows to exploit both increased frequency and time diversity as well as enhanced StatMux.

Intra-frame TFS is suited for long NGH frames (e.g., 150–250, 250 ms being the maximum frame size for both DVB-T2 and DVB-NGH), as it may not be possible to guarantee a correct frequency hopping among RF channels within shorter frames with a single tuner. However, this mode of operation also depends on the number of RF channels, the number of sub-slices, the MODCOD, and peak rate for the PLP. For example, the lower the MODCOD and the higher the selected peak bit rate, the larger will the total amount of cells be in the frame for the PLP. Increasing the number of RF channels also leads to lowering the time interval for frequency hopping.

With inter-frame TFS, each frame is transmitted in a different RF channel, and frequency hopping is performed on a frame basis (see Figure 13.3). Inter-frame TFS can be used for both Type 1 and Type 2 PLPs. It is the only TFS operation mode possible for short frames and Type 1 PLPs. Inter-frame TFS *requires* inter-frame time interleaving in order to jointly exploit the time and frequency diversity. It can

---

* The use of TFS is linked to the PLP concept introduced in DVB-T2. A PLP is a logical channel at the physical layer that may carry one or multiple services, or service components [3]. Each PLP can have different bit rates and error protection parameters (modulation, code rate, and time interleaving configuration, MODCODTI). PLPs of Type 1 are transmitted in a single burst (slice) within a logical frame. PLPs of Type 2 are transmitted in at least two sub-slices within each logical frame. Both types of PLPs allow multi-frame time interleaving over several logical frames.

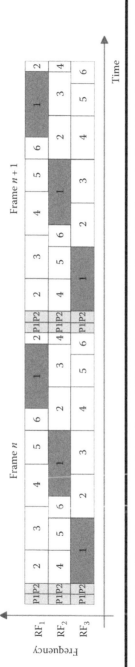

Figure 13.2 Example of two frames with intra-frame TFS for six PLPs over three RF channels. For this logical channel type (LC Type D) each set of time synchronized frames (RF1, RF2, RF3) contains exactly one logical frame.

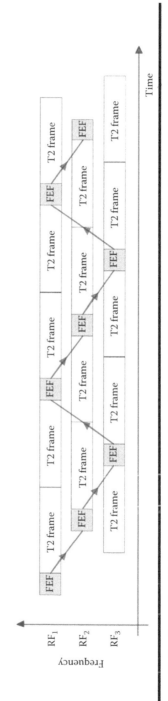

Figure 13.3 Example of inter-frame TFS transmission with DVB-NGH services in FEFs of DVB-T2 over three RF channels.

exploit a somewhat longer time interleaving durations than intra-frame TFS, but the drawback is that it is not possible to improve the efficiency of the StatMux. On the other hand, the time constraints between frames are usually more relaxed.

13.2.2 Statistical Multiplexing Gain

TFS can provide a capacity gain due to a more efficient StatMux [1,4]. StatMux exploits the fact that video codecs produce streams of VBR depending on the encoded content. Without StatMux, the capacity of a multiplex should be divided among the different services in a fixed way. This implies a constant bit rate (CBR) video encoding, which does not guarantee an optimum bandwidth usage. StatMux takes advantage of the fact that, statistically, and for a given video quality level, the instantaneous overall peak bit rate of all video streams together is significantly lower than the sum of the peak bit rates of each individual video stream, assuming a central control unit that dynamically allocates capacity to each service while trying to keep the quality of all services constant and potentially the same. Figure 13.4 illustrates the difference between CBR encoding and VBR encoding with StatMux.

The so-called StatMux gain is defined as the percentage reduction of the required bit rate compared to CBR encoding for a given quality. The StatMux gain depends on the number of services jointly encoded and multiplexed. Obviously, there is no gain for a single service, but the gain increases asymptotically as a function of the number of services until there is a point where it saturates. For example, in DVB-T2, the maximum StatMux gain for HDTV is reached when multiplexing 18–24 programs, in which case the StatMux gain is about 32% [1], corresponding to a (virtual) capacity gain of 47% ($1/(1 - 0.32) = 1.47$). In NGH, the capacity per multiplex is expected in general to be significantly lower than that in DVB-T2, but also the bit rate per service may be correspondingly lower, resulting in approximately the same number of services and a similar StatMux gain. Using, for example, an NGH configuration with 10 Mbps and 20 services, the bit rate per service becomes about 500 kbps, which may be adequate for a mobile service.

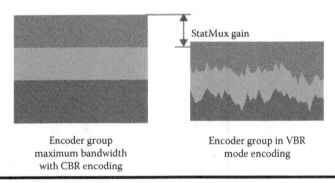

Encoder group
maximum bandwidth
with CBR encoding

Encoder group in VBR
mode encoding

Figure 13.4   CBR encoding vs. VBR encoding with StatMux.

It should be noted however that whereas the earlier T2 figures are derived using the MPEG-4/AVC coding standard, the new HEVC standard is a likely candidate for coding with NGH, and the StatMux properties of HEVC are not yet known.

13.2.3 Coverage Gain

TFS provides a coverage gain due to increased frequency diversity. In general, the signal of each RF channel is affected by different propagation conditions that cause imbalances in the received strength at each location although the same effective radiated power is transmitted in all channels [8]. The imbalances depend not only on the characteristics of the particular propagation scenario, but also on the frequency-dependent behavior of some physical elements of the transmission chain (e.g., antenna radiation patterns and ground echo), and the presence of interferences from other networks.

In a traditional DTT network, the perceived coverage of a set of services at a given location is determined by the channel with the worst signal level in each location. Receiving this worst channel assures correct reception of all the multiplexes, which is a natural commercial requirement. With TFS, on the contrary, the coverage is more likely to be determined by the *average* signal strength among all RF channels used by the TFS transmission. Figure 13.5 shows an illustrative example of the effect in the coverage of using TFS in a DTT network with three different RF channels. It should be noted that TFS enlarges the area where all services are correctly received, but reduces the area where at least one service is received (that corresponds to the RF channel with the best signal level). It should be noted that commercially, when competing digital TV delivery media exist, a missing multiplex is likely to make the user unsatisfied with the DTT service and may trigger a move to alternative delivery media.

The TFS coverage depends on the number of RF channels involved in the transmission and the frequency spacing among them. In general, the gain increases

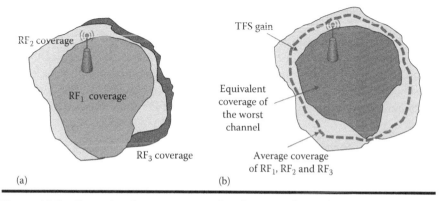

Figure 13.5 Example of area coverage for three RF channels in a DTT network without TFS (a) and with TFS (b).

with the number of RF channels and the frequency spacing. The potential gain is very high because with appropriate coding and interleaving, even a fully lost RF channel may still allow correct reception of all services. This is similar to the situation with conventional COFDM where a 0 dB echo may totally cancel, for example, every fourth OFDM carrier, but reception only be slightly degraded by this (a few dBs of *C/N* loss). What matters in both cases for reception is the "global" quality and not so much the quality of a particular RF channel (or OFDM carrier). When there are large variations in signal strength between different RF signals, the "good" RF channels tend to compensate for the bad ones. Rotated constellations (RCs) [9] can further improve the performance of TFS by means of increasing diversity [10].

13.2.3.1 Rotated Constellations and TFS

With conventional non-rotated constellations (non-RCs), the in-phase (I) and quadrature (Q) components of one constellation symbol are transmitted in a single cell at a given time and frequency (carrier). Hence, the loss of the cell implies the complete loss of the constellation symbol. Rotating the constellation in the complex plane with an appropriate angle, each transmitted component (I and Q) contains enough information by itself to allow the receiver to know which symbol was transmitted. Thus, if one of the components is lost (e.g., affected by a deep selective fading of the channel), the other can be used to recover the complete symbol. To be able to exploit this additional gain for TFS, the I and Q components of each cell should be transmitted in different RF channels, since these may have a very different fading. In DVB-T2, a cyclic delay is applied on the Q component after the constellation rotation, such that the I and Q components of each symbol are transmitted in consecutive cells. Then, a cell interleaver is applied to randomize the position of each component within the FEC frame. After time interleaving, I and Q of the same original cell appear in quasi-random positions in the T2 frame. This implies that I and Q sometimes appear in the same RF channel, which is undesirable. With, for example, two RF channels, every second I component has its original Q component in the same RF channel [10]. In DVB-NGH, a new component interleaver has been introduced to guarantee that the transmission of the component of the same symbol does not occur on the same RF channels. The resultant I/Q cells from the component interleaving with four RF channels and one TFS cycle over which a FEC block is time interleaved are depicted in Figure 13.6. For details on the bit-interleaved coding and modulation module of DVB-NGH, see Chapter 11.

13.2.4 Interference Reduction

From the interference point of view, TFS can also provide a gain as the interferences from other transmitters are usually frequency dependent [1]. Such interference reduction can be exploited to improve the coverage in interference-limited areas or to allow tighter frequency reuse patterns such that more DVB-NGH networks can

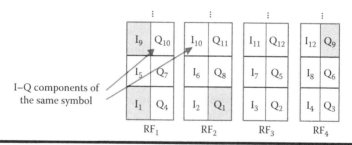

Figure 13.6   IQ cell transmission after interleaving. The I and Q components of the same symbol are transmitted in different RF channels.

fit within a given spectrum. In the T2 standardization process, interference gain was shown to exist in networks based on the frequency allocation plan GE'06, but it was not quantified.

TFS may also be beneficial to reduce potential interferences caused by the deployment of long-term evolution (LTE) cellular services. LTE will use the upper part of the UHF band (channels 61–69) as the result of the digital dividend after the analogue switch-off. These transmissions may have an adverse effect on broadcast reception on RF channels close to LTE. However, using TFS, only a part of the signal (corresponding to the RF channels close to LTE) would be affected, and reception could still be successful, with only minor degradation in terms of required signal-to-noise ratio (SNR), thanks to the reception of the other parts.

13.3 TFS Implementation in DVB-NGH

13.3.1 Gateway

At the transmitter side, the NGH gateway should assure that receivers can follow the TFS transmission with a single tuner. A correct scheduling of the data from each PLP is also necessary in order to exploit the combined frequency and time diversity. Ideally, an FEC code word should be uniformly distributed among all RF channels and properly spread in time.

However, time diversity is tightly linked to zapping time. Zapping time involves the inevitable delay produced when a user receiving a particular audio/video service decides to switch to a different service carried on another stream as the new service cannot be presented immediately to the user, unless the receiver performs decoding in parallel of both services (e.g., if they are both transmitted in the same PLP). The main factor that involves zapping time performance is the interleaving depth that is linked to the TFS cycle time (the time interval between the reception of two consecutive data bursts in the same RF channel). For intra-frame TFS (without using convolutional interleaving), the average zapping time, when zapping from another PLP of the same NGH signal, is 1.5 times the TFS cycle time (one-and-a-half frame), which corresponds to the time needed to receive the start of the frame (in average 0.5 frame) and the time before data can be reproduced (1 frame, in order to read the signaling parameters and to receive the time-interleaved data cells contained in the frame). For inter-frame TFS, it mainly depends on the time interleaving depth (the number of frames where an FEC code word is spread). A detailed analysis on the trade-off between time diversity and zapping time can be found in Chapter 12.

13.3.1.1 Data Scheduling for Intra-Frame TFS

For intra-frame TFS, the NGH specification defines a deterministic scheduling process that ensures the necessary time interval between sub-slices for sequential reception with a single tuner. The algorithm leads to a regular distance between sub-slices with a constant hopping time between slots. An illustrative example is depicted in Figure 13.7. The starting point is a set of consecutive data cells from the different PLPs that are going to be transmitted (in the example, there are a total of six PLPs). The total number of PLP cells is, then, divided into the number of sub-slices to be transmitted in each RF channel (two sub-slices per RF channel in our example). The previous division establishes the so-called $Subslice_{interval}$, which is the time interval (or the number of cells) between two data bursts of the same PLP in an RF channel:

$$Subslice_{interval} = \frac{n_{Cells}}{n_{Subslices}} \tag{13.1}$$

where
 n_{Cells} is the total number of cells of all PLPs
 $n_{Subslices}$ represents the total number of sub-slices that depends on the number of RF channels and the number of sub-slices in each one

The resultant number of sub-slices is divided into the number of RF channels to ensure an even distribution of the data from PLPs over the set of RF channels. To achieve feasible transmission with a single tuner, a suitable time interval, $RF_{Shift,n}$, is applied to the cells from channel to channel defined as follows:

$$RF_{Shift,n} = i \cdot \frac{Subslice_{interval}}{n_{RF}} \ (i = 0, \ 1,...,n_{RF}) \tag{13.2}$$

where
 n_{RF} is the number of RF channels
 n is defined from 0 to $n_{RF} - 1$

The final folding of the cells that exceed the last sub-slice (of the last PLP) of the RF channel that has not been shifted leads to a frame structure in which all PLPs are regularly spaced with time for frequency hopping.

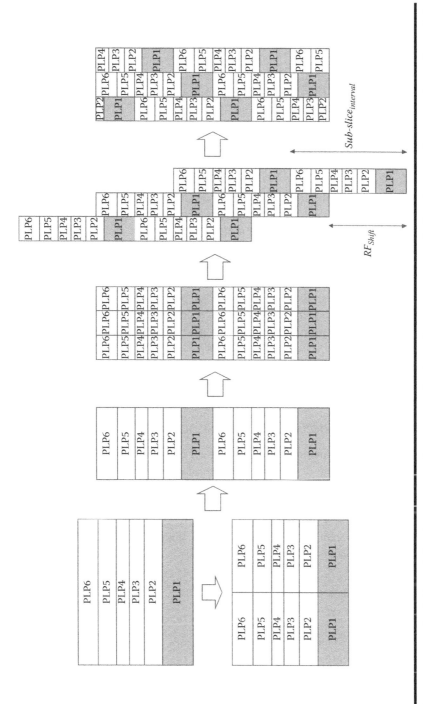

Figure 13.7 Scheduling of the data PLPs at the gateway for intra-frame TFS. Example for six PLPs, three RF channels, and six sub-slices.

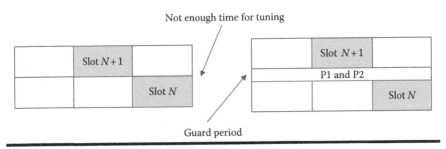

Figure 13.8 GI insertion between data slots.

It should be noted that the previous process defines the scheduling; however, scheduled cells have not yet been filled with the data of the services. Only positions in the frame have been defined. Time-interleaved PLP cells are introduced into sub-slices in the natural time sequence, independently of the RF channel. The first time-interleaved cell is, therefore, introduced in the first cell position of the first sub-slice of the PLP (independently of the RF channel in which it appears).

A proper scheduling for intra-frame TFS guarantees frequency hopping internally in a frame with a single tuner. However, a guard period is needed to allow enough tuning time between the last PLP sub-slice in one frame and the first PLP sub-slice in the following frame (see Figure 13.8). The guard period can be achieved either by the use of an FEF part between the frames or by the use of Type 1 PLPs for a long enough time period to allow the frequency hopping.

13.3.1.2 Data Scheduling for Inter-Frame TS

Inter-frame TFS operation across the FEFs of a T2 signal relaxes data scheduling since frequency hopping is not performed inside a frame, having longer time intervals between data bursts. Although timing is less critical for inter-frame TFS, a trade-off between the convenient time interleaving and zapping time must be reached for inter-frame TFS operation.

For regularly spaced T2 FEFs according to Figure 13.9c, inter-frame TFS scheduling must meet the following equation:

$$N \cdot (FEF_{length} + Gap_{length}) = FEF_{length} + k \cdot T_{frame} \qquad (13.3)$$

where
 N is the number of interleaved T2 FEFs
 FEF_{length} is the length of a T2 FEF
 Gap_{length} is the gap between adjacent T2 FEFs (but on different frequencies)
 T_{Frame} is the length of the T2 frame
 k is the number of T2 frames between T2 FEFs

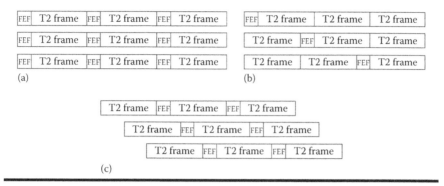

Figure 13.9 **Interleaving over FEFs with frequency hopping between RF channels using co-timed T2 frames (a), time-shifted FEFs but co-timed super frames (b), and time-shifted super frames (c).**

This equation defines the TFS cycle time for inter-frame TFS that mainly depends on the spacing among FEFs and the number of FEFs throughout data is spread. In general, zapping time increases with the spacing among FEFs. However, large-spaced FEFs increase time interleaving on the channel that also increases time diversity. The different schemes presented in Figure 13.9 deal with this issue.

Although the first two schemes provide the best performance for frequency and time diversity, they provide long zapping times. To guarantee the possibility of using long interleaving without seriously affecting zapping time, the use of time-shifted super frames is proposed. With all three options, some synchronization needs to be implemented in the TFS multiplex among the different RF channels.

13.3.2 Modulator

The basic modulator block diagram of an NGH transmitter is depicted in Figure 13.10. It includes several OFDM generation chains, one for each RF channel of the TFS multiplex. The most important blocks involved in the TFS operation are the frame mapper and, when using RCs, the component interleaver. The frame mapper allocates the interleaved NGH services in their corresponding frame according to the TFS configuration parameters. The component interleaver ensures that the dimensions of each constellation point are transmitted over all possible combinations of RF channels, increasing the frequency diversity as much as possible (see Chapter 12). In this case, the relevant parameters are the number of RF channels and the number of TFS cycles over which an FEC block is time interleaved.

13.3.3 Receiver

The implementation of TFS at the receiver is one of the most challenging issues on TFS implementation since receivers need to perform frequency hopping between two RF channels with a single tuner.

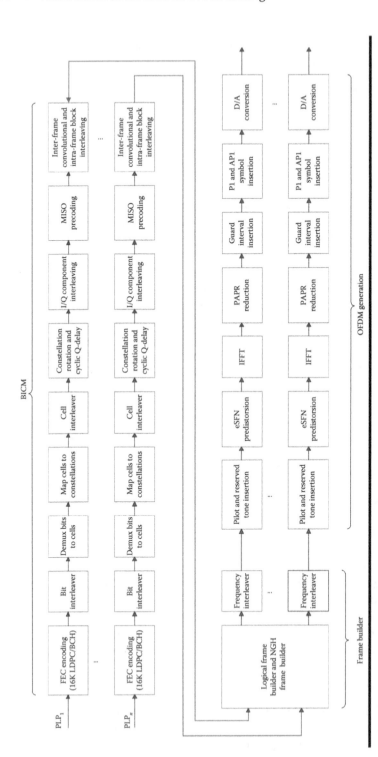

Figure 13.10   DVB-NGH modulator block diagram with TFS.

13.3.3.1 TFS Frequency Hopping Time

The NGH specification includes a receiver model for frequency hopping. Only those transmission modes that are receivable with receivers following this model are allowed. The receiver is assumed to have a tuning time, T_{tuning}, that is 5 ms or lower. The minimum frequency hopping time between two data slots, T_{FH}, can be calculated as follows:

$$T_{FH} = 2 \cdot T_{ChE} + T_{tuning} \tag{13.4}$$

where

T_{ChE} is the time required for fine frequency synchronization and channel estimation

T_{tuning} is the Phase Locked Loop (PLL) and Automatic Gain Control (AGC) tuning. Figure 13.11 depicts this timing

In the T2 standardization process, it was acknowledged that a skilful design of the PLL and AGC would require only 5 ms, and since then, receiver manufacturers have reported that even lower values are in fact possible. Not only the time gap between the same PLP on two different RF channels does however need to be somewhat larger than 5 ms because the channel estimation process has to be performed before starting to receive any data slot in the new RF channel, but also there should be some extra time in order to finish the channel estimation in the current RF channel before hopping. T_{ChE} depends on the FFT size, the OFDM guard interval (GI), and the pilot pattern. For example, for a typical mobile configuration for the UHF band with FFT 8K GI = 1/4* with pilot pattern PP1 (which requires time interpolation involving three future and past OFDM symbols), the required time gap is increased to 12 ms (11 symbols). For FFT 16K[5] GI = 1/4 with pilot pattern PP2, which requires time interpolation over one future and past symbol, the time gap is about 11 ms (five symbols).

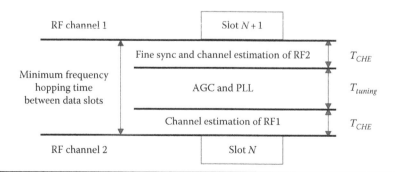

Figure 13.11   Minimum frequency hopping time between data slots.

* Maximum symbol duration with GI = 1/4 is 1.12 ms for 8K and 2.24 ms for 16K.

13.4 TFS Coverage Analysis Based on Field Measurements

13.4.1 Measurement Settings

The Swedish DTT network operator Teracom performed indoor and outdoor measurements in different areas of the DVB-T network in Sweden. Both sets of measurements consist of samples of the signal strength of four and six multiplexes that are recorded cyclically. The power level measured in each channel is represented as the relative level of each one to the average level of all the channels (see Figure 13.12a). Outdoor measurements only take into account variation in the

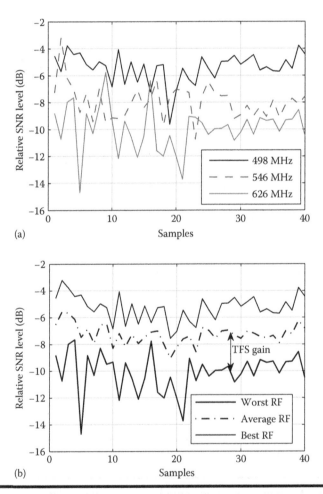

(a)

(b)

Figure 13.12 Samples of the outdoor measurements of three RF channels (a) and maximum, average, and minimum signal level among them (b).

signal level caused by path loss. The variability of the signal cannot be evaluated over time, and no fast-fading effect can be analyzed. Indoor measurements take into account variations in time. The signal differences in each measured RF channel come from path loss and fast-fading.

13.4.2 TFS Coverage Gain Definition

The TFS coverage gain at a particular location may be defined as the difference between an effective SNR, which represents the average SNR value of the RF channels, and the minimum instantaneous SNR among all RF channels (that corresponds to the one that would limit coverage in a traditional transmission) [1]:

$$TFS_{gain} = \text{mean}_N \left(SNR_i \right)\big|_{dB} - \min_N \left(SNR_i \right)\big|_{dB} \qquad (13.5)$$

where

SNR_i is the SNR of the i RF channel in linear scale

and mean(·) and min(·) are the average and minimum operators, respectively

In Figure 13.12b, the values of the average, the minimum, and the maximum level of each channel are depicted. TFS gain is also shown as the difference between the average and minimum values.

This approximation has been validated in [10], where the TFS gain is analyzed from an information theory point of view. The comparison between the two methods shows that computing the effective SNR as the average signal of the RF channels represents an upper bound for the TFS gain. The information theory method approaches the upper bound for low SNR values, but there are some differences in the effective SNR for high SNR values. Therefore, the results obtained can be considered as an upper-bound value.

13.4.3 Field Measurement Results

13.4.3.1 Outdoor Measurements

First, we investigate the TFS gain when only two RF channels are used in the TFS multiplex and evaluate the influence of the frequency spacing among them. Figure 13.13a shows the average TFS gain as a function of the frequency spacing. Figure 13.13b shows the probability density function (PDF) of the TFS gain for six different frequency spacing values. In Figure 13.13a, it can be observed that, in general, the TFS gain increases linearly with the channel spacing between two RF channels for all the transmitter areas. A 3 dB gain is achieved for 112 MHz (14 channels in-between). In Figure 13.13b, we can see that there are some locations where there is no TFS gain. But the percentage of locations is below 6% already for a separation of six channels.

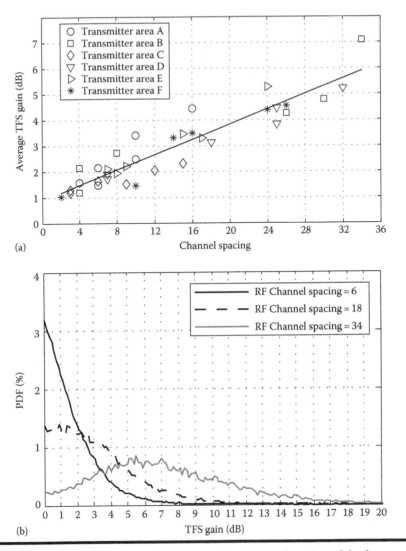

(a)

(b)

Figure 13.13 **Average TFS gain for two RF channels as a function of the frequency spacing (a) and PDF of the TFS gain (b) for different frequency spacing.**

Figure 13.14 shows the PDF of the TFS gain for two, three, and four RF channels in the TFS multiplex using all the possible combinations of frequencies available from the measurements. We can see that very important gains can be achieved. The PDF of the TFS gain for three and four RF channels resembles a Rayleigh distribution, whereas for two RF channels, it resembles an exponential distribution. The average TFS gain is 2.5, 4.3, and 6.2 dB, for two, three, and four RF channels, respectively.

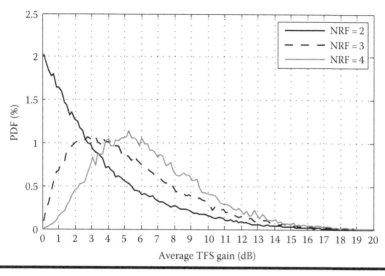

Figure 13.14   PDF of the TFS gain with different RF channel aggregations.

13.4.3.2 Indoor Measurements

Figure 13.15 shows the average TFS gain of a TFS multiplex with two RF channels as a function of the distance to the window for three different frequency spacing values: 8, 112, and 264 MHz, (1, 14, and 33 channels respectively). Indoor reception presents the particularity that the signal level varies strongly with frequency due to penetration losses into buildings. Hence, larger gains can be expected for indoor reception than for outdoor reception. With a very narrow spacing among the two channels (8 MHz), the TFS gain is reduced when moving away from the window because the building penetration loss evenly reduces the signal level of the two channels. But with a wide spacing (33 MHz), the TFS gain increases with distance to the window. The reason is that the signal level of the worst channel decreases significantly faster.

13.4.3.3 Optimal Spectrum Allocation with TFS

The presented results reveal the importance of spreading the RF channels of a TFS transmission as evenly as possible over the UHF band. In general, the coverage level is higher at lower frequencies, but the coverage requirement for all services is normally the same. This means that when several independent TFS transmissions are used, they should have the same coverage and therefore RF frequencies "equally" spread over the UHF band. If, for example, two TFS transmissions would be placed in the lower and upper half of the UHF spectrum, the differences between the RF channels of each TFS transmission would be lower than if these frequencies were

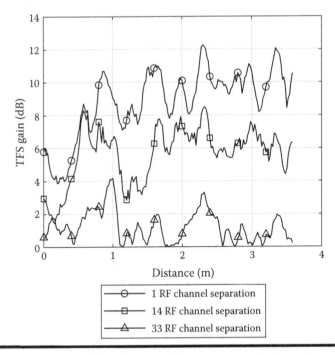

Figure 13.15 **TFS gain and loss for two RF narrow-spaced channels as a function of the distance inward the building.**

distributed over the full UHF band, which is in itself an advantage. However, the global coverage would be quite different between the lower and the upper half of TFS signals. In such a multiple TFS transmission scenario, it is recommended to spread the frequencies of each TFS signal as evenly as possible over the UHF band. This will maximize frequency diversity and will ensure the same coverage of these TFS transmissions. Also from a point of view of "equality between countries," it is natural to allocate frequencies to all countries as evenly as possible over the UHF band.

13.5 TFS Performance Based on Physical Layer Simulations

13.5.1 Simulation Settings

The performance of TFS has been evaluated with physical layer simulations to investigate the effect of the code rate, RCs, and the power imbalances among the different RF channels. The simulator has been calibrated within the NGH standardization process. Table 13.1 shows the main configuration parameters of the simulations.

Table 13.1 PHY Simulation Configuration for the TFS Performance Evaluation

FEC configuration			
Code rate	1/3 7/15 2/3	**FEC blocks per frame**	49
Bits per cell (modulation)	QPSK	**RCs (2D)**	Enabled–disabled
FEC type	16,200		
Time interleaving configuration			
Sub-slices per frame	1,620	**Time interleaver**	Activated
Frame interval	1	**Component interleaver**	Activated
Convol. interleaving length	Num. of RF channels	**Convol. interleaving type**	Uniform

OFDM generation configuration					
FFT size	8K (6817 carriers)	**GI**	¼	**Bandwidth**	8 MHz

Channel configuration					
Channel model	Typical Urban 6	**Doppler frequency**	33.3 Hz	**Antenna configuration**	SISO

System configuration	
Number of frames	500
Frame length	250 ms

Although in a real transmission the reception conditions in each RF channels might vary dynamically from frame to frame, in our simulations, it is considered that some channels of the TFS multiplex suffer from static imbalances. The operation of TFS is emulated at the receiver by means of applying a different carrier-to-noise ratio (CNR) level to each frame. Figure 13.16 represents a transmission using TFS with three RF channels in which one frame is received with a lower CNR level. The number of interleaved frames is equal to the number of RF channels.

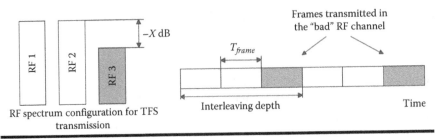

Figure 13.16 Emulation of the TFS operation for the PHY simulations.

13.5.2 Simulation Results

13.5.2.1 Influence of the Code Rate on the TFS Performance

In this section, the performance of TFS is evaluated for different code rates. The code rate defines the redundancy applied to the information data to cope with disturbances in the received signal, and thus, it plays an important role in the TFS operation because it determines the capability to recover from "bad" RF channels.

Figure 13.17 shows the frame error rate (FER) as a function of the CNR for different transmission configurations and different code rates (1/3, 7/15, and 2/3). The TFS multiplex is assumed to have four RF channels, with two of them having an imbalance (attenuation) of –6 dB with respect to the other two. In the figure, the solid lines refer to the performance of the TFS transmission. The dotted lines represent the performance of the RF channel with the lowest CNR value (in this particular case, –6 dB).

In the figure, we can see that the TFS gain increases when reducing the code rate (i.e., increasing the transmission robustness). The gain values at FER 10^{-4} are 2, 2, and 1.5 dB, for code rates 1/3, 7/15, and 2/3, respectively. It should be noted that these values are lower than the average TFS gain obtained using the Equation 13.5, which is 3.9 dB. The optimum code rate with TFS is linked to the quality of the channels in order to obtain the best trade-off between robustness and capacity. For example, we can see that QPSK 7/15 with TFS achieves a similar performance as QPSK 1/3 without TFS.

13.5.2.2 Influence of Rotated Constellations on the TFS Performance

In this section, the performance of TFS is evaluated together with two-dimensional (2D) RCs for two different code rates (1/3 and 2/3). The number of RF channels in the transmission has been set to three. One channel is attenuated with respect to the others by –6 and –12 dB. Figure 13.18 depicts the results of the simulations with and without RCs. The solid curves refer to the performance of the TFS transmission, and the dotted curves represent the performance of the attenuated RF channel.

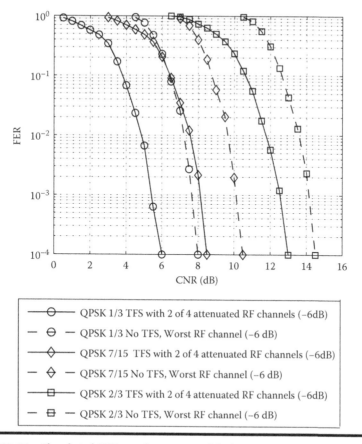

Figure 13.17 Simulated TFS performance with four RF channels for different code rates. Two channels are attenuated −6 dB with respect to the other two channels.

In the figure, it can be seen that there is no gain with RC for a code rate 1/3, whereas there is a noticeable gain for a code rate 2/3. This is consistent with the potential gain of RCs without TFS, providing its maximum gain for low-order constellations and high code rates [9].

Although there is some dependence with the code rate, the important parameter is the amount of erasures in comparison with the redundancy percentage. Using, for example, code rate 1/2 (50% redundancy), the required CNR will necessarily approach infinity without RCs when the percentage erasures approach 50%. Note that anything else is theoretically impossible. This is applicable to a situation with two RF channels, with one being erased. RCs, on the other hand, will however have only a limited penalty in this situation, so there is an infinite gain in such circumstances. Although it may not be so common to have full erasures, there may be very large differences in the signal strength between two RF channels, so there

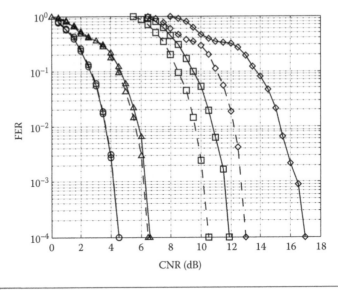

Figure 13.18   Simulated TFS performance with and without 2D RCs with different code rates. TFS multiplex with three RF channels, and one of them attenuated with respect to the other two.

will be a big gain also in practice when one RF channel is much attenuated. With more RF channels, we may have the same reasoning, for example, with code rate 1/3 and RC, it will, in principle, cope with up to 67% (2/3) of erasures, but again this will necessarily require an infinite CNR with non-RC. In general, when the proportion of erasures approach (or even exceed) the percentage of redundancy of the FEC code, RC will always show a big gain in comparison with non-RC.

13.6 Conclusions

DVB-NGH will be the first digital TV broadcasting system to transmit the services across several channels with TFS. In addition to a potential capacity gain due to enhanced StatMux, TFS can provide very important coverage gains for both outdoor and indoor reception conditions due to improved frequency diversity and considerably improve the robustness of the transmitted signal. Without TFS, the coverage level at a given location is limited by the multiplex with the lowest signal strength. With TFS, the reception at a particular location is more likely to be determined by the average signal strength of the RF channels. With appropriate coding and interleaving, even a fully lost RF channel may still allow correct reception of all services. The use of RCs improves the capability to recover from "bad" RF channels.

References

1. European Broadcasting Union, Frequency and network planning aspects of DVB-T2, EBU Tech 3348, May 2012.
2. ETSI EN 302 755 v.1.3.1, Digital video broadcasting (DVB); Frame structure channel coding and modulation for a second generation digital terrestrial television broadcasting system (DVB-T2), November 2011.
3. L. Vangelista et al., Key technologies for next generation terrestrial digital television standard DVB-T2, *IEEE Communications Magazine*, 47(10), 146–153, 2009.
4. M. Rezaei, I. Bouazizi, and M. Gabbouj, Statistical time-frequency multiplexing of HD video traffic in DVB-T2, *International Journal of Digital Multimedia Broadcasting*, 2009, Article ID 186960, 12 pages, 2009.
5. ETSI EN 300 468 v1.12.1, Digital video broadcasting (DVB); Specification for service information (SI) in DVB systems, October 2011.
6. ETSI TS 102 773 v1.3.1, Digital video broadcasting (DVB); Modulator interface (T2-MI) for a second generation digital terrestrial television broadcasting system (DVB-T2), January 2012.
7. DVB commercial module sub-group on next generation handheld, Commercial requirements for DVB-NGH, CM-1062R2, June 2009.
8. J. J. Giménez, D. Gozálvez, D. Gómez-Barquero, and N. Cardona, Statistical model of signal strength imbalance between RF channels in DTT network, *Electronics Letters*, 48(12), 731–732, June 7, 2012.
9. C. A. Nour and C. Douillard, Rotated QAM constellations to improve BICM performance for DVB-T2, *Proceedings of the IEEE International Symposium on Spread Spectrum Techniques and Applications (ISSSTA)*, Bologna, Italy, 2008.
10. M. Makni, J. Robert, and E. Stare, Performance analysis of time frequency slicing, *Proceedings of the ITG Conference on Electronic Media Technology*, Dortmund, Germany, 2011.

Chapter 14

DVB-NGH Logical Frame Structure and Bundling DVB-T2 Future Extension Frames

Jordi Joan Giménez, David Gómez-Barquero, and Alain Mourad

Contents

417

14.1 Introduction

The widespread success of DVB-T2 has been a key stimulus for the development of DVB-NGH. A primary deployment scenario envisaged for DVB-NGH is in-band with DVB-T2, hence promoting the coexistence of DVB-NGH with DVB-T2 in the same RF channel and allowing the reuse by DVB-NGH of the existing DVB-T2 network infrastructure. Along this line, the DVB project has opted for an alignment of DVB-NGH with DVB-T2 (e.g., OFDM waveform, LDPC+BCH coding, physical layer pipe (PLP) concept, etc.). Furthermore, the DVB has set an important commercial requirement that it shall be possible to combine DVB-NGH and DVB-T2 signals in the same RF channel. Such a requirement can be fulfilled thanks to the future extension frame (FEF) mechanism defined in DVB-T2.

Thanks to the FEF concept, DVB-T2 makes it possible to accommodate different technologies in the same multiplex in a time-division manner. Each FEF starts with a preamble OFDM symbol known as P1 [1], which, among other basic signaling information, identifies the frame type. The positions of the FEFs in time and their duration are signaled in the physical layer-1 (L1) signaling in the T2 frames. This way, DVB-T2 legacy receivers, not able to decode the FEFs, simply ignore the transmission during that time while still receiving the DVB-T2 signal. Terminals can switch off their RF front ends during the transmission of FEFs, saving power, like in a stand-alone DVB-T2 discontinuous transmission. Figure 14.1 illustrates the combined transmission of DVB-T2 with FEFs in the same multiplex. It should be pointed out that the FEF concept is relative to the system of focus, so that a system (e.g., DVB-NGH) sitting the FEF of DVB-T2 will see the T2 frames as their respective FEFs (e.g., of DVB-NGH).

DVB-NGH is the third type of transmission specified in the FEF of DVB-T2. The first type of transmission is used for transmitter identification in single-frequency

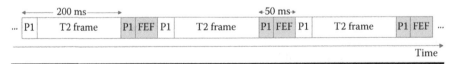

Figure 14.1 Coexistence of T2 frames and FEFs in a single multiplex. Each T2 frame and FEF starts with a preamble P1 OFDM symbol that identifies the type of frame.

networks (SFNs) [2], while the second type of transmission is used for the mobile profile of DVB-T2 known as T2-Lite (for more details on T2-Lite, see Chapter 6). A typical deployment scenario for DVB-NGH or T2-Lite in a shared DVB-T2 multiplex is the one depicted in Figure 14.1: one FEF of reduced size (e.g., 50 ms duration) after every T2 frame with a considerably longer duration (e.g., 200 ms). This configuration devotes most of transmission time to DVB-T2 services, which are supposed to be the main services for the broadcast operator, but it allows introducing few mobile services (e.g., four to six multimedia services at 500 kbps, including video and audio). Moreover, since the FEFs are rather short, the zapping time of the DVB-T2 services is practically not affected by the introduction of any new technology in the FEF.

In the case of T2-Lite, there is a one-to-one relation between FEFs and T2-Lite frames. That is, each DVB-T2 FEF is a T2-Lite frame. DVB-NGH allows for a more flexible and efficient allocation between FEFs and NGH frames, and several FEFs can belong to the same NGH frame. DVB-NGH has defined a logical framing concept suited to the transmission of DVB-NGH signals in DVB-T2 FEFs. The logical framing concept has taken particular care to include time-frequency slicing (TFS). TFS is a transmission technique adopted in DVB-T2 for transmitting one service across several RF channels with frequency hopping and time-slicing (i.e., discontinuous transmission). It can provide an important gain in coverage due to increased frequency diversity and also a gain in capacity due to improved statistical multiplexing (StatMux) for variable bit rate (VBR) services. TFS also offers the possibility of finding spectrum more easily and in a more flexible way, since it is possible to combine different RF channels. For more details on TFS, interested readers are referred to Chapter 13.

The new logical frame (LF) structure of DVB-NGH allows combining RF channels with different FEF sizes and transmission intervals, bandwidth, frequency band, transmission modes, etc., which can be seen as a generalization of TFS. Without the LF structure, it would be also required that the different DVB-T2 multiplexes are time synchronized to employ TFS in DVB-NGH. The LF structure also enables the possibility of hybrid satellite–terrestrial reception with a single demodulator (tuner). For more details on the satellite profile of DVB-NGH, see Chapter 22.

Another important advantage of the LF structure of DVB-NGH is that there is no need to transmit all L1 signaling in each FEF, which reduces the signaling overhead compared to DVB-T2 and T2-Lite. In DVB-T2 (and T2-Lite), the physical layer L1 signaling is transmitted in every frame. In addition to the preamble P1 symbol, each frame has so-called preamble P2 symbols that carry the rest of the physical layer L1 signaling at the beginning of each frame. For very short FEFs, which may be the most representative use case for the initial transmission of DVB-NGH services (like the example shown in Figure 14.1 of 50 ms), the amount of physical layer signaling overhead could then become significant. The LF structure also avoids any limitation in the number of PLPs used in the system, as the L1 signaling capacity is no longer constrained to a fixed number of preamble P2 symbols. For more details on physical layer signaling in DVB-NGH, see Chapter 15.

This chapter describes the frame structure of DVB-T2 and details the operation of a DVB-T2 multiplex with FEFs in Section 14.2. Section 14.3 presents the LF structure of DVB-NGH. Finally, the chapter is concluded with Section 14.4.

14.2 Physical Frame Structure and Future Extension Frames in DVB-T2

14.2.1 Super Frames, Frames, and OFDM Symbols in DVB-T2

The physical frame structure of DVB-T2 consists in super frames, frames, and OFDM symbols, as illustrated in Figure 14.2. Super frames comprise an integer number of frames that may be of two types: T2 frames and FEFs. Each frame is formed by an entire number of OFDM symbols (which contain preamble symbols P1 and P2, and data symbols).

14.2.1.1 Super Frames

Super frames can carry a configurable number of T2 frames and FEF parts. The maximum number of T2 frames in a super frame is 255, and all the frames within a super frame must have the same length, which can be up to 250 ms. Hence, the maximum length of a super frame is 63.75 s, if FEFs are not used. The pattern insertion scheme of FEFs is configurable in each super frame. The maximum number of FEFs in a super frame is 255, reached when one FEF is inserted after every T2 frame. Figure 14.2 illustrates a frame structure in which there is one FEF every four T2 frames. Super frames must start with a T2 frame and should end with an FEF, if they are used.

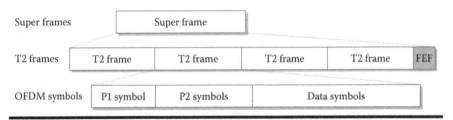

Figure 14.2 Physical frame structure of DVB-T2 consists of super frames carrying T2 frames and optionally FEFs. T2 frames consist of one P1 OFDM symbol, one or several P2 OFDM symbols, and a configurable number of data OFDM symbols. The maximum T2 frame duration is 250 ms.

14.2.1.2 Frames

T2 frames are intended to carry DVB-T2 services and related signaling by means of OFDM symbols. They consist of one P1 symbol, one or several P2 symbols (depending on the FFT size), and a configurable number of data symbols.

14.2.1.3 OFDM Symbols

The initial symbol of each T2 frame and FEF is a preamble OFDM symbol known as P1. The P1 symbol has a constant 1K FFT size, with guard intervals at both ends. It allows fast identification in the initial scan for detecting the presence of DVB-T2 signals on a given frequency. It carries some basic transmission parameters, such as the frame type (e.g., T2, T2-Lite, or NGH), and it enables the reception of the P2 symbol(s).

The subsequent OFDM symbols of a T2 frame are also preamble symbols called P2 symbols. The P2 symbols have the same FFT size as the data OFDM symbols, but have an increased number of pilots. The number of P2 symbols depends on the FFT size used (e.g., two symbols for 8K FFT). The P2 symbols carry the rest of static, configurable, and dynamic physical layer L1 (Layer 1) signaling information, which enables the reception of the PLPs that contain the actual DVB-T2 services in the data OFDM symbols. The P2 symbol(s) can also contain some data if there is any space left. For more information on the physical layer signaling, see Chapter 15.

14.2.2 Future Extension Frames in DVB-T2

FEFs provide a method for expanding the DVB-T2 standard by means and ways not known at the time of writing the original specification. The use of FEFs is optional. They are inserted, when needed, between T2 frames in such a way that they enable a flexible mixing of services within a single multiplex in a time-division manner. FEFs may also be empty or contain no data. The only defined attributes of the FEFs are [3] as follows:

- They shall begin with a preamble P1 symbol.
- Their positions in the super frame and duration in time are signaled in the L1 signaling in the T2 frames.

Several consecutive FEFs may be combined into a so-called FEF part. The maximum length of an FEF part is 250 ms for the T2-base profile, whereas for T2-Lite and DVB-NGH, it has been extended up to 1 s. The existing DVB-T2 receivers are not expected to decode FEFs,* but they should be able to detect and correctly handle FEF parts so that the reception of T2 frames is not disturbed.

---

* One particular exception is the mobile profile of DVB-T2 known as T2-Lite. T2-Lite is an integral part of the DVB-T2 v1.3.1 specification. Therefore, all DVB-T2 receivers supporting this version are supposed to be able to receive T2-Lite as well.

Although FEFs were designed to enable future extensions of the DVB-T2 standard, other use cases are possible. It should be noted that the source of the FEF can be either the T2 network or a different network with a different service coverage area. But the T2 network is always responsible for inserting the P1 symbol at the beginning of all FEF parts such that all T2 receivers can detect the FEF parts correctly. FEFs can be used, for example, to share spectrum with other networks, such as fourth-generation Long-Term Evolution–Advanced (LTE-A) cellular networks, with different sites and infrastructure. The use of FEFs also enables the use of bidirectional applications within the same frequency channel. In this case, the T2 network would leave an empty FEF, which would be used for the upstream traffic by the T2 receivers.

14.2.3 FEF Signaling

14.2.3.1 Preamble P1 OFDM Symbol

The use of FEFs is signaled in the preamble P1 and P2 symbols. The P1 symbol has two signaling fields S1 and S2, with three and four signaling bits, respectively. The S1 field is used to distinguish the preamble format and, hence, the frame type (see Table 14.1). If no FEFs are inserted in the T2 transmission, the S1 field is either 000 (single-input single-output SISO format) or 001 (multiple-input single-output MISO format). Other values denote the presence of FEFs in the multiplex. Currently, there are three types of DVB-T2 FEFs defined: the transmitter signature standard, T2-Lite, and DVB-NGH.

In the S1 field, there is one combination devoted for non-T2 applications. Currently, there is only one combination specified, which signals that the preamble corresponds to an FEF part intended for professional use. The transmitter signature standard [2] includes a method that uses FEFs for identifying transmitters in an SFN and measuring their channel impulse response. It is meant for professional receivers capable of decoding such FEFs.

T2-Lite makes use of two combinations of the S1 field as for normal DVB-T2: one for SISO and the other for MISO. DVB-NGH also makes use of two combinations for its base sheer terrestrial profile: one for SISO and another for MISO. For the terrestrial multiple-input multiple-output (MIMO) and the hybrid terrestrial–satellite profiles (SISO and MIMO) of DVB-NGH, there is an additional P1 (aP1) symbol. As shown in Table 14.1, when the S1 field of the P1 symbol signals the presence of an escape code, the associated S2 field indicates whether terrestrial MIMO, hybrid SISO, hybrid MISO, or hybrid MIMO are being used, as depicted in Table 14.2. For more information on the aP1 symbol, see Chapter 15.

For DVB-T2, T2-Lite, and the base profile of DVB-NGH, the S2 field of the preamble P1 symbol indicates the FFT size and gives partial information about the guard interval for the remaining symbols in the T2 frame, and it has one bit

Table 14.1   S1 Field of the Preamble OFDM Symbol P1

S1 Field	Preamble P2 Format	Description
000	T2 SISO	The preamble is a T2 preamble and the P2 part is transmitted in its SISO format.
001	T2 MISO	The preamble is a T2 preamble and the P2 part is transmitted in its MISO format.
010	Non-T2	Intended for professional applications.
011	T2-Lite SISO	The preamble is a preamble of a T2-Lite signal. The P2 part is transmitted in its SISO format.
100	T2-Lite MISO	The preamble is a preamble of a T2-Lite signal. The P2 part is transmitted in its MISO format.
101	NGH SISO	The preamble is a preamble of an NGH signal. The P2 part is transmitted in its SISO format.
110	NGH MISO	The preamble is a preamble of an NGH signal. The P2 part is transmitted in its MISO format.
111	ESC	General escape code. The current P1 may be followed by another symbol with additional signaling information.

Table 14.2   S2 Field for Escape Code in Preamble Symbol P1

S1 Field	S2 Field	Description
111	000X	NGH MIMO signal. The P1 symbol is followed by an aP1 symbol.
111	001X	NGH hybrid SISO signal. The P1 symbol is followed by an aP1 symbol.
111	010X	NGH hybrid MISO signal. The P1 symbol is followed by an aP1 symbol.
111	011X	NGH hybrid MIMO signal. The P1 symbol is followed by an aP1 symbol.
111	100X–111X	Reserved for future use.

Table 14.3   S2 Field of the Preamble OFDM Symbol P1

S1 Field	S2 Field	Meaning	Description
XXX	XXX0	Not mixed	All preambles in the current transmission are of the same type
XXX	XXX1	Mixed	Preambles of different types are transmitted

Source: EN 302 755 v1.3.1, *Frame Structure Channel Coding and Modulation for a Second Generation Digital Terrestrial Television Broadcasting System (DVB-T2)*, November 2011.

relevant for FEFs. This bit indicates whether the preambles are all of the same type or not, that is, whether more than one type of frame exists in the super frame (see Table 14.3). The bit is valid for all values of S1 and the rest bits of S2. This speeds up the scan as all receivers immediately know if there is a need to wait for another type of P1 symbol.

14.2.3.2 Preamble P2 OFDM Symbol and L1 Signaling

T2 super frames that contain T2 frames and FEF parts always begin with a T2 frame and end with an FEF part. The locations of the FEF parts are described with the following fields in the L1 signaling (see Figure 14.3):

- *FEF_LENGTH*: This 22-bit field (24 bits for T2-Lite) indicates the length of the FEF parts as the number of elementary time periods (samples in the receiver),* T, from the start of the P1 symbol of the FEF part to the start of the P1 symbol of the next T2 frame.
- *FEF_INTERVAL*: This 8-bit field indicates the number of T2 frames between two FEF parts and also the number of T2 frames at the beginning of a super frame before the beginning of the first FEF part.

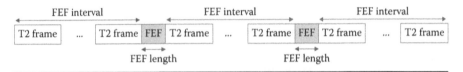

Figure 14.3   T2 multiplex operation with T2 frames and FEF parts.

---

* The value of the elementary time period depends on the bandwidth, for example, 7/64 μs for 8 MHz or 7/48 μs for 6 MHz.

The receiver can compute the number of FEFs in a super frame, N_{FEFs}, from the number of T2 frames in the super frame $N_{T2frames}$, which is signaled in the L1, and the FEF interval:

$$N_{FEFs} = \frac{N_{T2\,frames}}{FEF\_INTERVAL} \qquad (14.1)$$

The total super frame duration, T_{SF}, is determined by

$$T_{SF} = N_{T2\,frames} \times T_{frame} + N_{FEFs} \times T_{FEF} \qquad (14.2)$$

where
  T_{frame} is the duration of the T2 frame
  T_{FEF} is the duration of the FEF part signaled by *FEF_LENGTH*

In DVB-NGH, there is one additional L1 field named *FEF_PREAMBLES*, which signals the presence of a preamble of a given signal (i.e., T2 SISO/MISO, T2-Lite SISO/MISO, NGH SISO/MISO/MIMO, or hybrid SISO/MISO) that is carried in the FEF part of the NGH signal.

14.2.4 Examples of DVB-T2 Transmission with FEFs

When a signal of a different profile is transmitted within the FEF part of the T2 signal (e.g., T2-Lite or DVB-NGH), each signal (T2 signal and the other) will appear as it is being transmitted in the FEF part of the other and shall be signaled accordingly. T2-Lite and DVB-NGH super frames have the same restrictions as DVB-T2 (e.g., super frames may contain more than one FEF and, if FEFs are being used, must finish with an FEF). But it should be taken into account that the maximum duration of an FEF part of a DVB-T2 signal is 250 ms, whereas for T2-Lite and DVB-NGH, it is 1 s.

Figure 14.4 illustrates the transmission of DVB-T2 with other technology in the same multiplex using FEFs. The T2 frames are labeled "T2," and the frames of

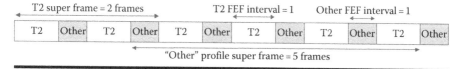

Figure 14.4   T2 multiplex with T2 frames and FEF parts with another technology (e.g., T2-Lite or DVB-NGH).

the other T2-Lite or DVB-NGH signal are labeled "Other." In this example, there is one FEF part after every T2 frame, so the *FEF_INTERVAL* for both technologies is 1. In the figure, the minimum length possible for the T2 super frame (2 frames) is assumed. For the other profile, the super frame consists of five frames, to be able to increase the time interleaving duration with multi-frame time interleaving. It should be pointed out that it is possible to combine more than two different technologies within the same multiplex.

14.3 Frame Building in DVB-NGH

Frame building in DVB-NGH is performed in two stages: LF building and NGH frame building, as depicted in Figure 14.5. LFs are carried in NGH frames that represent the physical containers of the NGH system.

14.3.1 NGH Frame Building

14.3.1.1 NGH Frames

A (physical) NGH frame is composed of one preamble P1 OFDM symbol, which may be followed by one additional aP1 symbol for the MIMO and the hybrid profiles, one or more P2 symbols (the number of P2 symbols is given by the FFT size as in DVB-T2), and a configurable number of data symbols (the maximum frame size is as in DVB-T2 250 ms). According to the combination of the FFT size, guard interval, and pilot pattern, the NGH frame may be closed by a frame closing symbol.

In addition to the aP1 symbol, the main difference compared to DVB-T2 is that not all L1 signaling information is necessarily transmitted in all NGH frames. The L1 signaling is divided into two fields, known as L1-pre and L1-post. The L1-pre is always transmitted in all NGH frames in the preamble P2 symbol(s).

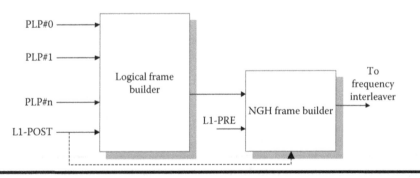

Figure 14.5 Two stages of DVB-NGH frame building.

An NGH frame provides capacity for carrying the L1-pre signaling followed by the contents of the LFs, which includes the L1-post signaling.

14.3.1.2 NGH Super Frames

An NGH super frame can carry NGH frames and also FEF parts. The maximum length of a super frame without FEFs is 63.75 s as in DVB-T2, equivalent to 255 NGH frames of 250 ms. The only difference compared to DVB-T2 is that the maximum length of a FEF part in DVB-NGH is 1 s instead of 250 ms.

14.3.2 Logical Frame Structure in DVB-NGH

The LF structure of DVB-NGH defines how to carry NGH services into DVB-T2 FEFs. The FEFs can be transmitted in a single RF channel (time-domain bundling) or in multiple RF channels with TFS (time-frequency domain bundling). The LF structure provides a lot of flexibility to TFS, as it relaxes constraints such as having the same length for the FEFs in all RF channels or synchronizing the different T2 multiplexes. Hence, a DVB-NGH network could be managed independently from the DVB-T2 network(s).

The LF structure defined in DVB-NGH is formed by the following elements:

- LFs
- Logical super frame
- Logical channels (LCs)
- LC group

14.3.2.1 Logical Frames

An LF is a data container that also carries L1-post signaling. Each LF starts with L1-post signaling followed by the PLPs. The capacity of the LF is defined in terms of the number of cells (constellation symbols), which comprises the L1-post cells, common and data PLP cells, auxiliary streams, and dummy cells.

The LF structure also avoids any limitation in the number of PLPs used in the system, since the L1-post signaling capacity is not constrained by a fixed number of preamble P2 symbols, as opposed to DVB-T2. The L1-post signaling must be transmitted after the L1-pre, but it does not have to follow immediately after. The L1-post may occupy any chunk of cells after L1-PRE in the NGH frame (see Figure 14.6).

14.3.2.2 Logical Super Frames

LFs are grouped in logical super frames. The maximum number of LFs in a logical super frame is 255. The number of LFs in a logical super frame must be chosen correctly in order that for every data PLP there is an integer number of interleaving frames in a logical super frame.

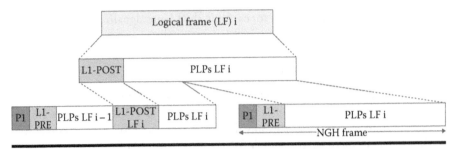

Figure 14.6   L1 signaling structure in DVB-NGH. The aP1 symbol is transmitted only in the MIMO and hybrid profiles.

14.3.2.3 Logical Channel

An LC is defined as a sequence of LFs that are transmitted over 1 to N RF frequencies available in the NGH network. The logical NGH channels are introduced to account for the possibility that more than one FEF is simultaneously transmitted in different RF channels. Each LC is an independent entity with its own transmission mode (e.g., FFT, guard interval, pilot pattern, etc.) in which all the necessary signaling is being transmitted.

Depending on how the LF is mapped into the NGH frames and the number of RF channels used, it is possible to distinguish four types of LCs. There may be different LCs in the same NGH network.

The basic LC Type A uses a single RF channel and the NGH frames contain cells from only one LF. That is, as in DVB-T2, there is a one-to-one relationship between LFs and NGH frames, and the L1-post is transmitted in every NGH frame (see Figure 14.7).

LC Type B uses a single RF channel, but each LF is mapped to several NGH frames (see Figure 14.8). Thus, each NGH frame may contain cells from multiple LFs. Compared to the reference LC Type A, this configuration has a reduced L1 signaling overhead since the L1-post is not transmitted in all NGH frames. It should be pointed out that LCs of Type B do not degrade the zapping time, provided that NGH frames are transmitted with a sufficient high rate.

LCs of Type C and Type D employ multiple RF channels. An LC of Type C corresponds to the case where each LF is mapped onto multiple NGH frames that are

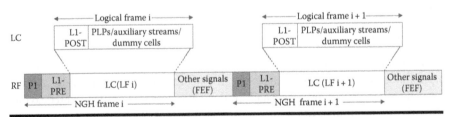

Figure 14.7   Logical channel Type A.

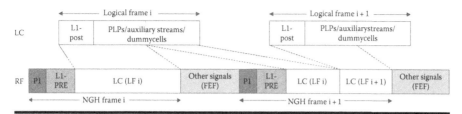

Figure 14.8   Logical channel Type B.

transmitted over a set of RF channels (see Figure 14.9). Each NGH frame may carry cells from multiple LFs. The NGH frames are separated from channel to channel in order to allow LC reception with a single tuner (for more details on the required time for frequency hopping, see Chapter 13) and may be of different length.

For LCs Type D, each LF is mapped one-to-one to multiple time-synchronized NGH frames of the same length that are transmitted in parallel over a set of RF channels (one NGH frame in each RF channel; see Figure 14.10). In this case, the different RF channels of the TFS multiplex shall be synchronized.

Type C and Type D LCs are also referred to as intra-frame and inter-frame TFS, respectively (see Chapter 13). LC Type C with inter-frame TFS benefits of an improved frequency diversity and higher StatMux gain, with L1 overhead reduction thanks to the transmission of L1-POST in only one physical container (NGH frame) of the LF. This configuration is also very flexible, since it allows finding spectrum for DVB-NGH joining several independent T2 multiplexes without restriction on perfect alignment of all FEF parts available in the different T2 multiplexes. LC Type D with intra-frame TFS provides higher frequency diversity and higher StatMux gain than LC Type C, but it is not that flexible because the length of the FEFs has to be identical in all T2 multiplexes, and the multiplexes have to be time synchronized.

Table 14.4 compares the four types of LC in terms of overhead reduction, time and frequency diversity, StatMux gain, and system flexibility. The reference configuration is LC Type A (without time-frequency bundling and without TFS).

14.3.2.4 Logical Channel Group

LCs are grouped such that the NGH frames of one LC can be separated in time from the NGH frames that carry the information of another LC within the same group such that it shall be possible to receive all the LCs of the same group with a single tuner. This concept was developed in order to allow parallel reception of terrestrial and satellite transmissions in hybrid multifrequency networks (MFNs) with a single demodulator. Figure 14.11 illustrates an example in which LFs are mapped to the terrestrial and satellite transmissions.

Figure 14.12 shows an example of two LCs that are members of the same group, a first logical channel LC1 of type C over RF1 and RF2 and a second logical channel LC2 of type A over RF3.

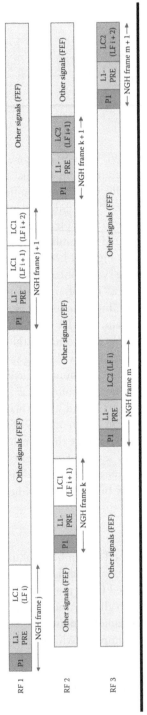

Figure 14.9   Logical channel Type C.

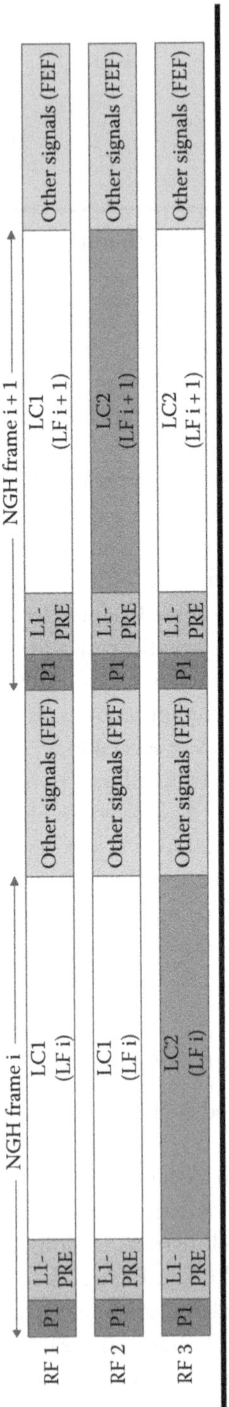

Figure 14.10   Logical channel Type D.

Table 14.4 Comparison between Different Types of Logical Channel

Logical Channel	Description	Overhead Reduction	Time Diversity	Frequency Diversity	StatMux Gain	System Flexibility
Type A	No bundling, no TFS	0	0	0	0	0
Type B	Time bundling, no TFS	+	+	0	0	0
Type C	Time + Freq. bundling TFS inter-frame	+ +	+	+	+	+
Type D	Time + Freq. bundling, TFS intra-frame	+	+	+ +	+ +	−

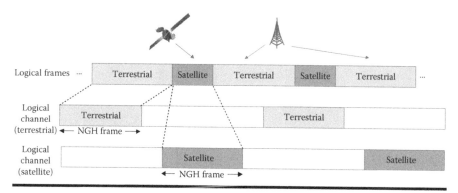

Figure 14.11 Logical frame mapping for hybrid terrestrial–satellite MFN networks. Parallel reception of both signals is possible for terminals with a single demodulator.

14.4 Conclusions

DVB-NGH has defined a new logical frame structure in order to guarantee flexible and efficient exploitation of the FEFs of DVB-T2. In DVB-NGH, one logical frame can be transmitted across a number of physical frames or FEFs, with the possibility for the associated L1-post signaling to appear in only one physical frame or FEF, yielding a significant overhead reduction, especially in the context of multiple RF channels. The logical frame structure enables the combination of FEFs of different length, bandwidth, frequency band, etc., without tight restriction on perfect

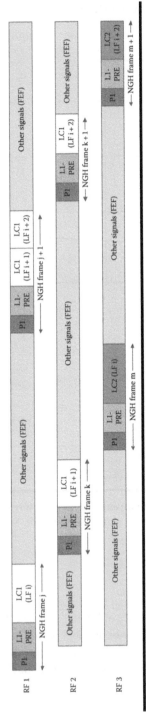

Figure 14.12 Logical channel groups with type C LC and type A LC transmitted in different RF channels.

time alignment. Hence, it can be seen as a generalization of TFS known from DVB-T2, hence preserving the high frequency diversity and high StatMux gains achieved by TFS. This is in addition to the possibility to allow for TFS reception with one single tuner, enabling, for example, hybrid terrestrial–satellite reception with relatively low complexity receivers.

References

1. EN 302 755 v1.3.1, Frame structure channel coding and modulation for a second generation digital terrestrial television broadcasting system (DVB-T2), April 2012.
2. ETSI TS 102 992 v1.1.1, Structure and modulation of optional transmitter signatures (T2-TX-SIG) for use with the DVB-T2, September 2010.
3. ETSI TS 102 831 v1.2.1, Digital video broadcasting (DVB), implementation guidelines for a second generation digital terrestrial television broadcasting system (DVB-T2), April 2012.

Chapter 15

Overview of the Physical Layer Signaling in DVB-NGH

José M. Llorca, David Gómez-Barquero, Hongsil Jeong, and Alain Mourad

Contents

15.1 Introduction

The physical layer signaling in the second-generation digital terrestrial TV standard Digital Video Broadcast–Terrestrial Second Generation (DVB-T2) has two main functions. First of all, it provides a means for fast signal detection, enabling fast signal scanning. Second, it provides the required information for accessing upper layers, that is, the layer-2 (L2) signaling and the services themselves. As the physical layer signaling enables the reception of the actual data, it should naturally be more robust against channel impairments than the data itself. It is generally recommended that the physical layer signaling is 3 dB more robust than the data. Furthermore, in order to maximize the system capacity, it should introduce as little overhead as possible.

The physical layer signaling of DVB-T2 is transmitted in preamble OFDM symbols at the beginning of each T2 frame, known as P1 and P2 symbols (see Figure 15.1). The preamble carries a limited amount of signaling information in a very robust way. The T2 frames begin with a preamble consisting of 1 P1 symbol and 1–16 P2 symbols. The preamble is followed by a configurable number of data symbols. The maximum length of a T2 frame is 250 ms.

The physical layer signaling is structured into the P1 signaling carried in the P1 symbol and the layer-1 (L1) signaling carried in the P2 symbol(s). The P1 symbol is used in the initial scan for detecting the presence of DVB-T2 signals. It carries some basic transmission parameters, such as the frame type, and it enables the reception of the P2 symbol(s). The L1 signaling is divided into two parts: L1-pre and L1-post signaling. The L1-pre signaling provides general information of the transmission related to the network topology, configuration, and the transmission protocols used. It enables the reception and decoding of the L1-post signaling, which in turn contains the information needed for extracting and decoding the data physical layer pipes (PLPs) from the frames.

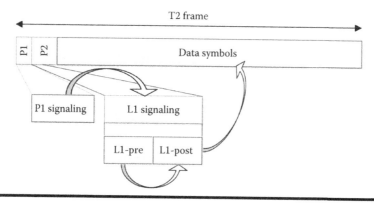

Figure 15.1   Physical layer signaling in DVB-T2 transmitted in preamble P1 and P2 OFDM symbols.

The physical layer signaling of DVB-T2 was designed such that it can always be made more robust than the data path. The transmission and detection of the preamble P1 symbol are very robust, and it can be correctly received even at negative signal-to-noise ratios (SNRs) under mobility conditions [1]. The transmission of the rest of the physical layer signaling in the P2 symbol(s) can be configured for enough robustness in rather static reception conditions. However, in mobile reception conditions, the robustness of the L1 signaling, in particular the L1-post field, may not be high enough due to the lack of time diversity [2]. DVB-T2 implements a flexible time interleaver at the physical layer in order to improve the robustness of the data signal against impulse noise and exploit the time diversity in mobile channels [3], though such a time interleaver is not used for L1 signaling. Since the L1 signaling is not time interleaved (only spread across one or few OFDM symbols) as opposed to data, it can be less robust than data in mobile channels despite having a more robust modulation and code rate (MODCOD).

One possibility to increase the robustness of the L1-post is to reduce the modulation order such that it can be received in all circumstances. This increases the signaling overhead and reduces the number of PLPs that can be used in the system. But this approach is not valid for the most robust data transmission mode: QPSK modulation with code rate 1/2 [2].

DVB-T2 specifies two optional mechanisms capable of improving the robustness of the L1 signaling, known as in-band L1 signaling and L1-repetition, but they are not useful when the receivers try to connect the very first time or when switching the radio frequency channel or re-scans for portable and mobile receivers. In these cases, the full signaling from the preamble needs to be received. With L1-repetition, the preamble of each T2 frame carries signaling information about the current and next frame. This increases the probability of correctly receiving the L1 signaling information after receiving two frames at the expense of increased signaling overhead (the maximum number of PLPs that can be signaled is also reduced). With in-band signaling, the L1 signaling is transmitted in-band together with the data, such that it has the same robustness and the receivers do not need to decode the preamble symbols for continuous reception.

Following the recommendations of its study mission and its commercial requirements [4], the next-generation mobile broadcasting standard DVB-NGH (Next Generation Handheld) has enhanced the physical layer signaling of DVB-T2 in three different aspects:

■ Reduced overhead
■ Higher capacity
■ Improved robustness

DVB-NGH has restructured the L1 signaling structure of DVB-T2 in order to reduce the signaling overhead. Instead of signaling the configuration of each PLP (MODCODTI, modulation, code rate, and time interleaving configuration), PLPs' configurations are classified into categories with the same settings, hence reducing the required signaling information when a number of PLPs have the same configuration (i.e., belong to the same category). Furthermore, DVB-NGH makes it possible to split across a number of frames some signaling information, which is in practice static, as opposed to DVB-T2, where all signaling information is always transmitted in every frame. DVB-NGH has also introduced a new concept of logical framing where the L1 signaling is further split into logical and physical parts, with the logical part (typically dominant) transmitted only once within the logical frame and so may appear in only one of the sequences of physical frames carrying the logical frame.

On the other side, the new logical frame structure avoids any limitation on the maximum number of PLPs that can be used in the system due to signaling constraints. For the preamble P1, DVB-NGH has also increased the signaling capacity. In DVB-T2, the preamble P1 symbol provides seven signaling bits. In DVB-NGH, for the terrestrial multiple-input multiple-output (MIMO) profile (see Chapters 19 and 21) and the hybrid terrestrial–satellite profiles of DVB-NGH (see Chapters 22 and 24), an additional preamble P1 (aP1) symbol has been introduced in order to cope with the required amount of signaling capacity of the P1 symbol. The P1 symbol signals the presence of the aP1 symbol through an escape code.

The improvement in the signaling robustness is also particularly relevant, because DVB-NGH adopts for the data path code rates more robust than that in DVB-T2 (i.e., 1/5 instead of 1/2). DVB-NGH adopts for L1 signaling new mini LDPC codes of size 4320 bits (4k) with a code rate 1/2. Although 4k LDPC codes have in general a worse performance than 16k LDPC codes, the 4k LDPC codes turn out to outperform 16k LDPC for L1 signaling thanks to the reduced amount of shortening and puncturing. In DVB-T2, LPDC codewords with L1 signaling are shortened (i.e., padded with zeros to fulfill the LDPC information codeword) and punctured (i.e., not all the generated parity bits are transmitted), which decreases the LDPC decoding performance. The adopted 4k LDPC codes have similar parity check matrix structure to the 16k LDPC codes used for data protection. This allows for efficient hardware implementations at the transmitter and receiver sides efficiently sharing the same logic.

DVB-NGH has also adopted two new mechanisms to improve the robustness of the L1 signaling known as additional parity (AP) and incremental redundancy (IR). The AP mechanism consists of transmitting punctured bits in the previous frame. When all punctured bits have been transmitted and there is still a need for more parity bits to be added, the IR mechanism extends the original 4k LDPC code into an 8k LDPC code of 8640 bits. The overall code rate is thus reduced from 1/2 down to 1/4. L1-repetition can be used to further improve the robustness of the L1 signaling as a complementary tool to AP and IR. It should be pointed out that the robustness improvement of the L1 signaling in DVB-NGH can be translated into a reduction of the signaling overhead for the same coverage.

Another improvement of the physical layer signaling in DVB-NGH compared to the first two releases of DVB-T2 (v.1.1.1 and v.1.2.1) is related to a reduction of the peak-to-average power ratio (PAPR) of the L1 signaling. After the first transmitters were manufactured, it was observed that large numbers of data PLPs resulted in peaks in the time domain signal during the P2 symbols. The reason was the lack of energy dispersal scrambling for the L1 signaling data. In the second release, several improvements were done to overcome this situation, based on a combination of the PAPR reduction mechanism of DVB-T2 active constellation extension (ACE) and tone reservation (TR), and using reserved bits for future use and additional bias balancing cells [5]. DVB-NGH, as well as the third release of DVB-T2 (v.1.3.1, whose main novelty is the profile T2-Lite, see Chapter 6), simply scrambles the L1 signaling based on the mechanism employed to scramble the data.

This chapter provides an overview of the physical layer signaling in the new-generation mobile broadcasting standard DVB-NGH. The rest of the chapter is structured as follows. Section 15.2 briefly reviews the physical layer signaling in DVB-T2. Section 15.3 deals with the L1 overhead reduction in DVB-NGH. Section 15.4 describes the signaling capacity improvements of the physical layer signaling in DVB-NGH. Section 15.5 focuses on the robustness enhancements of the L1 transmission in DVB-NGH, and it includes illustrative results comparing the robustness of the physical layer signaling and the data. Finally, the chapter is concluded with Section 15.6.

15.2 Physical Layer Signaling in DVB-T2

The physical layer signaling information in DVB-T2 is transmitted at the beginning of each T2 frame in preamble OFDM symbols known as P1 and P2. The physical layer signaling structure is illustrated in Figure 15.2.

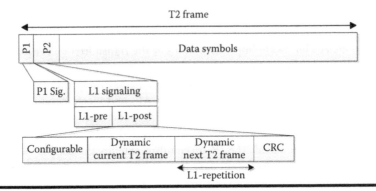

Figure 15.2 Physical layer signaling structure in DVB-T2.

15.2.1 Preamble Symbol P1

The preamble symbol P1 signals the basic parameters of the T2 frame in a very robust way with a very low overhead. It was designed for fast identification of T2 signals during the initial scan. In the first-generation digital terrestrial TV standard DVB-T (Terrestrial), initial scans when receivers are switched on for the first time are very time-consuming because receivers need to blindly test a high number of transmission parameters (e.g., FFT sizes, guard intervals) and frequency offsets of the transmitted signal, which may be applied by the broadcast network operator in order to avoid adjacent channel interference. In DVB-T2, P1 symbols are transmitted at most every 250 ms, and the P1 symbol can be received without knowing the data transmission parameters, which makes the initial scan much faster. Furthermore, its structure also allows signal detection at the nominal frequency, supporting up to 0.5 MHz frequency offsets. The P1 symbol improves robustness against spurious signals that could be superimposed on the received signal and consequently reduce the problem of false detections with respect to DVB-T.

15.2.1.1 P1 Signaling

The P1 symbol provides seven signaling bits that are carried by the modulation signaling sequences S1 and S2, with three and four signaling bits, respectively. The S1 signaling field is common for all DVB-T2-based systems and defines the frame type (e.g., T2, T2-Lite). For DVB-T2 (and T2-Lite), it also signals the transmission mode single-input single-output (SISO) or multiple-input single-output (MISO). The S2 field signals the FFT size and provides some hint about the guard interval used.

15.2.1.2 P1 Structure

The P1 symbol is a 1k FFT OFDM symbol that is differential binary phase shift keying (DBPSK) modulated in the frequency domain by the signaling sequences S1 and S2. The structure of the P1 symbol is shown in Figure 15.3.

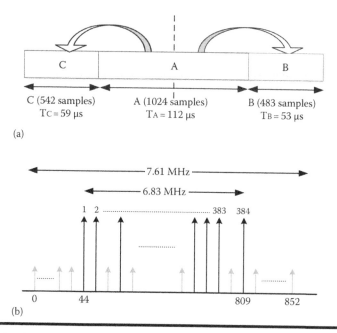

Figure 15.3 **DVB-T2 P1 symbol in time domain (a) and frequency domain (b) for 8 MHz channel bandwidth.**

In the time domain, the P1 symbol has three parts: one main part and two frequency-shifted cyclic prefixes with different lengths, which minimize the false detection risk. Its duration is fixed, and it equals 224 μs for 8 MHz channelization. In the frequency domain, the P1 symbol is narrowband, and it has a bandwidth of 6.83 MHz for 8 MHz bandwidth. This allows coping with frequency offsets up to 500 kHz in the first signal scan. The P1 symbol uses only 384 carriers with a boosted power in order to normalize the power of the P1 symbol with the data OFDM symbols, which enhances the robustness of the reception. The standard defines a pseudorandom carrier distribution sequence to determine the subset of used carriers. The transmission and detection of the preamble P1 symbol are very robust, and it can be correctly received even at negative SNR under mobility conditions [5].

15.2.2 Preamble Symbol(s) P2

The L1 signaling is transmitted in the P2 OFDM symbol(s) located in the beginning of the T2 frames after the P1 symbol. The number of the P2 symbols depends on the FFT size used in the system (P2 symbols use the same FFT size as the data symbols). In particular, for 1k, 2k, 4k, and 8k FFT modes, 16, 8, 4, and 2 P2 symbols are used, respectively, while for 16k and 32k FFT modes, only one P1 symbol is used. The P2 symbol(s) have a denser pilot pattern than the data symbols in order to allow frequency-only interpolation in the channel estimation process. They may also carry data in case the L1 signaling information leaves some carriers available, reducing the signaling overhead of the P2 symbol(s).

15.2.2.1 L1 Signaling

The L1 signaling transmitted in the P2 symbols is divided into two parts: L1-pre and L1-post signaling. The main objective of L1-pre field is to signal basic transmissions parameters that enable the reception and decoding of the L1-post signaling field. The L1-post signaling carries the information needed to identify and decode the data PLPs within the frame.

The amount of L1-pre signaling data is fixed to 200 bits, and it signals, among other things, the modulation, code rate, and size of the L1-post signaling field.

The L1-post signaling parameters are divided into two parts based on their frequency of change: configurable and dynamic. The configurable parameters change only when the network transmission parameters are changed (e.g., when a multiplex reconfiguration occurs). The changes are applied only at the border of a so-called super frame, which consists of configurable number of T2 frames. The dynamic information, on the other hand, can change from frame to frame due to dynamic multiplexing of the data PLPs. The amount of the L1-post information depends on the transmission system parameters, mainly the number of PLPs used in the T2 multiplex. The configurable field carries information with the features of each PLP (e.g., MODCODIT, type, position within the T2 frame, etc.). The L1-post signaling finishes with a cyclic redundancy check code that is applied to the entire L1-post including the configurable, the dynamic, and the optional extension field, and padding (if necessary).

15.2.2.2 L1 Signaling Transmission in DVB-T2

The L1-pre and L1-post signaling fields are independently coded and modulated, and they have different levels of robustness. Figure 15.4 shows the encoding and modulation process for L1 signaling.

The L1-pre field is protected with the most robust MODCOD possible: BPSK with code rate 1/5. The LDPC size used for L1 signaling in DVB-T2 is 16k, and because of the reduced size of the L1-pre, the LPDC codeword needs to be shortened and punctured. The standard defines optimized shortening and puncturing patterns in order to minimize the degradation in the LDPC decoding performance. The overall code rate including BCH after the shortening and puncturing for the L1-pre is 1/9.

The L1-post information is protected with a 16k LDPC with code rate 4/9 concatenated with a BCH code. Shortening and puncturing are also applied, but the effective LDPC code rate is variable according to the size of L1-post field, which depends on the transmission system parameters (mainly the number of PLPs in the

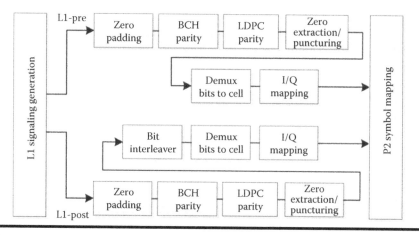

Figure 15.4   Error coding and modulation for L1 signaling in DVB-T2.

multiplex). The effective code rate decreases as the signaling information decreases (the minimum value is 1/4 and the maximum 4/9). This rate control compensates the performance degradation of LDPC decoding due to padding and puncturing, and it ensures the preservation of the coverage area.

The modulation order of the L1-post is the only parameter of the signaling preamble that can be adjusted by the broadcast network operator. The possible schemes are BPSK, QPSK, 16QAM, and 64QAM. When 16QAM or 64QAM is used, the LDPC codeword is bit interleaved, as it is done for the data path. The choice of the modulation order represents a trade-off between robustness and signaling overhead. Obviously, the L1 signaling needs to be more robust than the data because it carries information that allows the receivers decoding the data PLPs. For static reception conditions, the L1-post can be more robust than the data path by simply selecting one modulation order lower than the data. This way, the signaling overhead is also minimized. The problem is that the time interleaving duration is only several milliseconds, which is not sufficient in mobile scenarios. In case there is more than one P2 symbol per T2 frame, the L1-pre and L1-post signaling data are evenly distributed over all P2 symbols in order to maximize the time diversity. The L1-pre is always more robust than the data, but the L1-post is not robust enough for the most robust transmission mode for the data (QPSK 1/2) with time interleaving [2].

15.2.2.3 Additional L1 Signaling Robustness Mechanisms in DVB-T2

15.2.2.3.1 L1-Repetition

The mechanism known as L1-repetition consists of transmitting in each T2 frame the dynamic part of L1-post signaling of the current and the following T2 frame, as depicted in Figure 15.2. The use of L1-repetition is optional, and it is signaled in the L1-pre field. The probability of receiving the L1-dynamic information increases, since the correct reception of one T2 frame provides information for the next one. Furthermore, the time diversity is increased. However, it does not provide any benefit for the reception of the first T2 frame. L1-repetition introduces a delay in the transmission of one T2 frame, but it does not result in an increase in the channel change time. It also increases the signaling overhead and reduces the maximum number of PLPs that can be used in the T2 multiplex.

15.2.2.3.2 In-Band L1 Signaling

In-band signaling consists of transmitting L1 signaling information in the data path, such that there is no need for the receivers to decode the preamble P2 symbol(s). This signaling information is inserted in the first base band frame of the interleaving frame (time interleaving unit), and it carries L1-dynamic information about the next interleaving frame. In-band signaling reduces the time the receivers need to stay on, saving power. Another advantage is that the signaling information has the same robustness as the data. In a similar way to L1-repetition, in-band signaling increases the signaling overhead and the end-to-end delay (by one interleaving frame), but it cannot be used until the receiver is already synchronized.

15.3 L1 Signaling Overhead Reduction in DVB-NGH

The L1 signaling overhead in DVB-T2 is usually at reasonable values around 1%–5% [5]. However, the overhead increases with the higher number of PLPs, the higher robustness required for the signaling, and the shorter frame length. Further analysis of the L1 signaling in DVB-T2 shows that the configurable signaling (L1-config) represents the dominant part of the amount of signaling information, though it does not bring any useful information once the receiver is in its steady state (i.e., semi-static). In DVB-T2, all L1 signaling fields, that is, L1-pre, L1-config, and L1-dynamic, are transmitted in every frame. The values of the L1-dynamic can change frame by frame, whereas the L1-pre and L1-config may only change on a super frame basis. In practice, however, they change only when the multiplex of the RF channel is reconfigured, which occurs rather seldom. For these reasons, DVB-NGH allows the transmission of the L1-pre and L1-config signaling fields to be split into several frames (n-periodic L1-pre signaling (see Section 15.3.1), and self-decodable L1-config partitioning (see Section 15.3.3). The drawback is that the channel scanning time increases when the receiver is switched on for the very first time because the signaling information is not completed after the reception of the first frame.

The different techniques adopted in DVB-NGH to reduce the L1 signaling overhead can achieve an increase in the total system capacity higher than

1% (up to 1.5% in realistic settings). Nevertheless, the actual saving depends on the number of PLPs, the number of different PLPs configurations, the actual PLP settings and system features used, and finally the number of frames used to transmit the n-periodic L1-pre and the self-decodable L1-config partitions.

Next, the different mechanisms to reduce the signaling overhead of the L1-pre and L1-config are briefly described. For more details about the reduction of the L1 signaling overhead in DVB-NGH compared to DVB-T2, see Chapter 17.

15.3.1 L1-Pre Overhead Reduction

The L1-pre signaling field can be divided into n blocks of the same size, such that the signaling overhead is reduced by a factor of n. This is known as n-periodic transmission of L1-pre. The selection of the parameter n represents a trade-off between channel scanning time and signaling overhead. A superposed correlation sequence is used to detect the value of n and the order of the portions. The detection is very robust, being possible even at negative SNR. n-Periodic transmission of L1-pre cannot be used with time-frequency slicing (TFS) (see Chapter 13).

15.3.2 L1-Post Overhead Reduction

DVB-NGH has introduced a new logical frame structure compared to DVB-T2 specially suited to the transmission of DVB-NGH in DVB-T2 future extension frames (FEFs) (see Figure 15.5).

From an L1 signaling overhead point of view, in DVB-NGH, the L1-post signaling field does not need to be transmitted in every (physical) NGH frame, but only in the first NGH frame that carries the logical frame. One logical frame can span across several NGH frames, thus reducing the L1 signaling overhead. The overhead reduction is achieved at the expense of a larger zapping time because receivers may need to wait for more than one frame to start decoding the data PLPs. However, if the NGH frames are transmitted frequently, the impact can be negligible. For more details on the logical frame structure of DVB-NGH, see Chapter 14.

15.3.3 L1-Configurable Overhead Reduction

In DVB-T2, the L1 signaling allows each PLP to have completely independent parameters [5]. However, in realistic scenarios, there will be only a very limited number of different PLP configurations. This means that several PLPs would use exactly the same settings. For this reason, DVB-NGH has changed the L1 signaling paradigm for multiple PLPs, categorizing PLPs with the same features and reducing the signaling overhead of the L1 configurable field. The adopted solution allows for a totally general case, with up to 255 unique PLP settings, but in typical scenarios with few configurations of PLPs, the required amount of signaling information is radically reduced.

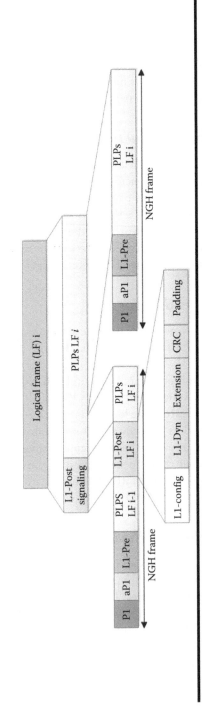

Figure 15.5 L1 signaling structure in DVB-NGH. The aP1 symbol is transmitted only in the MIMO and hybrid profiles.

Another improvement to reduce the L1-config overhead in DVB-NGH is the introduction of flags to signal the availability of some optional features that are not commonly used. The L1-config signaling format for DVB-T2 is very generic and supports a lot of features, such as TFS, auxiliary streams, reserved for future use fields, etc. The amount of required signaling information can be significant in some cases. However, in practical use cases, only a few of these features may be used, and thus the corresponding fields could be removed. In DVB-NGH, at the beginning of the L1-config field, one bit flags are introduced for some optional features to indicate whether the features are available or not.

The L1-config signaling cannot be divided into *n* blocks of the same size because the frame configuration (e.g., T2 frame length, FEF length) would not be known until the reception of *n* frames. Compared to n-periodic L1-pre transmission, the adopted approach for L1-config provides a flexible trade-off between overhead reduction and zapping delay and consists of transmitting the PLP configurations in L1-config not in every frame but in a carousel-like manner with controllable delay. The delay is controllable on a PLP basis, so that PLPs that cannot tolerate any delay can be sent with zero delay. This way, zero delay can be guaranteed for the constant signaling information that cannot tolerate any delay. The amount of L1-config information transmitted per frame is constant and self-decodable, but the signaling overhead is reduced because all the information of all PLPs is not transmitted in every frame.

15.4 Increased Physical Layer Signaling Capacity in DVB-NGH

DVB-NGH has improved the capacity of the physical layer signaling compared to DVB-T2 in two different aspects. On one hand, the adoption of an additional preamble P1 symbol has doubled the P1 capacity from 7 to 14 bits. On the other hand, the L1-post signaling capacity has been increased because it is not constrained to the preamble P2 symbol(s).

15.4.1 Additional Preamble aP1 OFDM Symbol for the Terrestrial MIMO and Hybrid Profiles of DVB-NGH

In DVB-T2, the P1 symbol provides seven signaling bits that define some essential transmission parameters such as the preamble format, the FFT size, and the guard interval. The preamble format, and hence the frame type, is signaled with three bits. These three bits are not sufficient to signal all profiles of DVB-T2, T2-Lite, and DVB-NGH. Therefore, additional bits are required. DVB-NGH has introduced an additional preamble P1 (aP1) symbol to identify the terrestrial MIMO and the hybrid terrestrial–satellite SISO and MIMO profiles. The presence of the aP1 symbol is signaled in the P1 symbol. For the basic profiles of DVB-NGH, as well as for DVB-T2 and T2-Lite, there is no aP1 symbol. The three bits of the P1 symbol used to identify the preamble format are used in the following way:

■ Two combinations are used for DVB-T2 (SISO and MISO).
■ Two combinations are used for T2-Lite (SISO and MISO).
■ One combination is used for non-T2 applications (e.g., identification of transmitters in single-frequency networks).
■ Two combinations are used for the terrestrial SISO and MISO profiles of DVB-NGH.
■ One combination is used as escape code to signal the presence of the aP1 symbol. This is used for the terrestrial MIMO and for the hybrid SISO and MIMO profiles of DVB-NGH.

The aP1 symbol provides seven signaling bits, and with this information, the receiver is able to receive the preamble P2 symbols and access the L1 signaling. The aP1 has the same structure as the P1 symbol and hence the same properties of the P1 symbol: robust signal discovery against false detection and the resilience to continuous interference. But in order to avoid interferences with the P1 symbol, the aP1 is scrambled with a different pseudorandom sequence, uses a different frequency offset value for the prefix and postfix, and employs a different set of carriers (see Figure 15.6).

The three design parameters mentioned earlier were carefully chosen to make aP1 have the same performance as P1 in both detection and decoding performance on the receiver side. Compared to DVB-T2, the aP1 symbol improves the performance of detecting and correcting frequency and timing synchronization on the receiver side.

15.4.2 L1-Post Signaling Capacity

DVB-T2 allows up to 255 PLPs to be used per multiplex. However, the maximum number of PLPs is in practice lower because the L1 signaling capacity is limited to the preamble P2 symbols. The L1 signaling capacity, and hence the maximum number of PLPs that can be used, is limited by the FFT size, the modulation used for the L1-post, and whether L1-repetition is used or not. For example, for any FFT size other than 32K, BPSK modulation, and with L1-repetition, the maximum number of PLPs is limited to 14 [5].

DVB-NGH has improved the L1-post signaling capacity compared to DVB-T2 because the L1-post is no longer constrained by a fixed number of preamble P2 symbol(s). The new logical frame structure allows transmitting the L1-post signaling outside the P2 symbol(s) (see Figure 15.5). In DVB-NGH, only the L1-pre signaling is transmitted within the P2 symbol(s) with a fixed length, modulation, and coding. The L1-post signaling may be transmitted after the L1-pre in the P2 symbol(s), but it can expand to the data symbols if needed.

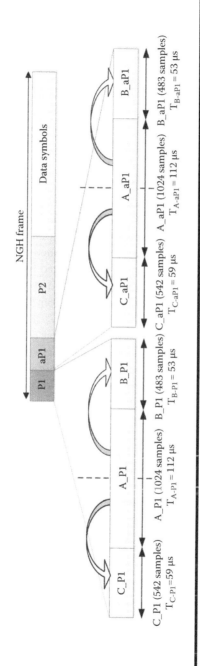

Figure 15.6 Preamble P1 and aP1 symbol in DVB-NGH.

15.5 Improved L1 Signaling Robustness in DVB-NGH

DVB-NGH adopts a similar L1 signaling structure to DVB-T2 and keeps the possibility of using L1-repetition, but it introduces three new mechanisms to improve the L1 robustness:

- 4k LDPC codes
- AP
- IR

15.5.1 4k LDPC Codes

The L1 signaling information of DVB-T2 does not generally fill one 16k LDPC codeword. Hence, codewords need to be shortened and punctured, which degrades the performance. For this reason, DVB-NGH adopts for L1 signaling new 4k LDPC codes of size 4320 bits. The reduced size of 4k LDPC codes is more suitable for L1 signaling, as the amount of shortening and puncturing can be significantly reduced (see Table 15.1). The LDPC code rates adopted for L1-pre and L1-post in DVB-NGH are 1/5 and 1/2, respectively. The modulation scheme used for L1-pre is always a BPSK, whereas L1-post is configurable between BPSK, QPSK, 16-QAM, and 64-QAM, as in DVB-T2.

The 4k LDPC codes have been created with similar parity check matrix structure to the 16k LDPC in order to efficiently share the hardware decoder chain at the receivers. For example, a quasi-cyclic structure in the information part and a staircase structure in the parity part are used. The 4k LDPC codes have a parallel factor of 72 (i.e., number of parallel computations to get the input from the memory for processing of the bit node and check node by the decoder), which is a divisor of the number used for 16k LDPC codes (360) to enable efficient sharing of the memory access. The maximum number of degrees of the parity check matrix, related at the decoder to the number of input values for variable node and check node processor, is also smaller than the one in 16k LDPCs.

Due to its lower size, 4k LDPC codes generally perform worse than 16k LDPC codes without padding and puncturing. This performance degradation depends on the code rate and the channel (i.e., it is larger for higher code rates and multipath fading channels), but it is less than 1 dB. However, for L1 signaling, 4k LDPC codes outperform 16k LDPC codes in the order of 1–2 dB because of the reduced size of the L1 signaling information. This performance improvement is also translated into a lower number of iterations and hence lower power consumption and faster convergence. Other advantages of 4k LDPC codes are lower complexity and higher efficiency at the receivers, reduced latency, and finer granularity when choosing the code rate.

Table 15.1   4k LDPC Codes in DVB-NGH vs. 16k LDPC Codes in DVB-T2 for L1 Signaling

L1 Signaling	LDPC Code	Code Rate	Information Bits	Parity Bits	Signaling Bits	Padded Bits	Punctured Bits
L1-pre	4k	1/5	1080	3,240	280	800	2,120
	16k	1/5	4000	12,000	200	3200	11,200
L1-post	4k	1/2	2160	2,160	640	1520	1,520
	16k	4/9	7200	9,000	640	6560	8,200

15.5.2 *Additional Parity*

The AP technique consists of transmitting punctured LDPC parity bits of the L1-post in the previous NGH frame. This technique improves the time diversity of the L1 signaling in a similar way to L1-repetition without affecting the zapping time, but it obtains a better performance because instead of just repeating the information, the effective code rate is reduced.

DVB-NGH employs optimized puncturing patterns for L1 in a similar way to DVB-T2 [5]. The code rate for L1-pre is fixed and is equal to 1/5. For L1-post, the mother code is 1/2, but a code rate control is also applied like in DVB-T2 to compensate the LDPC decoding performance degradation due to padding and puncturing such that the coverage does not depend on the amount of signaling information.

AP allows transmitting the punctured parity bits in the previous frame. Figure 15.7 shows the structure of the LDPC codeword with AP. The amount of AP bits is determined by the configuration parameter AP_RATIO, whose possible values are 0 (i.e., AP is not used), 1, 2, or 3. The proportion of AP bits transmitted in the previous frame is increased by a factor of 35% for each AP block. That is, AP1, AP2, and AP3 imply 35%, 70%, and 105% more parity bits.

Figure 15.8 shows the effective code rate for L1-post in DVB-T2 and in DVB-NGH with AP. We can see how the effective code rate decreases as the signaling information decreases (for DVB-NGH, the minimum value is 1/4 and the maximum 1/2). The main limitation of AP is the available number of punctured bits. In some cases, some AP configuration cannot be applied because there are not enough punctured bits. For this reason, the technique IR was adopted.

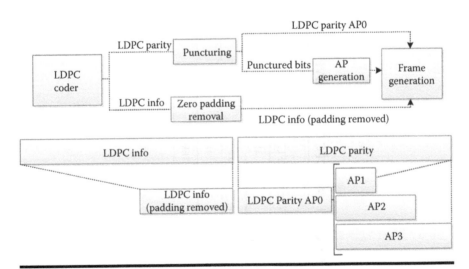

Figure 15.7   The resulting LDPC code word with AP bits.

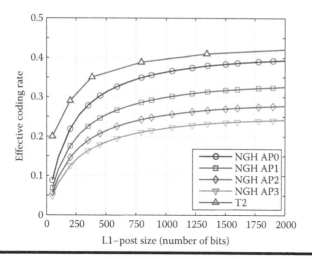

Figure 15.8   Effective code rate for L1-post in DVB-T2 and DVB-NGH with AP.

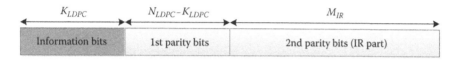

Figure 15.9   Extended 8k LDPC codeword with IR. K_{LDPC} 2160 bits, N_{LDPC} 4320 bits, M_{IR} 4320 bits.

15.5.3 Incremental Redundancy

The technique IR allows generating AP bits for AP when the amount of punctured bits is not enough for the selected AP configuration for L1-post. IR extends the 4k LDPC codeword into an 8k LDPC codeword, yielding an effective code rate of 1/4. The resulting codeword has two parts: the first part corresponds to the basic 4k LDPC codeword, and the second part corresponds to the additional IR parity bits (see Figure 15.9).

Figure 15.10 shows the structure of the IR parity check matrix. It can be noted that the parity check matrix of the 4k LDPC code is maintained.

The use of IR is determined by the code rate control mechanism depending on the AP configuration and the length of the L1-post signaling information. In particular, IR is applied when the amount of required parity bits is larger than the original 2160 bits generated by the 4k LDPC 1/2 code.

15.5.4 L1 Signaling Robustness Comparison in DVB-T2 and DVB-NGH

Figure 15.11 compares the robustness of the L1-post signaling in DVB-T2 and DVB-NGH. The performance has been evaluated with physical layer simulations

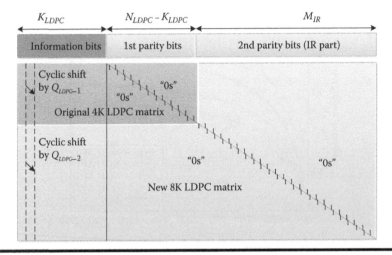

Figure 15.10   8k LDPC parity check matrix with IR.

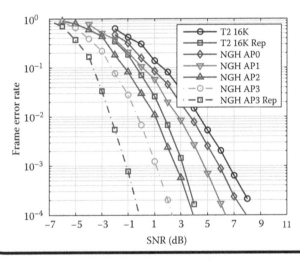

Figure 15.11   Robustness comparison of the L1-post in DVB-T2 and DVB-NGH. Modulation is QPSK. TU6 channel with 80 Hz Doppler. Bandwidth 8 MHz, FFT 8K.

for the mobile channel TU6 (6-Tap Typical Urban) for a Doppler frequency of 80 Hz. The modulation is QPSK, bandwidth is 8 MHz, and the FFT size is 8 K. The size of the L1-pre is representative for eight PLPs. In the figure, it can be noted that DVB-NGH considerably improves the robustness of the L1-post compared to DVB-T2. The gain due to the use of 4k LPDC (i.e., without repetition and without AP) is around 1 dB. The difference in SNR between the most robust configurations in DVB-T2 and DVB-NGH is around 4 dB.

15.5.5 L1 Signaling Robustness Performance Evaluation

The main objective in this section is to investigate the robustness of the DVB-NGH signaling path in terrestrial channels. As it has been explained before, it is important that the signaling path is transmitted at least 3 dB more robust than the data path in order to ensure the correct reception of the services. In DVB-T2, the data path with the most robust MODCOD QPSK 1/2 can be more robust than the L1 signaling path in mobile channels due to the short time interleaving of the L1 signaling. The use of L1-repetition can improve the robustness of the transmission (although recall that it is not useful in the reception of the first frame), but is not sufficient to ensure a 3 dB advantage for the L1-post signaling when the data path is transmitted with QPSK (see Chapter 6). In this case, the advanced decoding technique described next is needed.

15.5.5.1 Advanced Decoding of L1-Post

Advanced receivers can improve the robustness of the L1 signaling by reusing the information that has been decoded successfully in previous frames and that is also transmitted in the current frame when decoding the LDPC codewords. This is the case of the L1-config part, which generally remains the same from frame to frame. If L1-repetition is used, this technique can also be applied to the L1-dynamic part of the previous frame that is transmitted in the current frame.

This technique can be applied to both DVB-T2 and DVB-NGH. It should be pointed out that none of these techniques improves the robustness of the L1 signaling during the reception of the first frame.

15.5.5.2 Simulation Results

Table 15.2 shows the simulation parameters for the terrestrial TU6 channel model. The data transmission mode is QPSK 1/3. The modulation assumed for the L1 signaling is BPSK. Eight PLPs have been considered to compute the L1 signaling information. The effective code rate depends on the configuration of the AP ratio: AP0, AP1, AP2, and AP3.

Table 15.2   Simulation Parameters for the Terrestrial Profile of DVB-NGH

Parameter		Value	Parameter		Value
OFDM	FFT size	8192	Channel	Model	TU6
	BW	8 MHz		Doppler	33.3 Hz
	GI	1/4	Data	MODCOD	QPSK 1/3
	Pilot pattern (PP)	PP1		Time interleaver	100 ms Type 1 PLP
Time slicing	NGH frame	50 ms	L1	Modulation	BPSK
	Cycle time	300 ms		Nr. PLP	8 PLPs

Figure 15.12 shows the performance of the L1 signaling transmission in the terrestrial TU6 mobile channel with 4k LDPC, AP, and IR, with and without L1-repetition and advanced L1 decoding. The performance of the data path for QPSK 1/3 is also shown for comparison. The required SNR to achieve a frame error rate (FER) of 1% is 1.4 dB.

In Figure 15.12a, we can see that without L1-repetition and without advanced L1 decoding, only the most robust AP3 configuration is more robust than QPSK 1/3 at FER 10^{-2} (and only 1.4 dB). Advanced L1 decoding provides an SNR gain between 2.2 and 3 dB, yielding together with AP2 and AP3, 2.5 and 4.5 dB more robustness than QPSK 1/3, respectively (see Figure 15.12b). L1-repetition provides a similar gain to advanced decoding (between 2.5 and 3 dB) (see Figure 15.12c). The gain with both advanced decoding and repetition is in the order of 4 and 4.5 dB. In this case, AP0 is already 3 dB more robust than the data path. There is thus room for using the most robust MODCOD for the data path QPSK 1/5.

15.6 Conclusions

DVB-NGH has enhanced the physical layer signaling of DVB-T2 in three aspects: reduced overhead, higher capacity, and improved robustness. The overhead improvements allow reducing the signaling overhead and increasing the system capacity between 1% and 1.5% without affecting the system performance. The capacity enhancements allow signaling all five profiles of DVB-NGH (sheer terrestrial SISO, MISO, and MIMO; and hybrid terrestrial–satellite SISO and MIMO) without any restriction on the number of PLPs used in the system. Finally, the robustness improvements allow supporting extremely robust MODCODs for the data path such as QPSK 1/5 for both terrestrial and satellite mobile channels.

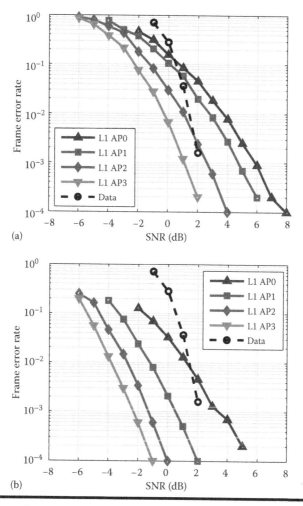

Figure 15.12   Robustness of the L1 signaling in the terrestrial TU6 mobile channel. (a) Reference case without L1-repetition and without advanced L1 decoding. (b) Advanced L1 decoding.

(*continued*)

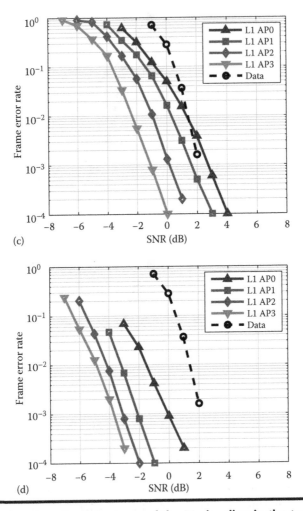

Figure 15.12 (continued) Robustness of the L1 signaling in the terrestrial TU6 mobile channel. (c) L1-repetition. (d) L1-repetition and advanced L1 decoding.

References

1. L. Vangelista, On the analysis of P1 symbol performance for DVB-T2, *Proceedings of IEEE SARNOFF*, Princeton, NJ, 2009.
2. T. Jokela, M. Tupala, and J. Paavola, Analysis of physical layer signaling transmission in DVB-T2 systems, *IEEE Transaction on Broadcasting*, 56(3), 410–417, September 2010.
3. D. Gozálvez, D. Gómez-Barquero, D. Vargas, and N. Cardona, Time diversity in mobile DVB-T2 systems, *IEEE Transactions on Broadcasting*, 57(3), 617–628, September 2011.
4. DVB Commercial Module, Commercial requirements for DVB-NGH, CM-NGH0 15R1, June 2009.
5. ETSI TR 102 831 v1.2.1, Implementation guidelines for a second generation digital terrestrial television broadcasting system (DVB-T2), 2012.

Chapter 16

Overview of the System and Upper Layers of DVB-NGH

Jani Väre, David Gómez-Barquero, and Alain Mourad

Contents

16.1 Introduction

Digital Video Broadcasting—Next Generation Handheld (DVB-NGH) is a transmission system for broadcasting multimedia contents to handheld devices, which allows for user interaction via the return channel of mobile broadband communication systems. The interoperation between DVB-NGH and mobile broadband networks to create an effective interactive user experience requires a full end-to-end Internet Protocol (IP) system, such that it becomes possible to deliver the same audio/visual content over the two networks in a bearer-agnostic way. Therefore, DVB-NGH includes full support of an IP transport layer, in addition to the most common transport protocol used in broadcast systems: MPEG-2 transport stream (TS). DVB-NGH supports two independent transport protocol profiles, one for TS and another one for IP, each with a dedicated protocol stack. The profiles have been designed as mutually transparent in order to improve bandwidth utilization. It should be noted that this approach is possible in DVB-NGH thanks to a physical layer packet unit, known as base band frame (BB frame), which is independent of the transport layer (i.e., TS or IP). This is a major difference compared to the first-generation mobile broadcasting DVB standards DVB-H (Handheld) and DVB-SH (Satellite to Handheld), where the physical layer unit in both is MPEG-2 TS packets. Hence, to transmit IP over DVB-H and DVB-SH, it is necessary to introduce an additional encapsulation protocol known as Multi Protocol Encapsulation (MPE). However, in DVB-NGH, this is no longer needed, and IP can be delivered directly over a generic encapsulation protocol known as Generic Stream Encapsulation (GSE), reducing the packet encapsulation overhead compared to MPE over TS by up to 70% (see Chapter 17).

Currently, the only available upper layer solution for the IP profile of DVB-NGH is based on Open Mobile Alliance Mobile Broadcast Services Enabler Suite (OMA-BCAST). However, it should be noted that the specification also allows the possibility to use any other upper layer solution, which may be defined in the future. OMA-BCAST is an open global specification for mobile TV and on-demand video services that can be adapted to any IP-based mobile delivery technology. OMA-BCAST specifies a variety of features including content delivery

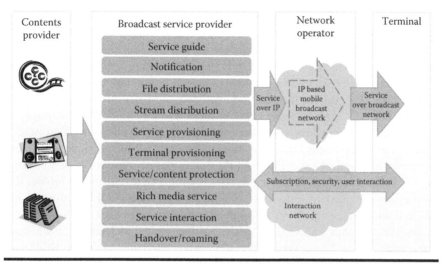

Figure 16.1 **OMA-BCAST overview: end-to-end hybrid cellular and broadcasting system.**

protocols for streaming and file download services, electronic service guide (ESG) for service discovery, service and content purchase and protection, terminal and service provisioning (e.g., firmware updates), interactivity, and notifications as shown in Figure 16.1.

OMA-BCAST is thus an interworking platform that enables the convergence of services from broadcast and cellular domains, combining the strengths of broadcast networks to efficiently deliver very popular content with the personalization, interaction, and billing features of cellular networks. In this sense, the mobile broadcasting network is a complement of the cellular system. It should be pointed out that there is no convergence in networks, since both broadcasting and cellular networks are used separately and independently. Convergence is realized in services, platforms, and multimode terminals that support both radio access technologies. OMA-BCAST specifies separate set of adaptation specifications for different physical layer bearers, such as 3GGP, DVB-H, and now DVB-NGH. The set of specifications consist of OMA-BCAST specification and the Adaptation specification(s), which are typically developed jointly with the standardization organization, which develops the specifications for the transmission bearer.

Both TS and IP profiles in DVB-NGH have been designed with the commercial requirement of minimizing the overall signaling overhead and increasing the bandwidth efficiency [1]. Also the latency in the service reception, and hence the end user experience, is emphasized in the commercial requirements. In order to minimize the redundancy and latency on the signaling, and hence also the complexity and signaling overhead, the OMA-BCAST adaptation for DVB-NGH has taken a radical approach to split the majority of the signaling between the physical layer and on the top of IP, more specifically for the latter inside the OMA-BCAST SG structures. Only minimal signaling is defined at the link layer (L2), for optimizing the receiver latency and power saving. One of the most important conclusions of the study performed prior to the development of DVB-NGH was that a significant data net throughput is today wasted in signaling overheads such as packet headers and metadata, and that the optimization of the signaling and packet encapsulation could yield to important throughput improvements without losing any functionality [2]. In DVB-NGH, two header compression mechanisms have been adopted for both TS and IP packets. For the TS profile, for Physical Layer Pipes (PLPs) that carry only one service component, the TS header can be reduced from 3 bytes to only 1 byte. For the IP profile, the unidirectional mode of the RObust Header Compression (ROHC) protocol [3] has been adopted alongside an external adaptation to allow for a more efficient compression without implication on the zapping time and latency. In addition to the new TS and IP packet header compression mechanisms, DVB-NGH has improved the physical layer signaling in terms of reduced signaling overhead.

The two transport protocol profiles of DVB-NGH have also been designed with the commercial requirement of transmitting layered video, such as scalable video coding (SVC) [4], with multiple PLPs. Layered video codecs allow for extracting different video representations from a single bit stream, where the different sub-streams are referred to as layers. SVC is the scalable video coding version of H.264/Advanced Video Coding (AVC). It provides efficient scalability on top of the high coding efficiency of H.264/AVC. SVC encodes the video information into a base layer, and one or several enhancement layers. The base layer constitutes the lowest quality, and it is a H.264/AVC compliant bit stream, which ensures backward compatibility with existing H.264/AVC receivers. Each additional enhancement layer improves the video quality in a certain dimension. SVC allows up to three different scalability dimensions: temporal, spatial, and quality scalability. SVC utilizes different temporal and interlayer prediction methods for gaining coding efficiency while introducing dependencies between the different quality layers. Due to these dependencies, parts of the bit stream are more important than others. In particular, the enhancement layer cannot be decoded without the base layer due to missing references.

The combination of layered video codec such as SVC with multiple PLPs presents a great potential to achieve a very efficient and flexible provisioning of mobile TV services in the DVB-NGH system [5]. By transmitting the SVC base layer using a heavily protected PLP and the enhancement layer in one PLP with moderate/high spectral efficiency, it is possible to cost-efficiently provide a reduced quality service with a very robust transmission, while providing a standard/high quality service for users in good reception conditions.

The benefits of using SVC compared to simulcasting the same content with different video qualities in different PLPs with different robustness are twofold. First of all, SVC has reduced bandwidth requirements compared to simulcasting.

Compared to single-layer AVC coding, the SVC coding penalty for the enhancement layer can be as little as 10% [4]. Second, with SVC, it is possible to provide a graceful degradation of the received service quality when suffering strong channel impairments with seamless switching between the different video qualities.

The DVB-T2 specification states that DVB-T2 receivers are only expected to decode one single data PLP at a time [6]. But they must be able to decode up to two PLPs simultaneously when receiving a single service: one data PLP and its associated common PLP, which is normally used to transmit the information shared by a group of data PLPs. This feature cannot be used to deliver SVC with multiple PLPs and differentiated protection in DVB-T2 because the behavior of the receivers when only one PLP is correctly received is not specified [5]. In DVB-NGH, it is possible to receive up to four PLPs simultaneously [7]. Compared to DVB-T2, DVB-NGH has enhanced the handling of multiple PLPs belonging to the same service updating the receiver buffer model (RBM) and the synchronization requirements, which determine the necessary time delays at the transmitter in order to avoid buffering overflow or underflow in the receivers, and defining a new signaling to provide mapping between service components and PLPs and the scheduling of the PLPs at the physical layer.

This chapter provides an overview of the system and upper layers of DVB-NGH, as depicted in Figure 16.2. The rest of the chapter is structured as follows. Section 16.2 provides background information on the DVB-NGH physical layer, such as physical layer signaling, physical layer adaptation, and SVC delivery with

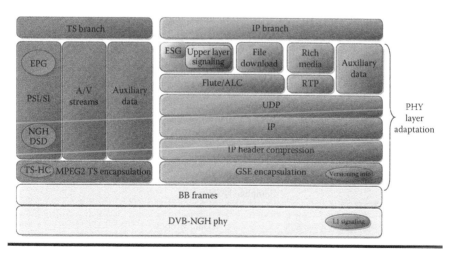

Figure 16.2 TS and IP profiles of DVB-NGH, including header compression mechanisms. Both profiles may coexist in the same multiplex. The TS signaling and service discovery is based on the PSI/SI tables with a new NGH delivery system descriptor. The upper layer IP signaling is carried within the ESG for better convergence with IP-based systems (cellular networks) and is compatible with OMA-BCAST.

multiple PLPs. Section 16.3 provides an overview of the TS profile of DVB-NGH. It analyzes the protocol stack, including the null TS packet deletion mechanism, the TS signaling and the service discovery, and the delivery of SVC over TS. Section 16.4 provides an overview of the IP profile of DVB-NGH. It describes the protocol stack, the adopted ROHC header compression with external adaptation, the IP signaling and the service discovery, the end-to-end functionalities provided by OMA-BCAST, and the delivery of SVC over IP. Finally, the chapter is concluded with Section 16.5.

16.2 DVB-NGH Physical Layer Background

16.2.1 DVB-NGH Physical Layer Signaling

Layer 1 (L1) signaling data include the necessary information for the receivers to identify the DVB-NGH signal, find the different data PLPs transmitted in the frame, and allow their decoding. The L1 signaling is split into two parts, a physical part and a logical part. The physical part is carried in the preamble OFDM symbols of each frame, known as P1, aP1, and P2 symbols, whereas the logical part is carried in the data OFDM symbols (see Figure 16.3). The additional aP1 preamble symbol is transmitted only for the MIMO and hybrid terrestrial–satellite profiles of DVB-NGH.

The signaling detection starts with the P1 symbol (and optionally the aP1 symbol), which enables fast access to the targeted signal. After successful decoding of the P1 symbol and upon a detection of a desired signal, the process continues with the decoding of the P2 symbol. The P2 symbol includes the L1-pre, which provides some basic signaling parameters such as bandwidth, peak-to-average power ratio (PAPR) techniques; it establishes the structure and location for the L1-post decoding. Further on, L1-post signaling contains parameters that provide enough information for the receiver to decode the desired PLP(s). The TS and IP profiles share the same L1 signaling. For more details on the physical layer signaling of DVB-NGH, see Chapter 15.

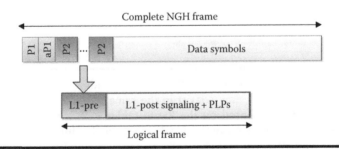

Figure 16.3 Physical layer signaling in DVB-NGH.

Prior to starting the decoding process of any PLPs, the receiver needs to access the upper layer signaling in order that it can discover which services are carried in which PLPs. This can be done by accessing the common PLP, which carries all upper layer signaling information. Depending on the profile, TS or IP, a different set of upper layer signaling is provided. In the case of TS profile, the upper layer signaling consists of PSI/SI tables (including NGH delivery system descriptor). The IP profile, in turn, has dedicated upper layer signaling elements inside the OMA-BCAST SG and versioning signaling information, which is carried within the logical link control (LLC) defined in [8]. The upper layer signaling is further elaborated in Section 16.3.4.

16.2.2 DVB-NGH Physical Layer Adaptation

In the second-generation DVB standards, the physical layer unit is the so-called BB frame. All input data from the link layer (L2) (i.e., MPEG-2 TS packets and GSE packets with IP datagrams) has to be adapted first for delivery over BB frames. In DVB-T2, the physical layer adaptation is performed by means of two processes, the mode adaptation and the stream adaptation. Basically, most of the physical layer adaptation is performed in the mode adaptation, which puts the upper layer packets on the BB frame payload.

16.2.2.1 Mode Adaptation

DVB-T2 defines two adaptation modes: the normal mode (NM) and the high efficiency mode (HEM) [6]. In the NM, the upper layer packets are transmitted over BB frames with a data structure that includes signaling information to make the decoding at the receiver easier. These signaling and synchronizing methods represent more overhead for the system. On the other hand, if the transmission is configured in a HEM, it is possible to save overhead by means of less signaling transmission. In this mode, it is assumed that the receiver knows a priori some transmission information signaled by other ways (L2 signaling), avoiding a repeated transmission of the data flow.

In DVB-NGH, a new definition of the BB header is performed in order to improve the overhead optimization. The signaling has been reorganized in order to separate different concepts that in DVB-T2 were duplicated or misallocated. The reorganization in the signaling aims to separate the PLP-specific information of the BB frame information. In this manner, specific PLP information is present only in the L1 signaling, whereas the BB frame information appears only in the BB frame header. With this new structure, it is possible to reduce the BB frame header from 10 bytes in DVB-T2 to 3 bytes in DVB-NGH, only with the essential parameters of the BB frame if TS header compression (TS-HC) is not used (see more details in Section 16.3.4). This way, ambiguities concerning signaling are avoided, an overhead reduction is achieved, and the receiver implementation is also simplified.

The mode adaptation module in DVB-NGH can process input data for a PLP in one of three modes [7]: the ISSY-LF mode (i.e., one ISSY field per logical frame carried as part of the mandatory in-band signaling type B), the ISSY-BBF mode (i.e., one ISSY field per BB frame carried in the BB frame header), or the ISSY-UP mode (one ISSY field attached to each user packet). The Input Stream Synchronizer (ISSY) is the synchronization mechanism used in the TS profile of DVB-T2. The field carries the value of a counter clocked at the modulator clock rate and can be used by the receiver to regenerate the correct timing of the regenerated output stream.

16.2.2.2 Stream Adaptation

The stream adaptation is composed by the scheduler, a frame delay block, padding or in-band signaling inserter, and BB scrambler, as shown in Figure 16.4.

The scheduler maps data BB frames into constellation cells. Working with multiple PLPs (MPLPs), the scheduler is the most important agent in order to perform the service multiplexing. To do that, the scheduler generates the physical layer signaling information, the L1, and assigns for each PLP the cell number to fill

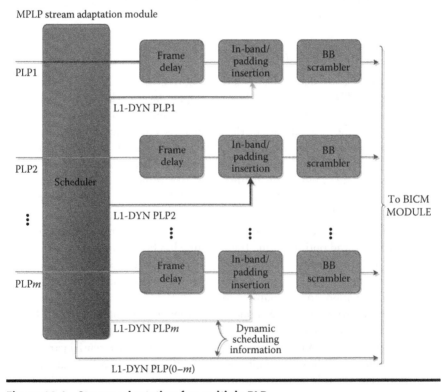

Figure 16.4 Stream adaptation for multiple PLP.

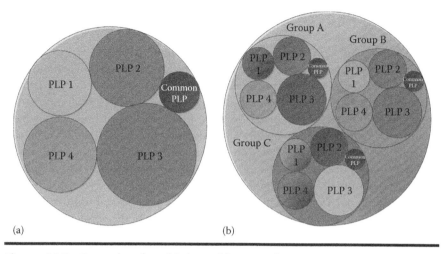

Figure 16.5 Example of multiplex with several PLP groups. (a) One group of PLPs with a single common PLP. (b) Three groups of PLPs with a common PLP for each.

with data. After the scheduler, there is a frame delay block and the in-band and/or padding insertion. In this block, it is possible to fill the BB frames with in-band L1 or simply complete the BB frame with padding. And finally, the last stream adaptation block is a scrambler for the complete BB frame to increase information diversity. This scrambling is applied separately BB frame per BB frame.

16.2.3 Common PLP

DVB-T2 defines a common PLP that is an entry point for the all upper layer signaling. In DVB-NGH, this common PLP concept has been enhanced to work with different groups of PLPs within a single multiplex. In this case, the common elements of multiple PLPs shall be transported centrally in order to achieve overhead reduction. This configuration allows having one common PLP per group of PLPs, and the content transported in the common PLP applies to all data PLPs of the same group in the same way (see Figure 16.5).

16.3 Transport Stream Profile of DVB-NGH

The TS profile of DVB-NGH is dedicated for the transmission of the TS multiplexes. The DVB-NGH TS profile is based on DVB-T2 and follows the same protocol stack structure, taking advantage of the methods to achieve an efficient transmission such as null packet deletion (NPD). In addition, DVB-NGH includes novelties to increment the flexibility of the system and the efficiency, such as SVC delivery with multiple PLPs and TS-HC. In addition, an adaptation of the PSI/SI signaling has been performed by adding a new descriptor, the NGH delivery system descriptor, containing the new necessary information for DVB-NGH.

MPEG-2 imposes strict constant bit rate requirements on a TS multiplex. However, a TS multiplex may contain a number of variable bit rate (VBR) services with null TS packets that do not contain any useful data. Depending on the bit rate characteristic of the multiplexed services, it is possible that null packets may constitute a large percentage of the TS multiplex bit rate. On the other hand, statistical multiplexing algorithms can exploit the properties of the VBR video content by means of real-time analysis of the encoded content. This analysis allows the adaptation of the transmission to the set of service bit rates. Nevertheless, this improvement is not free. Statistical multiplexing requires a real-time analysis of the video content and processing to allocate appropriately the optimal amount of information of each service in each time interval. This processing complexity implies usually extra buffering in the transmitter, hence increasing the end-to-end delay.

In DVB-NGH, it is possible to transmit with different physical layer configuration by means of MPLPs. In this case, the service multiplexing needs to be performed by the scheduler at the physical layer in the stream adaptation, managing the cell number per PLP and allowing a most accurate mechanism to optimize the transmission. In any case, the benefits of statistical multiplexing are substantial in terms of bandwidth optimization allowing more service transmission per multiplex.

16.3.1 TS Protocol Stack Overview

The protocol stack of the TS profile keeps the same structure as that of the previous first generation of DVB standards (see Figure 16.6). The elementary streams (ESs) containing audio and video information are encapsulated in packetized elementary streams (PESs) and at lower lever to TS packets. Finally, this flow of TS packets undergoes both mode and stream adaptation to achieve BB frame flows, as described in Section 16.2.

16.3.2 TS Null Packet Deletion

DVB-NGH adopts the TS NPD mechanism of DVB-T2, which avoids the transmission over the air of null packets. The removal of null packets at the transmitter side is performed in a way that at the receiver side the removed null packets can be reinserted in the TS at their original positions (see Figure 16.7). NPD is optional, and when used, useful TS packets are transmitted and null TS packets are removed. A deleted null packet (DNP) counter of 1 byte is defined that is first reset and then incremented at each DNP. This DNP counter is sent instead of the null packets, allowing an overhead reduction. For more details on the null TS packet deletion mechanism, see Chapter 17.

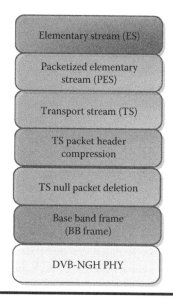

Figure 16.6 **DVB-NGH TS profile protocol stack.**

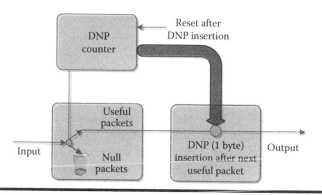

Figure 16.7 **TS null packet deletion process at the transmitter.**

16.3.3 TS Packet Header Compression

TS-HC is a new technique introduced in DVB-NGH that allows reducing the TS header signaling information for PLPs with only one packet identifier (PID), for example, video. The TS/HC scheme assumes that the PID values can be transmitted by some other means such as L1 signaling or PSI/SI. TS packets that implement the compression have another structure of header that provides a more efficient transmission and, in addition, it is possible to perform TS decompression purely on L1 with negligible complexity. With this method, it is possible to achieve overhead saving by about 1% within the overall system performance. However, it is not easy to manage contents with a dynamic PID where it is necessary to synchronize the changes with the PMT. To solve this complication, in DVB-NGH, the PID value is transmitted in the BB header, and a new pointer (SYNCD-PID) is defined to indicate the TS packet where the PID changes. These new fields in the BB header, in combination with the new consistent signaling described in Section 16.2, represent an overhead saving of 4 bytes in the BB frame in comparison with DVB-T2, in addition to the TS-HC performed. With this solution, when a particular PLP uses TS-HC, all the TS packets are compressed and have a constant length, and receivers can be correctly synchronized working with dynamic PID. For more details on the TS packet header compression mechanism, see Chapter 17.

16.3.4 PSI/SI Signaling

As in other previous TS-based DVB systems, the signaling of the upper layers within DVB-NGH TS profile is based on the PSI/SI tables. Traditionally, all network-related signaling is kept in the NIT, including in addition to network information everything related to the transmission characteristics. In a similar way, all PID mapping is done by the PMT. The system provides a demodulation mechanism at the receiver to output correct MPEG2-TS packets without first parsing PSI/SI tables.

The PSI/SI signaling consists of tables that are carried over TSs. The main PSI/SI tables are as follows: Network Information Table (NIT), Program Association Table (PAT), Program Map Table (PMT), Service Description Table (SDT), Event Information Table (EIT), and Time and Date Table (TDT). Each table, in turn, excluding PAT and TDT, carries a number of different descriptors that contain most of the actual information that is carried within the tables. The majority of the descriptors and PSI/SI tables are common for different DVB technologies. However, the *NGH_delivery_system_descriptor* is specific to the TS profile of NGH. This descriptor maps the services with the network information and PLPs (see Table 16.1).

The descriptor follows the same philosophy adopted in DVB-T2, being used for one of the TS carried by a particular *network_id/ngh_system_id* combination. The other TSs of the same combination carry only the upper part in order to save overhead. With the same *network_id/ngh_system_id* combination, it is possible to get a one-to-one mapping between *transport_stream_id* (16 bits) and *ngh_stream_id* (8 bits), the latter used in the L1 signaling. In addition, the PMT table also contains, as usual, the *service_id/component tag*, that is, the PID for all service components. On the other hand, as opposed to DVB-T2, the MIMO/SISO field goes to the component loop because of the possibility to have different schemes per service component. Other fields in the lower part are assumed to be the same for a complete service.

Otherwise, it should be noted that there are two key differences in the receiver behavior for the TS profile of DVB-NGH, one in the initial scanning and the other one in the zapping case, as presented later.

Table 16.1 NGH Delivery System Descriptor

Syntax	# of Bits	Identifier
NGH_delivery_system_descriptor() {		
descriptor_tag	8	uimsbf
descriptor_length	8	uimsbf
descriptor_tag_extension	8	uimsbf
ngh_system_id	16	uimsbf
ngh_stream_id	8	uimsbf
service_loop_length	8	uimsbf
for (i = 0;i<N,i++){		
service_id	16	uimsbf
component_loop_length	8	uimsbf
for (j = 0;j<N,j++){		
component_tag	8	uimsbf
plp_id	8	uimsbf
anchor_flag	1	uimsbf
SISO/MIMO	2	bslbf
reserved_for_future_use	5	uimsbf
}		
}		
if (descriptor_length - service_loop_length > 5) {		
bandwidth	4	bslbf
reserved_future_use	3	bslbf
guard_interval	4	bslbf
transmission_mode	3	bslbf
other_frequency_flag	1	bslbf
tfs_flag	1	bslbf
common_clock_reference_id	4	bslbf
for (i = 0;i<N,i++){		
cell_id	16	uimsbf

(continued)

Table 16.1 (continued) NGH Delivery System Descriptor

Syntax	# of Bits	Identifier
`if (tfs_flag== 1){`		
`    frequency_loop_length`	8	uimsbf
`    for (j=0;j<N;j++){`		
`        centre_frequency`	32	uimsbf
`    }`		
`}`		
`else {`		
`    centre_frequency`	32	uimsbf
`}`		
`subcell_info_loop_length`	8	uimsbf
`for (k=0;k<N;k++){`		
`    cell_id_extension`	8	uimsbf
`    transposer_frequency`	32	uimsbf
`    }`		
`   }`		
`  }`		
`}`		

16.3.4.1 Initial Scanning

The initial scanning procedure in the receiver is described as follows:

- The receiver reads the configurable part of L1-post signaling (L1-config) and parses the *stream_id* loop. For each *stream_id* entry, it will therefore see the collection of *plp_ids* carrying the TS (or other stream).
- Later, the receiver can demodulate the relevant PLPs and access the TS packets (or other packets) of each demodulated PLP.
- In the TS case, the receiver can easily identify which TS packets are desired by checking the *stream_id* byte of the packets in the relevant PLPs (see Section 16.3.5).
- All packets having the same *stream_id* byte are merged and will together form a complete TS after replacement of *stream_id* with sync byte.

- When the common PLP is used, it is included in the merging. Packets in the common PLP do not use the *stream_id* feature (since the packets are merged into all TSs).
- The TS is output from the demodulator and parsed by the backend.
- Other TSs are output in the same way or via help from already parsed PSI/SI tables of so-far received TSs.

16.3.4.2 Zapping Case

Here is the receiver procedure in a zapping case:

- When a new service is selected, the receiver finds the *ts_id*, *service_id*, and *component_tags* via the stored SDT
- It then finds the access parameters of the relevant service components of the TS via the stored NIT, for example, frequency, *stream_id* and relevant *plp_ids*
- The receiver can then demodulate the relevant *plp_ids* and merge the TS packets having the right *stream_id* to form an output TS
- The receiver then reads the PAT/PMT of the output TS and finds the PIDs for the selected service

16.3.5 TS Signaling

Once the L1 signaling has provided the required information to start the PLP decoding and has access to the TS flow, the receiver needs to identify which TS packets belong together to the desired TS by checking the *stream_id* byte (the byte replacing the TS sync byte in transmission (see Figure 16.8) of the packets in the relevant PLPs). With this identification, all the packets with the same *stream_id* are merged and together form a complete TS stream after replacing the *stream_id* byte by the sync byte. It should be noted that this sync byte replacement is not performed in the common PLP as the packets here are merged into all TS flows.

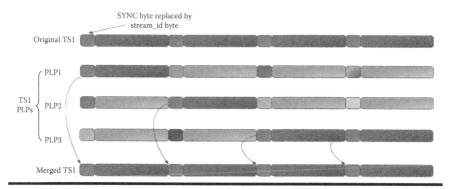

Figure 16.8 *Stream_id* **byte utilization example for PLP splitting and merge.**

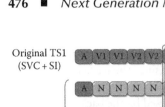

Figure 16.9 **SVC delivery with multiple PLPs in the TS profile. The original TSs include audio (A), SVC video (base layer V1 and enhancement layer V2), and PSI/SI signaling information.**

16.3.6 SVC Delivery in the TS Profile

The synchronization between PLPs that carry the different layers of the same NGH service is a critical aspect. The synchronization depends on the profile chosen. TS specifies a container format encapsulating PESs (video, audio, and other information), with error correction and stream synchronization features for maintaining transmission integrity when the signal is degraded. The synchronization mechanism used in the TS profile is ISSY. The required signaling for SVC with multiple PLPs in DVB-NGH is based on the delivery system descriptor present in the NIT of the PSI/SI, where the different PLPs are signaled, allowing the corresponding selection in the receiver (see Figure 16.9).

16.3.7 TS Profile Receiver Buffer Model

A typical DVB-NGH receiver should be composed of an RF input and followed by a number of physical layer stages in order to recover BB frames [9]. The data field bits of decoded BB frames belonging to a PLP are then converted to a canonical form, independent of the mode adaptation options in use. The resulting bits are written into a de-jitter buffer (DJB). Bits are read out from the buffer

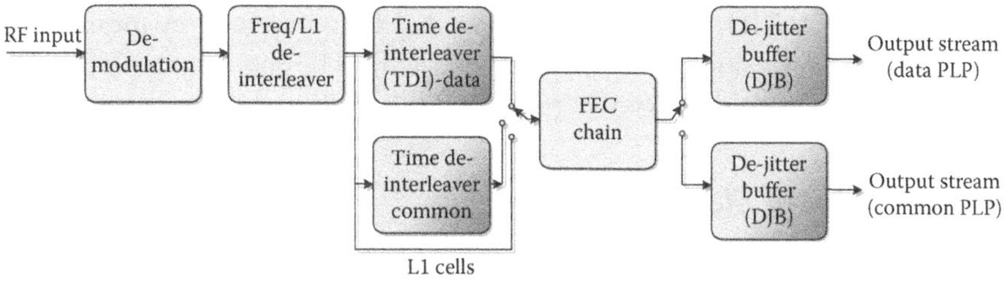

Figure 16.10   DVB-NGH RBM. DVB-NGH terminals are capable of simultaneously receiving up to four PLPs including the common PLP.

according to a read clock; removed sync bytes and DNPs are reinserted at the output of the DJB. When the receiver is decoding the data PLPs together with its associated common PLP, it shall be assumed that the time de-interleaver and DJB are duplicated as shown in Figure 16.10. It should be noted that there is a single FEC HW chain and that the TDI memory is shared among the data PLPs and the common PLP.

After this process, it is necessary to have a TS packet reconstruction mechanism in order to recover the original TS flow. Basically, this mechanism has to reverse partially the mode adaptation performed at the transmitter. This process includes sync byte reinsertions, CRC-8 removals (if present), and DNP analysis in order to reinsert null packets deleted at the transmission side. All these processes make use of the synchronization mechanisms introduced by the gateway at the transmission side such as the SYNCD, SYNC, and UPL (to regenerate the user packets), the TTO (to set the initial occupancy of the DJB), the DNP (to reinsert the DNPs), and the ISCR (to calculate the output bit rate and for fine adjustment of the relative timing of data and common PLPs).

16.4 Internet Protocol Profile of DVB-NGH

The IP profile of DVB-NGH is dedicated for the transmission of IP data flows. IP protocols, as part of the so-called Internet protocol suite, are designed to ensure reliable interchange of data over a communication channel with multiple hops, requiring routing. DVB-NGH is designed to deliver a wide range of multimedia services over its IP profile in a bearer-agnostic way, so that these services can be made accessible through mobile cellular networks.

DVB-T2 is not optimized for the transmission of IP content delivery especially as it lacks a corresponding signaling to deliver the information properly. In addition, the amount of overhead present in the DVB-T2 IP encapsulation chain is significantly high and not effective for consideration in DVB-NGH specification. In DVB-T2, the IP profile uses several protocols that are not specifically designed for broadcasting, so they introduce big amounts of overhead through the added headers. In DVB-NGH, the bandwidth usage has been improved by means of an overhead reduction module (ORM) based on the ROHC specification [3], which allows reducing the overhead connected with RTP, UDP, and IP headers.

16.4.1 Protocol Stack Overview

In the IP profile of DVB-NGH (see Figure 16.11), the SG together with the upper layer signaling is carried on the top. On the same level with the SG, audio and video streams and files are provided as contents. FLUTE/ALC is used for the encapsulation of the SG and files, while RTP/RTSP is used for the delivery of audio and video streams on the top of User Datagram Protocol (UDP). Below UDP, there is IP, which can be compressed by using ROHC, which follows after the IP. Next, GSE is used as encapsulation protocol for ROHC and/ or IP streams. The LLC signaling element inside the GSE carries also the versioning information of the upper layer signaling. The BB frames are next used for presenting the GSE streams into frames, which are then encoded into the DVB-NGH physical layer as NGH frames. The DVB-NGH physical layer also contains the L1 signaling.

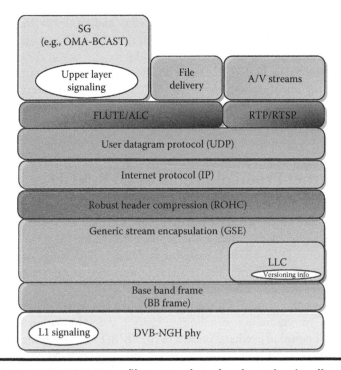

Figure 16.11 DVB-NGH IP profile protocol stack, where the signaling elements are marked as ellipsoids.

16.4.2 ROHC

In DVB-NGH IP profile, the useful data are encapsulated over several protocols such as RTP, UDP, or IP. All these protocols were not designed for broadcasting content delivery as they include a lot of header information that is not relevant for the broadcasting use case. In addition, most of these headers' fields have a static behavior or change in a more or less predictable way. A method to optimize the transmission of headers over DVB-NGH was deemed necessary in order to improve the transmission efficiency. DVB-NGH has thus considered methods such as ROCH for unidirectional transmission (ROHC-U). ROHC aims to provide a RTP/UDP/IP header compression that allows overhead reductions about 1% of the total transmission capacity.

Basically, ROHC algorithms use the interdependencies between packets to determine a correct static and dynamic context on the receiver side. However, standard ROHC algorithms introduce undesirable effects in terms of zapping delay and error propagation due the interdependency utilization. To solve that, DVB-NGH adds an adaptation module to the conventional ROHC-U standard module in order to reduce or avoid the mentioned problems, forming the ORM (see Figure 16.12).

With this structure, the ROHC module of the ORM is responsible for the IP flow overhead reduction, whereas the adaptation module is in charge of mitigating the effects of zapping delay and error propagation. Thanks to this scheme, backward compatibility with the stand-alone ROHC framework is guaranteed, while minimizing the impact on zapping delay and error propagation.

The way through which ROHC provides compression is an algorithm based on state machines in transmission and reception that establish different packet-type deliveries depending on the context (see Figure 16.13). The state machine determines whether it is necessary to send all the information (static and dynamic) or simply a set of modified fields (only dynamic information).

The ORM block is introduced in the DVB-NGH system to be transparent for the rest of the components as it is shown in Figure 16.14. It should be noted that the adaptation module sends the essential header fields of the ROHC process in a common PLP. This improves the zapping delay and error propagation.

More detailed information about the ROHC header compression mechanism adopted in DVB-NGH can be found in Chapter 17.

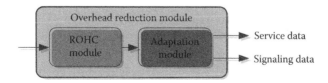

Figure 16.12　IP ORM in DVB-NGH.

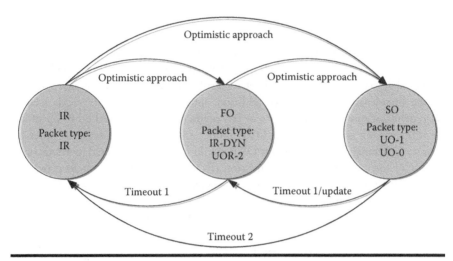

Figure 16.13   ROHC compressor states and transition between states.

16.4.3 Signaling in the DVB-NGH IP Profile

The upper layer signaling in the DVB-NGH IP profile consists of the network information together with signaling information for mapping services and service components onto PLPs, and related versioning information. The upper layer signaling is carried inside the service guide delivery descriptor (SGDD) in OMA-BCAST, and it contains the mapping between the services and PLPs, ROHC context information as well as the mapping between the services available within the current frequency/multiplex and within the neighboring multiplexes. The versioning information, however, is a dedicated descriptor defined within the LLC information at the link layer (L2), and it contains versioning information of the upper layer signaling tables, SG, and possible bootstrap information. Finally, it also provides information of the upper layer type that is being used, even though OMA-BCAST is currently the only upper layer defined so far.

16.4.4 OMA-BCAST Service Guide

OMA-BCAST is the "default" upper layer solution considered for the delivery of the SG over the DVB-NGH bearer (see Figure 16.15). As opposed to the previous generation of mobile DVB standards, DVB-H, so far only OMA-BCAST solution has been adopted in DVB-NGH as upper layer solution. However, as stated before, the versioning information within the LLC signaling enables future additions of other upper layer solutions as well.

In OMA-BCAST, the SGDD plays a very important role to deliver the SG efficiently. SGDD is transported on the SG Announcement Channel and informs the terminal about the availability, metadata, and grouping of the fragments of

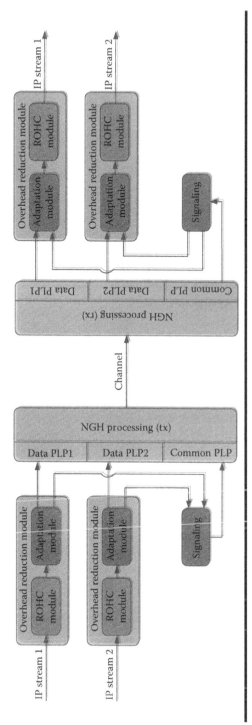

Figure 16.14 IP overhead reduction method in DVB-NGH transmission–reception chain.

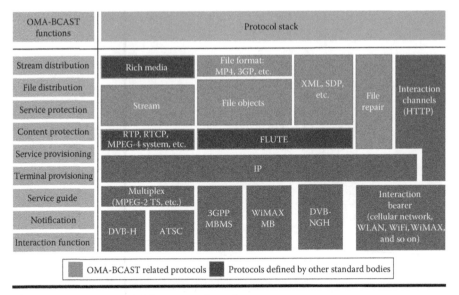

Figure 16.15   OMA-BCAST protocol stack.

the SG in the SG discovery process. This allows for a quick identification of the SG fragments that are either cached in the terminal or being transmitted. For that reason, the SGDD is preferably repeated for distribution over a broadcast channel. In addition, the SGDD also provides the grouping of related SG fragments and thus a means to determine completeness of such a grouping.

The SGDD is especially useful if the terminal moves from one service coverage area to another. In this case, the SGDD can be used to quickly check which of the SG fragments that have been received in the previous service coverage area are still valid in the current service coverage area, and therefore do not have to be re-parsed and reprocessed.

This solution reports benefits as the BCAST SG delivery protocol was tested, implemented, and commercialized, and so it is a stable solution. In addition, BCAST SGDD can deliver BDS-specific entry information with the element *"BDSSpecificEntryPointInfo."* This element is used in OMA-BCAST adaptation specification to DVB-H for the provision of DVB-H network information such as network ID and frequency. It is also adapted and used in OMA adaptation specification to DVB-NGH for the transport of bearer-specific upper layer signaling.

16.4.5 Service Discovery

The service discovery process in DVB-NGH starts from the search of the P1 symbol while the receiver is trying to synchronize to signals available within the frequency range. This so-called signal scan is a traditional procedure in all other broadcast systems. Once, the P1 symbol is found, the receiver can detect whether the found signal is desirable, that is, if it is a DVB-NGH signal the receiver is capable to decode. Once the P1 phase is complete, the receiver utilizes additional tuning information, particularly FFT and GI information, which it can use for fast reception of the next signaling element, the P2 symbol. The P2 symbol carries the L1 signaling information that the receiver needs in order to find out the necessary parameters needed to decode L1-post signaling and desired PLPs from the physical layer frames. Prior to the start of the decoding process of desired PLPs, the receiver needs to resolve which services are available on which PLPs. This can be done by accessing the common PLP, which carries all signaling information. The terminal checks whether there is an IP stream with an IP multicast address 224.0.23.14 or FF0X:0:0:0:0:0:0:12D available within the data streams carried in the common PLP defined in [7]. If either or both of the IP addresses are available, then the entry point for one or more OMA-BCAST SGs can be found through SGDD according to [10]. The upper layer signaling of DVB-NGH is carried on the top of IP layer, inside the SGDD. Hence, the upper layer signaling can be discovered by using "*BDSSpefificEntryPointInfo*." The encapsulation and the signaling parameters for the upper layer signaling elements of the BCAST 1.2 are defined in [11] and in [8].

In addition, the LLC information contains versioning information of the upper layer signaling, which may also be used for checking whether the signaling information needs to be updated.

16.4.6 SVC Delivery in the IP Profile

The main critical issue for SVC delivery is the synchronization of the multiple streams carrying multiple associated PLPs. The IP profile takes advantage of the protocol stack defined, where the RTP protocol is able to synchronize different multimedia streams by means of the time stamp, a field designed for this purpose. Furthermore, the IP profile also uses the ISSY field in the same way as in the TS profile for the synchronization of the multiple associated PLPs delivering the SVC components.

16.4.7 IP Profile Receiver Buffer Model

The mission of the RBM is to ensure continuous reception of RTP stream packets of an RTP stream that is delivered over DVB-NGH [9]. This model is designed to avoid buffer underflows or overflows at the receiver for a specific service component. This receiver model is implemented at the sender side, in order to mimic the receiver behavior when handling the receiver service component. The sender signals required minimum buffering time to correctly and continuously receive and decode the corresponding service component. In this manner, the receiver buffers are sized according to this minimal buffering time, taking into account that the consumptions of the corresponding service will not result in buffer overflows or underflows.

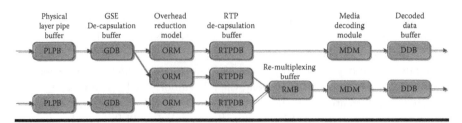

Figure 16.16 IP profile RBM example. Service with one audio and two video flows (base layer and upper layer). Base layer and audio are allocated in the same PLP.

The RBM is composed of Physical Layer Pipe Buffer (PLPB), GSE De-capsulation Buffer (GDB), RTP De-capsulation Buffer (RTPDB), Re-Multiplexing Buffer (RMB), Media Decoding Module (MDM), and Decoded Data Buffer (DDB) (see Figure 16.16). The RBM also takes into account the operations of ORM, if present. It should be noted that there is one PLPB and GDB per each receiving PLP, one ORM and RTPDB per each transport flow, and one RMB, one MDM, and one DDB per each service component.

In a first term, the PLPB extracts the generic stream carrying GSE packets. Then, the GDB output gives one or more transport flows or one or more compressed transport flows. If the GDB output flows are compressed, the next block has to be the ORM; otherwise, the flows are directly passed to the corresponding RTPDB. The ORM uses De-Adaptation Module and the ROHC Decoding Module to extract the datagram of the transport flow to the block output. Later, the RTPDB model extracts from the datagrams at its input the RTP packets that are passed to the RMB. If needed, a re-multiplexing is performed to join different service components transmitted in different PLPs (e.g., SVC with multiple PLPs) and then passed to the MDM to be decoded. The MDM is a decoding module that includes buffering of both coded and decoded data. Finally, the DDB aims to smooth out differences in the buffering delays of different service components until the DDB.

16.5 Conclusions

The system and upper layers in DVB-NGH split across the entire protocol stack, from Layer 1 (Physical Layer) up to Layer 7 (Application Layer). The L1 signaling solution is the same for both TS and IP profiles. The key difference between the IP profile and the TS profile is in the contents and transport of the upper layer (above Layer 1) signaling. The IP profile carries the whole upper layer signaling (including both service and system signaling information) above the IP layer (Layer 3) (e.g., in the SG), whereas the TS profile stays with the traditional approach for conveying PSI/SI at the link layer (i.e., below the IP). The IP profile is defined with the

OMA-BCAST as the "default" upper layer solution, even though it remains open for other upper layer solutions in the future. The encapsulation of the IP packets at the link layer is done using the GSE protocol and in accordance with the GSE Link Layer Control specification.

References

1. DVB Commercial Module sub-group on Next Generation Handheld, Commercial requirements for DVB-NGH, CM-1062R2, June 2009.
2. DVB Technical Module sub-ground on Next Generation Handheld, NGH study mission report, TM-H 411, June 2008.
3. C. Bormann et al., RObust Header Compression (ROHC): Framework and four profiles: RTP, UDP, ESP, and uncompressed, ITEF RFC 3095, July 2001.
4. H. Schwarz, D. Marpe, and T. Wiegand, Overview of the scalable video coding extension of the H.264/AVC standard, *IEEE Transactions on Circuits and Systems for Video Technology*, 17(9), 1103–1120, September 2007.
5. C. Hellge, E. Guinea, D. Gómez-Barquero, T. Schierl, and T. Wiegand, HDTV and 3DTV services over DVB-T2 using multiple PLPs with SVC and MVC, *Proceedings of the IEEE Broadcast Symposium*, Alexandria, VA, 2012.
6. ETSI EN 302 755 v.1.3.1, Frame structure channel coding and modulation for a second generation digital terrestrial television broadcasting system (DVB-T2), November 2011.
7. ETSI EN 303 105 v.1.1.1, Digital video broadcasting (DVB); next generation broadcasting system to handheld, physical layer specification (DVB-NGH), in press.
8. ETSI TS 102 606 v1.1.1, Digital video broadcasting; generic stream encapsulation protocol (GSE); part 2 logical link control (LLC), October 2007.
9. C.R. Nokes and O.P. Haffenden, DVB-T2 receiver buffer model (RBM): Theory & practice, *BBC White Paper*, July 2012.
10. Open Mobile Alliance v1.0, Service guide for mobile broadcast services, February 2009.
11. Open Mobile Alliance v.1.0, BCAST distribution system adaptation—DVB next generation handheld, in press.

Chapter 17

Overhead Reduction Methods in DVB-NGH

Lukasz Kondrad, David Gómez-Barquero,
Carl Knutsson, and Mihail Petrov

Contents

17.1 Introduction

One of the recommendations of the study mission on next-generation mobile broadcasting standards carried out prior to the standardization of Digital Video Broadcast—Next Generation Handheld (DVB-NGH) was to improve the bandwidth usage by designing low-overhead protocol stacks [1]. This recommendation was subsequently adopted as part of the DVB-NGH commercial requirements [2]. DVB-NGH targets the delivery of a wide range of multimedia services, having been designed to support Internet Protocol (IP)- and Transport Stream (TS)-based network-layer solutions. The support for IP services enables DVB-NGH to deliver in a bearer-agnostic way the same multimedia services that are available on mobile cellular networks, while the support for TS services allows for sharing the same infrastructure with other DVB broadcast systems, such as DVB-T2 (Terrestrial Second Generation). The TS and IP profiles have been optimized independently in order to improve the bandwidth utilization. For a detailed description of the system and upper layers of DVB-NGH, see Chapter 16.

This chapter presents an overview of the different overhead reduction methods introduced in DVB-NGH, targeting the physical (layer 1, L1) and network layers. The physical layer overhead is reduced through a more efficient physical layer adaptation and a more compact physical layer signaling. In addition, dedicated packet header compression methods have been adopted for the TS and IP profiles, which further reduce the protocol overhead.

The rest of the chapter is structured as follows. Section 17.2 is devoted to the overhead reduction at the physical layer through more compact adaptation and signaling. Section 17.3 describes the two overhead reduction mechanisms defined in the TS profile: the null packet deletion (NPD), as specified in DVB-T2, and a new packet header compression scheme. Section 17.4 describes the overhead reduction for the IP profile, illustrating the benefits of using the Generic Stream Encapsulation (GSE) protocol for encapsulating IP data. The overhead reduction is achieved by compressing the IP packet headers using a mechanism known as Robust Header Compression (ROHC). Finally, the chapter is concluded with Section 17.5.

17.2 Overhead Reduction at the DVB-NGH Physical Layer

17.2.1 Physical Layer Adaptation

In the second-generation DVB standards, the basic physical layer payload unit is the so-called baseband (BB) frame. All input data (i.e., MPEG-2 TS packets and GSE packets with IP datagrams) have to be adapted for delivery over BB frames.

DVB-T2 defines two adaptation modes: the normal mode (NM) and the high efficiency mode (HEM) [3].

In the NM, the user packets transmitted over BB frames contain signaling and synchronization information that facilitates the timing recovery in the receiver. This additional information increases the signaling overhead. The HEM enables a more efficient transport of the user packets by moving some parameters from the user packet headers to the BB frame headers. But there is room for further improvements.

In DVB-NGH, a new physical layer adaptation mode is defined in order to reduce the overhead and prevent ambiguities. The signaling has been reorganized, separating the PLP-specific information from the BB frame information. In DVB-NGH, the PLP-specific information is present only in the L1 signaling, whereas the BB frame information appears only in the BB frame headers. Thanks to this new signaling structure, information duplication is avoided, thus reducing the overhead and improving the consistency. Moreover, the BB frame header is reduced in size from 10 to 3 bytes, as illustrated in Figure 17.1. It should be pointed out that the for TS packet header compression, the BB frame header is increased by 3 bytes (i.e., 6 bytes in total).

17.2.2 Physical Layer L1 Signaling

The L1 signaling data include the necessary information for the receivers to identify the DVB-NGH signal, find the data PLPs transmitted in the frame, and enable their decoding. The L1 signaling is divided into two parts: L1-pre and L1-post signaling. The L1-pre signals basic transmission parameters that enable the extraction and decoding of the L1-post. The L1-post carries the information needed to identify and decode the data PLPs within the frame. The L1-post signaling parameters are further divided into two parts based on their frequency of change: configurable and dynamic. The configurable parameters change only when the network transmission parameters are changed (e.g., when a multiplex reconfiguration occurs), which is a rather infrequent event. The changes are applied only on the border of a so-called super frame, which consists of configurable number of NGH frames. The dynamic parameters, on the other hand, can change from frame to frame due to the dynamic nature of the frame allocation of the PLPs. For more details on the physical layer signaling, see Chapter 15.

In DVB-T2, the relative overhead of the L1 signaling is usually at reasonable values around 1%–5%. However, as the number of PLPs and the signaling robustness increase, the relative overhead can become quite significant, especially for short frame lengths. A further analysis of the L1 signaling in DVB-T2 shows that the L1-config represents the main part of the signaling information. Like the L1-pre, it does not convey any relevant information once the receiver is in normal operation (i.e., it is semi-static).

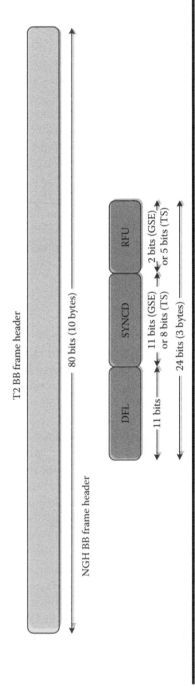

Figure 17.1   Comparison of DVB-T2 BB frame header and DVB-NGH BB frame header. DFL, data frame length in bits; SYNCD, distance to the start of next entire packet in bits; RFU, reserved for future use.

17.2.2.1 L1 Logical Frame Structure

DVB-NGH introduces a new logical frame structure compared to DVB-T2, which is especially suited to the transmission using DVB-T2 Future Extension Frames (FEFs). In order to reduce the L1 signaling overhead, the L1-post signaling is not transmitted in every NGH frame, but only in the first NGH frame that corresponds to one logical frame. One logical frame can span several NGH frames, thus reducing the overhead. The overhead reduction is achieved at the expense of increased zapping and handover times. However, if the NGH frames are transmitted frequently, the impact can be negligible. In any case, it should be noted that there is a flexible trade-off between the L1 signaling overhead and zapping and handover times. For more details on the logical frame structure of DVB-NGH, see Chapter 14.

17.2.2.2 L1-Configurable Structure

In DVB-T2, there is a PLP signaling loop in the L1-config, which fully defines the properties of each PLP, such as the PLP identification number, modulation, code rate, time interleaving configuration, and the PLP location inside the frame.

In DVB-NGH, the PLP signaling loop has been optimized in order to reduce the signaling overhead. There are now two loops instead of one. The first loop defines the so-called PLP configuration modes, each mode being identified through a 6-bit configuration ID. The second loop is the actual PLP signaling loop. Unlike DVB-T2, only the configuration ID is specified for each PLP. Thus, if there are several PLPs with exactly the same parameters, a significant overhead reduction can be achieved since the parameters can now be defined only once, in a single PLP configuration. The solution adopted in DVB-NGH retains the full flexibility of DVB-T2. Up to 255 PLPs are supported, each with its unique configuration. However, since the number of PLP configurations is typically very low, the signaling overhead is significantly reduced.

A further improvement for reducing the L1 signaling overhead in DVB-NGH consists in the introduction of flags to signal the presence of optional features in the L1-config section. The L1-config signaling format in DVB-T2 is very generic and supports a lot of options, such as auxiliary streams, reserved for future use fields (both inside and outside the PLP loop), time-frequency slicing (TFS), and more than one PLP group. The amount of signaling information can be significant in some cases. However, in practical use cases, many of these features may be unused, so the corresponding fields can be removed. In DVB-NGH, these optional features are signaled through flags located at the beginning of the L1-config field.

It should be pointed out that the overhead reduction mechanisms described in this section do not affect the zapping time.

17.2.2.3 N-Periodic L1-Pre and Self-Decodable L1-Config Transmission

Since L1-pre and L1-config convey information that is required only for the initial channel scanning and after a multiplex reconfiguration, they do not need to be fully transmitted in every frame.

The DVB-NGH standard introduces the concept of *n*-periodic signaling, which means that the signaling information is transmitted interleaved over *n* frames instead of being repeated every frame. Thus, the overhead is reduced by *n*, while the time diversity is improved. The drawback is an increased channel scanning time when the receiver is switched on because at least *n* frames need to be received before the signaling information can be decoded. The selection of the parameter *n* represents therefore a trade-off between channel scanning time and signaling overhead.

In DVB-NGH, the L1-pre signaling field can be divided into *n* blocks of the same size, as illustrated in Figure 17.2. A superposed correlation sequence is used to detect the value of *n* and the indices of the L1-pre blocks.

The L1-config cannot be transmitted using the same approach, since the frame configuration (e.g., T2 frame length, FEF length) cannot be known until *n* frames have been received. This is a major problem when DVB-NGH shares the same multiplex with DVB-T2 or when TFS is used. For TFS, *n*-periodic signaling is not even possible for L1-pre because the receiver needs to know the frequencies of all RF channels in order to extract L1-pre. However, these frequencies are signaled only in L1-config.

The approach adopted for L1-config provides a more flexible trade-off between overhead reduction and zapping time, and consists of transmitting the PLP configurations in L1-config not in every frame but in a carousel-like manner with configurable delay. The delay is configurable on a PLP-by-PLP basis, in order to enable those PLPs that cannot tolerate any delay to be sent with zero delay. The amount of L1-config information transmitted per frame is constant and self-decodable, but the signaling overhead is still reduced because each frame carries only a part of the signaling information.

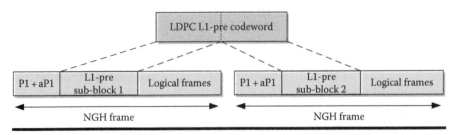

Figure 17.2 **N-periodic transmission of the L1-pre with *n* equal to two frames.**

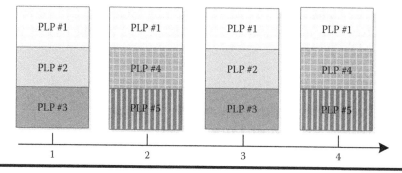

Figure 17.3 *N*-periodic transmission of L1-config with one PLP transmitted every frame and four PLPs transmitted every two frames.

Figure 17.3 shows a generic example of self-decodable L1-config transmission. Each PLP has associated a repetition interval that indicates how frequently the PLP is signaled in the L1-config. In this example, PLP #1 is transmitted every frame, and the other four PLPs are transmitted every two frames. It should be noted that each frame carries exactly the same amount of L1-config information. The PLPs are sorted according to their repetition interval, so that PLPs with lower repetition intervals are transmitted first. When several PLPs have the same repetition rate, PLPs with lower IDs are transmitted first.

17.2.2.4 L1 Signaling Overhead Comparison between DVB-T2 and DVB-NGH

The various techniques adopted in DVB-NGH for reducing the L1 signaling overhead can result in an increase in the total system capacity in excess of 1% (as high as 1.5% in realistic settings). The actual figure depends on the number of PLPs and PLP configurations, on the PLP settings and the system features used, and on the number of frames *n* used for the *n*-periodic transmission of the L1-pre and the self-decodable L1-config. Reducing the number of PLP configurations can provide savings of up to 0.85% with respect to the total frame capacity, especially for configurations with many PLPs. The lower the ratio of the number of PLP configurations to the number of PLPs, the higher the overhead saving. The main advantage of this proposal is that the zapping time is not affected. Introducing a flag for each of the available options of the L1-config can achieve remarkable savings (between 35% and 18%) for such a small change, without any drawback. The *n*-periodic transmission of L1-pre can already provide a capacity increase of around 0.5% for a value of *n* equal to four frames, at the expense of a proportionally longer zapping time. The self-decodable L1-config can provide another 0.5% capacity increase. Moreover, if the number of frames that contain the L1-config information of a given PLP does not exceed the number of frames that carry the L1-pre, the zapping time is not affected for that PLP.

17.3 Overhead Reduction in the Transport Stream Profile of DVB-NGH

17.3.1 Null Packet Deletion Mechanism

The TS profile of DVB-NGH is dedicated for the transmission of TS multiplexes [4]. MPEG-2 imposes strict constant bit rate (CBR) requirements on a TS multiplex. However, the individual services and service components in the TS multiplex may be of variable bit rate (VBR). In order achieve a CBR for the TS multiplex, so-called null packets are inserted, which carry no information. Depending on the bit-rate characteristics of the multiplexed services, the null packets may constitute a significant percentage of the TS multiplex bit rate. Therefore, in order to reduce the overhead, it is desirable to remove the null packets prior to transmission.

Like DVB-T2, the physical layer of DVB-NGH must be transparent for the TS multiplex and must provide a constant end-to-end delay. The null packets must be therefore reinserted by the receiver in their initial positions.

DVB-NGH adopts exactly the same NPD mechanism specified in DVB-T2 to reduce the overhead in its TS profile. NPD is a configurable feature of each PLP. When used, payload TS packets are transmitted, whereas null TS packets are removed. A Deleted Null Packet (DNP) counter of 1 byte is defined, which indicates the number of null TS packets preceding each payload packet (see Figure 17.4). This DNP counter is sent after each payload packet, and the null packets themselves can be discarded. This mechanism is sufficient for the receiver to restore the null packets in their initial positions. It should be noted that if a run length of 256 null packets is encountered, one null packet will still be transmitted, with a DNP value of 255.

17.3.2 TS Packet Header Compression

DVB-NGH introduces an optional header compression method for TS packets, which allows reducing the header size from 4 bytes to only 1 byte. The compression is performed on the transmitter side, and the information needed for restoring the header in the receiver is signaled in the BB frame header and the L1 signaling, such that the compression and decompression processes are transparent. This technique increases the overall capacity by 1.1%, but it is applicable only to PLPs that carry one program component, identified by one

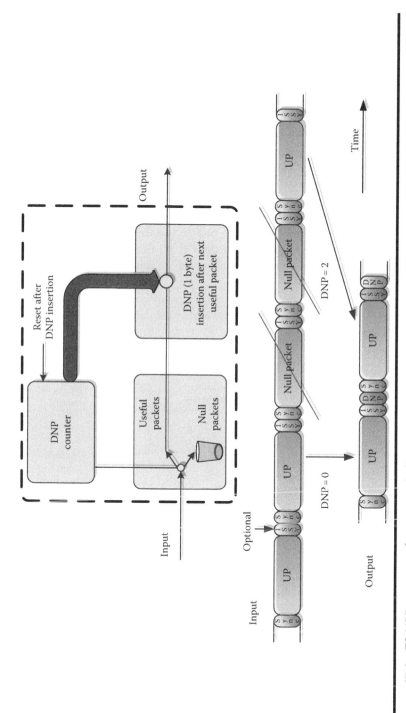

Figure 17.4 TS NPD process. The counter indicates the number of null packets preceding each payload packet. (From ETSI EN 302 755 v.1.3.1, digital video broadcasting (DVB); frame structure channel coding, and modulation for a second-generation digital terrestrial television broadcasting system (DVB-T2), November 2011.)

Program ID (PID). As mentioned before, a few null packets can still be present even if the NPD is enabled.

The TS packet header is compressed as follows (see Figure 17.5):

- The sync byte is removed
- The 1-bit transport priority indicator is removed
- The 13-bit PID field is replaced by a single bit
- The 4-bit continuity counter is replaced with a 1-bit duplication indicator

The *sync byte* is a fixed field (0×47) used for detecting the packet boundaries in systems where the packet starts are not signaled. In DVB-NGH, the packet starts are signaled in the BB frame headers, making the sync byte unnecessary. Like in DVB-T2, this byte is not transmitted. The *transport error indicator* signals that at least one uncorrectable error exists in the TS packet. It is usually set by the receiver if the error correction mechanism fails, so there is no need to transmit it. Also not transmitted is the *transport priority indicator*, which is now signaled in the BB frame header.

A further header field that can be efficiently compressed is the 13-bit PID, which uniquely identifies the elementary stream that the data carried in the packet

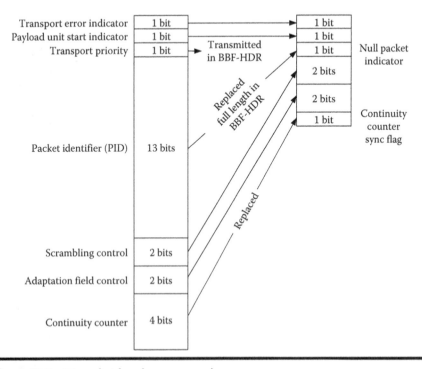

Figure 17.5   TS packet header compression.

payload belongs to. For those PLPs that carry only one elementary stream, the PID is no longer needed. However, since null packets can still be present even when the null-packet deletion is enabled, the PID is replaced by a null-packet flag, which identifies the null packets.

The original PID value, which is needed in the receiver, is signaled in each BB frame header. Since the PIDs change very rarely, this is perfectly acceptable. When a change does occur, however, its position must be signaled accurately in order to enable a seamless reception. For this purpose, the BB frame header also contains a pointer that indicates the first TS packet with the new PID. These two new fields require 3 bytes in the BB frame header. But thanks to the new consistent signaling described in the previous section, the BB frame header still has 4 bytes less compared to DVB-T2.

The *continuity counter* is a further field that can be compressed. It is incremented for each TS packet with the same PID. The TS syntax allows the transmission of duplicate packets, in which case the duplicated packet (having the same PID) is sent with the same continuity counter. In DVB-NGH, the transmitted continuity counter is replaced by a duplication flag, which identifies the duplicated packets. The counter can be regenerated in the receiver using a 4-bit counter that is incremented whenever the duplication flag is false. Since only the relative values matter, it is not important to have exactly the same continuity counter values like that in the transmitter. In order to support the decoding of multiple PLPs in parallel, the receiver needs to implement several counters.

17.4 Overhead Reduction in the Internet Protocol Profile of DVB-NGH

The IP profile of DVB-NGH is dedicated to the transmission of services that are supported by the Internet Engineering Task Force (IETF) protocols, such as IP [5,6] on the network layer, UDP [7] on the transport layer, and RTP [8] on the application layer. The link layer encapsulation used for IP packet transmission over DVB-NGH is the GSE protocol.

The IETF protocols are intended to operate in the Internet world and are designed to ensure reliable data transfers over multi-hop, heterogeneous packet networks. As a consequence, these protocols are quite verbose, and the packet headers are relatively large. The header sizes of the common protocols' packets are 12 bytes for RTP, 8 bytes for UDP, 20 bytes for IPv4, and 40 bytes for IPv6 packets. If the average protocol data unit (PDU) size is assumed to be 1000 bytes, the resulting overhead would amount to roughly 4% (6% for IPv6). In broadcasting, where no routing is necessary, some of the header fields are no longer needed. Other header fields do not vary randomly from packet to packet during the transmission, but in a more or less predictable way. Others may not change at all. Moreover, some of the fields are redundant since the information they carry can also be extracted from the lower layers of the protocol stack. These specific aspects facilitate the use of an efficient overhead reduction mechanism. One of the new features of DVB-NGH is the use of the ROHC protocol [9] as the mechanism for reducing the over-head associated with the IETF protocols. ROHC, however, introduces inter-packet dependencies in a transmitted stream, which can be seen as a drawback. Due to these inter-packet dependencies, the zapping time increases and error propagations become more likely. Therefore, in order to reduce the negative impact of ROHC on the DVB-NGH performance, an NGH adaptation layer for ROHC has been intro-duced. The NGH adaptation layer for ROHC loosens the constraints of ROHC in terms of zapping delay and error propagation. Consequently, compared to stand-alone ROHC operation, the overhead reduction performance is maintained with no impact on the zapping time and error propagation. Moreover, the NGH adap-tation layer for ROHC guarantees backward compatibility with the stand-alone ROHC framework, which allows the reuse of existing software implementations of the ROHC protocol.

17.4.1 Generic Stream Encapsulation

In the first-generation DVB standards, IP-based services were transmitted over MPEG2-TS [4] using Multiprotocol Encapsulation (MPE) [10]. However, MPEG2-TS is a legacy technology optimized for media broadcasting rather than for IP service transmission. The main drawback of IP delivery over MPEG2-TS\MPE is the additional overhead, which reduces the efficiency of the channel bandwidth utilization.

In the second-generation DVB standards, IP datagrams can be transmitted using the GSE protocol [11], which was designed to carry IP datagrams. GSE provides efficient IP datagram encapsulation over variable-length link-layer packets, which are then directly scheduled on the link-layer BB fames, rather than transmitted over MPEG2-TS\MPE. Additionally, GSE has less overhead. Only 7–10 bytes of GSE header are needed per PDU. When using MPE over MPEG2-TS, the absolute overhead is 2 bytes of MPE header plus 4 bytes of CRC per PDU plus 4 bytes of TS header for every 184 bytes of TS payload. The relative overhead depends on the average PDU size, but approximately up to 70% overhead reduction can be achieved. See Table 17.1 for concrete values.

17.4.2 Robust Header Compression

The ROHC framework was created by ITEF and was designed to be general and extensible [9,12,13]. ROHC supports several profiles, such as RTP, UDP, ESP, or Uncompressed. Each profile specifies a set of procedures for a specific protocol use case. Moreover, the profiles define the state machine transitions for the compressor and the decompressor and also describe procedures for the start of transmission and error recovery. ROHC classifies the protocol header fields depending on their

Table 17.1   Comparison of the Overhead Caused by Encapsulation Protocols GSE and MPEG2-TS with MPE for IP Delivery

Average PDU Size (Bytes)	Overhead	
	GSE (%)	MPEG2-TS (%)
100	9.09	9.09
200	4.76	6.54
400	2.44	4.31
600	1.64	3.54
800	1.23	3.15
1000	0.99	2.91
1200	0.83	2.76
1400	0.71	2.64

changing pattern between consecutive packets. A simplified view of this classification consists of dividing the fields into three types: inferred, static, and dynamic. An ROHC flow is defined by its static fields (such as IPv4/IPv6 address and UDP port numbers). Flows are separated into different compressor contexts and given unique context identifiers (CID). Inferred are the fields that contain values that can be inferred from other protocol header fields or from lower-level protocols (e.g., IPv4 header checksum, IP length, and UDP length). These values are removed from the compressed header and are not transmitted by the compression flow. Static are the fields that are (a) expected to be constant throughout the lifetime of the packet flow (e.g., IP destination address) and therefore must be communicated to the receiver only once; or (b) expected to have well-known values (e.g., IP version) and therefore do not need to be communicated at all. Dynamic are the fields that are expected to vary during the transmission of the packet flow (e.g., IPv6 hope limit, UDP checksum, RTP timestamp). The classification of the protocol header fields by ROHC into static, dynamic, and inferred fields based on IPv6, IPv4, UDP, and RTP examples is shown in Figure 17.6.

ROHC can be modeled as an interaction between two state machines, one compressor machine and one decompressor machine, each instantiated once per context. The compressor and the decompressor have three states each that, in many ways, are related to each other even if the meanings of the states are slightly different for the compressor and the decompressor. Both machines start in the lowest compression state and transition gradually to higher states. The transitions

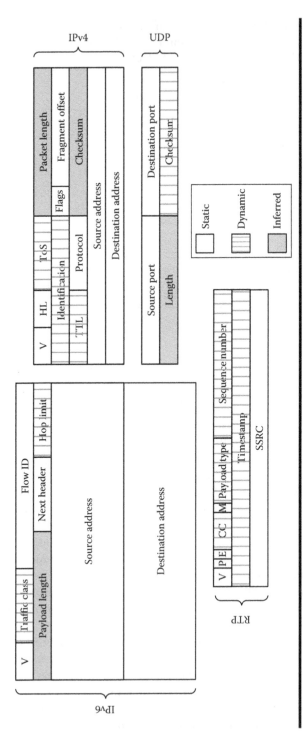

Figure 17.6 **Classification of protocol header fields by ROHC into static, dynamic, and inferred fields, based on IPv6, IPv4, UDP, and RTP examples.**

do not have to be synchronized between the two machines. In normal operation, only the compressor can temporarily transition back to lower states. The decisions about transitions between the various compression states are taken by the compressor on the basis of periodic time-outs or variations in packet headers. In the decompressor, transitions between states occur when a context damage is detected.

The compressor can generate two groups of packet types: IR (initialization and refresh) and CO (compressed header type). The IR header format contains all the original static header fields and/or the dynamic header fields of the uncompressed header plus ROHC context information. The IR packet, which includes both static and dynamic header fields, is used at the start of the transmission to indicate when to reestablish or to refresh the decompressor context due to error detection or time-out. An IR-DYN is sent upon the detection of a dynamic context damage or following a time-out, in order to refresh the dynamic part of the decompressor context. CO header formats compress the complete header and communicate as few dynamic fields as possible. CO header formats include UOR-2, UO-1, and UO-0. An example of the generic format of IR, IR-DYN, and UOR-2 packets in the ROHC framework is shown in Figure 17.7.

In the compressor, the three states are IR (initialization and refresh), FO (first order), and SO (second order). In each state, the compressor creates different

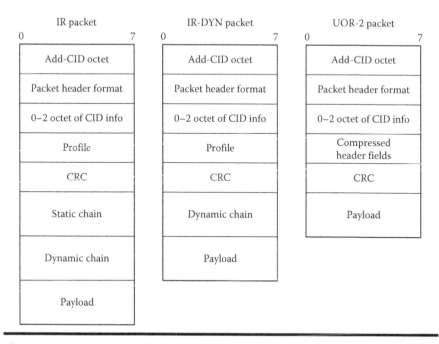

Figure 17.7 **An example of the generic format of IR, IR-DYN, and UOR-2 packets in the ROHC framework.**

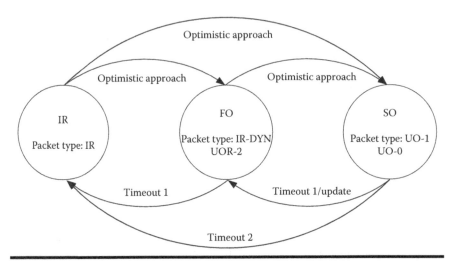

Figure 17.8   Three ROHC compressor states, IR, FO, and SO, and packet types generated in each state as well as transitions between the states when an ROHC compressor operates in U-mode.

packet types to provide information to the decompressor. The operation of the compressor, as well as the packet types generated in each state, is shown in Figure 17.8. In the IR state, the compressor initializes the static and dynamic parts of the context at the decompressor. The IR state is also used when recovering from static context damage caused by packet loss. In this state, the compressor sends the complete header information using IR packets. The purpose of the FO state is to efficiently communicate irregularities in the packet stream. When in the FO state, the compressor rarely sends information about all the dynamic fields, and the information sent is usually compressed, at least partially. The compressor enters the FO state from the IR and the SO state whenever the headers of the packet stream do not conform to their previously established pattern and stays in the FO state until it is confident that the decompressor has acquired all the parameters of the new pattern. The compressor enters the SO state when the header to be compressed is completely predictable and leaves the SO state to go back to the FO state when the header no longer conforms to the uniform pattern and cannot be independently compressed on the basis of previous context information.

The three states of the decompressor are "No Context," "Static Context," and "Full Context." The operation of the decompressor and the packet types accepted in each state are shown in Figure 17.9. The decompressor starts in its lowest compression state, "No Context," and gradually transitions to higher states. Initially, while in the "No Context" state, the decompressor attempts to decompress a first packet. Once a packet has been successfully decompressed, the decompressor can

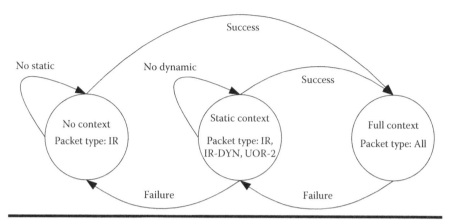

Figure 17.9 The ROHC decompressor state transition diagram and packet types accepted in each state.

transition directly to the "Full Context" state, but not before it receives at least one IR packet. Normally, the decompressor state machine never leaves the "Full Context" state once it reaches it. It transitions back to lower states only in case of decompression failure. When this happens, the decompressor first transitions back to the "Static Context" state. In this state, the reception of any packet generated in the IR or FO state of the compressor (IR, IR-DYN, UOR-2) is sufficient to enable the transition back to the "Full Context" state. If the decompression of the packets generated in the FO state of the compressor fails, the decompressor goes back to the "No Context" state.

ROHC has three modes of operation: unidirectional (U-mode), optimistic (O-mode), and reliable (R-mode). The U-mode is designated for unidirectional channels, i.e., with no feedback from the decompressor. As a consequence, in U-mode, the transition between the compressor states is based on an optimistic approach where a number of IR packets are sent to a decompressor to establish the context. Additionally, in order to ensure context synchronization, the context information is updated periodically using time-out mechanisms that force the compressor to return back to FO state or IR state. As a result, the U-mode, due to limited error recovery and context synchronization procedures, is characterized by a lower compression gain and transmission robustness compared to the other two modes. In the O-mode and R-mode, a feedback channel between the compressor and the decompressor exists. The decompressor is able to send requests for error recovery or indications of successful context update. The O-mode and R-mode differ from each other in the frequency with which they use the feedback channel. The R-mode uses the feedback channel quite often to improve the robustness of the transmitted data. Due to the lack of a feedback channel, only the U-mode is used in DVB-NGH.

17.4.3 NGH Adaptation Layer for ROHC

DVB-NGH introduces an NGH adaptation layer for ROHC in order to minimize the dependencies among ROHC packets, since these dependencies may cause error propagation and increased zapping time.

The channel zapping time is the cumulative sum of several components. In DVB-NGH, these components include the delay of synchronizing to an NGH frame, acquiring the L1 signaling, acquiring the L2 signaling, refreshing the media decoders to produce correct output samples, compensating for the VBRs of media (e.g., video buffering), and synchronizing associated media streams (e.g., audio and video). The ROHC may introduce an additional delay, caused by the time needed for the decompressor to initialize; in ROHC, it would be the time to receive at least one IR packet.

The DVB-NGH channel is prone to errors, which means that some of the ROHC compressed packets may be received incorrectly. If any of these packets is an IR packet, all packets subsequently received, even if not corrupted, would be marked as unusable until the next IR packet is received. Therefore, from the zapping time and the error propagation perspective, it would be desirable to maximize the IR packet transmission frequency. On the other hand, from a compression efficiency perspective, it is desirable to make the refresh as seldom as possible.

The NGH adaptation layer for ROHC allows for a more flexible trade-off by providing full refresh with IR-DYN packets, which can be transmitted more frequently than IR packets because they have a smaller impact on the compression efficiency. To achieve this, the NGH adaptation layer for ROHC splits the transmission of the static and the dynamic information fields of the compressed ROHC stream into two logical channels. The dynamic header information is sent in-band (over the same PLP as the payload data) without any modification, while the static header information is sent out-of-band using dedicated signaling fields.

The adaptation module extracts the static chain information from IR packets and converts these IR packets to IR-DYN packets by changing the corresponding header fields accordingly and recalculating the CRC values. The extracted static chain data are then transmitted over designated signaling bytes, and the newly created IR-DYN packets are transmitted together with the rest of the compressed data flow.

Due to splitting the transmission into two logical channels, the receiver possesses all static information for all available services as soon as the signaling data are processed. The receivers are able to convert any IR-DYN packet into an IR packet using the signaled static chain data. The conversion can be done, after a CRC validation, by adding the static chain data to IR-DYN packet, changing the corresponding header fields accordingly, and recalculating the CRC value. As a result, the ROHC decompressor may start the decompression of the ROHC flow upon the reception not only of an IR packet but also of an IR-DYN packet.

17.4.4 IP Profile Header Compression in DVB-NGH

In DVB-NGH, the header compression is optional in the IP profile and, if used, is applied to IP streams carrying service data. The header compression in the IP profile comprises the ROHC framework and an NGH-specific adaptation layer. An example deployment of the header compression in DVB-NGH is depicted in Figure 17.10. The ROHC modules implement the ROHC framework, while the adaptation modules implement the NGH adaptation layer for ROHC operation.

The detailed operation of ROHC framework is specified in [9,12,13], where multiple header compression algorithms, called profiles, are defined. Each profile is specific to the particular network layer, transport layer, or upper layer protocol combination. In DVB-NGH, the profiles presented in Table 17.2 are supported.

Moreover, DVB-NGH imposes additional constraints on the operation of the ROHC framework. If any change in the static fields of the input IP stream is detected by the ROHC compressor, then the ROHC compressor performs context re-initialization, and a new CID value is assigned to the compressed IP stream. The new value should be unique for the DVB-NGH system, and it should not be used by any other instance of ROHC compressors within the system.

The adaptation layer reduces the impact of the ROHC mechanism on the zapping delay and improves the robustness of the compressed flow. Loosening the constraints imposed on the ROHC compressor allows it to operate with an increased compression efficiency. In the transmitter, the adaptation module extracts the static fields from the IR packets produced by the ROHC compressor, converts the IR packets to IR-DYN packets, and transmits the extracted fields out of band. In the receiver, the adaptation module detects the IR-DYN packets and converts at least one such packet to an IR packet using the static fields received out of band. In DVB-NGH, the static fields are transmitted in a common PLP, as indicated in Figure 17.10. As the common PLPs are transmitted in every NGH frame, the adaptation block receives the static bits once per frame. The ROHC decompressor can therefore transition to the "Full Context" state with a delay of at most one NGH frame.

17.4.5 Header Compression Results

The efficiency of the ROHC scheme depends on the setup of the compressor, the characteristic of the transmitted IP flow, and whether or not the NGH adaptation layer for ROHC is used. In this section, the header compression efficiency is discussed based on the overhead of the RTP/UDP/IPv4 stack, which is 40 bytes per PDU.

Figure 17.11 shows the theoretical RTP/UDP/IPv4 overhead after ROHC compression for different average PDU sizes and average service bit rates in bytes per packet and relative to the data. It is assumed that the ROHC compressor

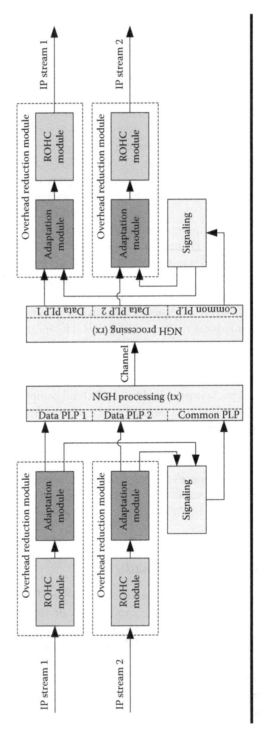

Figure 17.10 Block diagram of DVB-NGH including header compression stage.

Table 17.2   ROHC Profiles Supported in DVB-NGH

Profile Identifier	Usage	Reference
0 × 0001	RTP/UDP/IP	RFC 3095, RFC 4815
0 × 0002	UDP/IP	RFC 3095, RFC 4815
0 × 0003	ESP/IP	RFC 3095, RFC 4815
0 × 0004	IP	RFC 3843, RFC 4815

transitions back to the FO state every 0.1 s and to the IR state every 1 s. This assumption is valid from the DVB-NGH perspective, as the goal of DVB-NGH is to provide the same zapping time regardless of the service bit rate and the PDU size.

In Figure 17.11a, it can be seen that the ROHC compression efficiency is a function of the bit rate and the PDU size. The compression efficiency is higher for smaller packets and higher service bit rates, and smaller for larger packets and lower bit rates. For a PDU size of 100 bytes and a bit rate of 1000 kbps, the compressed overhead is close to 4 bytes per packet, which is 10 times smaller than the uncompressed overhead. At the other extreme, when the PDU size is 1400 bytes and the service bit rate is 100 kbps, the overhead is close to 20 bytes per packet per second after compression.

The highest compression gain of the ROHC is achieved for small PDU sizes and high bit rates. However, when looking from the system perspective and the overall percentage of the overhead in transmitted data, the combination of small PDU size and high bit rate does not yield optimal results. In Figure 17.11b, it can be seen that in order to reduce the relative overhead, the PDU size should be as high as possible regardless of the service bit rate.

So far, an ROHC compressor with fixed transition times to the FO and IR states was discussed. In Figure 17.12, theoretical RTP/UDP/IPv4 overhead sizes in bytes per packet per second after ROHC compression for different transition times to the IR state and to the FO state are shown, for a PDU size of 800 bytes and an average service bit rate of 800 kbps. It can be seen that the transition time to the IR state has the biggest impact on the compression efficiency. In order to increase the compression efficiency, an ROHC compressor should transition back to the IR state as rarely as possible. However, in a broadcast system such as DVB-NGH, the transition time to the IR states has a direct impact on the zapping time. Therefore, the transitions to the IR state should be frequent in order to minimize the zapping time to a service. Based on Figure 17.12, it can also be observed that the transition to the FO state has a rather minor impact on the compression efficiency. Here, the NGH adaptation layer comes into play, which allows any IR-DYN packet generated in the FO state to be converted into an IR packet based on the static fields already received via out-of-band signaling.

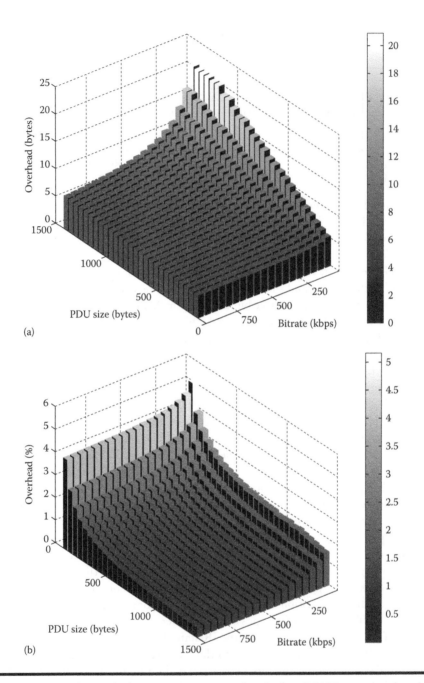

(a)

(b)

Figure 17.11 Theoretical RTP/UDP/IPv4 overhead after ROHC compression, in bytes per packet per second (a) and relative to the data (b) for different average PDU sizes and average bit rates. The ROHC compressor transitions back to the FO state every 0.1 s and to the IR state every 1 s.

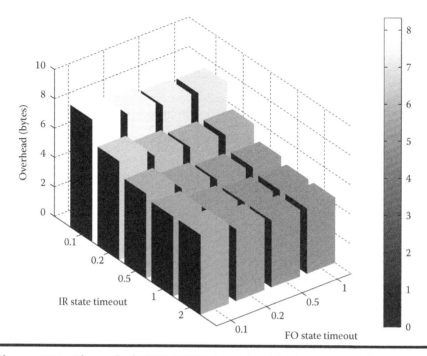

Figure 17.12 Theoretical RTP/UDP/IPv4 overhead in bytes per packet per second after ROHC compression, for different transition times of the compressor to the IR state and to the FO state. The PDU size is 800 bytes and the average service bit rate is 800 kbps.

As an example, thanks to the NGH adaptation layer, the zapping time with an IR transmission time and an FO transmission time of only 0.1 s is the same as that of a stand-alone ROHC with an IR transmission time of 2 s and an FO transmission time of 0.1 s. When looking at this from the compression efficiency perspective, the adaptation layer reduces the overhead from 8.3 to 5.4 bytes, which corresponds to a 40% overhead reduction.

Table 17.3 presents ROHC compression efficiency results for three use cases using real data flows described in Table 17.4. For RTP packetization mode 0 (use cases 1 and 3), the timestamp field of the RTP header is monotonically increasing over time, but the timestamp difference between two consecutive packets is not constant. For RTP packetization mode 1 (use case 2), the RTP timestamp is not monotonically increasing over time, but the timestamp difference between two consecutive packets is constant. The ROHC TS_STRIDE parameter is the expected increase in the timestamp value between two RTP packets with consecutive sequence numbers. When the RTP timestamp does not increase by an arbitrary number from packet to packet, then it shall be possible to represent the increase as an integral multiple of TS_STRIDE [13].

Table 17.3 Packet Header Overhead in the Tested Use Cases

	ROHC	*Uncompressed*
Use case 1		
Ice	10.10 bytes (0.92%)	40.00 bytes (3.47%)
City	10.08 bytes (0.92%)	40.00 bytes (3.47%)
Use case 2		
Ice	9.91 bytes (1.27%)	40.00 bytes (5.00%)
City	9.67 bytes (1.38%)	40.00 bytes (5.00%)
Use case 3		
Live	10.65 bytes (1.25%)	40.00 bytes (4.40%)

Table 17.4 Use Cases

Parameter	*Use Case 1*	*Use Case 2*	*Use Case 3*
Video sequences	Ice, city		Live video content
Resolution	4CIF (704 × 480) at 30 fps		VGA at 30 fps
Bit rate	1500 kbps		800 kbps
Average NAL unit	1150 bytes	800 bytes	900 bytes
Encapsulation	RTP/UDP/IP		
RTP packetization	Mode 0	Mode 1	Mode 0
ROHC TS_STRIDE	90	3600	1
Refresh packet rate	IR 380 ms	IR 270 ms	IR-DYN 550 ms

Finally, in order to test how the error propagation may degrade the video quality, a simple scenario is assumed, in which a BB frame carrying an IR packet is corrupted during transmission. The transmission of the same two sequences (Ice and City) was simulated. In the compressed ROHC stream, IR packets were sent every 250 ms, and IR-DYN packets were sent every 50 ms. Thus, for a frame rate of 30 fps, eight video frames are affected in the stand-alone ROHC mode, but only two frames are affected when the NGH adaptation layer was used. Table 17.5 shows the impact of the loss of one IR packet on the video quality in terms of the average PSNR value of the luma component.

Table 17.5 Luma PSNR in dB of the Tested Sequences When ROHC Is Used in Stand-Alone Mode and in Conjunction with the NGH Adaptation Layer

	Ice	*City*
ROHC stand-alone	37.191	32.859
ROHC + NGH adaptation layer for ROHC	39.706	34.131
Error free	40.346	34.509

17.5 Conclusions

Compared to the state-of-the-art digital terrestrial TV standard DVB-T2, DVB-NGH has considerably improved the bandwidth utilization efficiency without any compromise to the functionality of the system. The improved physical layer adaptation reduces the size of the BB frame headers from 8 to 3 bytes, while avoiding signaling ambiguities. The reduction of the L1 signaling overhead provides an increase in the system capacity between 1% and 1.5%. The newly adopted TS packet header compression method allows the size of the TS packet header to be reduced from 4 bytes to only 1 byte, providing a 1.1% system capacity increase. With ROHC, the IP packet overhead can be reduced to approximately 1% of the transmitted data. All these improvements result in a combined capacity increase of 2.5%–3.5%. Compared to first-generation mobile broadcasting systems, the encapsulation protocol overhead for IP delivery is reduced by up to 70%, thanks to the use of GSE as link layer encapsulation protocol instead of the older MPE over MPEG-2 TS.

References

1. DVB Technical Module sub-ground on Next Generation Handheld, NGH study mission report, TM-H 411, June 2008.
2. DVB Commercial Module sub-group on Next Generation Handheld, Commercial requirements for DVB-NGH, CM-1062R2, June 2009.
3. ETSI EN 302 755 v.1.3.1, Digital video broadcasting (DVB); frame structure channel coding and modulation for a second generation digital terrestrial television broadcasting system (DVB-T2), November 2011.
4. ISO/IEC 13818-1, Information technology—Generic coding of moving pictures and associated audio information: Systems, 2007.
5. IETF RFC 791, Internet Protocol, September 1981.
6. S. Deering and R. Hinden, Internet protocol, version 6 (IPv6), IETF RFC 2460, December 1998.
7. J. Postel, User datagram protocol, IETF RFC 768, August 1980.
8. H. Schulzrinne et al., RTP: A transport protocol for real-time applications, IETF RFC 3550, July 2003.

9. C. Bormann et al., Robust header compression (ROHC): Framework and four profiles: RTP, UDP, ESP, and uncompressed, ITEF RFC 3095, July 2001.
10. ETSI EN 301 192 v1.5.1, Digital video broadcasting (DVB); DVB specification for data broadcasting, November 2011.
11. ETSI TS 102 606 v1.1.1, Digital video broadcasting (DVB): Generic stream encapsulation (GSE) protocol, October 2007.
12. L.-E. Jonsson and G. Pelletier, Robust header compression (ROHC): A compression profile for IP, ITEF RFC 3843, June 2004.
13. K. Sandlund et al., The robust header compression (ROHC) framework, IETF RFC 5795, March 2010.

Chapter 18

Local Service Insertion in DVB-NGH Single-Frequency Networks

Jan Zöllner, Jaime López-Sánchez,
David Gómez-Barquero, Samuel Atungsiri,
and Erik Stare

Contents

513

18.1 Introduction

One of the main advantages of digital terrestrial television (DTT) networks is the possibility of deploying single-frequency networks (SFNs). In contrast to a multi-frequency network (MFN), in which neighboring transmitters employ a different transmitting frequency, in an SFN, all transmitters operate with the same frequency. Transmitters are synchronized in both time and frequency using, e.g., a GPS reference signal. SFNs are enabled by the use of orthogonal frequency division multiplex (OFDM) technology, which is particularly well-suited to the needs of the terrestrial broadcasting radio channel. With the introduction of a cyclic prefix in each OFDM symbol (also known as guard interval, GI), receivers can combine signals coming from several transmitters without inter-symbol interference if the GI duration is longer than the relative delay of all the multipath components. The duration of the cyclic prefix, T_g, defines the maximum transmitter distance within the SFN, with no self-interference from adjacent transmitters. Its value depends on the FFT size and the OFDM symbol GI fraction chosen. The maximum GI duration in the first-generation European digital terrestrial TV standard Digital Video Broadcasting–Terrestrial (DVB-T) [13] is given by the FFT 8K, GI 1/4 mode, which provides a T_g of 224 µs (8 MHz bandwidth), which corresponds to a maximum transmitter distance of about 67 km [1] to guarantee no self-interference from adjacent transmitters. The second-generation DTT standard DVB-T2 (second-generation terrestrial) allows for much bigger SFNs. In particular, it allows a transmitter distance of up to 159 km for 8 MHz bandwidth (FFT 32K, GI 19/128, T_g 532 µs) [2] and even larger using 7 MHz bandwidth.

SFNs achieve improved spectrum efficiency compared to MFNs, which normally require a higher-frequency reuse factor to bring co-channel interference down to an acceptable level. It should be noted that SFNs also require more than one frequency for an international frequency plan of large SFNs. But SFNs also exhibit other important advantages such as improved coverage* (or less required transmission power), no need for frequency handover within the SFN (but at borders to other SFNs) and may also allow for less interference and higher reliability [4]. The drawback of SFNs is that all users within the coverage area receive the same content, causing inefficient insertion of local services.

The TV services transmitted in a DTT network can be classified depending on their target area. Some services are consumed by many users throughout the whole network, e.g., nationwide TV programs covering a whole country. These services are usually called *global services*. Contrariwise, some services attract enough users for efficient broadcasting only in a certain subregion of the network, for example, in a city. Thus, broadcasting those services is reasonable only within this subregion, and they are referred to as *local services*. There may also be regional services, which are, e.g., unique for a region in a country. The area where such a local service is broadcast is called the *local service area* (LSA). Each LSA may consist of one transmitter or several transmitters transmitting identical and synchronized signals forming an SFN. SFNs are ideally suited for services where the service area matches one or more SFN areas (e.g., global or regional services), but they are not efficient for transmitting local services where this condition is not fulfilled, because the local service would then have to be transmitted in a wider area than needed from a service point of view. Hence, local content is transmitted in areas where it is not desired, making thus an inefficient use of the spectrum because the available channel capacity has to be shared among all local services. On the other hand, using the MFN approach, the full channel capacity is available for the content requested within each cell. The drawback is that more frequency spectrum is required and the network cost is higher compared to the SFN approach.

---

* A coverage gain is achieved due to a statistical gain by exploiting the signal diversity and a power gain by the combination of the received signal strengths (so-called *SFN gain*) [3].

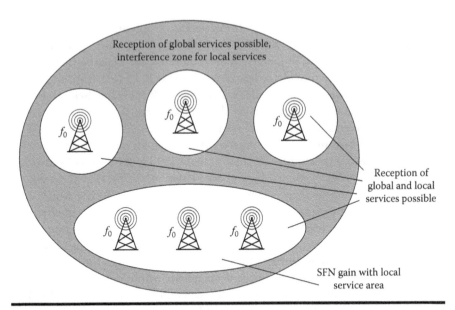

Figure 18.1   Principle of LSAs in SFNs.

Following is a list of desirable characteristics of a DTT broadcast network deployment for both global and local services (see Figure 18.1) [5]:

■ Only a single transmission frequency is used for the whole network (national or regional).
■ Local services may be transmitted using any subset of transmitters of the network.
■ For global services, SFN gain exists, and they are not affected by the local services.
■ For local services, the coverage area should be large enough to cover the specific areas where local content is to be consumed, which for some use cases is the same area as the global coverage area.

Generally speaking, it is beneficial to decouple global services from local ones. This means that global services are transmitted according to the SFN principle with all its advantages and are not affected by the local services. For the transmission of local services, the SFN principle has to be broken partially or for a short period. The main problem to be solved is that different local services transmitted in a single frequency would interfere. Thus, in areas where the signals of one or more interfering transmitters transmitting different local services are strong, reception would not be possible at all. However, for local services, an interference zone between different LSAs may be acceptable for some use cases if most of the users for the local services are within the core area of the LSA, for example, a city.

One of the commercial requirements of the next-generation mobile multimedia broadcasting standard DVB-NGH (next generation handled) was to efficiently transmit local contents in SFNs with minimum increase to network overhead [6]. The two most important local content types are probably local news and local advertising. Both are already widely implemented using temporal windows in the national DTT service in many countries. But the exponential growth of digital content platforms has also fostered the development of multimedia content and interactive services targeted at specific geographic communities. These services have the potential to create significant citizen and consumer benefits, as well as new revenue streams for broadcasters. Such local content may be potentially relevant for future broadcast services, serving audience needs that are not fully met by the current national and local broadcasting. However, the technical infrastructure of broadcast networks and the economic viability of this type of local services have not been clearly established, and thus the audience demand for them has not been adequately assessed. DVB-NGH will allow exploring the viability of inserting "classical" as well as new types of local services in SFNs in a way that has not been possible before.

DVB-NGH has adopted two complementary techniques to transmit local content in SFNs. The first technique uses hierarchical modulation [7]. The hierarchical modulation feature consists of combining two independent data streams into a single stream with different robustness. One stream, called the low priority (LP) stream, is embedded within another stream known as the high priority (HP) stream. The most significant bits (MSBs) of each modulation symbol are used for the HP stream (e.g., the first two bits identify the quadrant of the constellation). The remaining bits, less robust, are used by the LP stream to determine the exact position of the symbol in the complex plane. This way, the effective modulation order of the HP stream is lower than the LP stream, and thus the HP stream is more robust.

Hierarchical modulation can be used to transmit local services on top of the global services in areas close to the transmitters, by transmitting the local services in the LP stream and the global services in the HP stream [7]. The hierarchical modulation cannot assure the same coverage for local and global services. This means that, in some areas, it is not possible to receive any local service at all. One drawback of the hierarchical modulation is that it suffers from inter-layer interference, because each stream acts as noise to the other. That means that the coverage of the global services is reduced when local services are transmitted on top of them. The susceptibility to noise of the HP stream can be reduced by increasing the spacing between HP constellation states or using a lower code rate, at the expense of degrading the performance of the LP stream and reducing the available total capacity, respectively.

In order to avoid interferences between neighboring local areas, the LP stream can be shared in a time-division manner, such that each local area transmits local services at different instants of time. If there are two local areas, each of them could transmit during 50% of the time, which implies a reduction in the available capacity for each local service. Local areas far away from each other can then transmit at the same time. As in MFNs, it is possible to define a spatial reuse depending on the distance between the corresponding transmitters.

The second technique adopted in DVB-NGH to transmit local services in SFNs is known as orthogonal local service insertion (O-LSI). The idea is that a set of OFDM subcarriers within the NGH frame structure are allocated to transmit local services, and the transmitter of each LSA transmits local services only on its allocated set of subcarriers. The main benefits of this technique are twofold. First, the coverage of the global services is not affected by the local services. Second, the coverage of the local services is potentially the same as the coverage of the global services. It should be noted that local services do not fully benefit from the SFN gain (except within an LSA containing several transmitters), and hence there is no power gain. But there is still a statistical network gain.

In a similar way like hierarchical modulation, the transmission of local content can be done in such a way that different local areas can reuse the same OFDM carriers as soon as they do not interfere with each other. The number of different (and orthogonal) sets of carriers is equivalent to a frequency reuse figure, which plays the same role as the frequency reuse factor in conventional frequency planning. This means that the transmission capacity depends on the frequency reuse factor. For O-LSI, this capacity loss can be partially compensated by increasing the modulation order, since it is possible to transmit the carriers devoted to local services with higher power because for each transmitter, only some carriers within one OFDM symbol are active. It should be noted that the total transmission power of each OFDM symbol is the same, but at each transmitter, only one set of local carriers is active.

This chapter provides an overview of the two mechanisms adopted in DVB-NGH to efficiently insert local content in SFNs. The rest of the chapter is structured as follows. Section 18.2 provides some background information. It briefly introduces the different DTT network topologies SFN and MFN, describes use cases for local services in DTT networks, and reviews the state of the art. Section 18.3 describes in detail the use of hierarchical modulation for local service insertion (H-LSI). Section 18.4 describes in detail the O-LSI technique. Section 18.5 presents illustrative performance evaluation results of both techniques H-LSI and O-LSI. Finally, the chapter is concluded with Section 18.6.

18.2 Background

18.2.1 DTT Broadcast Network Topologies

18.2.1.1 Multi-Frequency Network

In an MFN, each transmitter employs a dedicated frequency that is not used in any other adjacent cell, using a cellular approach (see Figure 18.2). Cells that use the same transmitting frequency are called co-channel cells. The co-channel distance

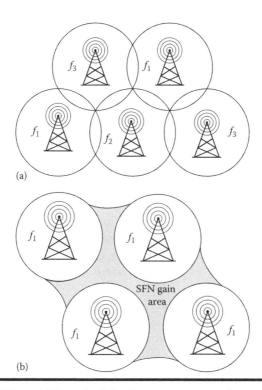

Figure 18.2 MFN and SFN principle. In an MFN (a), adjacent cells use differ-ent transmission frequencies. In an SFN (b), all cells use the same transmission frequency.

depends on the cell sizes and the frequency reuse factor, and the latter determines the number of unique frequency sets used for a cell cluster. The inter-cell interfer-ence can be ignored with a sufficiently large co-channel distance. In an MFN, the transmitters operate independently, and hence different contents can be transmit-ted in each cell. Local services may be transmitted in a subset of the cells while the transmission capacity may be used for other services in the other cells. The drawback is that more of the scarce frequency spectrum is required because frequencies can-not be reused between cells whose proximity is closer than the co-channel distance.

18.2.1.2 Single-Frequency Network

SFNs make use of only one frequency for all transmitters in the SFN, although adjacent SFNs need to use other frequencies, so frequency reuse is required also in the case of, e.g., a national SFN, but on an international level (see Figure 18.2). All transmitters are synchronized in time and frequency and transmit the same content. SFNs are very spectrum efficient for the transmission of global services.

Furthermore, due to the so-called SFN gain, the coverage level is enhanced due to mutual support of the signals from the different transmitters. However, if local services are transmitted, they have to be transmitted across the whole network including in regions where they are not required. This leads to a significant waste of capacity (for global services) if the proportion of local content is large.

18.2.2 Use Cases for Local Content Insertion

18.2.2.1 Local Service Content Classification

There are three main use cases for the transmission of local content in DTT SFNs: *local services*, *local advertising*, and *local emergency information*.

Local services can offer audiences access to deeper and richer news and information about local issues, events, and developments. Generally, news is identified as the main local content service, and it is the primary driver of audience interest in local services. Other areas of particular interest include local language programs and local sports and events.

Local content insertion in SFNs allows broadcasters to segment its viewers for local advertisers, such that they may target their commercials at a particular portion of the broadcaster's audience. From the advertiser's perspective, they can therefore reach more effectively viewers who are more likely to be interested in their product at a lower delivery cost since they do not have to purchase the rights to advertise in the entire network. The broadcaster may attract new advertisers who would not have wanted, or been able to afford, to advertise on the whole network, achieving an overall higher income.

Local content insertion in SFNs may allow the transmission of local emergency information without the need of transmitting it across the whole network. This may include providing information in emergency situations, such as local alerts or local traffic information.

18.2.2.2 Local Service Delivery Classification

Local services can be classified as a function of their time on air and the target geographic area (see Table 18.1). If the content must be broadcast all the time (e.g., local entertainment), these services are referenced as continuous services. On the contrary, if the content is emitted for short time slots or at different times of day, they are called intermittent services. In this case, the network has to dynamically signal the availability of the services. This is a very common case when a global service is substituted by a local service, either planned (e.g., local advertisements) or not (safety-related services). Local content may be inserted by all transmitters of the SFN (with different contents), but in some cases, it may be meaningful that only a subset of the transmitters in the SFN transmit local content (e.g., a single transmitter inserts local content).

Table 18.1   Classification of Local Services as a Function of the Time on Air and Target Localization

	Intermittent Services	*Continuous Services*
Localized geographically (one or some transmitters of the SFN insert local content)	Emergency alarm	Local radio service
Available around all transmitters in the SFN (with different content)	Local news or advertising as temporal window in a global service	Local service with same coverage as global services. The same local service bit rate must be allocated from all TXs in the SFN

18.2.3 State-of-the-Art Review

18.2.3.1 Hierarchical Modulation

Hierarchical modulation builds each QAM symbol from two bit streams with different robustness levels [7]. This is achieved by a QAM modulation, which is subdivided into a robust stream consisting of a particular amount of the HP bits of each QAM symbol, and a second less robust stream that corresponds to the remaining LP bits. For instance, a 16-QAM modulation carrying 4 bits per symbol can be fed by two bit streams, each providing 2 bits per QAM symbol, as depicted in Figure 18.3. An equivalent description of hierarchical modulation is the arithmetical addition of two QPSK streams with different magnitudes also resulting in a hierarchically modulated 16-QAM constellation.

The HM suffers from inter-layer interference because the LP stream acts as noise to the HP stream (and vice versa). To adjust the robustness of both LP and HP bit streams, the distance of the constellation points can be adapted by the magnitude of their QPSK constellation. This is characterized by means of the hierarchical parameter α that describes the ratio of the lowest distance between the constellation points carrying different HP bits (b) to the distance between the LP bits (a) (see Figure 18.3). The usual 16-QAM constellation corresponds to $\alpha = 1$, whereas increasing values of α reduce the robustness of the LP stream. The limiting value $\alpha = \infty$ corresponds to a QPSK constellation, i.e., no hierarchical modulation is used. Another possibility to adapt the robustness of the HP and LP stream is by the selection of the forward error correction (FEC) code rate. Of course, the concept of hierarchical modulation is not limited to 16-QAM, but can be extended to larger QAM sizes. Hierarchical modulation was adopted for the first time for DVB-T allowing for the transmission of two streams with individual robustness levels [1]. Later, it has been also adopted for Media FLO, a mobile

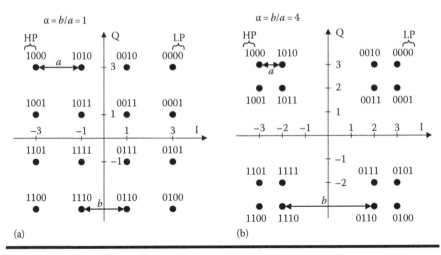

Figure 18.3 **Hierarchical 16-QAM constellation with $\alpha=1$ (a) and $\alpha=4$ (b). The HP stream is mapped on the two most significant bits. The LP stream is mapped on the two least significant bits.**

broadcasting system developed by Qualcomm [8], and the first-generation hybrid terrestrial–satellite European mobile broadcasting standard DVB-SH (Satellite to Handhelds) [9].

The idea of local service insertion (H-LSI) utilizes the concept of hierarchical modulation to transmit global content within the HP bit stream, whereas the local content is inserted into the LP stream of the hierarchical modulation. In the case of DVB-SH, the hierarchical modulation can be used for the insertion of local services in terrestrial repeaters [10]. However, hierarchical modulation has never been commercially deployed yet, neither with DVB-T, Media FLO, nor DVB-SH.

18.2.3.2 Time-Slicing

The coexistence of global and local services in an SFN is in general possible if each type of service is transmitted in different time slots. That is, the SFN principle is violated within the time slot of the local services. Figure 18.4 shows the principle of this idea.

This technique allows global services to be transmitted by all transmitters of the network like in a classical SFN, exploiting the SFN gain. The coverage area of the local services is, however, smaller compared to the MFN approach, due to interference between different LSAs [5]. This technique was originally proposed for DVB-H (Handheld), which employs a time-slicing technology, but it can also be applied to DVB-T [8]. In both cases, in order to decouple the transmission of global and local services, it is necessary to insert an adaptation interval with padding just before and after the local service slot. Such adaptation intervals ensure that no data

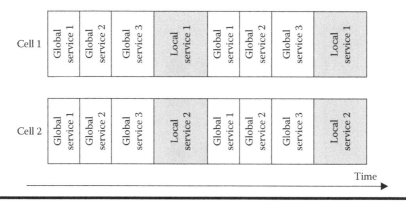

Figure 18.4 **Time-slicing to insert local services in SFNs.**

from the global services is mixed with data from the local services at the physical layer, so that the local services do not interfere with the global services. These adaptation intervals waste capacity.

18.3 Hierarchical Modulation for Local Services Insertion in DVB-NGH SFNs

The idea of H-LSI is directly related to the concept of hierarchical modulation introduced in the previous section. A local service stream is embedded in the LP bits of the QAM constellation, whereas the global stream is transmitted in the HP bits. Since not all transmitters carry local services, a transmitter containing the global service only is using normal QAM constellations. Transmitters also transmitting the local service then add an additional QPSK constellation on top of the existing QAM constellation. This leads to the constellation diagrams depicted in Figure 18.5, where both global and local services are using QPSK, resulting in a hierarchical 16-QAM modulation. Besides a hierarchical 16-QAM, DVB-NGH also supports hierarchical 64-QAM modulation for the insertion of local services wherein the LP stream consists of the two least significant bits (LSBs) per QAM cell, i.e., the local stream is carried in a QPSK modulation on top of a global 16-QAM modulation.

Depending on the distance of the receiver to the local service inserting transmitter, it is possible to decode the global service only or both local and global services. Naturally, the required signal-to-noise ratio (SNR) for successful decoding of the global service is degraded close to the local service inserting transmitter. This coverage reduction can be expressed by means of a higher SNR threshold [10]. The coverage reduction of the global service further depends on the choice of α. The highest degradation occurs when a hierarchically modulated 16-QAM modulation with $\alpha = 1$ is used, and the receiver is close to the local service inserting transmitter. For high values of α, the degradation is practically negligible,

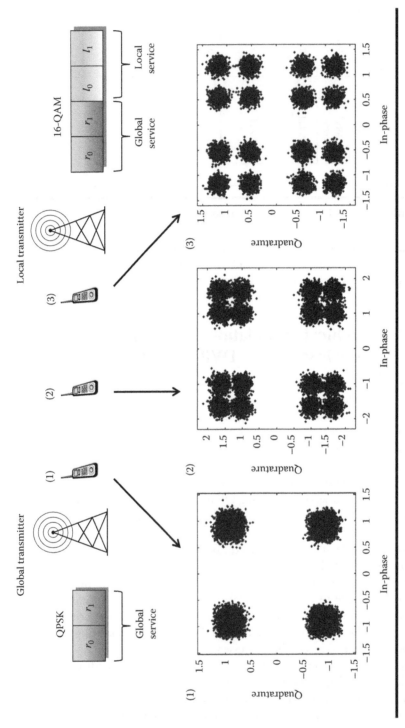

Figure 18.5 **Hierarchically modulated 16-QAM; left: receiver is distant from a local transmitter; right: receiver is close to a local transmitter.**

but the coverage level of the local services is highly reduced. The selection of α is therefore a trade-off between the coverage reduction of the global stream and the achieved coverage of the local stream. In DVB-NGH, the allowed values for α are 1, 2, and 4 for 16-QAM, and 1 and 3 for 64-QAM, respectively. Since the local stream is embedded into the LP bits, the reception of the local service requires a much higher SNR compared to the global stream carried by the HP bits. But this may be achieved when most receivers interested in the local service are in the vicinity of the local transmitter. Within the coverage area of adjacent SFN cells, the hierarchically modulated QAM symbols "look" like noise, requiring an increase in SNR. This SNR penalty diminishes with distance from the local service inserting transmitter.

A second penalty for global and local service decoding arises from the violation of the SFN principle. Since all OFDM carriers that carry hierarchically modulated local services transmit different complex values from the global and local transmitters, interference between the global and local OFDM carriers occurs when propagation channel characteristics from the two transmitters are different at the receiver. This causes imperfect equalization in the receiver and causes a performance penalty for the global and local services that depends on the channel characteristics at the receiver. The penalty can be reduced with an iterative equalization and decoding scheme. This scheme is discussed in detail in Section 18.3.2.2.

18.3.1 Network Topologies

Consider two neighboring SFN transmitters. In the absence of local insertion, both transmitters carry the same global SFN service in the QAM symbols. When there is need for one transmitter to insert a local service, there may also be need for the other to do the same. The insertion of local services is carried out on a physical layer pipe (PLP) basis with H-LSI. A PLP in the DVB-NGH context consists of a GSE or TS input stream that is FEC encoded, QAM modulated, time interleaved, and afterward mapped to the NGH frame structure. Thus, the PLP concept allows assigning an individual robustness to each service or service component of the corresponding input stream that are carried by a given PLP. If then both transmitters insert different local services over the same PLP, then doing it simultaneously would cause severe interference within their respective coverage areas for the QAM symbols concerned.

The solution is to time share the PLP carrying QAM symbols, e.g., on a frame-by-frame basis. For example, the first transmitter can hierarchically modulate all the QAM symbols of the corresponding PLP with its local content during even physical layer frames, while the other transmitter does it only on odd physical layer frames. If there are more neighboring transmitters to insert different local services on the same PLP, then the time sharing increases. Time share slots can be reused between transmitters that are sufficiently far apart. Increased time sharing (high reuse factors) also leads to a reduced capacity of local content that

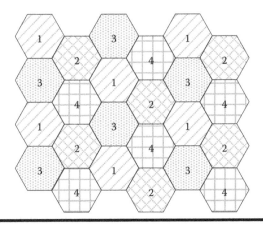

Figure 18.6　Spatial reuse pattern with different time slots or PLPs.

can be inserted at each local transmitter. Another approach is to use different PLPs for the local service insertion instead of time slots. The first transmitter would insert its local service over PLP 1, whereas the neighboring transmitters use different PLPs for the local service insertion. The used PLPs are then spatially reused within the network in the same fashion as described for the time slot reuse. In terms of spectral efficiency, both approaches are the same, but the usage of different PLPs instead of time sharing achieves better uniformity of the local service data rates.

Figure 18.6 illustrates local insertion with a time slot reuse factor of 4. In this example, each transmitter uses all the QAM symbols available of the corresponding PLP in the current physical frame if the frame number modulo 4 corresponds to the number shown in the cell. In this case, the capacity of each of the locally inserted services would be at most a quarter of the PLP capacity. During each physical layer frame, therefore, only one out of four transmitters in a cluster would insert local content. The spatial reuse on PLP basis is carried out in the same way. The number in each cell then corresponds to the number of the PLP transmitting the local service.

18.3.2 H-LSI Implementation in DVB-NGH

18.3.2.1 Transmitter Side

The block diagram of the H-LSI transmitter is shown in Figure 18.7. The local and global service bits are processed separately in stages, both containing the typical blocks of the DVB-NGH signal generation such as LDPC encoder, time interleaver, and frequency interleaver. The processing path of the local PLP comprises a burst builder, which groups the coded local service bits of an integer number of FEC frames to a single burst that is carried in a local service (LS) frame.

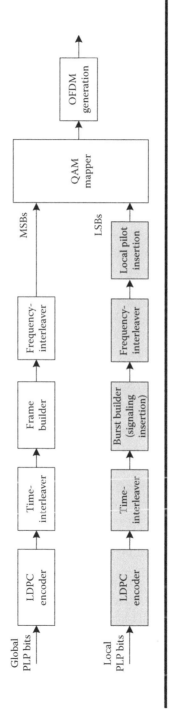

Figure 18.7 H-LSI transmitter block diagram.

Before the local payload burst, each LS frame starts with a 64 bit synchronization header to ease decoding in the receiver. The synchronization header also carries the signaling information for local stream decoding. Before the hierarchical QAM mapping takes place, local pilots are inserted that are required for correct channel estimation and equalization in the receiver. Afterward, both streams are jointly mapped by the QAM mapper to the MSBs and LSBs of the hierarchical QAM constellation.

Just like DVB-T2, DVB-NGH allows for the use of rotated constellations to improve the robustness of the transmission. If constellation rotation is used for the global service PLP, the rotation angle applied to the hierarchically modulated QAM symbols is the same as is used in the global service. Thus, if the global service is carried, for example, with 16-QAM, then the hierarchical modulation symbols of the 64-QAM would be rotated with the rotation angle for 16-QAM.

However, some combinations of different technologies in DVB-NGH with H-LSI are excluded for reasons of incompatibility. This concerns the combination of H-LSI with nonuniform QAM constellations (see Chapter 11) and the peak-to-average power ratio (PAPR) reduction scheme active constellation extension (ACE) [2]. The combination of hierarchical modulation and nonuniform QAM constellations is excluded because the densely packed inner constellation points would be strongly impacted by hierarchal modulation compared to uniform QAM constellations. Decoding of the local stream would be more complex as nonuniform constellations cannot be considered as the sum of two smaller QAM constellations, which complicates iterative sliced decoding (ISD; see Section 18.3.2.2). The usage of the PAPR reduction scheme ACE is excluded because the algorithm cannot be applied in the same way to hierarchical QAM modulations.

18.3.2.1.1 H-LSI PLPs

In DVB-T2, there are two types of PLPs, known as Type 1 and Type 2 [2]. PLPs of Type 1 are transmitted in a single burst (slice) within each frame, whereas PLPs of Type 2 are transmitted in at least two subslices within each frame. In addition, there may be so-called common PLPs carrying common information of a group of input streams. Common PLPs are transmitted in a single burst within the frame as Type 1 PLPs, and they appear first in the frame followed by Type 1 PLPs and then Type 2 PLPs.

In DVB-NGH, two new data PLPs are defined for local content insertion in SFNs, known as Type 3 and Type 4, which are used for O-LSI and H-LSI, respectively. Figure 18.8 shows the NGH logical frame (LF) structure with the different types of PLPs. H-LSI PLPs are transmitted on top of data PLPs of Type 1, and hence they appear after the common PLPs. The O-LSI PLPs are transmitted at the end of the frame, after data PLPs of Type 2. Auxiliary streams or padding cells may exist in between Type 2 and Type 3. The LF structure of DVB-NGH is described in detail in Chapter 14.

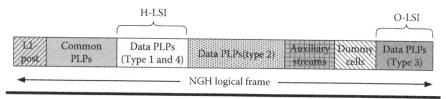

Figure 18.8 **Structure of the DVB-NGH LF with the different types of PLPs.**

18.3.2.1.2 H-LSI Signaling

Whereas DVB-T hierarchically modulates all payload carriers, the local services are introduced in DVB-NGH on a PLP basis, i.e., are inserted only on a subset of the carriers of a global PLP. The synchronization on the corresponding carriers of the global PLP is achieved by means of the 64 bit synchronization header at the beginning of each LS frame, by correlating with a known Walsh–Hadamard sequence contained in the header. The header furthermore carries the signaling information of the local PLP, which is protected by a (32, 16) Reed–Muller channel code with code rate 1/2. The signaling information in the synchronization header describes the parameters of the local PLP, such as its LDPC code rate and the number of FEC frames in the LS frame, etc. The value of the hierarchical parameter α and the ID of the global PLP carrying the local PLP are signaled in the global layer 1 (L1) signaling, since this information is required to extract the local stream. A receiver interested in a local service thus first reads the PLP ID of the national PLP carrying the local service and then obtains synchronization by means of the synchronization header inserted in each LS frame.

18.3.2.1.3 Pilot Patterns for H-LSI

For successful decoding of the local stream, knowledge of the local channel properties is required. For that reason, local pilots are hierarchically modulated on top of the global QAM symbols with a regular spacing allowing for the channel estimation of the local path only. The pilot spacing of the local pilots is always the same as the global PLPs, as shown in the example of Figure 18.9 for a pilot spacing of $D_x = 3$ and $D_y = 4$. However, the local pilots are shifted by one carrier compared to the global pilots to avoid the distortion of the global pilots. To reduce the impact of the local pilots on the global payload QAM symbols, a complex pilot sequence $(1/2 \pm j/2)$ is used for the local pilots instead of a real sequence.

18.3.2.1.4 Transmission Capacity for Local Services with H-LSI

The maximum number of hierarchically modulated OFDM carriers is naturally limited to the amount of available QAM symbols of the global PLP carrying the local PLP. Thus, the data rate of the local PLP in bits per second, B_{local}, depends

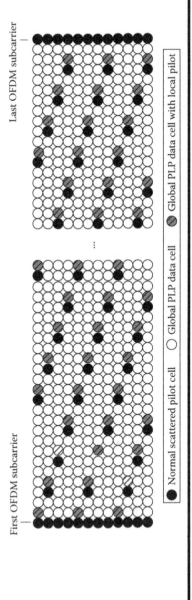

Figure 18.9 Pilot pattern PP1 of local PLP ($D_x = 3$, $D_y = 4$).

on the carrier rate of the global PLP in payload OFDM carriers per second, C_{global}, which is directly related to its data rate B_{global}, the spectral efficiency of the global modulation, μ_{global} (i.e., 2 bits/carrier for QPSK and 4 bits/carrier for 16-QAM), and its code rate, CR_{global}:

$$C_{global} = \frac{B_{global}}{\mu_{global} \cdot CR_{global}}. \tag{18.1}$$

The effective payload bit rate of the local PLP is furthermore influenced by the LDPC code rate of the local PLP, CR_{local}. Taking into account the overhead of the local pilot carriers with horizontal spacing D_x and vertical spacing D_y, the data rate of the local service is approximately

$$B_{local} \approx 2 \cdot CR_{local} \cdot C_{global} \cdot \frac{D_x D_y - 1}{D_x D_y}. \tag{18.2}$$

The factor "2" in Equation 18.2 reflects that each carrier is modulated using a local QPSK modulation on top of the global constellation, carrying two bits per carrier. For typical pilot patterns like PP1 ($D_x = 3$, $D_y = 4$) and PP3 ($D_x = 6$, $D_y = 4$), the resulting local pilot overhead is 8.3% and 4.2%, respectively.

18.3.2.2 Receiver Side

For initial acquisition to a local service, the receiver first decodes the global L1 signaling to find out which global PLPs might contain local PLPs within the network and for each such local PLP, which Walsh–Hadamard sequence to use for its synchronization. For these global PLPs, the receiver tries to find the synchronization sequence and to decode the additional signaling information in the local signaling. Afterward, decoding of the payload of the local PLP is carried out, which requires a particular equalization scheme.

In SFNs, the channel equalization is carried out based on the resulting channel transfer function (CTF) of the different transmitters, which is typically determined by known OFDM pilot carriers. In an SFN with one transmitter Tx1, transmitting the complex value S_g of the global service at OFDM carrier k, and transmitter Tx2, transmitting both the global service S_g and the local service S_l with the CTFs H_l and H_g, respectively, the received OFDM symbol R at OFDM carrier k is

$$R(k) = [H_g(k) + H_l(k)]S_g(k) + H_l(k)S_l(k). \tag{18.3}$$

However, if H_l and H_g differ, which is typically the case in practice, equalization with the joint estimated CTF ($H_l + H_g$) is not possible. For successful decoding of the local service, the global service needs to be subtracted from the received signal

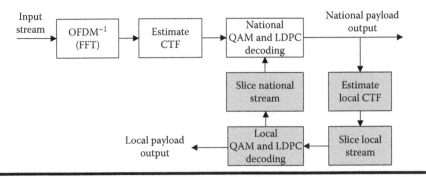

Figure 18.10   Simplified block diagram of the ISD receiver.

and vice versa. This operation requires knowledge of the CTF of the local transmitter. By means of the inserted OFDM pilots in the frequency domain, which are only present in the local signal, it is possible to estimate the local CTF.

As the initial decoding of both global and local services is distorted since neither the local nor the global streams for subtraction from the received signal are known, the decoding is carried out in an iterative manner [11]. In the first iteration, imperfect reception of the global service is performed, and the local service is decoded by subtracting the global service from the received signal. The corresponding equations result from rearranging Equation 18.1 to

$$S_l(k) = \frac{R(k) - S_g(k)[H_g(k) + H_l(k)]}{H_l(k)}, \tag{18.4}$$

$$S_g(k) = \frac{R(k) - S_l(k)H_l(k)}{H_l(k) + H_g(k)}. \tag{18.5}$$

If decoding is not fully successful, a second iteration is carried out, and so on. This decoding algorithm is called ISD, as the first decoding iteration of the global signal is carried out by means of hard decision in a plain receiver implementation. The simplified block diagram of an ISD decoder is shown in Figure 18.10. The ISD decoding is not to be mistaken for iterative decoding between QAM demapper and LDPC decoder, which can be used mutually within each ISD iteration.

18.3.3 System Configuration Examples

H-LSI is especially suited to the transmission of geographically localized services, since neighboring SFN transmitters do not require any multiplex or parameter reconfiguration. This is especially beneficial for intermittent local services like local advertisement, traffic information, or emergency alarms (see again Table 18.1).

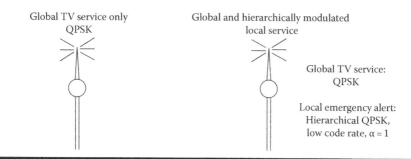

Figure 18.11 **H-LSI system configuration example in a two-transmitter SFN, one transmitter introducing a local emergency alert.**

A system configuration example of an SFN with two transmitters, one transmitter locally inserting an emergency alert, is shown in Figure 18.11. Since the coverage of the global service is slightly reduced in areas where hierarchically modulated local services are inserted, the robustness of the global service can be compensated by means of increasing the transmission power for transmitters that insert local services. Since most transmitters are operating with the maximum allowed output power, an alternative is to decrease the code rate of the global stream, reducing the useful bit rate of the global service in the corresponding OFDM carriers. Since the local emergency alert has a very high priority, α is chosen to 1 and a low code rate is selected, achieving a very high robustness of the local service.

18.4 Orthogonal Local Services Insertion in DVB-NGH SFNs

18.4.1 O-LSI Technique

18.4.1.1 Concept

The O-LSI technique adopted in DVB-NGH to efficiently insert local content in SFNs uses a similar concept as the auxiliary stream solution specified in the DVB-T2 transmitter signature standard [12]. With O-LSI, a set of dedicated OFDM carriers on dedicated OFDM symbols are reserved for the transmission of local services. Within the same OFDM symbol, the transmitters of different LSAs employ a different subset of carriers for broadcasting local services, whereas the other OFDM carriers used by the other transmitters are set to zero. This orthogonality obtained by using dedicated carriers for each local service ensures no interference between adjacent transmitters.

Figure 18.12 shows the general concept of the O-LSI technique for the insertion of local services in an SFN with four LSAs. For the sake of clarity, the picture shows the allocated data subcarriers in one OFDM symbol before frequency interleaving.

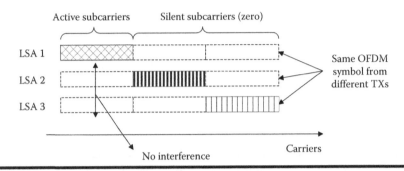

Figure 18.12 General concept of O-LSI technique. Each local transmitter employs a specific set of OFDM carriers.

After the frequency interleaving, each set is spread across the complete bandwidth to achieve good frequency diversity, but still no interference occurs between transmitters of different LSAs. LSAs separated by a long distance could use the same carriers, in a similar way as the spatial reuse pattern shown in Figure 18.6 for H-LSI.

The main advantage of the O-LSI technique is that there is no interference between transmitters from different LSAs because the inserted local content is orthogonal to each other. This way, local services can potentially have the same coverage area as global services, such that it is possible to receive one local service at any point in the network. Furthermore, the insertion of local services does not affect the coverage of the global services. Compared to the H-LSI technique, the drawback of O-LSI is that the capacity available for global services is reduced, since the local services are not transmitted on top of the global services. With O-LSI, global and local services share all the available OFDM data carriers.

18.4.1.2 OFDM Carrier Power Boosting

With O-LSI, the available data rate for local content in each LSA depends on the fraction of resources devoted to local services and the cluster size of the network, which corresponds to the minimum value among the number of LSAs and the frequency reuse factor in the network. It should be pointed out that with O-LSI, the same data rate is reserved for each LSA to insert local content. Compared to the conventional approach based on simulcasting the local services, the main difference is that instead of transmitting all the local services across the whole network, which would require an unnecessary overhead, the local content is transmitted only in the particular area (LSA) of interest using a specific subset of OFDM carriers. A further advantage of O-LSI compared to simulcasting comes from the fact that the power of the OFDM carriers devoted to local services can be boosted in order to keep constant the OFDM symbol power in the transmitters. This power boost at the transmitter side can yield improved area coverage for local services. But it can be also translated into a capacity increase using a transmission mode with higher spectral efficiency (higher code rates and/or higher modulation order). For example, if there are four LSAs in the network, a gain of 6 dB in the link budget can be achieved due to the transmission of four times more power in each subset of cells. This allows using, e.g., 16-QAM modulation instead of QPSK for the O-LSI PLPs, doubling the spectral efficiency or using 2/3 rate coding instead of 1/3. Therefore, the total transmission capacity in an SFN for global and local services with O-LSI is higher compared to the conventional simulcasting approach while keeping a similar coverage level.

18.4.2 O-LSI Implementation in DVB-NGH

18.4.2.1 O-LSI PLPs

PLP data using O-LSI are transmitted as Type 3 after any preceding Type 1 and Type 2 PLPs in specific OFDM symbols, as shown in Figure 18.8. O-LSI requires a one-to-one mapping between logical and physical frames (i.e., only logical NGH channels Type A and D are supported with O-LSI, see Chapter 14). All Type 3 PLP data in a logical frame is transmitted in a number of consecutive OFDM symbols, which are disjoint from those using other PLP types. The first and the last O-LSI symbols are denoted as O-LSI starting symbol and O-LSI closing symbol, respectively, and they have a denser pilot pattern.

18.4.2.2 Filling of O-LSI Symbols with Data Cells

The cells with local content data are introduced into the O-LSI part of the NGH frame symbol by symbol. The orthogonality among the insertion of different local contents is obtained by dividing the available number of data carriers in each O-LSI symbol into n_{LSA} parts, n_{LSA} being the frequency reuse factor in the SFN. Only one part is transmitted from a particular transmitter. For each symbol, data cells are introduced following their carrier index and excluding all positions already filled with pilot cells. When all O-LSI data cells have been introduced, frequency interleaving is performed symbol by symbol. Before frequency interleaving, the different parts appear one after the other, as shown in Figure 18.12. After frequency interleaving, the data cells of a particular part are quasi-randomly distributed over the full bandwidth without losing the orthogonality with the rest of O-LSI parts.

18.4.2.3 O-LSI Pilot Patterns

In a similar way to H-LSI, O-LSI PLPs need to add dedicated pilot patterns for channel estimation. Since the receivers need to demodulate each transmitter independently of the other transmitters, each transmitter, or set of adjacent transmitters using the same carriers, must therefore include its own pilot pattern to allow for channel estimation. These additional pilots reduce the useful data capacity and depend on the frequency reuse factor n_{LSA} in the SFN.

The continual pilots are the same for all transmitters in the network, but the scattered pilots for the transmitters of an LSA are based on the corresponding regular scattered pilot pattern, with a number of cells being frequency shifted, such that the different patterns are orthogonal (see Figure 18.13). Since the scattered pilot pattern is repeated for each LSA, the densest patterns PP1 and PP2 with a density of 8.3% of pilots are not used for O-LSI.

18.4.2.4 Power Level of the O-LSI Cells

The reserved O-LSI data and pilot cells in each LSA are transmitted with an amplitude boosting factor equal to $\sqrt{n_{LSA}}$, followed by a normalization factor K. The value of K is such that the O-LSI starting and closing symbols have the same expected average power as the preamble P2 symbol. The power level of any intermediate symbols is slightly lower due to a lower pilot density.

18.4.2.5 O-LSI Signaling

O-LSI signaling is transmitted as part of the physical Layer 1 (L1) signaling. The L1 signals the pilot pattern, the starting and total number of OFDM symbols in the current NGH frame, the number of frequency chunks used for O-LSI denoted as reuse factor and the frequency chunk used by the current PLP.

18.4.3 System Configuration Examples

The main advantage of O-LSI is that it is possible to provide local services across the whole SFN network with a similar coverage to the global services, not necessarily only in the vicinity of the transmitters as with H-LSI. Hence, this technique is suitable, for example, for local news or advertising as temporal window in a global service. Figure 18.14 depicts the coverage level for global and local services in an SFN network with O-LSI. In the overlapping zones between adjacent transmitters, global services experience an SFN gain, but the receivers can decode more than one local service in addition to the global service because they do not interfere with each other.

18.5 Performance Evaluation

18.5.1 H-LSI Performance

18.5.1.1 Coverage Degradation for Global Services

Due to inter-layer interference, the hierarchically modulated local service adds a penalty to the global service. This can be expressed by means of an increased effective SNR for the global service depending on the value of the hierarchical parameter

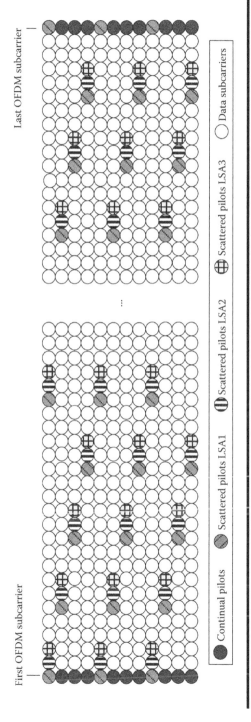

Figure 18.13 Example of O-LSI pilot pattern with three LSAs based on PP7.

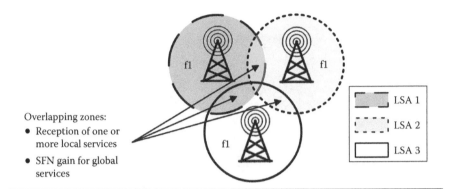

Figure 18.14 Coverage areas for global and local services in an SFN with O-LSI.

α [10]. In Figure 18.15, the effective SNR of the global service is shown for the different values of α if the receiver is only receiving a signal from the local transmitter. The effective SNR values show the worst case values, since the inter-layer interference decreases with greater distance from the local service inserting transmitter. The dashed line shows the required SNR for successful decoding of the global stream with LDPC code rate 7/15 in the Rayleigh P1 multipath channel [1]. Error-free decoding is achieved when the effective SNR exceeds the required SNR threshold (dashed line). The penalty in this scenario is 0.5 dB for $\alpha = 4$, 1.3 dB for $\alpha = 2$, and 3.1 dB for $\alpha = 1$,

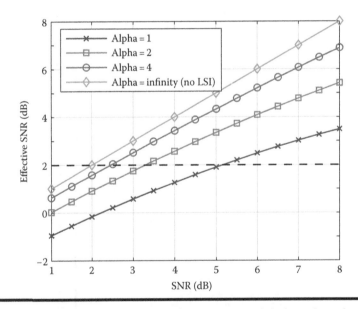

Figure 18.15 Effective SNR of hierarchical 16-QAM global service. The dashed line is the required SNR for successful decoding with code rate 7/15 in the Rayleigh P1 channel.

respectively. The penalty increases for code rates that require higher SNR for successful decoding. Naturally, without LSI, i.e., $\alpha = \infty$, no penalty occurs.

The second penalty due to the violation of the SFN principle cannot be calculated in such a general way since it depends on the channel characteristics experienced by the receiver. Figure 18.16 depicts the overall penalty on the global stream decoding, taking also the effect of the SFN violation into account. For the simulations, a delay between the global and local channel impulse responses of the utilized Rayleigh P1 model has been assumed, causing the violation of the SFN principle. The four curves correspond to different locations within the network. It can be observed that the penalty is relatively low in the vicinity of the global transmitter, but is getting larger with greater distance to the global transmitter and reaches its highest value just in-between both transmitters. With $\alpha = 1$, the penalty reaches up to 4 dB, whereas in case of $\alpha = 4$, it is in the order of 2 dB.

18.5.1.2 Iterative Sliced Decoding Performance for Local Services

The effective SNR for the local service can be expressed in the same way compared to the global service if the receiver is only receiving a signal from the local transmitter. This is depicted in Figure 18.17 for the different values of α. It is clear that the effective SNR of the local stream is much lower compared to the global stream due to the reduced Euclidian distance between the LSBs of the hierarchical QAM constellation. To achieve the required effective SNR for successful decoding of the local stream with LDPC code rate 7/15 in the Rayleigh P1 channel, 9.0 dB SNR is required with $\alpha = 1$. With $\alpha = 2$ and $\alpha = 4$, 12.0 and 16.2 dB are required, respectively.

To achieve a good performance, the subtraction of the global stream from the local stream by means of ISD decoding is required. The performance of the local stream decoding with and without ISD decoding in the Rice F1 multipath channel [1] is shown in Figure 18.18 for different positions between the global and the local transmitter. Despite the low delay spread of the Rice channel, decoding is not possible without ISD even in the presence of a modest regional transmitter signal power. However, with ISD, successful decoding is possible even if the receiver is located just in-between the local and global transmitters. Although, as the distance from the local transmitter increases, the required SNR for successful decoding also increases, since the ratio of the local reception power to the overall power is decreasing, thereby decreasing the effective SNR of the local stream in the receiver.

18.5.1.3 Capacity Performance

In this section, the additional capacity provided by H-LSI compared to a conventional SFN, where the local services are transmitted across the whole network, is

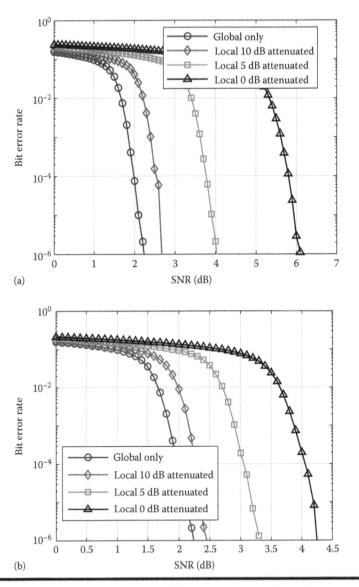

(a)

(b)

Figure 18.16 Performance of global QPSK stream decoding with α = 1 (a) and α = 4 (b) at different locations between the global and local transmitters with code rate 7/15 in Rayleigh P1 channel.

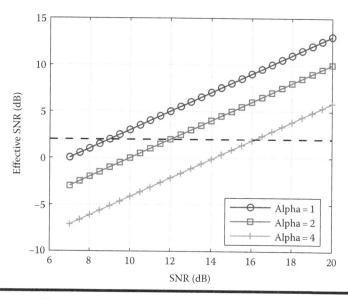

Figure 18.17 Effective SNR of hierarchical 16-QAM local service. The dashed line is the required SNR for successful decoding with code rate 7/15 in the Rayleigh P1 channel.

analyzed. It is assumed that one frequency is available for the whole network area for both global and local services. Hence, MFNs are not considered. The gain is defined as the ratio between the available data rate for the transmission of global and local services in each LSA according to the new approaches, R_{LSI}, and the data rate provided by the conventional approach, R_{SFN}:

$$G_{LSI} = \frac{R_{LSI}}{R_{SFN}}. \tag{18.6}$$

The gain depends on the percentage of the data rate of local services relative to the whole data rate of both global and local services ρ_{local}, and the number of LSAs in the network, n_{LSA}.

For a classical SFN, the available data rate may be calculated accordingly as [5]

$$R_{SFN} = R_{Mux}\left((1-\rho_{local})+\frac{\rho_{local}}{n_{LSA}}\right), \tag{18.7}$$

where R_{Mux} is the multiplex data rate. In this case, the data rate for local services is shared among the different LSAs in the network.

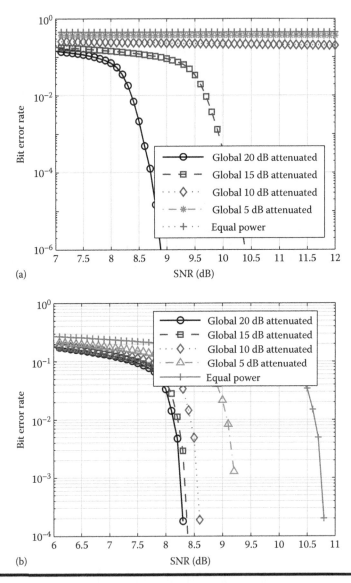

Figure 18.18 Performance of local decoding with α = 1 and code rate 7/15 without ISD decoding (a) and with ISD decoding with 1 iteration (b). Rice F1 channel.

With H-LSI, local services are transmitted on top of global services with a hierarchical modulation. Assuming that local PLPs are transmitted in all OFDM carriers, the available data rate can be calculated as

$$R_{H-LSI} = R_{mux}\left(1 + \frac{1}{n_{LSA}} \times \frac{2 \cdot CR_{local}}{\mu_{global} \cdot CR_{global}} \times \frac{\rho_{local}}{1 - \rho_{local}}\right), \qquad (18.8)$$

where

CR_{local} and CR_{global} are the code rate of the local and global PLPs, respectively

μ_{global} is the modulation efficiency of the global PLP

It should be noted that the maximum percentage of local services for H-LSI when all carriers are used to transmit local content depends on the MODCODs used for the global and local PLPs. It can be computed as

$$\rho_{local, \max H-LSI} = \frac{2 \cdot CR_{local}}{\mu_{global} \cdot CR_{global}}. \qquad (18.9)$$

When the global PLP employs QPSK modulation, and the same code rate is used for the global and local PLPs, the maximum percentage of local services is 50%. In this case, in Equation 18.8, we can notice that with H-LSI, it is possible to double the available data rate in the network (i.e., 100% gain).

Figure 18.19 shows an illustrative example of the potential gains that can be achieved with H-LSI compared to the classical SFN approach as a function of the fraction of local services. It has been assumed that all OFDM carriers are used to insert local content. Two modulations have been considered for the global PLP (QPSK and 16-QAM). It has been assumed that the global and local PLPs employ the same coding rate. It can be observed that important gains can be achieved, but the percentage of local services is limited up to 50% and 33% when the global PLP uses QPSK and 16-QAM, respectively.

18.5.2 O-LSI Performance

18.5.2.1 Coverage Performance Discussion

With O-LSI, each transmitter that inserts local content employs a specific subset of OFDM carriers not used for global services or local services by other transmitters. From a coverage point of view, the benefits of O-LSI compared to H-LSI are two-fold. On one hand, the insertion of local services does not affect the coverage of the global services. On the other hand, it is possible to provide local services across the whole SFN network with a similar coverage to the global services, being possible to

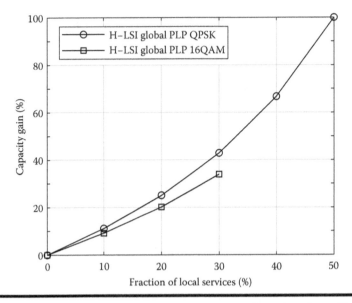

Figure 18.19 Capacity gain of H-LSI as a function of the fraction of local services. The same coding rate is assumed for the global and the local PLPs.

potentially receive one local service at any point in the network rather than only in the areas close to the transmitters. It should be pointed out that in the overlapping zones between adjacent transmitters not inserting the same local content, global services experience an SFN gain whereas local services do not [3].

18.5.2.2 Capacity Performance

In this section, the capacity of O-LSI is analyzed in a similar way to the H-LSI capacity in Section 18.5.1.3. With O-LSI, the available data rate at each LSA depends on the proportion of OFDM carriers devoted for local services and the number of LSAs in the SFN. In a similar way to H-LSI, the transmission capacity available for local services in each LSA is reduced by a factor $1/n_{LSA}$. Compared to a classical SFN, O-LSI provides a capacity gain because it is possible to employ transmission modes for local services with a higher spectral efficiency thanks to the fact that the active carriers are transmitted with a higher power in order to keep the average OFDM symbol power constant. The available data rate with O-LSI can be computed as

$$R_{O-LSI} = R_{Mux}\left((1-\rho_{local}) + \Delta C \cdot \frac{\rho_{local}}{n_{LSA}}\right), \qquad (18.10)$$

Table 18.2   Power Boost of the O-LSI Cells as a Function of the Number of LSAs in the SFN

Number of LSAs	1	2	3	4	5	6	7
Power boost (dB)	0	3	4.8	6	6.9	7.8	8.5

where ΔC is the capacity gain due to power boosting. The boosted power is equal to $\sqrt{n_{LSA}}$, and it can be directly translated into a link budget gain as shown in Table 18.2. Obviously, the higher the number of LSAs, the higher the power increase.

The actual capacity gain cannot be easily derived because it depends on the minimum signal level and the reference channel model considered. But the gain is higher for robust transmission modes. As an example, it is easier to double the spectral efficiency for QPSK than for 16-QAM, which requires moving up to 16-QAM and 256-QAM, respectively. And the required link margin gain also depends on the channel model. Generally speaking, the more severe propagation conditions, the larger is the required gain (e.g., TU6 requires more than AWGN).

Figure 18.20 shows the capacity gain provided by O-LSI compared to the classical SFN approach as a function of the fraction of local services and the number of LSAs in the network. It has been also assumed that the global PLP uses QPSK 1/2 and that the reference channel is AWGN. It can be observed that the gain depends not only on the percentage of local services, but also on the number of LSAs (cluster size).

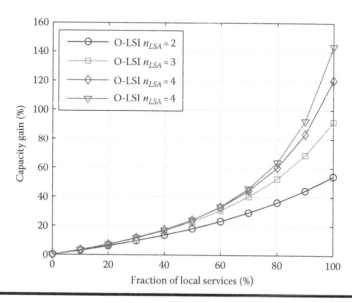

Figure 18.20   Capacity gain of O-LSI as a function of the number of LSAs and the fraction of local services.

18.6 Conclusions

One of the commercial requirements of DVB-NGH was to efficiently transmit local content in SFNs with minimum increase to network overhead. Hence, two complementary techniques were adopted, known as H-LSI (based on hierarchical modulation) and O-LSI (orthogonal insertion). Each technique addresses different use cases with different coverage-capacity performance trade-off, such that DVB-NGH will allow exploring the viability of inserting local services in SFNs in a way that has not been possible before. For example, DVB-NGH will allow inserting local content from a single transmitter in the network or providing local services with the same coverage as global services. The optimum transmission technique depends on the target use case and the particular scenario considered (location and power of the transmitters, distribution of the LSAs, etc.).

References

1. ETSI EN 300 744 v.1.6.1, Digital video broadcasting (DVB); framing structure, channel coding and modulation for digital terrestrial television (DVB-T), January 2009.
2. EN 302 755 v1.3.1, Frame structure channel coding and modulation for a second generation digital terrestrial television broadcasting system (DVB-T2), November 2011.
3. W. Joseph, P. Angueira, J. A. Arenas, L. Verloock, and L. Martens, On the methodology for calculating SFN gain in digital broadcast systems, *IEEE Trans. Broadcast.*, 56(2), 331–339, September 2010.
4. A. Mattson, Single frequency network in DTV, *IEEE Trans. Broadcast.*, 51(4), 413–422, December 2005.
5. G. May and P. Unger, A new approach for transmitting local content within digital single frequency broadcast networks, *IEEE Trans. Broadcast.*, 53(4), 732–737, December 2007.
6. DVB Commercial Module, Commercial requirements for DVB-NGH, Document CM-NGH015r1, June 2009.
7. H. Jiang and P. Wilford, A hierarchical modulation for upgrading digital broadcast systems, *IEEE Trans. Broadcast.*, 51(2), 223–229, June 2005.
8. M. R. Chari et al., FLO physical layer: An overview, *IEEE Trans. Broadcast.*, 52(1), 145–160, March 2007.
9. ETSI EN 302 583 v1.1.2, Digital video broadcasting (DVB); framing structure, channel coding and modulation for satellite services to handheld devices (SH) below 3G, February 2010.
10. H. Jiang, P. Wilford, and S. Wilkus, Providing local content in a hybrid single frequency network using hierarchical modulation, *IEEE Trans. Broadcast.*, 56, 532–540, 2010.
11. J. Zöllner, J. Robert, S. Atungsiri, and M. Taylor, Local service insertion in terrestrial single frequency networks based on hierarchical modulation, *Proc. IEEE International Conference on Consumer Electronics (ICCE)*, Las Vegas, NV, 2012.
12. ETSI TS 102 992 v1.1.1, Structure and modulation of optional transmitter signatures (T2-TX-SIG) for use with the DVB-T2, September 2010.
13. U. Reimers, *DVB—The Family of International Standards for Digital Video Broadcasting*, 2nd edn., Springer, Berlin, Germany, 2005.

NEXT GENERATION HANDHELD DVB TECHNOLOGY

Multiple-Input Multiple-Output (MIMO) Profile

III

Chapter 19

Overview of the Multiple-Input Multiple-Output Terrestrial Profile of DVB-NGH

Peter Moss and Tuck Yeen Poon

Contents

19.1 Introduction

MIMO stands for multiple-input, multiple-output and refers to a radio link employing at least two transmitters and two receivers. A conventional topology comprising a single transmitter and a single receiver is referred to as SISO, with SIMO and MISO having corresponding meanings. In addition, we use the term MIXO to indicate the presence of multiple transmitting antennas together with one or more receive antennas.

This chapter provides an overview of MIMO technology as employed within the MIMO terrestrial profile of DVB-NGH. The fact that there is a MIMO profile at all contrasts with the situation in DVB-T2, where the need to make changes to both transmission plant and domestic antennas in order to use MIMO for roof-top reception led to its exclusion from that standard. This problem is eased, at least at the user terminal end of the chain, for DVB-NGH since portable handsets integrate the antenna with the decoder and are changed relatively often by consumers. This allows MIMO decoder antennas to be included along with matching decoder technology in new handset designs as they emerge.

However, it is an unfortunate fact that the generally lower signal-to-noise ratios (SNRs) of portable reception compared to fixed rooftop substantially reduces the available MIMO gain that may be exploited; indeed, portable indoor reception gains nothing from MIMO. For this reason, the DVB-NGH MIMO profile is best suited for outdoor medium/high signal use cases such as tablet PCs and automotive reception. The schemes included are Alamouti, offering diversity gain only, enhanced single-frequency network (eSFN), again diversity gain, and enhanced spatial multiplexing plus phase hopping (eSM + PH), a "rate 2" scheme (see Section 19.5) offering increased data throughput. These schemes need not be present in isolation; indeed, integration with existing SISO services via new or upgraded is a key issue that will be discussed.

When the study of MIMO within the DVB-NGH standardization process commenced, it was decided at the outset that the cross-polar configuration was to be employed, and transmission frequencies of most interest would be those within the ultra high frequency (UHF) band. So a suitable UHF MIMO channel model was required in order that candidate MIMO schemes could be evaluated and compared. Double-directional channel models based on ray-tracing were available from other projects focusing on mobile communications, such as 3GPP [1] and WINNER II [2], but these were aimed at somewhat higher frequency bands than UHF. They also require antenna details to be embedded to create a resultant end-to-end electrical path suitable for simulation, necessitating detailed antenna evaluation work. The BBC Guildford model, from which Figure 19.1 was derived, might have been a reasonable candidate but did not employ representative receive antennas of small form factor, even for the case described as portable, and so may have been overoptimistic in terms of attainable performance. In addition, the BBC campaign did not include any indoor reception measurements. For these reasons, it was decided to create a new simple channel model based on measurement campaign carried out at a frequency near 500 MHz and including the effects of a "typical" handset/portable device antenna placed in both indoor and outdoor scenarios. The net result would be an end-to-end channel models inclusive corresponding to an "average" antenna that could be embedded directly into simulations designed to compare MIMO schemes. In practice, two measurement campaigns took place, both in Helsinki in the summers of 2009 and 2010, the latter employing five

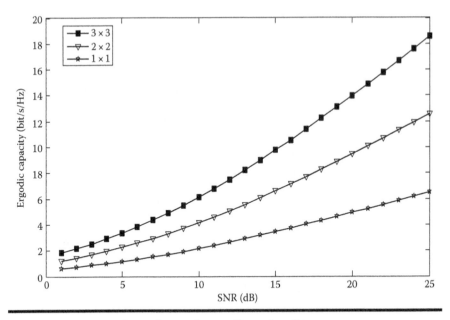

Figure 19.1   MIMO channel capacity vs. dimensionality.

test antennas from different sources to allow an informed decision about typical behavior. The final indoor and outdoor channel models (known as the "Helsinki2" models within DVB-NGH) were based predominantly on the 2010 campaign, with the analysis of the data producing a simple model for both outdoor and indoor scenarios of the antenna-inclusive 2-by-2 channel.

At the time of writing, a potential addition to the DVB-T2 standard in the form of a MIMO profile is under discussion. The MIMO techniques developed within DVB-NGH provide a possible starting point for the possible use of MIMO in the context of fixed rooftop reception. We shall present a short discussion of what could be expected and which other techniques might be used in such a high-capacity T2 profile.

The rest of the chapter is structured as follows. Section 19.2 introduces basic MIMO concepts focusing on issues central to its use in DVB-NGH. Sections 19.3 and 19.4 describe a channel sounding campaign and resulting channel model describing the UHF cross-polar channel to a portable receiving device. Section 19.5 briefly introduces the rate 1 and rate 2 MIMO codes adopted within DVB-NGH. Section 19.6 describes the pilot patterns used for MIMO. Section 19.7 describes how MIMO may actually be deployed in practice, bearing in mind that the proposal to use MIMO within NGH was met with a number of technical, economic, and regulatory constraints, including how it may be integrated into new and existing networks that continue to also support SISO transmission. Finally, the chapter is concluded with Section 19.8, which includes a discussion of the possibilities for a high-capacity T2 profile.

19.2 MIMO Background

19.2.1 MIMO Overview and Benefits

These more elaborate transmission and reception arrangements necessitated by MIMO offer a payback in the form of potentially increased *diversity* (leading to greater robustness in fading channels) and/or *multiplexing* (to offer greater average throughput than a SISO system). In doing so, the fundamental Shannon capacity limit of a SISO channel occupying the same bandwidth and employing the same transmitted power can be exceeded. As an example, Figure 19.1 compares the ergodic capacity of SISO (i.e., 1×1) and MIMO channels of dimension 2×2 (i.e., two transmitting antennas, two receiving antennas) and 3×3 for the case of independent Rayleigh fading on each path from a transmitter antenna to a receiver antenna. It is important to note that the total transmitted power is held constant as the dimensionality increases, in order to give a fair comparison, and that the channel is unknown to the transmitter. That is, each channel path has mean square value ½ in this plot, rather than one as in most texts, to allow direct comparison with the cross-polar case to be discussed later. This does not affect the general form of the plot.

As suggested by the plot, it turns out that for the chosen cases, the ergodic channel capacity increases almost pro rata with the number of transmitters (assuming that this is equal to the number of receivers). Furthermore, it can be shown that in the limit of N_t (the number of transmitters and receivers) becoming large, the capacity gain becomes exactly pro rata with N_t. The plot was derived from the well-known formula introduced by Telatar [3] for the capacity of a MIMO channel described by a matrix **H**, which assumes the channel is unknown to the transmitters and it may be written as

$$C = \log_2\left(\det\left[\mathbf{I} + \frac{P}{\sigma^2 n_T} \mathbf{H}\bar{\mathbf{H}}^T \right] \right), \qquad (19.1)$$

where

C is the MIMO channel capacity
P is the total power of all the transmitters
n_T the number of transmitters
σ^2 the variance of the i.i.d. noise at each antenna
$\bar{\mathbf{H}}^T$ is the Hermitian transpose of $\mathbf{H} \cdot \mathbf{I}$ is the identity matrix

19.2.2 Broadcast MIMO: Co-Polar or Cross-Polar?

Typical circumstances where the MIMO channel may approximate independent Rayleigh fading are a combination of co-polarized transmit and receive antennas and a "rich-scattering" environment where many multipath components may be

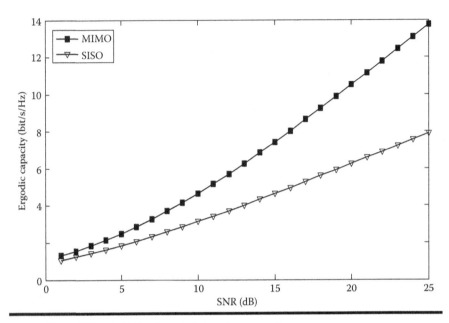

Figure 19.2 BBC Guildford model ergodic capacity.

identified. A practical example might be indoors in a laboratory or office with many physical "scatterers." Although tending to maximize diversity in a cluttered environment, a disadvantage of such a co-polar antenna configuration is the loss of matrix condition number in predominantly line-of-sight (LOS) propagation conditions. For this reason, earlier experiments with MIMO for broadcast, e.g., [4], employed instead a cross-polar 2×2 configuration that retains full matrix rank, and hence potentially high capacity, in LOS conditions through the essentially orthogonal nature of the cross-polar channel. The channel model derived from the measurements made in [4] leads to a predicted capacity behavior as shown in Figure 19.2. Here, the MIMO configuration refers to 2×2 cross-polar.* For both this plot and the previous, the average squared Frobenius norm of the 2×2 channel matrix has the common value 2.0 and hence the two may be directly compared.

19.2.3 Decoding Aspects

Although optimal bit-error performance is obtained with a maximum-likelihood decoder (MLD), it has the disadvantage that the complexity grows exponentially with the number of transmit antennas. Although the number of transmit antennas specified for MIMO in DVB-NGH standard is relatively small, complexity might

---

* For this plot, Ricean K-factors were set to 4 and 1 for the co-polar and cross-polar coupling, respectively.

still be an issue for low-cost portable devices, and it is worth investing in new techniques to reduce its complexity.

The first step in reducing the complexity of the decoder is to simplify the log-likelihood ratio (LLR) calculation for soft decision by using the max-log approximation on the LLR. It was shown that the performance penalty is very small, less than 0.05 dB, for a 2-by-2 16QAM system in the context of the DVB NGH channel model. This is significantly lower than a typical hardware implementation margin.

Second, there has been significant research in reducing the complexity of MLD by first decomposing the MIMO channel matrix using the QR decomposition $\mathbf{H} = \mathbf{QR}$, which results in $\mathbf{Q}$, an orthonormal matrix, and $\mathbf{R}$, an upper triangular matrix with real diagonal values. The QR decomposition can form the basis of reduced-complexity decoding as follows:

It is well known that finding the ML solution is equivalent to solving

$$\hat{\mathbf{s}}_{ML} = \arg \min_{s \in D} \left\| \mathbf{x} - \mathbf{Hs} \right\|^2 , \qquad (19.2)$$

where

D is the search-space
$\mathbf{x}$ the received vector
$\mathbf{H}$ is the channel matrix
$\mathbf{s}$ is the transmitted vector

The QR-based decoder will first decompose $\mathbf{H}$ into $\mathbf{Q}$ and $\mathbf{R}$, hence the ML solution will now be solving

$$\hat{\mathbf{s}}_{ML} = \arg \min_{s \in D} \left\| \mathbf{y} - \mathbf{Rs} \right\|^2 , \qquad (19.3)$$

where $\mathbf{y} = \mathbf{Q}^H \mathbf{x}$ and the squared norm remains unaltered.

An indication of the complexity can be done by calculating the number of multiplication and addition operations required by the decoders as shown in the following:

1. MLD (2-by-2 MIMO)
   a. (13 multipliers and 15 adders) $\times 2b1$
   b. (13 multipliers and 15 adders) $\times 2b2$
2. QR-based MLD (2-by-2 MIMO)—excluding QR decomposition
   a. (5 multipliers and 6 adders) $\times 2b1$
   b. (11 multipliers and 11 adders) $\times 2b2$
   c. 16 multipliers and 12 adders

The $b1$ and $b2$ correspond to the number of bits in the QAM constellation for the first and second symbols, respectively. The number of multipliers/adders required

Table 19.1 Resource Usage for Different QAM Combination

Resource / Decoder	16QAM/16QAM		16QAM/64QAM		64QAM/64QAM	
	Multipliers	*Adders*	*Multipliers*	*Adders*	*Multipliers*	*Adders*
MLD	416	480	1040	1200	1664	1920
QR-based MLD	272	284	512	572	1040	1100
Savings	144	196	528	628	624	820

Calculations do not include resources needed for QR decomposition.

by the QR decomposition depends on implementation and known to be small for a 2-by-2 matrix. Table 19.1 illustrates three possible QAM combinations for the 2-by-2 MIMO system.

This shows that the QR-based MLD saves around 35%–51% multipliers and 41%–52% adders on first inspection before taking into account the resources required for QR decomposition. It is worth noting that the QR decomposition is only done once for every received vector.

The sphere decoding technique is another way to reduce the complexity of the decoder. The sphere decoders can be classified as a QR-based decoder, and it has been known by different names throughout the research community because of its slight variant. The MLD decoding structure can be illustrated in a tree diagram as shown in Figure 19.3, and the search space is represented by the points at the lowest level (Layer 1).

The hard-decision MLD searches over the entire search space for the most probably transmitted symbol based on the received vector while the sphere decoder searches over a fraction of the search space by using an iterative process and boundary conditions. The sphere decoding concept can also be extended to soft decision, and it has become a good choice for MIMO decoding. However, there are still challenges in implementing sphere decoder in very large scale integration (VLSI) because of practical trade-offs and general assumptions.

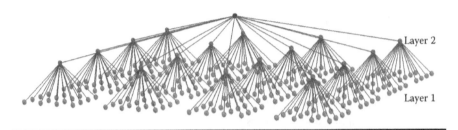

Layer 2

Layer 1

Figure 19.3 MIMO 2 × 2 tree diagram (16QAM, 16QAM).

It is worth mentioning that basic linear decoders such as zero-forcing and MMSE equalizers are simple and easy to implement in hardware but produce sub-optimal bit-error performances. The complexity of such decoders is not determined by the size of the QAM modulation like that in MLD, and the resource usage is just a very small fraction compared to MLD.

19.3 MIMO Channel Sounding Campaign

In this section, we shall discuss some details of the channel sounding campaign that took place in July 2010 in Helsinki, Finland, focusing on cross-polar UHF transmission and reception, and employing a number of practical designs for antennas suitable for a handheld terminal.

First, we briefly describe the excitation signal and test arrangements before introducing the form of the channel model. Finally, the actual parameters of the model that have resulted from the measurements are discussed and related to the observable behavior of the NGH cross-polar signal. The model describes fast fading as seen during antenna displacements of up to 50 m rather than slow fading caused by shadowing or large-scale path loss. Two variants are provided, defining outdoor pedestrian and indoor reception, respectively. A 4×2 model or SFN model may be derived from two uncorrelated 2×2 models of the type described, the excess delays of one model being increased to be close to the defined system tolerance of the system under test (for instance, to 90% of the guard interval).

19.3.1 Channel Measurement System Overview

The channel sounding campaign was carried out in Helsinki, Finland, in a collaborative group comprising Amphenol, BBC R&D, Digita Oy, Elektrobit, Nokia, Tampere University, and Turku University of Applied Sciences. The sounding data were transmitted from a 146 m high TV tower in Pasila, situated to the north of Helsinki city center. Two separate transmit antennas, horizontally polarized (HP) and vertically polarized (VP), were fixed on the transmitting tower, and a series of cross-polarized patch antennas were mounted on the mobile receiving station to create a representative 2-by-2 MIMO system. A modified DVB-T OFDM-based signal containing only pilots, no data, was used to sound the channel.

19.3.2 Transmitter

The transmitter YLE transmission tower (shown in Figure 19.4) used during this channel sounding campaign was located in Pasila, in the north of Helsinki city, and

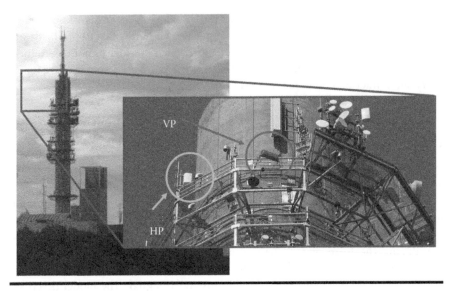

Figure 19.4   YLE transmission tower at Pasila and the transmitting antennas.

it is 146 m high. Two standard UHF antenna panels, one HP and the other VP, were used, and these are shown in Figure 19.4 along with the transmitting tower.

The transmit antennas were standard horizontally polarized UHF panels. One of the panels was rotated through 90° to act as a vertically polarized transmit antenna. The panels comprised four stacked dipoles, which means that the beamwidth of the vertical radiation pattern was considerably narrower than the horizontal radiation pattern. This restricted beamwidth meant that the VP antenna was unable to cover all the measurement locations. Therefore, the VP antenna employed a mechanism to steer it, ensuring that the sounding location has similar field strength levels for both polarizations. Note that, in a practical network, a bespoke antenna design would ensure equal H/V coverage within about 2 dB at any point in the service area.

19.3.3 Receiver

The portable receiving station was constructed as shown in Figure 19.5 and consisted of the Elektrobit channel recorder (connected to the control laptop), BBC MIMO receiver with real-time channel display, cross-polar application antenna, and batteries to power up all the equipment. All these were mounted on a push trolley, and the application antenna was mounted on an "extended arm" to minimize interference from the equipment. The antenna was about 1.5 m above ground.

The real-time channel display was very useful to give an instant feedback on the channel response and to ensure the integrity of the captured data.

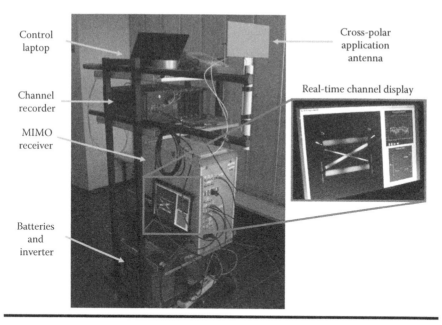

Figure 19.5 Portable receiving station used in the field measurement campaign.

19.3.4 Measurement Campaign

The channel sounding was done at five different locations covering indoor and outdoor scenarios, and over 100 measurements were recorded with 280 GB of data. Five cross-polar application antennas designed by various partners were used in this campaign with the intention that the channel model extracted from this campaign is representative of a practical antenna. The locations where the measurements were taken are shown in Figure 19.6 along with the location of the YLE transmission tower in Pasila.

The type of location and measurement was carefully planned to capture as many scenarios as possible. The indoor capture scenarios include measurements along a corridor, inside a conference room, at reception area, and inside an office while the outdoor scenarios include LOS and non-LOS (NLOS) environment. Every care was taken during the trial, and calibration runs were done throughout the campaign to ensure that all equipment on the portable receiving station was working.

19.4 NGH MIMO Channel Model

19.4.1 General Approach

Data from the 2010 Helsinki campaign were analyzed and allowed construction of the channel model that will now be described. The aim is to provide a time-domain model of paths shown as b_{11}, b_{12}, b_{21}, b_{22} in Figure 19.7, where $Tx1$ and $Tx2$

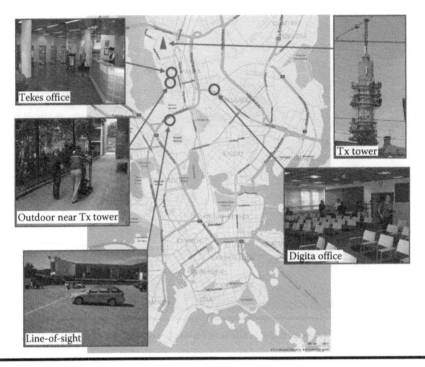

Figure 19.6   Locations where measurements were taken.

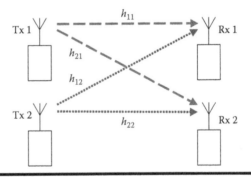

Figure 19.7   2-by-2 MIMO system.

represent twin antenna elements at a terrestrial transmitter site and $Rx1$ and $Rx2$ the two elements of the MIMO receive antenna. It is important to note that the model is antenna inclusive so it is intended to represent the full transmitter output to receive terminal input path. The model describes fast fading as seen during antenna displacements of up to 50 m rather than slow fading caused by shadowing or large-scale path loss. This allows receiver algorithm testing to be carried out satisfactorily as the path/loss shadowing is implicitly included in the choice of test SNR.

An eight-tap model of the propagation paths was proposed, to include the contribution of a "representative" antenna, with each parameter based on an average or typical behavior of those made available for the trial. The model has two variants, "indoor" and "outdoor," customized to represent those scenarios by suitable choices of parameters. The delays and relative powers of the paths are tabulated later.

19.4.2 Mathematical Form of the Model

A simplifying assumption underpinning the model is symmetry of the horizontal/ vertical fading components. This was largely confirmed by observation, although differences in the transmit antenna H/V patterns made it difficult to establish the precise balance. As a working assumption for modeling and simulation, the approximation was considered adequate. Similarly, the cross-polar coupling is taken as symmetrical.

The channel model for both indoor and outdoor variants has the following common form:

$$\text{vec}(\mathbf{H}^T(t,\tau)) = \mathbf{L}_1(t)\delta(\tau - \tau_1) + \sum_{j=1}^{8} \mathbf{R}_j^{1/2}\mathbf{x}_j(t)\delta(\tau - \tau_j), \qquad (19.4)$$

where

t is the time index

τ the delay-time

τ_j the delay-time position of the jth tap (as tabulated in Tables 19.2 and 19.3).

Table 19.2 Tap Values of the NGH MIMO Indoor Model

Tap Number	Excess Delay (μs)	P_j (h_{11}, h_{22}) (dB)	wP_j (h_{12}, h_{21}) (dB)
1	0	−6.0	−8.5
2	0.1094	−8.0	−10.5
3	0.2188	−10.0	−12.5
4	0.6094	−11.0	−13.5
5	1.109	−16.0	−18.5
6	2.109	−20.0	−22.5
7	4.109	−20.0	−22.5
8	8.109	−26.0	−28.5

Table 19.3 Tap Values of the NGH MIMO Outdoor Model

Tap Number	Excess Delay (μs)	P_j (h_{11}, h_{22}) (dB)	wP_j (h_{12}, h_{21}) (dB)
1	0	−4.0	−10.0
2	0.1094	−7.5	−13.5
3	0.2188	−9.5	−15.5
4	0.6094	−11.0	−17.0
5	1.109	−15.0	−21.0
6	2.109	−26.0	−32.0
7	4.109	−30.0	−36.0
8	8.109	−30.0	−36.0

$$\mathbf{vec}(\mathbf{H}^T(t,\tau)) = \begin{pmatrix} h_{11}(t,\tau) \\ h_{12}(t,\tau) \\ h_{21}(t,\tau) \\ h_{22}(t,\tau) \end{pmatrix}. \tag{19.5}$$

A 2×2 time-variant channel matrix $\mathbf{H}$ is hence

$$\mathbf{H}(t,\tau) = \begin{bmatrix} h_{11}(t,\tau) & h_{12}(t,\tau) \\ h_{21}(t,\tau) & h_{22}(t,\tau) \end{bmatrix}. \tag{19.6}$$

The LOS term is given by

$$\mathbf{L}_1(t) = \sqrt{\frac{K_1 P_1}{1+K_1}} \begin{pmatrix} \exp(j\theta_{11}) \\ w\exp(j\theta_{12}) \\ w\exp(j\theta_{21}) \\ \exp(j\theta_{22}) \end{pmatrix}, \tag{19.7}$$

where K_1 is the tap 1 Ricean K-factor.

If the tap is pure LOS, the expression reduces to

$$\mathbf{L}_1(t) = \sqrt{P_1} \begin{pmatrix} \exp(j\theta_{11}) \\ w\exp(j\theta_{12}) \\ w\exp(j\theta_{21}) \\ \exp(j\theta_{22}) \end{pmatrix}. \tag{19.8}$$

$\mathbf{R}_j$ is the 4-by-4 covariance matrix of the four-element coefficient vector $\mathbf{c}_j = \mathbf{R}_j^{1/2}\mathbf{x}_j(t)$ at the jth tap, ordered as Equation 19.8. The terms $\mathbf{x}_j$ are (distinct) random vectors for each j with i.i.d. complex Gaussian components of unit variance. They are time varying with a power spectral density (PSD) in accordance with Section 19.4.7.

$\mathbf{L}_1$ defines the tap 1 LOS terms; it has uniformly distributed random phases θ_{nm} in the interval $[0, 2\pi]$. P_j is the total power of the jth tap weight associated with terms h_{11}, h_{22}, i.e., $P_1 = \mathbf{L}_1^2 + |\mathbf{c}_1(1)|^2 = \mathbf{L}_1^2 + |\mathbf{c}_1(4)|^2$ and $P_j = |\mathbf{c}_j(1)|^2 = |\mathbf{c}_j(4)|^2 \ldots j \neq 1$. The coefficient w adjusts the power level appropriately for the terms h_{12}, h_{21}. For the indoor model, $w = 0.562$ (2.5 dB); for the outdoor model, $w = 0.25$ (6 dB).

Since tap 1 has both an LOS and a Rayleigh component, the latter has a power level $1/1 + K_1$ of the tabulated tap power P_1 (see Section 19.4.2). The required square root of $\mathbf{R}_j$ satisfies $\mathbf{R}_j = \mathbf{R}_j^{1/2}\mathbf{R}_j^{H/2}$ and can be obtained from the Cholesky decomposition of $\mathbf{R}_j$. The tap covariance matrix is common to taps 2–8 and the Rayleigh part of tap 1 within each model variant.

19.4.3 Additional Antenna Rotation and Asymmetry Terms

In deriving the antenna-specific data from which the model parameters were derived, the raw data were rotated by up to ±40° to find the angle that maximized the cross-polar discrimination. This was to correct for both physical mounting differences and antenna axis differences with respect to the casing. It also ensured that averaging across antennas retained legitimacy in terms of cross-polar discrimination. However, in practice, this "ideal" alignment may not be representative, and so a further rotation matrix $\mathbf{W}$, with angle Ω chosen from the set $\{-45°, 0°, +45°\}$, is recommended with the angle fixed for a particular simulation run. In addition, an asymmetry matrix $\mathbf{\Gamma}$ is included to model observed H/V asymmetries that persist over many contiguous channel realizations ($\sim$10 m). $\mathbf{\Gamma}$ is also fixed for a particular simulation run and takes values from the set:

$$\left\{ \begin{bmatrix} 1.1074 & 0 \\ 0 & 0.8796 \end{bmatrix} \begin{bmatrix} 1 & 0 \\ 0 & 1 \end{bmatrix} \begin{bmatrix} 0.8796 & 0 \\ 0 & 1.1074 \end{bmatrix} \right\}. \tag{19.9}$$

The resulting channel matrix $\mathbf{H}_c(t, \tau)$ is hence derived from $\mathbf{H}$ as follows:

$$\mathbf{H}_c(t,\tau) = \mathbf{W}\mathbf{H}(t,\tau)\Gamma = \begin{bmatrix} \cos\Omega & -\sin\Omega \\ \sin\Omega & \cos\Omega \end{bmatrix} \begin{bmatrix} h_{11}(t,\tau) & h_{12}(t,\tau) \\ h_{21}(t,\tau) & h_{22}(t,\tau) \end{bmatrix} \begin{bmatrix} \Gamma_{11} & 0 \\ 0 & \Gamma_{22} \end{bmatrix}.$$

(19.10)

19.4.4 Tap Values

The tap values P_j are set so as to model the power delay profile of the channel. Two sets of tap values were derived from the data, corresponding to the indoor and outdoor scenarios. These are tabulated in Tables 19.2 and 19.3.

19.4.5 Ricean K-factor

The Ricean K-factor (i.e., ratio of the LOS power to total of the NLOS power) of each matrix element was estimated using the method of moments [5]. This technique first evaluates the second and fourth moments of the frequency response data along the t-axis.* From this, the K-factor of an assumed underlying Ricean distribution may be deduced. The K-factors of the model are tabulated in Table 19.4.

19.4.6 Tap Correlation Matrices

The tap correlation matrix for the indoor channel is given, except for tap 1 $(2 \leq j \leq 8)$, by

$$\mathbf{R}_j = P_j \begin{pmatrix} 1.00 & 0.15 & 0.10 & 0.15 \\ 0.15 & 0.56 & 0.06 & 0.04 \\ 0.10 & 0.06 & 0.56 & 0.15 \\ 0.15 & 0.04 & 0.15 & 1.00 \end{pmatrix}.$$

(19.11)

Table 19.4　Ricean K-Factors of the NGH MIMO Channel

Model	First Tap	Overall
Indoor	1.0	0.2
Outdoor	LOS only, $K=\infty$	1.0

---

* That is, we examine the evolution in time of each frequency bin and assign a K to each, from which an overall K is deduced.

For tap 1, we have

$$\mathbf{R}_1 = \frac{1}{1+K_1} P_1 \begin{pmatrix} 1.00 & 0.15 & 0.10 & 0.15 \\ 0.15 & 0.56 & 0.06 & 0.04 \\ 0.10 & 0.06 & 0.56 & 0.15 \\ 0.15 & 0.04 & 0.15 & 1.00 \end{pmatrix}. \tag{19.12}$$

The tap correlation matrix for the outdoor channel is given by (except for tap 1)

$$\mathbf{R}_j = P_j \begin{pmatrix} 1.00 & 0.06 & 0.06 & 0.05 \\ 0.06 & 0.25 & 0.03 & 0.05 \\ 0.06 & 0.03 & 0.25 & 0.06 \\ 0.05 & 0.05 & 0.06 & 1.00 \end{pmatrix}. \tag{19.13}$$

For tap 1, the energy is pure LOS, so there is no Rayleigh part to the fading. Hence $\mathbf{R}_1 = [0]_{4 \times 4}$.

19.4.7 Doppler Spectrum

While the experimental methodology of pushing a trolley at approximately constant speed was sufficient to produce an indicative spectrum, it was clearly not sufficiently accurate to produce quantitative results. Because of this, it was decided at the outset to use a "classical" PSD as a basis of the Rayleigh fading taps. This PSD (also known as Jakes Doppler spectrum) for $|f| \le f_d$ is given by

$$S(f, f_d) = \frac{1}{\pi f_d \sqrt{1 - (f/f_d)^2}} \tag{19.14}$$

and is zero elsewhere.

f_d is a parameter controlling the maximum Doppler width. It is chosen to be proportional to an assumed vehicle speed. The form of this PSD is illustrated in Figure 19.8 (with arbitrary scales), which is a plot from a simulation based on the method of exact Doppler spread (MEDS) [6].

In this model, the Jakes spectrum is specified to be used at a particular tap in conjunction with a fixed frequency offset (pure Doppler) proportional to f_d. Table 19.5 specifies the spectra of each tap.

Choice of Doppler width parameter f_d: To represent portable reception, both indoor and outdoor, receive terminal speeds of 0 and 3 km/h are specified. At 600 MHz carrier frequency, the latter corresponds to 1.667 Hz Doppler width f_d. For the 0 km/h case, see next section.

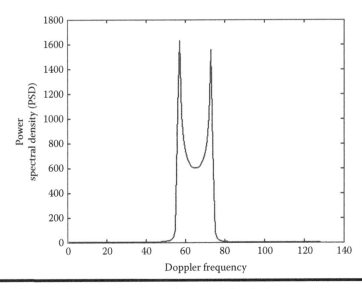

Figure 19.8   Jakes spectrum from MEDS model simulation.

Stationary channels ("snapshots"): If we simulate a time-invariant channel, i.e., $f_d = 0$, the Doppler shift for the LOS components and the Doppler spreads for the NLOS components are not modeled by the procedures described earlier. Instead, a number of independent realizations ("snapshots") should be generated in which the vectors $\mathbf{x}_j$ remain (distinct) for each j retain the prescribed intra-tap correlation properties, but are not time varying, so the MEDS procedure is not invoked.

19.4.8 Mobile Vehicular Outdoor Model

For the mobile case, the outdoor model is used but with $f_d = 33.3$ Hz and 167 Hz Doppler half-width. This corresponds to 60 and 300 km/h at 600 MHz.

19.4.9 Two-Tower 4×2 Model

A pair of uncorrelated 2×2 models can be used to represent a 4×2 transmission from two cross-polar towers. A time offset and level difference between them can also be introduced to represent the time-of-arrival and signal level differences at the receiver with respect to each tower.

19.4.10 SISO, MISO, and SIMO Systems

Appropriate transmission paths of the MIMO model may be selected to provide SISO, MISO, and SIMO models. For SISO, this amounts to using h_{11} or h_{22} (which have identical statistics); for SIMO, h_{11} and h_{21}; and for MISO, h_{11} and h_{12}.

Table 19.5 Tap Doppler Spectra

Tap Number, p	Spectrum
1 (Indoor model NLOS component only)	$S(f, f_d)$
2	$S\left(f - \dfrac{3f_d}{4}, \dfrac{f_d}{4}\right)$
3	$S\left(f - \dfrac{3f_d}{4}, \dfrac{f_d}{4}\right)$
4	$S\left(f + \dfrac{3f_d}{4}, \dfrac{f_d}{4}\right)$
5	$S\left(f + \dfrac{3f_d}{4}, \dfrac{f_d}{4}\right)$
6	$S\left(f + \dfrac{3f_d}{4}, \dfrac{f_d}{4}\right)$
7	$S\left(f + \dfrac{3f_d}{4}, \dfrac{f_d}{4}\right)$
8	$S\left(f + \dfrac{3f_d}{4}, \dfrac{f_d}{4}\right)$

Of course, it must be remembered that the model is appropriate only for the cross-polar transmission/reception case, so the MISO variant has limited practical application. The SIMO model does, however, represent a realistic case of dual-polarized reception of single-polarized transmission, for instance, with rotation-insensitive handset antenna designs.

19.4.11 Capacity Prediction from the NGH Channel Model

Once again the model allows us to produce a prediction of the available capacity as a function of SNR. The plot for the outdoor model is shown in Figure 19.9, and this may be compared with the corresponding plot for fixed rooftop reception given earlier as in Figure 19.2. We see that the capacity is somewhat lower in the portable outdoor model, but not greatly so, despite the fact that the Ricean K-factor is considerably lower and the cross-polar coupling substantially higher. This is quite encouraging for the prospects of MIMO in such channels.

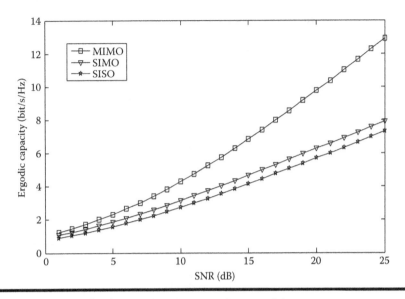

Figure 19.9 Capacity from NGH2 2 × 2 outdoor model.

19.5 MIMO NGH Codes

19.5.1 MIMO Rate 1 and Rate 2 Codes

19.5.1.1 Distinction between Rate 1 and Rate 2

The first family of techniques to be considered is MIMO rate 1 codes, which exploit the spatial diversity of the MIMO channel without the need for multiple antennas at the receiver side. They do not, however, offer any multiplexing gain as a consequence. As well as being applicable to an individual multiple-antenna transmitter site, they can also be applied across the transmitters of SFNs to reuse the existing network infrastructure (i.e., DVB-T and DVB-T2). This type of code is described in Chapter 20 and includes the well-known Alamouti coding scheme already featured in DVB-T2. In addition, eSFN, a novel transmit diversity scheme that utilizes a kind of delay diversity per OFDM subcarrier, is described.

The second family of techniques is known as MIMO rate 2 codes, which exploit both the diversity and multiplexing capabilities of the MIMO channel. As we have seen, spatial multiplexing techniques obtain very attractive capacity gains in favorable reception conditions, e.g., outdoor vehicular reception, for the provisioning of high data bitrate applications. Specifically, DVB-NGH adopts enhanced spatial multiplexing in combination with phase hopping (eSM-PH) as technical solution to combat high-correlation channel condition of LOS condition and specific harmful conditions. Detailed description of eSM-PH is presented in Chapter 21.

19.5.1.2 Alamouti Code

The most well known of rate 1 codes is Alamouti code [7], where the transmission conventionally characterized by the following matrix (with rows representing time, i.e., channel use, and the columns transmitter index):

$$\begin{pmatrix} s_0 & s_1 \\ -s_1^* & s_0^* \end{pmatrix}. \tag{19.15}$$

In DVB-NGH, as in DVB-T2, the pair of time indices is replaced by a pair of frequency indices to form an orthogonal space frequency block code where each row of the matrix corresponds to a complex cell at the input to the IFFT.

The particular symbols to be transmitted by this code word are S_0 and S_1, so it can be seen that these in effect appear twice in each channel use pair and so the net throughput is one symbol per channel use as it would have been for SISO. The benefit of Alamouti code is largely therefore that of increased diversity, as any on path to a particular receiver may be lost yet both symbols can be recovered. An additional feature is that Alamouti code creates an equivalent channel matrix (over two channel uses) that is orthogonal, and this leads to a particularly simple decoding.

19.5.1.3 Enhanced Spatial Multiplexing Equal (eSFN)

Surprisingly good results can be obtained in the cross-polar channel with simple spatial multiplexing, i.e., with the transmission matrix:

$$\begin{pmatrix} s_0 & s_1 \end{pmatrix}. \tag{19.16}$$

However, this scheme can be outperformed by the approach adopted in DVB-NGH that is enhanced spatial multiplexing (eSFN), particularly in highly correlated channels that often occur in LOS condition. eSFN processing starts with a vector of two constellation points **s** (not necessarily drawn from the same constellation) and applies diagonal and rotation matrices to give a general form:

$$\tilde{\mathbf{s}} = \mathbf{\Sigma}_1 \mathbf{\Theta} \mathbf{\Sigma}_2 \mathbf{s}, \tag{19.17}$$

where
  $\tilde{\mathbf{s}}$ is the eSM precoded output
  $\mathbf{\Theta}$ is a real rotation matrix
  $\mathbf{\Sigma}_1$ and $\mathbf{\Sigma}_2$ are real diagonal matrices of which only one has unequal terms in any particular variant of the scheme

The rotation matrix couples a proportion of each element of **s** to each output introducing diversity, whereas the diagonal matrices facilitate the use of disparate constellations making up the vector **s** and also allow unequal mean power in the elements of $\tilde{\mathbf{s}}$. Such "unequal power" codes provide for easier integration into real-world transmission scenarios as will be discussed in Section 19.7.

Variants of the scheme are specified for 6, 8, and 10 bpcu (i.e., QPSK+16QAM, 16QAM+16QAM, and 16QAM+64QAM, respectively) and for 0, 3, or 6 dB transmitted power imbalance. eSM will be discussed in greater detail in Chapter 21.

19.5.1.4 Phase Hopping

PH is used in conjunction with eSM and is applied after eSM processing, i.e., to vector $\tilde{\mathbf{s}}$, by introducing a pre-multiplying matrix of the form

$$\mathbf{X}(i) = \begin{pmatrix} 1 & 0 \\ 0 & e^{j\Phi_{PH(i)}} \end{pmatrix}. \tag{19.18}$$

The argument of the exponential term at the ith cell position increments the phase in with a period of nine cell positions in order to prevent a fixed phase relationship between the two transmitter output terms. In effect, due to the position in the chain immediately before the transmission, the equivalent channel is modified, and this prevents the LOS terms from each output forming a broad interference pattern. PH will be discussed in greater detail in Chapter 21.

19.6 Pilot Patterns for MIMO Channel Estimation

In MIMO systems, some means must be provided to estimate the channels from each transmission antenna to each receiving antenna in order that the receiver can have knowledge of the channel matrix **H**. In DVB-NGH, scattered pilots are inserted at certain positions in the OFDM ensemble to allow estimation in SISO operation in a manner similar to that adopted in DVB-T and DVB-T2, the latter also having support for multiple transmit antennas due to the presence of MISO as a transmission option. DVB-NGH makes use of these MISO pilots, the sign of which are based on a Hadamard matrix, for MIMO channel estimation. If we consider the 2×2 Hadamard matrix

$$\begin{pmatrix} 1 & 1 \\ 1 & -1 \end{pmatrix}, \tag{19.19}$$

and regard the columns as representing transmitter antennas and the rows as time (i.e., OFDM symbol index), then a "1" indicates no change to the radiated scattered pilots and a "−1" indicates a sign change. That is to say, during the first (and every other) OFDM symbol, both MIMO antennas radiate unchanged pilots, and during the second (and every other) OFDM symbol, the sign of the scattered pilots of

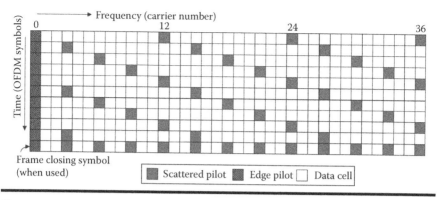

Figure 19.10   Scattered pilot pattern PP1 (SISO).

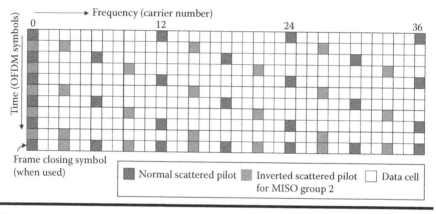

Figure 19.11   Scattered pilot pattern PP1 (MISO).

transmitter 2 is inverted. The receiver then directly obtains the sum and difference of the two channel paths and can hence deduce the paths themselves.

As an illustration of a pilot pattern for SISO, Figures 19.10 and 19.11 show one of the patterns used in DVB-NGH for the SISO and MISO cases, respectively. Notice that in Figure 19.11, the inverted scattered pilots are indicted as belonging to "MISO group 2," which corresponds to transmitter 2 in the earlier discussion.

As well as the patterns illustrated, alternatives are present in the standard with differing total pilot densities in the frequency and time domains, intended for instance to match shorter guard intervals where a lower rate of sampling in the frequency domain is sufficient. This is because the highest frequency component of the channel variation when viewed in the frequency domain is proportional to the delay of the latest component in the channel impulse response, and this is assumed to be less than the chosen guard interval.

19.7 MIMO Deployment Scenarios

19.7.1 Integration with SISO Broadcasts

19.7.1.1 Mixing SISO/MIXO Frames

As DVB-NGH is a standard in which MIMO features as an optional profile, it is quite possible that a transmission that has a MIXO component will also contain SISO (base-profile) frames. Of course, in terms of spectral efficiency, the use of full-time MIXO transmission is almost always the best choice, but it was recognized that mechanisms for mixed SISO/MIXO operation would greatly increase system flexibility.

A basic requirement for mixed operation is that the SISO frames should be decodable by a SISO receiver, and the presence of the MIXO component should not cause the SISO performance to be unduly compromised. In addition, as the MIMO transmission will be over a pair of dual-polarized antennas, the question also arises as to what signal to radiate from each element during SISO frames. The DVB-NGH MIMO profile is designed to address these issues as we shall now discuss.

19.7.1.2 Polarization Power Fluctuations

One option that could be considered during SISO frames would be simply to switch all station power* into the single SISO polarization in use. This might put more strain on power amplifiers, but they could be designed at the outset for these conditions so reliability need not be an issue. Potentially more serious would be the need for automatic gain control (AGC) in receivers to accept such power fluctuations without unwanted effects, including those intended for NGH-SISO, NGH MIMO, and receivers for different services in the adjacent channels. Similar AGC problems have already been observed when LTE signals (with strong envelope fluctuations) are present in the adjacent channel of some DVB-T receivers.

An alternative approach is to radiate a constant horizontal and vertical polarization power during SISO and MIXO frames, although not necessarily the same fixed value on each polarization. This avoids the AGC issues, but creates a new problem in that if the MIXO polarizations were set to have equal power, as desirable from a MIXO viewpoint, this would imply a 3 dB loss to existing SISO services, which is excessive and unlikely to be acceptable. For this reason, the proposed approach in NGH is to radiate the second MIXO polarization at –3 or –6 dB with respect to the SISO polarization. Assuming fixed total station power, the loss to SISO can be reduced to as little as 0.8 dB with the "–6 dB" option. The penalty

---

* The total station power is assumed fixed, as this seems a most likely scenario. The problem remains to a large extent if the SISO polarization power is fixed and the second MIXO polarization transmission allowed switching on and off.

now occurs in MIXO performance, where we have a polarization power imbalance, which particularly affects rate 2 codes. Capacity analysis suggests that this will be a penalty of 0.8 dB at 16 dB SNR in the NGH "Helsinki2" channel. In order to maximize remaining performance in the presence of this deliberate imbalance, specific transmission schemes were developed in DVB-NGH that have already been described.

19.7.1.3 Employing Unequal Power Codes and eSFN

The penalty to rate 2 MIMO arising from the unequal polarization transmission power can be minimized by using mixed constellation types in conjunction with unequal power codes. It is an intuitively reasonable result that the lower-power polarization of the MIXO pair has the smaller, more robust constellation. The MIMO loss seen in simulation for the chosen constellation pair and associated coding is typically in the range 0.5–1.0 dB with the –6 dB power imbalance case, which is broadly in line with the predicted capacity loss.

During SISO frames, the recommended approach is to employ eSFN and treat each polarization as a distinct transmitter. The eSFN technique, which prevents extended parts of the frequency spectrum being affected simultaneously by spatial fading in the context of multi-tower SFNs, shows an analogous benefit in the dual-polar case in controlling inter-polarization self-interference.

19.7.2 Attainable Signal-to-Noise Ratios in Typical Network Deployment

The following set of assumptions and deductions aims to give an illustration of what field strengths DVB-NGH may be able to enjoy based on signal-strength assumptions typical of a DVB-T2 network for fixed reception. The implication is that in order to deploy NGH within an existing network, the latter is upgraded for dual-polarization transmission with the constraint that the total number of transmitters, their powers, and their locations are unchanged. From these assumptions, we identify in particular the likely areas of applicability of the NGH MIMO profile.

19.7.2.1 Indoor Portable Use-Case Correction Factor

Starting from the assumptions of Table 19.6, we include additional correction terms as follows in order to estimate the available indoor carrier-to-noise ratio (CNR) as follows:

■ Antenna height factor (1.5 m compared to 10 m): –8 dB
■ Building penetration loss 8 dB
■ Antenna gain –10 dBd (i.e., 17 dB worse)
■ Total correction with respect to rooftop reception 33 dB

Table 19.6 Typical DVB-T2 ("Host Network") Planning Parameters

Parameter	Value	Notes
Assumed service mean field strength (inner coverage area)	65 dBμV/m	At 10 m AGL
Assumed service mean field strength (outer coverage area)	54 dBμV/m	At 10 m AGL
Carrier frequency	700 MHz	
Rooftop antenna gain	11 dBd	
Rooftop feeder loss	4 dB	
Effective antenna gain	7 dBd	
Location variation	5.5 dB	Implies required margin 7 dB for 90% coverage with log-normal distribution
Free space wavelength (λ)	0.429 m	
Dipole effective cross section	0.0146 m^2	$\lambda^2/4\pi$
Power flux density at 47 dBμV/m	$1.33 \cdot 10^{-10}$ W/m^2	7 dB below mean of 54 dBμV/m for 90% coverage
Received power, net of effective antenna gain	−80.1 dBm	
Required CNR (theoretical, DVB-T2 256QAM 3/5 and 2/3)	16.1 dB, 17.9 dB	
Receiver implementation margin	2 dB	
Receiver noise figure	5 dB	
Required practical CNR (256QAM 3/5 and 2/3) at receiver input	23.1, 24.9 dB	
Noise floor, 7.6 MHz bandwidth	−105.2 dBm	
Available CNR	−80.1 + 105.2 = 25.1 dB (i.e., exceeding 24.9 dB)	Hence system operates down to 47 dBμV/m as required

19.7.2.2 Outdoor Vehicular Use-Case Correction Factor

A similar set of correction factors may be estimated for the outdoor vehicular use case:

■ Antenna height factor (1.5 m compared to 10 m): −8 dB
■ Antenna gain 0 dBd
■ Feeder loss 1 dB
■ Net antenna losses with respect to rooftop: 8 dB
■ Total correction with respect to rooftop: 16 dB

19.7.2.3 Outdoor Portable Use-Case Correction Factor

The outdoor portable case lies between the two cases already presented; a correction figure of 24 dB is probably appropriate.

19.7.2.4 Summary

Now we can draw conclusions in terms of available SNR for each of these use cases. Table 19.7 summarizes the situation for indoor portable and outdoor vehicular scenarios:

From the table, we can conclude as follows:

■ For effective indoor operation, we need the equivalent of 65 dBμV/m or higher at 10 m.
■ With a single antenna, this still provides only −5 dB CNR.
■ For outdoor vehicular scenarios, this would equate to 16 dB CNR.

In practice, a 10 m field strength greater than 65 dBμV/m is probably necessary if indoor handheld reception is to be a serious possibility, perhaps a figure of 70 dBμV/m. By so providing, the outdoor vehicular CNR will be of the order of 20 dB, comfortable enough for high-bitrate rate 2 MIMO coverage.

Table 19.7   Available NGH CNR vs. 10 m Planning Field Strength

Reception Case	Receiver Antenna	54 dBμV/m	65 dBμV/m
Indoor	Single antenna	−15 dB	−4 dB
	Dual antenna	−12 dB	−1 dB
Vehicular	Single antenna	+2 dB	+13 dB
	Dual antenna	+5 dB	+16 dB

19.7.3 Concerns Raised and How These May Be Addressed

19.7.3.1 International Co-Ordination

In 2006, the RRC06 conference in Geneva resulted in a new international agreement on the use of broadcasting spectrum, and this included in some cases a limit not just on power radiated from a particular station but also on polarization use. It could be that, in some administrations, renegotiation to allow dual-polarized MIMO may be necessary. However, the use of unequal power codes may ease this problem and allow operation within existing interference limits.

19.7.3.2 Economic Considerations

Issues of capital cost and running costs were raised. There is certainly a need for capital expenditure to upgrade existing transmitters to MIMO, and this must be weighed against the benefits and revenues the higher data rate MIMO services could offer. In terms of operating costs, the total radiated power of a MIMO signal was assumed to be the same as the SISO signal during the evaluation phase, so that electricity costs should not rise as a result of this.

19.7.3.3 Polarization Power Fluctuations

The potential of MIMO to interfere with other services due to its pulsed nature was raised. This problem is completely overcome using the same fixed per-polarization power during MIXO and SIXO frames as has already been discussed in this chapter.

19.7.3.4 Polarization Pattern Asymmetries

Practical design and implementation constraints mean that a dual-polarized transmit antenna will not have exactly the same radiation pattern in the horizontal and vertical planes. This introduces a power imbalance in addition to that which may have been intentionally introduced by employing unequal power codes. The solution is good design, where this can be kept to 1–2 dB maximum variation or less. It should also be remembered that the SISO broadcast will itself be subject to performance variation due to the pattern irregularity.

19.8 Discussion for a High-Capacity T2 Profile Including MIMO

The MIMO techniques developed within DVB-NGH provide a possible starting point for the possible use of MIMO in the context of fixed rooftop reception. What follows is a short discussion of what could be expected and which other techniques might be used in such a high-capacity T2 profile.

19.8.1 MIMO Gain for Rooftop Reception

The potential benefit of MIMO is greater for rooftop reception than for the portable/mobile scenarios envisaged for DVB-NGH. This is mainly due to the higher signal levels available to a rooftop antenna, where CNRs of 20 dB are realistic. At this value, the MIMO capacity gain is around 66% over SISO as is illustrated in Figure 19.12, which shows the channel capacity derived from the BBC Guildford channel model [4]:

By writing the MIMO capacity equation in a somewhat different form to that of Equation 19.1, we can identify two contributions to the capacity:

$$C = \log_2 \left\{ 1 + \left| \det \mathbf{H} \right|^2 \left(\frac{\text{SNR}}{2} \right)^2 + \left\| \mathbf{H} \right\|_F^2 \left(\frac{\text{SNR}}{2} \right) \right\}. \qquad (19.20)$$

The first term depends on the matrix determinant, and the square of the SNR can be associated with the multiplexing gain of the channel—it falls to zero if the channel matrix is not full rank. The second term depends on the matrix Frobenius norm, i.e., the total "power" of the channel and the SNR. This can be interpreted as a contribution from the diversity of the channel. As SNR rises, the first term begins to dominate, and this is the behavior that is illustrated in Figure 19.10 and is the reason for the greater potential MIMO advantage with rooftop reception.

To optimize performance at these relatively high signal levels, constellation sizes larger than those for DVB-NGH are likely to be considered. An indication

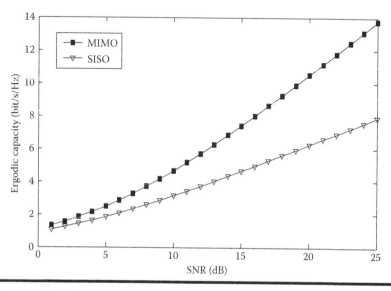

Figure 19.12 MIMO vs. SISO ergodic capacity with 10 m rooftop reception.

of the gain in throughput over DVB-T2 the T2-MIMO profile may offer, assuming the use of unequal power codes, mixed constellations and 6 dB power imbalance is shown below:

- 256QAM + 64QAM = +75% throughput on SISO 256QAM.
- More conservatively, 256QAM + 16QAM = +50% throughput on 256QAM.
- Intermediate values by choice of code rates.
- At rate 2/3 coding, these correspond to 9.33 bps/Hz and 8.0 bps/Hz achievable at SNR 20 and 18 dB, respectively (c.f. UK-mode DVB-T2 256QAM 2/3, which is 5.33 bps/Hz at 19 dB).
- Nonuniform constellations: these can be tailored to the failure-point SNR of a particular code rate and offer a useful gain of up to about 1 dB for larger constellations. See Chapter 11 (BICM).
- Time frequency slicing (TFS): data are cycled over different RF carrier frequencies thereby benefiting from frequency diversity. With four RF channels, benefits of several dB are possible. For details on TFS, see Chapter 13.

19.9 Conclusions

In this chapter, an overview of the application of MIMO technology in the DVB-NGH standard has been presented, including a description of the UHF cross-polar channel model used for code evaluation, the adopted MIXO codes and their performance, and an indication of the SNR range over which MIMO provides a worthwhile benefit. In addition, an indication of how in practice MIMO may be incorporated into new and existing networks has been provided, thereby answering some commonly raised questions about the practicality of deploying this technology. Following this, some speculation has been offered as to the possible benefits of MIMO in a future standard for rooftop reception, as well as indicating the potential of some potentially associated technologies such as nonuniform QAM and TFS.

References

1. ETSI TR 125.996 v10.0.0 (2011–04) Universal Mobile Telecommunications System (UMTS); Spatial channel model for Multiple Input Multiple Output (MIMO) simulations (3GPP TR 25.996 version 10.0.0 Release 10).
2. J. Meinilä, P. Kyösti, L. Hentilä, T. Jämsä, E. Suikkanen, E. Kunnari, and M. Narandžić, D5.3: WINNER+ final channel models, in *Wireless World Initiative New Radio— WINNER+*, P. Heino (ed.), CELTIC/CP5-026, 2010.
3. E. Telatar, Capacity of multi-antenna Gaussian channels, *European Transactions on Telecommunications (ETT)*, 10(6), 585–595, November 1999.

4. J. Boyer et al., MIMO for broadcast—Results from a high-power UK trial, *BBC White Paper*, October 2007.

5. A. Abdi, C. Tepedelenlioglu, M. Kaveh, and G. Giannakis, On the estimation of the K parameter for the rice fading distribution, *IEEE Communications Letters*, 5(3), 92–94, March 2001.

6. M. Patzold, U. Killat, F. Laue, and Y. Li, On the statistical properties of deterministic simulation models for mobile fading channels, *IEEE Transactions on Vehicular Technology*, 47(1), 254–269, February 1998.

7. S. Alamouti, A simple transmit diversity technique for wireless communications, *IEEE Journal on Selected Areas in Communications*, 16(8), 1451–1458, October 1998.

Chapter 20

Multiple-Input Single-Output Antenna Schemes for DVB-NGH

Jörg Robert and Jan Zöllner

Contents

20.1 Introduction

MISO stands for multiple-input, single-output and describes a radio link that uses at least two transmitters but only one receiver. A conventional topology comprising a single transmitter and a single receiver is referred to as single-input, single-output (SISO). This chapter gives an overview of MISO technology as employed within Digital Video Broadcasting–Next Generation Handheld (DVB-NGH). This comprises firstly the Alamouti scheme that is already well-known from DVB-T2 (terrestrial second generation) and the enhanced single-frequency network (eSFN) scheme.

The rationale of the adopted MISO schemes for DVB-NGH is to increase the robustness of the terrestrial transmission by utilizing the spatial diversity of multiple transmitters. For this purpose, the Alamouti encoding divides the available transmit antennas into two groups. The signals from the first group are transmitted without any modification. In contrast, the signals transmitted within the second group are modified within blocks of two QAM symbols. First, the symbols are swapped, and second, the symbols' real or the imaginary axes are inverted, respectively, resulting in an orthogonal coding between both transmitter groups. This can be considered as a space time block code (STBC) with rate-1, since two symbols are transmitted in two time slots. The Alamouti code is the only orthogonal STBC achieving rate-1 and reduces frequency-selective behavior in difficult reception environments like the center of a single-frequency network (SFN). However, decoding of an Alamouti signal requires the estimation of the channel characteristics of both transmitter groups, increasing the pilot overhead for channel estimation compared to SISO transmission.

The idea of the eSFN scheme is the avoidance of any regular fades for multiple transmitter antennas, by de-correlating the signals from the different transmitters using a specific linear phase-distortion algorithm. This de-correlation reduces the coherence bandwidth and, thus, reduces the presence of frequency-selective fades that span the complete signal bandwidth. The phase distortions are introduced in a way that makes them almost invisible for the channel estimation process within the receiver. Consequently, eSFN does not require any additional pilots compared to pure SISO transmission and can also be applied to existing OFDM-based standards, e.g., DVB-T2.

The rest of the chapter is structured as follows. Section 20.2 gives an overview of the background and the rationale for MISO transmission. Section 20.3.1 describes in detail the Alamouti-based MISO scheme and the eSFN scheme that have been adopted for DVB-NGH. In Section 20.4, the performance of the schemes is analyzed by link-level system simulations and field measurements that have been carried out in the scope of the DVB-T2 field trial in the northern part of Germany. Finally, the chapter is concluded with Section 20.5.

20.2 MISO Background

It is well known that frequency-selective fading leads to higher required signal-to-noise ratios (SNRs) for error-free decoding compared to flat channels. One source of frequency selectivity is a rich scattering environment with many strong echoes that arrive at the receiver. However, an additional source is the application of multiple transmit antennas that radiate identical signals.

One scenario for the application of multiple transmit antennas radiating identical signals are so-called SFNs. Figure 20.1 shows the SFN principle. Multiple transmitters radiating identical signals—synchronized in time and frequency—are typically used for covering larger areas. This offers several benefits compared to classical multifrequency networks (MFNs) where only one transmitter uses a specific frequency. The first benefit is the high spectral efficiency, as only one single RF channel is required for covering a large area, while multiple RF channels would be required in the case of an MFN. The second benefit is the so-called SFN gain [1], as multiple transmitters increase the diversity and, hence, reduce the probability of deep fades.

Figure 20.2 shows an impulse response and the corresponding channel transfer function measured within the DVB-T network in Braunschweig, Germany. The depicted impulse response clearly shows the four transmitters used within this SFN network. As the measurement took place in Braunschweig, the two local transmitters highly dominate the reception power. The other two transmitters are located close to Hannover, which is approximately 60 km distant from Braunschweig. Hence, their signals are already highly attenuated and delayed by more than 100 μs. The resulting channel transfer function, which is the Fourier transform of the impulse response [2], shows the relative attenuation of the received

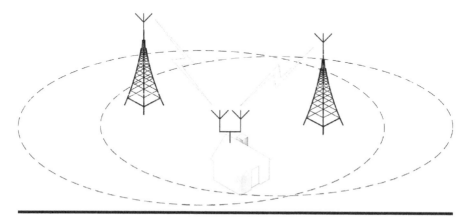

Figure 20.1   Example of an SFN. Multiple transmitters use the same RF frequency for covering a large area. The receiver may be equipped with a single or multiple reception antennas.

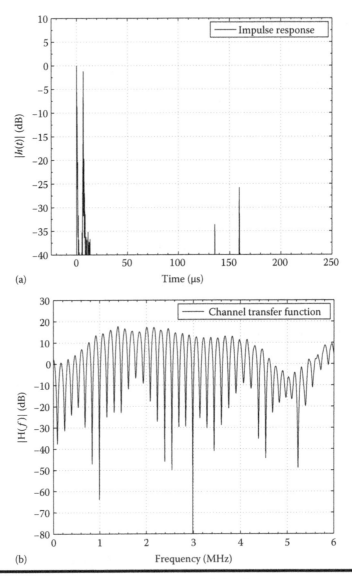

Figure 20.2 Impulse response (a) measured in the DVB-T SFN Hannover/ Braunschweig and the corresponding channel transfer function (b). The network consists of four transmitters that are clearly visible within the impulse response.

signal over the frequency axis. This figure shows the high frequency selectivity of the channel, where specific frequencies suffer attenuations of more than 30 dB compared to the average signal. Naturally, the OFDM subcarriers transmitted within these highly attenuated regions do not contribute to the data transmission and have to be compensated by means of the forward error correction (FEC).

The resulting loss compared to the flat channel strongly depends on the performance of the FEC.

The coherence bandwidth of the fades results from the delay of the individual impulses within the impulse response. Due to the properties of the Fourier transform, long delays will cause a low coherence bandwidth. However, due to a channel bandwidth of typically 8 MHz already, delays of 1 μs (corresponds to a distance of 300 m) lead to eight fading periods within the channel transfer function. Hence, the channel in an SFN can be more or less modeled as fast-fading Rayleigh channel, which has also been confirmed by measurements [3]. The SFN gain depends on the channel characteristics and can vanish, or even cause a negative SFN gain, in difficult reception environments [1]. This is typically the case in the core of the SFN, where the signals of the different transmitter are received with a similar power.

The second scenario for the application of multiple transmit antennas radiating identical signals is the mixed transmission of rate-1 and rate-2 signals in a DVB-NGH network. Figure 20.3 shows the principle of such a scenario. The transmission of the rate-2 signals requires the application of multiple antennas at the transmitter and receiver side. However, part of the signal is radiated as rate-1 signal, allowing also receivers equipped with a single antenna or located in bad reception conditions to decode the signal.

The simple switch-off of one transmit antenna during the transmission of the rate-1 signal may potentially disturb other receivers. Hence, it is preferred that both

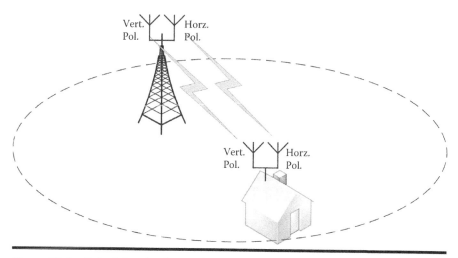

Figure 20.3 **Principle of colocated signal transmission. Multiple antennas with, e.g., different polarizations are for the radiation of the signals from a single transmitter site. A receiver equipped with a single RX antenna or multiple RX antennas that match the polarization of the transmitter antennas are used for the signal reception.**

transmit antennas are used for the transmission of the rate-1 signal. The simple approach to transmit identical signals on both antennas (SFN approach) again leads to frequency selectivity within the channel. In contrast to the distributed SFN, the delay between the signals of both antennas arriving at the receiver is almost zero. This leads to a very high coherence bandwidth that may reach up to several MHz. As a result, both signals may almost cancel each other completely, making successful reception almost impossible. This is exemplarily depicted in Figure 20.4, which shows the impulse response recorded in the core SFN with two echoes arriving with similar power, causing a massive fade in the channel transfer function.

20.3 DVB-NGH MISO Schemes

For overcoming the effects mentioned in the previous section, two rate-1 schemes have been specified for DVB-NGH. The first scheme uses space time coding based on the Alamouti approach, while the second uses an approach that is based on cyclic delay diversity [4] called eSFN.

20.3.1 MISO Alamouti

The transmission of identical signals from multiple transmit antennas results in a significant increase in the frequency selectivity. In contrast, the channel transfer function to the individual transmitters does not show this high frequency selectivity. Figure 20.5 shows the channel transfer functions to the two strongest transmitters of Figure 20.2. The channel transfer functions were extracted using the algorithm described in Reference 5.

The application of MISO encoding based on the Alamouti scheme [6] tries to avoid an increase in the frequency selectivity by means of orthogonal decoding of the signals radiated from the individual transmit antennas.

For this purpose, the Alamouti encoding divides the available transmit antennas into two groups. Using the scheme defined in DVB-NGH, the signals from the first group are transmitted without any modification. In contrast, the signals transmitted within the second group are modified within blocks of two QAM symbols. First, the symbols are swapped, and second, the symbols' real or the imaginary axes are inverted, respectively. Mathematically, we can describe this as follows:

- $x_1(t_0) = c_0$ and $x_1(t_1) = c_1$ for the first group
- $x_2(t_0) = -c_1^*$ and $x_2(t_1) = c_0^*$ for the second group

Where c_0 and c_1 are the QAM symbols prior Alamouti encoding, and * denotes the complex conjugation operation. For retaining orthogonality between the two transmitter groups, the channel must remain constant during the transmission

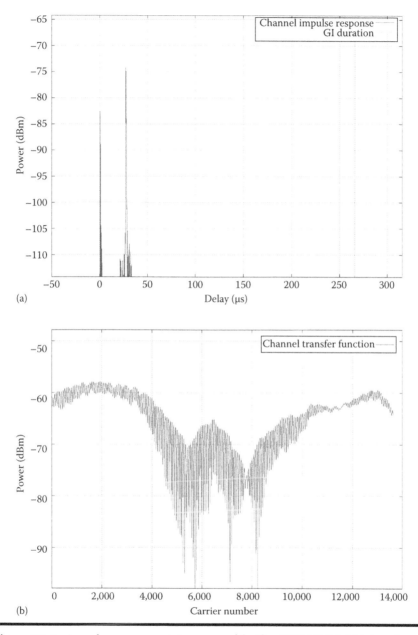

Figure 20.4 **Impulse response (a) measured in the DVB-T2 Trial Network near Winsen, Germany, and the corresponding channel transfer function (b). The network consists of two transmitters that are clearly visible within the impulse response.**

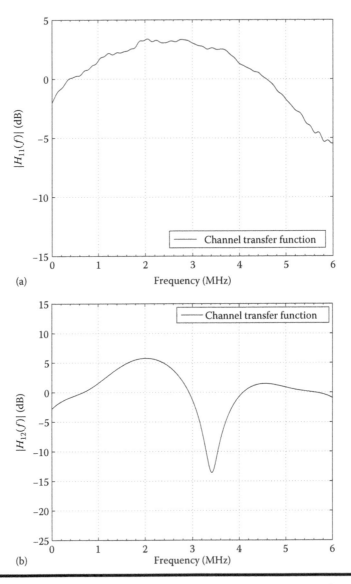

(a)

(b)

Figure 20.5 Separated channel transfer functions for the first (a) and second (b) transmitter signals of Figure 20.2.

of the jointly encoded QAM symbols, i.e., h_{11} from transmitter group 1 and h_{12} from transmitter group 2. Hence, DVB-NGH transmits both jointly encoded QAM symbols on the same OFDM symbol, and if possible, on adjacent OFDM subcarriers, since adjacent OFDM subcarriers have approximately identical channel characteristics.

We can use the theoretical channel capacity for estimating the potential gross gain of using Alamouti encoding. According to Reference 7, the achievable channel capacity for a classical SFN configuration in bps/Hz is

$$C_{SFN} = \log_2 \left(1 + \frac{1}{\sigma^2 N_T} |h_{11} + h_{12}|^2 \right),$$ (20.1)

where
σ^2 is the noise variance of the channel
the term N_T is the total number of transmit antennas for the normalization of the overall transmitted power to one

It gets clear that the complex channel coefficients from both transmitter groups add in amplitude and phase. The capacity equation for Alamouti encoding is the following:

$$C_{Alamouti} = \log_2 \left(1 + \frac{1}{\sigma^2 N_T} \left(|h_{11}|^2 + |h_{12}|^2 \right) \right).$$ (20.2)

In contrast to the previous SFN configuration, the absolute squared value of both channel coefficients is added. Hence, both transmitter groups always add constructively, and it is easy to show that $C_{Alamouti} > C_{SFN}$. Therefore, Alamouti encoding does not always provide a gain, but ensures that both signal components can never add destructively, i.e., they can never cancel each other completely.

The application of SFN with more than two transmitters may be desirable in some network configurations. However, it has been shown by Tarokh et al. [8] that such an orthogonal encoding does not exist for complex QAM constellations with more than two transmitter groups. Thus, orthogonality cannot be achieved between all transmitters if more than two transmitters are used in an SFN.

Another essential matter for the application of Alamouti is the channel estimation, since it requires the knowledge of the channel to both transmitter groups. It needs to be ensured by the transmission system that the receiver is able to estimate all channel characteristics. Therefore, the application of Alamouti requires the same pilot schemes as MIMO transmission that have been described in Chapter 19 of this book, causing additional pilot overhead compared to SISO transmission.

20.3.2 Enhanced Single-Frequency Network

The application of Alamouti has the drawback that it requires the estimation of the channel to both transmitter groups. In contrast, the eSFN approach does not try to reduce the frequency-selective fading, but tries to de-correlate the signals from the different transmitters using a specific linear phase-distortion algorithm.

This de-correlation reduces the coherence bandwidth and, thus, reduces the presence of frequency-selective fades that span the complete signal bandwidth. However, the phase distortions are created by means of a specific algorithm, which makes them almost invisible for the channel estimation process within the receiver. Consequently, the scheme does not require any additional pilots compared to pure SISO transmission and can also be applied to existing OFDM-based standards, e.g., DVB-T2. As the scheme shows certain similarities to SFN operation, it will be referred as eSFN.

20.3.2.1 Requirements for an "Invisible" Predistortion

The idea of the eSFN scheme is the avoidance of any regular fades for multiple transmitter antennas. As the signals of multiple transmit antennas have to be added in amplitude and phase at the receiver position, the signals may cancel each other in specific parts of the network. In case of high correlation between the signals, which is especially the case for the line-of-sight components, the signals may cancel each other over the complete bandwidth if both transmitters have similar field strength at the position of the receiver. Though, even if the signals do not cancel each other completely, significantly higher field strength for error-free reception may be required. Furthermore, the spatial location of the fades change over time if the transmitters have slight frequency offsets, which is practically the case in all SFNs.

In order to avoid this negative effects caused by the line-of-sight component, we can use a linear predistortion on the transmitter side. This predistorting has to be unique for each transmitter and has to change over the different OFDM subcarriers. Hence, we can express the transmitted signal as follows:

$$T_x(k) = S(k) \cdot P_x(k), \tag{20.3}$$

where

k is the OFDM subcarrier number

$S(k)$ is the non-distorted complex frequency-domain representation of the OFDM symbol at the transmitter (which we would transmit without predistortion)

$P_x(k)$ is the complex predistortion function of transmitter x on OFDM subcarrier k

$T_x(k)$ is the transmitted OFDM symbol for transmitter x in the frequency-domain representation

On the receiver side, we can describe the received signal as the sum of the transmitted signals multiplied by their corresponding channel transfer function:

$$R(k) = \sum_x T_x(k) \cdot H_x(k) = S(k) \cdot \sum_x P_x(k) \cdot H_x(k), \tag{20.4}$$

where $H_x(k)$ is the channel transfer function from transmitter x to the receiver, neglecting the additive noise to ease notation. Even if the channel transfer functions from the different transmitters to the receiver are highly correlated (e.g., caused by the line-of-sight terms), the predistortion term $P_x(k)$ ensures a de-correlation and, hence, avoids broad fades. However, the term $P_x(k)$ cannot be chosen freely. First, the term should not reduce the performance if only one transmitter is received, which means that the absolute value of the predistortion function should be one over all subcarriers k. Second, $P_x(k)$ should be almost invisible for the channel estimation within the receiver. We can see this effect if we describe the received signal in its time-domain representation (only a single TX is assumed for simplification):

$$R(k) = H_1(k) \cdot T_1(k) = H_1(k) \cdot P_1(k) \cdot S(k)$$
$$\Rightarrow r(t) = h_1(t) * p_1(t) * s(t) = h_{eq1}(t) * s(t)' \tag{20.5}$$

where
$h_{eq1}(t)$ is the equivalent impulse response seen by the receiver
$*$ denotes the convolution

Practically, this means that the length of our predistortion function $p_x(t)$ in its time-domain representation should be as short as possible, as the channel estimation unit of the receiver sees an equivalent channel impulse response, which is extended by the length of the predistortion function. A longer predistortion function would not decrease the usable guard interval length (as the predistortion does not have any effect on the orthogonality of the carriers) but would require a higher pilot density to sample the resulting channel.

Consequently, we have to find a function $P_x(k)$ that offers high de-correlation, which has an amplitude of one on all subcarriers k. Furthermore, the lengths of the corresponding time-domain representation should be as short as possible. However, except some trivial solutions, such a function—especially fulfilling the requirements 2 and 3 at the same time—is difficult to find.

20.3.2.2 Cyclic Delay Diversity

One solution that nicely fulfils the mentioned requirements is the application of cyclic delay diversity [4]. This technique delays the signal of a given transmitter by a delay coefficient Δ. Hence, we can rewrite Equation 20.5 as follows:

$$r(t) = h_1(t) * p_1(t) * s(t) = h_1(t) * \delta(t - \Delta_1) * s(t)$$
$$\Rightarrow R(k) = H_1(k) \cdot P_1(k) \cdot S(k) = H_1(k) \cdot e^{-j2\pi\Delta_1 k} \cdot S(k)' \tag{20.6}$$

where δ is defined as Dirac impulse. As only the delay term changes the phase of the signal, it is obvious that we do not change the amplitude of the transmitted signal.

On the other hand, the extension of the equivalent channel impulse response is limited to the delay term Δ. Using this approach, we can design our network such that the terms $H_x(k) \cdot P_x(k)$ in Equation 20.4 are as uncorrelated as possible at positions within the network where two or more transmitter may be received with similar field strength. A drawback of this approach is the fact that this technique is only able to move the difficult reception areas over the landscape, but it is not able to prevent them completely. If the delay of the channel plus the artificial delay due to the cyclic delay diversity match for two transmitters, still the effect of broad fades occurs.

20.3.2.3 Predistortion Using Raised Cosine Functions

This section describes an approach that does not suffer the drawbacks of cyclic delay diversity. Furthermore, it allows for a unique identification of the transmitters within the network, which can be used, e.g., for monitoring applications. This approach divides the N OFDM subcarriers into L segments of equal size. Each of these L segments is then modulated by means of a different phase (same phase in each segment). Additionally, the phases for the segments also differ between different transmitters. Consequently, the different phases decorrelate the signals and, second, the specific modulation of the phases allow for unique transmitter identification. In an example for the 8K FFT mode, we can divide the $N = 8192$ available OFDM subcarriers into $L = 16$ segments, as shown in Figure 20.6.

Naturally, a hard switch between the phases of two adjacent segments is not possible, as this would significantly extend the length of the equivalent impulse response. Therefore, we have to smooth the transition between two neighboring

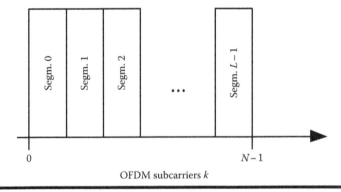

Figure 20.6 **Segmentation of an OFDM symbol in segments: The OFDM symbol in the frequency domain is divided into L segments; the phases of the subcarriers within each segment are multiplied by a common phase term. In case of $N = 8192$ and $L = 16$, each segment comprises 512 subcarriers.**

segments. Ideal candidates for this purpose are raised cosine functions. As shown later, this implicates that we have to accept slight variations of the amplitude of the final predistortion function $P_x(k)$, because we cannot change the phase without any amplitude variations. However, as the variations are of low amplitude, this has negligible effect on the performance.

The discrete frequency-domain description of the raised cosine functions is

$$H_{RC}(k) = \begin{cases} 1 & \text{if} & |k| \leq \dfrac{1-\alpha}{2T} \\ \cos^2\left[\dfrac{\pi T}{2\alpha}\left(|k| - \dfrac{1-\alpha}{2T}\right)\right] & \text{if} & \dfrac{1-\alpha}{2T} < |k| \leq \dfrac{1+\alpha}{2T}, \\ 0 & \text{else} \end{cases} \quad (20.7)$$

where T is a timing constant that has to be set as $T = L/N$. Furthermore, the roll-off factor α can be used for optimizing the final predistortion function. The values for α must be in the range of [0, 1], where $\alpha = 0$ results in a rectangular filter. A good compromise seems to be $\alpha = 0.5$.

We then obtain the final predistortion function $P_x(k)$ by a summation of the L raised cosine functions that are frequency shifted against each other by

$$P_x(k) = e^{-j2\pi\Delta_x k} \cdot \sum_{l=1}^{L-1} e^{j2\pi\Psi_x(l)} \cdot H_{RC}\left(k - l\frac{N}{L}\right), \quad (20.8)$$

where $\Psi_x(l)$ are the phase terms for transmitter x and segment l. Additionally, we still may use the possibility to cyclically delay each transmitter signal using the delay coefficient Δ, which is applied on the complete predistortion function.

Actually, only $L - 1$ raised cosine functions are used, because the edge carriers of the OFDM signal are not modulated. Figure 20.7 illustrates Equation 20.8 for $N = 8192$, $L = 16$, and $\alpha = 0.5$. The dashed line shows the sum of all raised cosine functions, which is equal to 1 if the phase between adjacent segments is kept constant.

20.3.2.4 Modulation of the Phase Term ψ

To ensure the de-correlation between different transmitters, the phase $\Psi_x(l)$ should vary between the different transmitters. Supplementary, we can modulate the phase to allow for unique transmitter identification. Therefore, we can use a differential encoding scheme with the transmitter ID sequence $C_x = \{c_{0,x}, \ldots, c_{L-1,x}\}$, which consists of $L - 1$ elements having the possible values $\{-1, 0, 1\}$. We cannot utilize the full amount of L elements, because we do not modulate the raised cosine

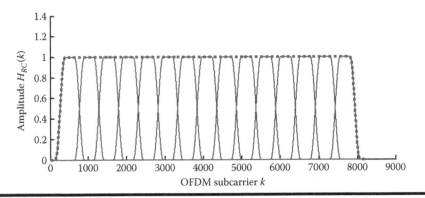

Figure 20.7 **Division of 8K OFDM signal into 16 raised cosine functions. The sum (dashed line) of the raised cosine functions is constant if the phase is kept constant between adjacent segments; only 15 raised cosine functions are shown as the OFDM subcarriers at the edges are not modulated.**

functions at the edges of our OFDM signal (see Figure 20.2). Finally, the phase term is defined as

$$\Psi_x(l) = \begin{cases} c_{l,x} = c_{0,x} & \text{if } l = 0 \\ \Psi_x(l-1) + c_{l,x}/8 & \text{else} \end{cases}. \tag{20.9}$$

The first coefficient absolutely defines the phase. The other coefficients realize a differential encoding of the transmitter sequence. Using Equation 20.9, we can differentially demodulate 14 values, leading to $3^{14} = 4{,}782{,}969$ possible transmitter IDs.

The phase difference between two adjacent components is limited to 1/8, which limits the phase differences in Equation 20.8 to $\pi/4$ between adjacent segments. This limitation is required because the different raised cosine functions do no longer add themselves to one if the phase changes between two adjacent segments. Figure 20.8 shows the resulting absolute value of a predistorting function $P_x(k)$ with a transmitter ID sequence that contains only one value that is nonzero. The resulting ripple is visible and has a depth of less than 0.7 dB, which is almost independent from the value $\alpha = 0.5$. Thus, this will have a negligible effect on the performance of the system, since the mean power of $P_x(k)$ can be normalized to one afterward.

20.3.2.5 Application to SIMO and MIMO

One main application scenario for eSFN is a mixed SISO/MIMO scenario. In case of cross-polar MIMO transmission, we have two transmit antennas on each transmitter site. Though, in case of SISO, only a single antenna is typically required.

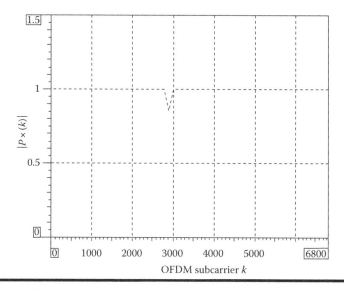

Figure 20.8 **Ripple in the absolute value of the predistortion function $P_x(k)$ caused by phase change between two adjacent segments, $C_x = \{0, 0, 0, 0, 0, 0, 1, 0, 0, 0, 0, 0, 0, 0, 0\}$ with $\alpha = 0.5$.**

Switching off the non-required antenna during SISO transmission would typically halve the overall transmitted power, besides causing possible issues for the amplifiers that have to switch from high power to zero power in very short time. On the other hand, simulation results show [9] that the transmission of identical data on both antennas during the SISO period significantly reduces the performance, as this may lead to a strongly selective channel with a very high coherence bandwidth.

Therefore, an enhanced SFN-like configuration for SISO transmission can be used by applying the described predistortion function, which avoids a highly selective channel and a very high coherence bandwidth (as the line-of-sight terms arrive almost at the same time).

Figure 20.9 shows the transmit antenna configuration that is used for further explanation. If an SFN-like configuration is used, we can express the transmitted signals as follows:

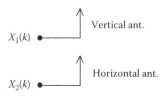

Figure 20.9 **Example for a transmitter antenna configuration.**

$$X_1(k) = X_2(k) = S(k), \qquad\qquad (20.10)$$

where $S(k)$ is the frequency-domain representation of our transmitted OFDM symbol. For avoiding the negative effects of this SFN-like configuration, we can now add our predistortion functions:

$$\begin{aligned} X_1(k) &= P_1(k) \cdot S(k) \\ X_2(k) &= P_2(k) \cdot S(k) \end{aligned} \qquad\qquad (20.11)$$

We still transmit identical data $S(k)$ on both antennas. However, each antenna has its own predistortion function. Thus, we do not remove the frequency selectivity, but we avoid a high coherence bandwidth. Consequently, we avoid that both line-of-sight components cancel each other completely, as they cannot interfere destructively over the complete OFDM symbol bandwidth. Additionally, we are able to identify and monitor both components separately to check the transmitted signals.

The application of this scheme to MIMO is done similarly to the SISO eSFN case:

$$\begin{aligned} X_1(k) &= P_1(k) \cdot S_1(k) \\ X_2(k) &= P_2(k) \cdot S_2(k) \end{aligned} \qquad\qquad (20.12)$$

The only difference to the SFN case is the presence of two different input signals. These input signals could be any kind of MIMO signal, e.g., SM, eSM, or MIMO rotated constellations.

20.4 Performance Evaluation

In the following, the performance of the Alamouti scheme and eSFN is discussed. In Section 20.4.1, their performance is analyzed based on system-level simulations, whereas Section 20.4.2 describes results obtained by field measurements during the DVB-T2 field trial in the northern part of Germany.

An important parameter for the performance evaluation is the MISO group power ratio (MGPR) that has been defined in the DVB-T2 measurement guidelines [10]. It describes the relative received power between the transmitters of the first and the second group and results from $MGPR = 10 \cdot \log_{10}[P_{Rx,GR1}/P_{Rx,GR2}]$. The MGPR is easily calculated in a receiver based on the channel estimates of both MISO groups and gives a good indication of the potential MISO gain. Low absolute values of the MGPR indicate that both transmitter groups are received at almost identical levels, leading to "high" potential gains. A high absolute value indicates that a single group dominates the reception, leading to almost no performance difference between SISO and MISO operation. In addition, the transmitter group with higher reception power can simply be obtained based on the sign of the MGPR.

20.4.1 System-Level Simulations

This section analyzes the performance of the presented MISO schemes by means of system-level simulations. For the simulations, the 8K FFT mode with guard interval 1/8 has been used. For all simulations, two transmit antennas have been assumed to ensure a fair comparison between the different schemes. On the receiver side, both one and two receive antennas have been simulated, to evaluate the impact of receive diversity on the system performance. Besides two curves for the presented Alamouti and the eSFN scheme, two additional curves denoted by "SISO" ("diversity" for the two receive antenna case) and "SFN" ("SFN+diversity" for the two receive antenna case) allow for a comparison with the known transmission modes. The "SISO" curves correspond to the case of a single transmit antenna, whereas the "SFN" case corresponds to two cross-polarized transmit antennas, transmitting the same signal synchronized in time and frequency. The utilized channel model for the system simulations is the NGH portable outdoor channel, which has been described in Chapter 19 in detail. For each OFDM symbol, a new channel snapshot has been generated. In this case, the time-varying fading effects that depend on the Doppler frequency have no impact. Thus, the Doppler frequency has been chosen to 0 Hz, to avoid inter-carrier interference between the OFDM subcarriers. Due to the generation of a new channel snapshot after each OFDM symbol, the simulation assumptions correspond to the case of ideal interleaving in time and frequency, i.e., ideal frequency interleaving and a time interleaver duration that is sufficiently larger than the coherence time.

Figure 20.10 shows the mutual information depending on the SNR in dB that can be achieved for the different schemes. With one receive antenna (Figure 20.10a), the difference between the performance of the schemes is very close in the lower SNR region. For higher SNR values, the Alamouti scheme shows the best perfor-mance. With two receive antennas (Figure 20.10b), the performance gain of the Alamouti scheme compared to SISO, SFN, and eSFN is reduced. However, all three schemes outperform the simple diversity case (SIMO) in this scenario.

Figure 20.11 shows the required SNR of the four schemes that is necessary to achieve a particular mutual information at the outage probability of 5%. That is, LDPC decoding would be successful in 95% of the channel snapshots. Again, the Alamouti scheme shows the best performance, followed by eSFN, SISO, and SFN. SISO outperforms the SFN scheme due to the high correlation between the two cross-polar paths, leading to high frequency selectivity. With two receive antennas, the simple diversity case performs worse, and the gap between Alamouti and eSFN gets smaller compared to the case with a single receive antenna.

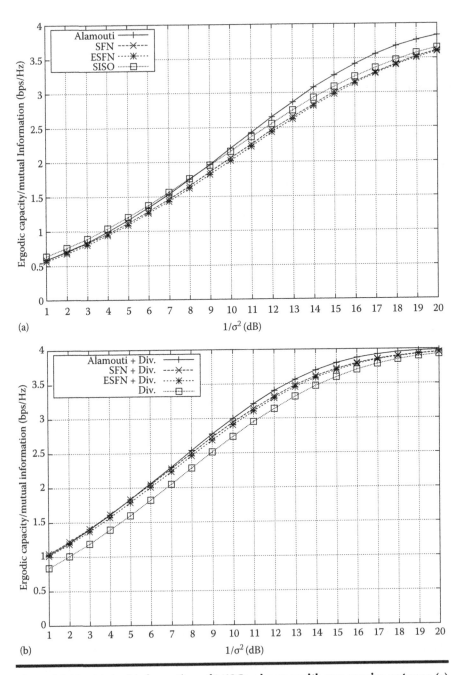

Figure 20.10   Mutual information of MISO schemes with one receive antenna (a) and two receive antennas (b) in the NGH portable outdoor channel, not taking pilot overhead into account.

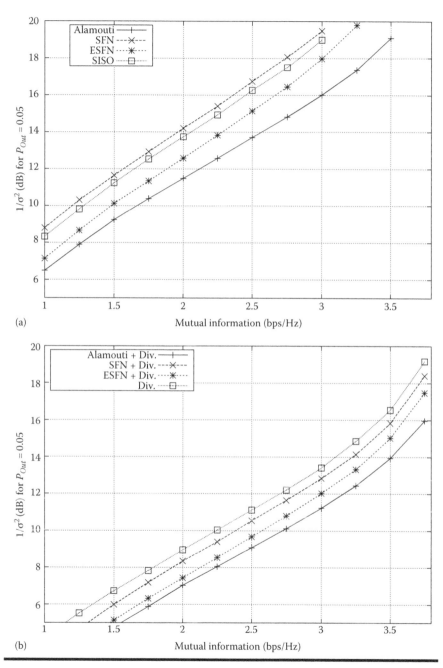

Figure 20.11 Outage probability of MISO schemes with one receive antenna (a) and two receive antennas (b) in the NGH portable outdoor channel, not taking pilot overhead into account.

So far, the pilot overhead for channel estimation has not been taken into account in the simulations. But since the Alamouti scheme requires two channel estimates for decoding, the spectral efficiency is reduced due to additional pilot overhead compared to the other schemes. The pilot overhead can be expressed as a penalty in terms of spectral efficiency and is considered by a shift of the mutual information curves by the relative pilot overhead. Figure 20.12 depicts the outage probability tanking the pilot overhead into account. For the modes except Alamouti, pilot pattern PP3 has been assumed, causing 4.18% pilot overhead. The Alamouti scheme requires twice the amount of pilots, leading to 8.3% pilot overhead. The additional loss caused by the pilot boosting is also reflected in the curves.

For the scenario with a single receive antenna, it can be observed that the gap between the Alamouti and the other schemes is reduced. However, the Alamouti scheme still requires slightly less SNR to achieve the same mutual information as the eSFN scheme. With two receive antennas, eSFN and Alamouti perform basically the same.

However, it is important to highlight that real channel estimation has not been taken into account in any of the system simulations in this section. If channel estimation is considered, the Alamouti scheme suffers stronger from the noise stemming from the channel estimate than the three other schemes, namely, SISO, SFN, and eSFN. This is caused by a high correlation of the channel estimation noise of two required channel estimates, since they are estimated by means of the same pilots. The impact of the channel estimation on MIXO decoding is analyzed in detail in Reference 9.

20.4.2 Alamouti Field Measurements

The project "Modellversuch DVB-T2 Norddeutschland" [11] was a German field trial for the evaluation of DVB-T2. Unlike other field trials, this field trial was focusing on portable and mobile reception of DVB-T2 signals, with special focus on the potential gain provided by techniques as rotated constellations and MISO based on Alamouti coding. As both DVB-T2 and DVB-NGH rely on the very same technologies, the obtained results also apply for DVB-NGH.

Figure 20.13 shows the covered area of the field trial network south of Hamburg, Germany. The two transmitters "Rosengarten" and "Lüneburg" are separated by approx. 45 km and were operated in SFN operation with and without additional MISO encoding based on the Alamouti scheme. Furthermore, the transmit frequency was 690 MHz with an ERP of approx. 10 kW at each transmitter site. The OFDM parameters for the SISO/MISO comparisons were 16K FFT with guard interval 19/128 and Pilot Pattern 1 (MISO) and Pilot Pattern 2 (SISO).

The performance was measured in mobile and portable reception scenarios using a single RX antenna. A special focus was on the area around "Winsen," where both transmitters are received at almost identical signal level and with minimal

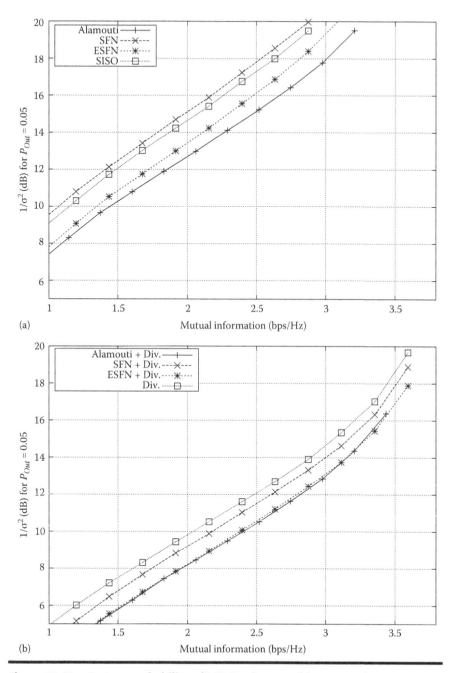

Figure 20.12 **Outage probability of MISO schemes with one receive antenna (a) and two receive antennas (b) in the NGH portable outdoor channel, taking pilot overhead into account.**

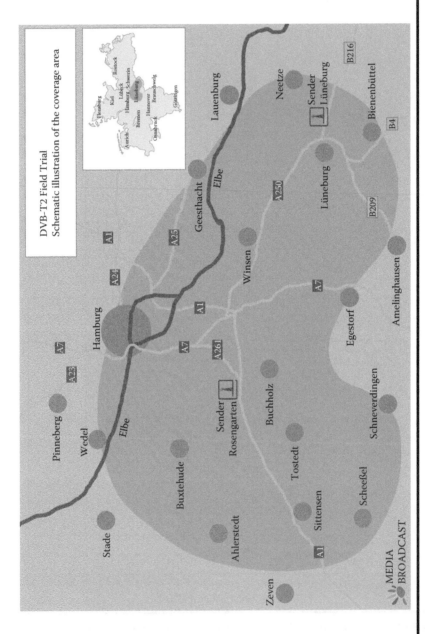

Figure 20.13 Approximate coverage area of the field trial (with roof-top antenna). (From http://www.dvb-t2-nord.de/netz/data/versuchsnetz.pdf, July 2012.)

delay. Thus, this is the region where the maximum gain by means of Alamouti can be expected.

In a first measurement step, several hundred impulse responses were measured in the area between both transmitters. For this purpose, both transmitters were operating in the SISO mode. The algorithm presented in Reference 5 was then used for separating the impulse responses of the two transmitter signals. Examples for the obtained impulse responses are given in Figure 20.14. The impulse responses were then used for simulating the theoretical MISO gain using system-level simulations. The gain was defined as the delta between the minimum SNR required for error-free reception between the SISO and MISO modes, where a positive gain indicates a benefit for MISO. Examples of such simulation results are also given in Figure 20.15.

The MISO gain was then analyzed on statistical bases. The results indicate that the gain mainly depends on the MGPR. Low absolute values indicate that both transmitter groups are received at almost identical levels, leading to "high" potential gains. A high absolute value, indicating that a single group highly dominates the reception, leads to almost no difference between SISO and MISO operations. Figure 20.15 shows the dependency of the achievable MISO gain in dependency of the MGPR for different code rates using 16-QAM. Detailed analyzes in Reference 3 show that the gain in dependency of the absolute value of the MPGR can be well approximated by means of an exponential function with a negative exponent. Furthermore, the maximum achievable gain is limited to the delta between the minimum SNR required for error-free reception in the Rayleigh compared to the minimum required SNR for error-free reception in the Ricean channel. Taking these values from Reference 12, it becomes clear that the achievable gain is limited to values in the order of 1 dB for the lowest code rate 1/2 and increasing to approx. 2 dB for the highest available DVB-NGH code rate 11/15. There theoretical gains are already halved if one transmitter group dominates the reception level in the order of 8 dB ($|\text{MGPR}| = 8$ dB).

However, the presented theoretical gains do not include the additional loss caused by the higher amount of pilots required for the extended channel estimation. Especially, in case of large SFNs utilizing the guard interval fractions 1/8 or 19/128, a significant reduction of the theoretical MISO gain can be expected. While SISO requires PP3 with approx. 4% pilot overhead, this value is extended to approx. 8% for PP1. This also leads to the fact that a net loss is present within the network in the areas with high absolute MGPR.

An additional item that reduces the theoretical gain is the channel estimation. Theoretical analysis in Reference 9 shows that noise caused by the realistic channel estimation affects the MISO mode twice as strong compared to the SISO mode. A channel estimation loss of 0.5 dB for SISO leads to a channel estimation loss of approx. 1 dB for MISO, without taking the additional pilot overhead into account.

These theoretical analyses using measured impulse responses are also confirmed by actual MISO measurements within the network using a commercially

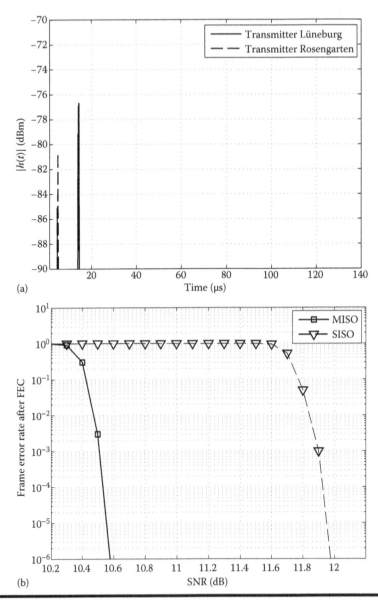

Figure 20.14 Examples for impulse responses with almost identical reception levels from both transmitters ((a), |MGPR| =5 dB) and with a single transmitter highly dominating the reception level ((c), |MGPR| =28 dB). The simulation curves show the required SNR for error-free reception at 16-QAM and code rate 3/4 for (b) and (d) MISO and SISO. A significant gross MISO gain is present only if both transmitters are received at almost identical reception levels. Please note that these results do not consider the additional MISO overhead required for channel sounding.

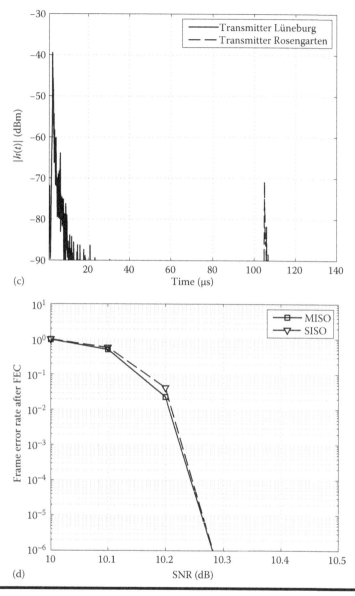

(c)

(d)

Figure 20.14 (continued) **Examples for impulse responses with almost identical reception levels from both transmitters ((a), |MGPR| = 5 dB) and with a single transmitter highly dominating the reception level ((b), |MGPR| = 28 dB). The simulation curves show the required SNR for error-free reception at 16-QAM and code rate 3/4 for (b) and (d) MISO and SISO. A significant gross MISO gain is present only if both transmitters are received at almost identical reception levels. Please note that these results do not consider the additional MISO overhead required for channel sounding.**

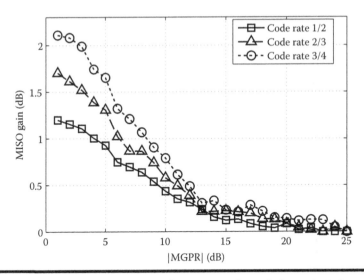

Figure 20.15 Relationship between the MISO gain and the MGPR for modulation mode 16-QAM. Simulated values based on hundreds of measured impulse responses.

available chip. The results presented in Reference 13 indicate a slight gain only in the very center SFN, where the MGPR approaches almost zero. In the other areas, a loss due to MISO has been measured. As these values do not include the additional overhead required for channel estimation, a net loss due to the application of MISO was present in all parts of the network. As these measurements were based on a single chip, one cannot finally conclude that MISO brings a net loss. However, also the theoretical analyses indicate that a gain due to MISO does not exist.

20.4.3 eSFN Field Measurements

The channel impulse responses used for the evaluation of the MISO gain were also used for theoretical analyses of the potential eSFN gain. Analyses of the correlation between the two transmitters indicate that both received signals are almost fully uncorrelated [3]. This result was not surprising for high delays between the signals from both transmitters, as discussed in Section 20.2 of this chapter. However, even for very short delays of far less than microseconds, both signals were fully de-correlated. Consequently, the application of eSFN was not able to further de-correlate the signals from the two transmitters. As a result, both simulation curves for normal SFN and eSFN operation ideally overlap. However, in contrast to the application of Alamouti, no additional loss is present, as the application of eSFN does not require any additional pilots or other overhead. It is important to highlight that these results are based on the measured impulse responses in the core SFN only, and further studies must be carried out in real SFN operation to draw final conclusions. This would require a modulator supporting eSFN operation, which is not available on the market yet.

20.5 Conclusions

In this chapter, the MISO technologies adopted for DVB-NGH have been discussed. This was first the Alamouti scheme, which is already well known from DVB-T2 and the eSFN scheme. Simulations have shown that both schemes can offer performance gain in particular regions of the network. For the Alamouti scheme, it has been observed that the pilot overhead and the correlated noise of the channel estimation play an important role for the overall performance, reducing its potential gain. Measurements have shown that the received power ratio of the different MISO transmitter groups, the so-called MGPR, gives a good indication for the gross MISO gain. Therefore, the pilot overhead of the Alamouti scheme is especially a drawback in regions of an SFN with a high MGPR, where a low Alamouti gain is expected.

System simulations have shown that eSFN can offer performance gains by decorrelating the channel components and, hence, reducing the frequency selectivity. The advantage of eSFN is its invisibility to the channel estimation of the receiver, therefore not requiring additional pilots. However, the potential gain of eSFN strongly depends on the channel characteristics, especially the extent of the channel correlation. First measurements in distributed SFNs indicate a weak correlation between the channel components, reducing the potential eSFN gain. However, the gain of both Alamouti and eSFN schemes still needs to be assessed by additional measurements.

References

1. D. Plets et al., On the methodology for calculating SFN gain in digital broadcast systems, *IEEE Transactions on Broadcasting*, 56(3), 331–339, September 2010.
2. J. Proakis, *Digital Communications*, McGraw-Hill, New York, 2000.
3. J. Qi, J. Robert, K. L. Chee, M. Slimani, and J. Zöllner, DVB-T2 MISO field measurements and a calibrated coverage gain predictor, *Proc. IEEE BMSB*, Seoul, Korea, 2012.
4. G. Bauch and J. S. Malik, Cyclic delay diversity with bit-interleaved coded modulation in orthogonal frequency division multiple access, *IEEE Transactions on Wireless Communications*, 5(8), 2092–2100, August 2006.
5. H. Meuel and J. Robert, MIMO channel analysis using DVB-T signals, *Proc. IEEE ISCE*, Braunschweig, Germany, 2010.
6. S. M. Alamouti, A simple transmit diversity technique for wireless communications, *IEEE JSAC*, 16(8), 1451–1458, October 1998.
7. C. Oesteges and B. Clerckx, *MIMO Wireless Communications: From Real-World Propagation to Space-Time Code Design*, Academic Press, Oxford, U.K., 2007.

8. V. Tarokh, H. Jafarkhani, and A. R. Calderbank, Space-time codes from orthogonal designs, *IEEE Transactions on Information Theory*, 45(5), 1456–1467, July 1999.

9. J. Robert, Terrestrial TV broadcast using multi-antenna systems, PhD dissertation, Technische Universität Braunschweig, Braunschweig, Germany, 2012.

10. DVB Document A14–2, *Digital Video Broadcasting (DVB); Measurement Guidelines for DVB Systems; Amendment for DVB-T2 System*, 2012.

11. Homepage of the field trial. http://www.dvb-t2-nord.de, July 2012.

12. ETSI TR 102 831 v.1.2.1, Digital video broadcasting (DVB); Implementation guidelines for a second generation digital terrestrial television broadcasting system (DVB-T2), 2012.

13. Terrestrik der Zukunft: Zukunft der Terrestrik, Projektbericht DVB-T2 Norddeutschland, final project report, Shaker Verlag, Aachen, Germany, 2012.

Chapter 21

Enhanced MIMO Spatial Multiplexing with Phase Hopping for DVB-NGH

David Vargas, Sangchul Moon, Woo-Suk Ko,
and David Gómez-Barquero

Contents

609

21.1 Introduction

Digital Video Broadcasting—Next Generation Handheld (DVB-NGH) is the first digital broadcasting standard to include *multiple-input multiple-output* (MIMO) antenna techniques to increase the transmission robustness and the system capacity without any additional bandwidth or transmit power. First multi antenna techniques* for broadcasting systems were adopted for the digital terrestrial TV standard DVB-T2 (Terrestrial Second Generation). However, in this case, a transmitter site distributed configuration was defined to exploit only diversity gain (see Chapter 20). Hence, DVB-NGH is the first broadcast system to employ pure MIMO as key technology exploiting all the benefits of the MIMO channel.

MIMO is the only technology to overcome the information-theoretic limits of single-input single-output (SISO) systems without any additional bandwidth or increased transmit power. MIMO provides three important gains, i.e., *array gain*, *diversity gain*, and *multiplexing gain* [1]. Array gain increases the received carrier-to-noise ratio (CNR) with coherent combination at the receive side (signal co-phasing and weighting for constructive addition). To do so, channel state information (CSI) is required, and it is commonly obtained by tracking the channel variations with the transmission of pilot signals. Diversity gain improves the reliability of the transmission by sending the same information

---

* The utilization of multiple antennas only at the receive side is known as single-input multiple-output (SIMO). Multiple antennas at the transmitter side is referred as multiple-input single-output (MISO).

through independently faded spatial branches to reduce the probability that all channels are in a deep fade. In addition to array gain and diversity gains, the MIMO channel can increase the system capacity by transmitting independent data streams across the transmit antennas.

MIMO spatial multiplexing (SM) is specified in DVB-NGH as MIMO rate-2 codes where the term "rate-2" stands for the transmission of two independent streams. MIMO rate-2 codes in DVB-NGH use cross-polar antenna arrangement (antennas with orthogonal polarization) with two transmit and two receive aerials. Compared with the co-polar counterpart (antennas with the same polarization), cross-polar antennas are feasible for small handset devices and provide higher multiplexing gains in line-of-sight (LOS) conditions due to orthogonal nature of the cross-polar channel (see Chapter 19). Moreover, in the ultrahigh frequency range, the antenna separation required in the co-polar case to provide sufficiently uncorrelated fading signal may exceed typical handheld device sizes.

MIMO rate-2 in DVB-NGH requires the implementation at the transmitter of two individually fed cross-polar antennas and the corresponding pair of cross-polar aerials at the receiver to decode the MIMO signal. Therefore, to transmit MIMO rate-2 signals, the current network infrastructure with single-antenna transmitters need to be upgraded. However, in DVB-NGH, transmission MIMO rate-2 codes are specified in two optional profiles to provide flexibility to network operators.

The two profiles in which MIMO rate-2 codes are specified are the MIMO profile and the hybrid MIMO profile. The former includes terrestrial single-frequency network (SFN) transmissions where all the transmitters in the network use the same signal. The latter adds an additional satellite component with two possible configurations in multifrequency network (MFN) and SFN modes (see Chapter 24). MIMO rate-2 codes are transmitted in MIXO frames with the appropriate pilot pattern structure to decode the incoming signal by each transmit antenna and time multiplexed with SIXO frames by means of Future Extension Frames (see Chapter 19). Here, in the case of time multiplexing of SIXO (i.e., one radiating element) and MIXO (i.e., two radiating elements) transmissions, the total transmit power needs to be distributed along the transmit antennas during SIXO and MIXO frames (see Chapter 19). In this case, one possible deployment scenario would be to transmit the same power at each antenna element at all times but lowering the power accommodated to the cross-polarized antenna. This deliberated transmitted power imbalance provides a reasonable coverage reduction for SIXO terminals while rate-2 codes maintain good performance due to their optimized performance to overcome this situation.

The main technology for MIMO rate-2 is enhanced spatial multiplexing with phase hopping (eSM-PH). It provides optimized performance in mobile/portable broadcasting scenarios with reasonable decoding complexity at the receiver side. eSM exploits a pre-coding matrix to increase the spatial diversity compared to simple

MIMO SM. In addition, PH avoids specific harmful channel conditions by a periodic phase rotation of the modulated symbols at one of the transmit antennas. Moreover, eSM-PH handles transmitted power imbalance situation with a set of optimized transmission parameters to reduce the performance loss in such conditions.

The implementation of MIMO rate-2 codes requires independent RF chains per antenna at the transmitter and the receiver. This embraces two independent and identical time (de-) interleavers, OFDM (de-) modulators at the (receiver) transmitter, as well as individually fed two cross-polar antennas at both sides of the communication link. The memory constraints imposed for MIMO are the same as for the SISO profile. Therefore, the TI memory assigned to each MIMO branch is halved compared to the base profile as to maintain this constraint. The frequency interleaver is also different for MIMO rate-2 where pair-wise frequency interleaver is performed.

The performance gains already obtained by rate-2 codes can be further increased by iterative MIMO decoding. Here, the MIMO demodulator and channel decoder exchange extrinsic information in an iterative manner. However, the complexity of optimal MIMO demapping together with iterative decoding is too high for many practical applications. To reduce the computational complexity, numerous suboptimal MIMO receivers have been proposed, e.g., linear zero-forcing (ZF) and minimum mean square error (MMSE) receivers. We note that iterative decoding is a receiving technique, and its implementation is a chip manufacturers' decision.

The MIMO demodulator and LDPC decoder are linked by the bit de-interleaving, which is specific for MIMO rate-2. It has been optimized to improve the performance with iterative decoding and to provide low complexity, low latency, and fully parallel design tailored for hardware implementation.

This chapter describes the structure of eSM-PH as well as the required elements at the transmitter and receiver terminals to decode a MIMO eSM-PH transmission. The rest of the chapter is structured as follows. Section 21.2 describes eSM-PH as solution for MIMO rate-2 for DVB-NGH. Then, Section 21.3 presents the transmit and receive MIMO rate-2 chains. Finally, illustrative performance simulation results are provided in Section 21.4, followed by the conclusions in Section 21.5.

21.2 MIMO Precoding: Enhanced Spatial Multiplexing with Phase Hopping

Most technical challenges for MIMO transmission in broadcasting comes from the rate-2 SM performance degradation in high correlation channel condition, which frequently happens in broadcasting due to LOS channel conditions. In the literature, there are a vast number of MIMO schemes to exploit all the potential of the MIMO channel. Among them, plain MIMO SM [2] and Golden Code (GC) [3] are two well-known schemes that provide low decoding complexity and

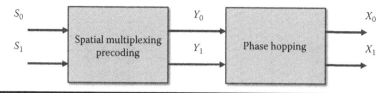

Figure 21.1   DVB-NGH MIMO precoding block diagram.

optimal performance in diversity-multiplexing trade-off sense, respectively. Plain MIMO SM is a rate-2 MIMO scheme that has most benefit from its simplicity. MIMO SM is regarded as lacking diversity because one symbol is transmitted only once from one of the transmit antennas. In addition, two symbols are transmitted at one frequency slot (OFDM carrier), thus it gets less diversity than GC using two slots.

Nevertheless due to its minimum decoding complexity among rate-2 MIMO schemes, MIMO SM is still attractive for practical wireless communication systems. The decoding complexity* of MIMO SM is proportional to $O(M^2)$ while that of GC to $O(M^4)$, where M is the cardinality of the signal constellation, e.g., $M = 4$ for QPSK. Among the full rate 2×2 square matrix codes, GC is known to achieve the diversity-multiplexing frontier with an encoding matrix providing nonvanishing determinant. This is a crucially advantageous property when we consider uncoded symbol-error-rate (SER) performance. But as far as bit-error-rate (BER) after error correction is concerned, the situation is different. The advantage of performance gain in uncoded SER domain becomes negligible after FEC due to powerful error correction capability. As such, the disadvantage of high decoding complexity becomes a more dominant factor as the cardinality of the QAM constellation increases.

The high decoding complexity of GC and the meaningless performance difference against MIMO SM [4,5] were the reasons to discard the adoption of GC in favor of low-complexity MIMO decoder more suited for mobile devices.

The next two sections describe the MIMO signaling scheme chosen for DVB-NGH and illustrated by the block diagram of Figure 21.1. The signal processing is divided into two separated blocks, SM precoding and PH.

21.2.1 Spatial Multiplexing Precoding: Enhanced Spatial Multiplexing

SM precoding processes the input symbols on each antenna stream "layer" to increase spatial diversity. This is known as eSM, and a block diagram is presented in Figure 21.2. The signal processing of the precoding block linearly combines the

---

* The term $O(\cdot)$ stands for the standard "big-O" notation and characterizes the rate of growth of a function that in our case refers to algorithm complexity.

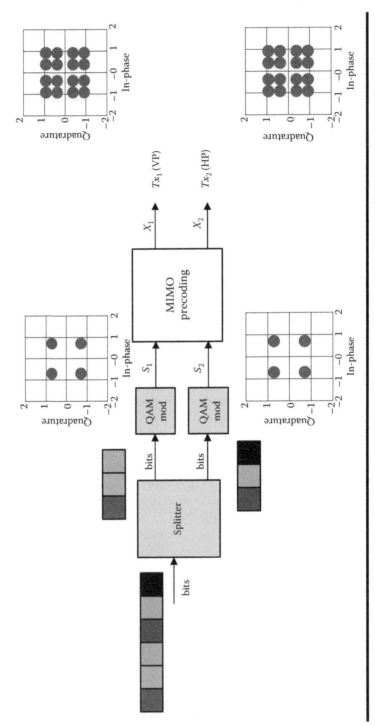

Figure 21.2 Example block diagram of pre-coded SM with gray labeled 4-QAM constellation on each antenna.

constellation symbols of each layer retaining the multiplexing gain of rate-2 schemes and at the same time improving the spatial diversity under spatially correlated channels.

eSM was designed under three criteria. The first goal is to make the final MIMO encoded symbols mapped onto independent constellation points according to the bits to be delivered. For instance, if a pair of QPSK symbols is encoded by a MIMO precoder, the output should have a total of 16 unique constellation points. The second goal is to reduce SER, which corresponds to maximization of the minimum Euclidean distance between constellation points. If we consider only this goal, the resultant precoded SM will produce uniform square QAM symbols capable of conveying total number of bits of two symbols. The third goal is to minimize the BER after symbol demapping. This can be achieved by maximizing the minimum Euclidean distance in connection with the corresponding Hamming distance. Gray bit labeling should also be an important factor to affect the resultant BER performance.

Considering all these three design metrics, the eSM precoding matrix was designed to have best compromise among them. High priority was put on the performance under high channel correlation while keeping the SM gain in rich scattering condition. In order to satisfy all these conditions, the precoding matrix should be a unitary matrix and as such can be expressed as a kind of rotation matrix with angle θ. Additionally, factors α and β are included for intentional deliberated power imbalance. The SM precoding processing is given in the following expression:

$$
\begin{bmatrix} X_{2i}(Tx1) \\ X_{2i+1}(Tx2) \end{bmatrix} = \sqrt{2} \begin{bmatrix} \sqrt{\beta} & 0 \\ 0 & \sqrt{1-\beta} \end{bmatrix} \cdot \begin{bmatrix} \cos\theta & \sin\theta \\ \sin\theta & -\cos\theta \end{bmatrix}
$$

$$
\cdot \begin{bmatrix} \sqrt{\alpha} & 0 \\ 0 & \sqrt{1-\alpha} \end{bmatrix} \begin{bmatrix} S_{2i}(Tx1) \\ S_{2i+1}(Tx2) \end{bmatrix}, i = 0,\ldots,\frac{N_{data}}{2} - 1, \quad (21.1)
$$

where

S_{2i} (*Tx*1) and S_{2i+1} (*Tx*2) are the input QAM constellation symbols

X_{2i} (*Tx*1) and X_{2i+1}(*Tx*2) are the processed QAM constellation symbols allocated to transmit antenna 1 and 2, respectively

i is the QAM symbol index

N_{data} is the number of QAM symbols corresponding to one LDPC codeword, for a given modulation order

The specified parameters for DVB-NGH are listed in Table 21.1, where N_{bpcu} stands for the total number of bits per cell unit. The term "cell" defines one subcarrier in the time-frequency OFDM grid. For instance, for MIMO rate-2 and with 8 bpcu

Table 21.1   eSM Parameter for DVB-NGH

Deliberate Power Imbalance between two Tx Antennas			0 dB			3 dB			6 dB		
n_{bpcu}	Modulation		β	θ	α	β	θ	α	β	θ	α
6	$S_q(Tx1)$	QPSK	0.50	45.0°	0.44	1/3	0.0°	0.50	0.20	0.0°	0.50
	$S_q(Tx2)$	16-QAM									
8	$S_q(Tx1)$	16-QAM	0.50	57.8°	0.50	1/3	25.0°	0.50	0.20	0.0°	0.50
	$S_q(Tx2)$	16-QAM									
10	$S_q(Tx1)$	16-QAM	0.50	22.0°	0.50	1/3	15.0°	0.50	0.20	0.0°	0.50
	$S_q(Tx2)$	64-QAM									

(bits per cell unit), a group of 4 bits is assigned to one cell on each transmit antenna. The power imbalance ratio was expressed as dB scale and also three modes of 0, 3, and 6 dB are possible.

21.2.2 Phase Hopping

The eSM performance can be further improved over various channel realizations in average sense if randomizing function is additionally used to avoid any possible loss caused by specific worst case channel condition. In DVB-NGH specification, a PH technique periodically changes the phase of symbols transmitted by one of two antennas within one FEC block by the multiplication with the term $e^{j\phi(i)}$, where

$$e^{j\phi(i)} = \cos\phi(i) + j\sin\phi(i), \quad \phi(i) = \frac{2\pi}{N}(i), i = 0,\dots\frac{N_{data}}{2} - 1, (N = 9). \quad (21.2)$$

Figure 21.3 illustrates a conceptual block diagram of SM with PH to the second transmit antenna. The phase of the resulting constellation is periodically rotated. The amount of phase change is uniformly distributed in the range 0–2π and the number of phase changes is nine. This number was confirmed through simulation to be sufficient for randomization purpose. Moreover, for every modulation, the number of cells corresponding to one LDPC codeword is a multiple of nine, providing an integer number of PH periods. Though using PH may slightly degrade the performance under some specific channel conditions, it improves overall performance by improving worst channel condition.

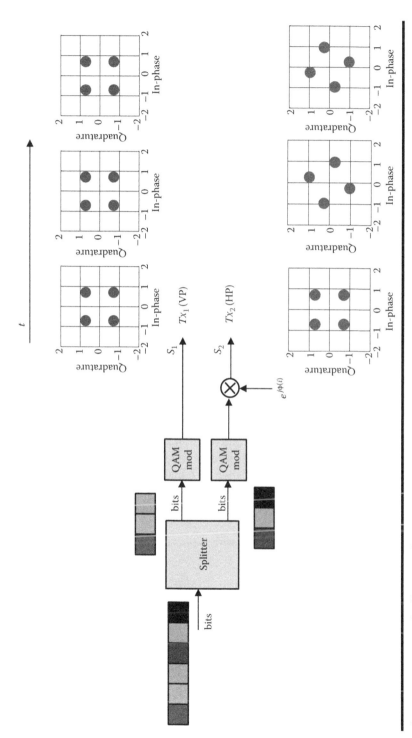

Figure 21.3   Example diagram of SM-PH with gray labeled 4-QAM constellation on each antenna.

21.2.3 Enhanced Spatial Multiplexing with Phase Hopping

The combination of eSM with PH is the so-called eSM-PH, the adopted rate-2 code MIMO for DVB-NGH. The resultant structure for eSM-PH is found in the following equation:

$$
\begin{bmatrix} X_{2i}(Tx1) \\ X_{2i+1}(Tx2) \end{bmatrix} = \sqrt{2} \begin{bmatrix} 1 & 0 \\ 0 & e^{j\phi(i)} \end{bmatrix} \cdot \begin{bmatrix} \sqrt{\beta} & 0 \\ 0 & \sqrt{1-\beta} \end{bmatrix} \cdot \begin{bmatrix} \cos\theta & \sin\theta \\ \sin\theta & -\cos\theta \end{bmatrix} \cdot \begin{bmatrix} \sqrt{\alpha} & 0 \\ 0 & \sqrt{1-\alpha} \end{bmatrix}
$$

$$
\cdot \begin{bmatrix} S_{2i}(Tx1) \\ S_{2i+1}(Tx2) \end{bmatrix}, \quad \phi(i) = \frac{2\pi}{N}(i), \quad i = 0,\dots, \frac{N_{data}}{2}-1, \quad (N=9).
$$

(21.3)

The conceptual block diagram of eSM-PH is illustrated in Figure 21.4.

21.2.4 MIMO Precoding Optimization: Rotation Angle and Power Imbalance Factors

Parameters of expression (Equation 21.3) α, β, and θ have different values according to the transmitted power imbalance and spectral efficiency as shown in Table 21.1. The specific values have been selected after optimization process based on physical layer simulations and detailed next.

The parameters α and β control the power imbalance factor applied to the transmitted streams. Specifically, α decides the minimum Euclidian distance for the given QAM-modulated symbols. For instance, in the case of using QPSK and 16-QAM (i.e., 6 bpcu) on first and second transmit antennas respectively, the power assigned to the constellation with higher minimum Euclidian distance (i.e., QPSK) can be lowered in favor to the constellation with higher cardinality (i.e., 16-QAM), hence increasing its minimum Euclidian distance. On the other hand, the factor β controls the output power and allows the independent optimization of the factor α and the rotation angle θ.

The first parameter optimized is α with expression:

$$
\begin{bmatrix} 1 & 0 \\ 0 & e^{j\phi(i)} \end{bmatrix} \cdot \begin{bmatrix} \cos\theta & \sin\theta \\ \sin\theta & -\cos\theta \end{bmatrix} \cdot \begin{bmatrix} \sqrt{\alpha} & 0 \\ 0 & \sqrt{1-\alpha} \end{bmatrix} \cdot \begin{bmatrix} S_1 \\ S_2 \end{bmatrix},
$$

$$
\phi(i) = \frac{2\pi}{N}(i), \quad i = 0,\dots, \frac{N_{data}}{2}-1, \quad (N=9). \tag{21.4}
$$

where the term β is not included because the Euclidean distance depends only on α. Given the power imbalance factor, $r = p_2/p_1$, where p_1 and p_2 are the transmit power in the first and second antennas respectively with expressions

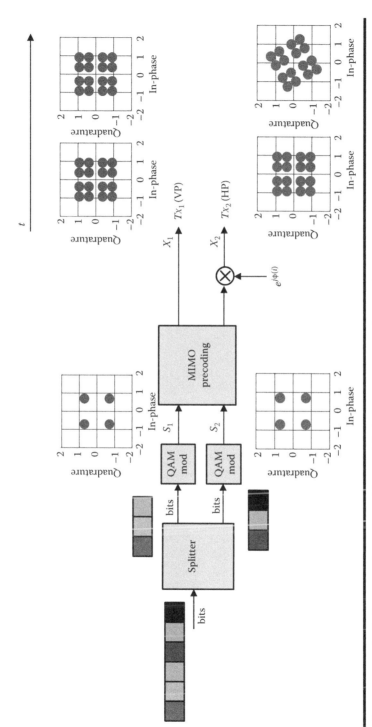

Figure 21.4 Block diagram of eSM-PH with gray labeled 4-QAM symbol constellation on each antenna.

$$p_1 = \alpha\cos^2(\theta) + (1-\alpha)\sin^2(\theta),$$

$$p_2 = \alpha\sin^2(\theta) + (1-\alpha)\cos^2(\theta).$$

(21.5)

The optimization was performed by computer simulations with the DVB-NGH MIMO rate-2 physical layer under the NGH MIMO channel model with optimization criteria of minimum coded BER. Figure 21.5 presents an example of

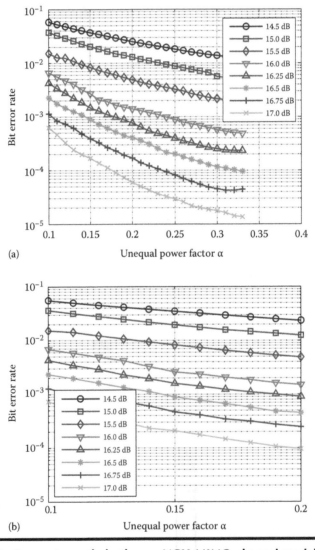

(a)

(b)

Figure 21.5 Parameter optimization α, NGH MIMO channel model with 3 dB (a) and 6 dB (b) of transmitted power imbalance.

the α term optimization for 10 bpcu with 3 and 6 dB of transmit power imbalance, Figure 21.5a and b respectively.

Euclidean distance is optimized by α, but rotation angle θ cannot be optimized because rotation angle is derived by α and given transmit power imbalance factor r, not by optimization. Rotation angle optimization is possible by introducing β. With the optimized value of α, θ, and given r, the term β is derived from expression

$$\beta = \frac{p_2}{p_1 r + p_2} \tag{21.6}$$

The final optimization is done over the rotation angle θ, given r and α. Here, again as for the parameter θ, the optimization is performed by computer simulations under NGH MIMO channel model finding the angle with minimum coded BER. Figure 21.6 shows the optimization over the rotation angle θ for 10 bpcu with 3 and 6 dB of transmit power imbalance, Figure 21.6a and b respectively.

21.3 MIMO Rate-2 Code Implementation in DVB-NGH

The implementation of MIMO rate-2 for the optional MIMO profile requires additional elements compared to the base profile such as extra cross-polarized antenna with the inferred hardware (i.e., cooling systems, RF feedings, power combiners, amplifiers, etc.), as well as a modified transmission chain. Figure 21.7 depicts a block diagram of a complete transmit-to-receive DVB-NGH MIMO rate-2 chain. We note that the receiver functionalities are not included in standard specification documents, leaving to the chipmakers the freedom to develop specific receiving signal algorithms. Hence, the receiver block diagram is just an example of a possible implementation.

21.3.1 Transmitter

The DVB-NGH MIMO rate-2 transmitter processes the incoming bits into a MIMO bit-interleaving coded modulation (BICM) chain to provide IQ samples to be radiated through two cross-polar aerials. In next subsections, we detail the new blocks and characteristics of the DVB-NGH MIMO rate-2 chain in comparison with the base profile.

21.3.1.1 New Bit Interleaver for MIMO

The code bits are passed through a bitwise interleaver that decorrelates the error events along the codeword. This new bit interleaver, different from the one for the basic profile, has been optimized for MIMO transmissions exploiting the

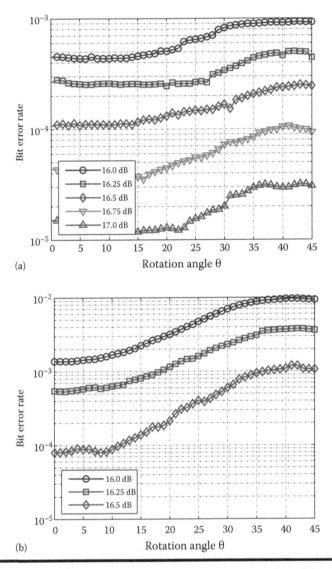

Figure 21.6 Parameter optimization θ, NGH MIMO channel model with 3 dB (a) and 6 dB (b) of transmitted power imbalance.

quasi-cyclic structure of the LDPC codes. The new bit interleaver for MIMO rate-2 consists of two components: a parity interleaver and a parallel bit interleaver. The parity interleaver is the same as that for the base profile and the parallel interleaver in turn comprises two stages: an adjacent quasi-cyclic block (QB) interleaver and a section interleaver and presented in Figure 21.8. It exhibits a low complexity, low latency, and fully parallel design that ease the implementation of iterative

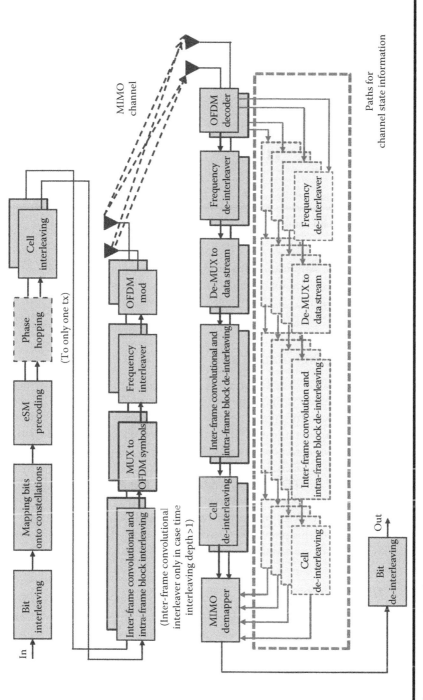

Figure 21.7 DVB-NGH MIMO rate-2 system.

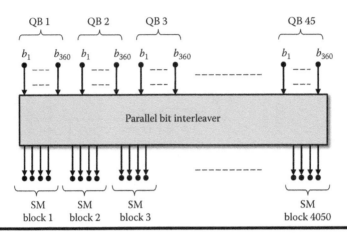

Figure 21.8   Parallel bit interleaver, QB, and section interleaver part.

structures that provide significant gains on the top of the significant MIMO gain. The QB interleaving is based on permutation sequences have been optimized for all combinations of N_{bpcu} and transmit power imbalance. Furthermore, these permutations optimize the gain achieved by iterative decoding receivers.

21.3.1.2 Mapping Bits onto Constellations

The incoming interleaved code bits are mapped to data symbols from a symbol alphabet of size 2^d, where d is the number of bits per channel use and antenna. For MIMO rate-2, the incoming bit data is split into two streams or layers, i.e., one stream or layer per transmit antenna. The number of bits allocated to the first and second layers is b and $m-b$, respectively, where (b, m) are $(2, 6)$ for 6 bpcu, $(4, 8)$ for 8 bpcu, and $(4, 10)$ for 10 bpcu. For MIMO rate-2, the first $m/2$ bits consecutively define the real part of the complex data symbol on the first and second layers, and in the same way the second $m/2$ bits define imaginary part. The modulated data streams are therefore processed by the SM precoding and PH blocks as described in Section 21.2.3. From this point, the transmission chain is duplicated compared with the SISO profile to process the information symbols on each transmit layer.

21.3.1.3 Time Interleaver

As described in Chapter 12, time interleaver is composed of two stages, i.e., inter-frame convolutional interleaver (see Section 12.3.1) and intra-frame block interleaver (see Section 12.3.2). However, for MIMO rate-2, the time interleaver process is implemented to each transmit layer. Due to the fact that same memory constraints apply for all DVB-NGH transmission schemes, i.e., MIMO and non-MIMO, the time interleaving depth for MIMO rate-2 is halved compared to non-MIMO transmissions.

21.3.1.4 Frequency Interleaver

For rate-2 MIMO scheme, both MIMO branches, i.e., the signal generation for both transmit antennas, shall use the same frequency interleaver configuration. In addition, since MIMO physical layer pipes (PLPs) can be multiplexed with MISO PLPs into one NGH frame, pair-wise interleaving scheme is necessary. As described in Chapter 20, this latter condition clearly comes from the use of Alamouti encoding for MISO in NGH system as Alamouti-encoded output cells pair should go through as similar transmission channel as possible for best performance. In OFDM system, the cells are allocated to frequency domain carriers therefore the best effort for the "similar channel" condition is to locate the encoded cell pair into adjacent carrier positions.

For this purpose and for simplicity, NGH system reuses the frequency interleaving scheme already adopted for SISO transmission. That is, the frequency interleaver in a given FFT mode is required to generate only a half of the interleaver addresses compared to when operating with SISO. The cell pair always shares the address and interleaved together into adjacent carrier positions.

21.3.1.5 Pilot Patterns

All the pilot patterns for MISO frame defined in Chapter 19 can also be used for MIMO frame. One special case for rate-2 MIMO arises when an intentional power imbalance of either 3 or 6 dB is introduced. When the power imbalance is applied, it automatically requires the pilot power level to be affected accordingly. This means that 3 and 6 dB power imbalance rate-2 MIMO scheme requires 3 and 6 dB lower pilot power than no power imbalance case.

When the MIMO and MISO PLPs are multiplexed into one NGH frame, the power of MIMO PLPs should also be lowered by 3 and 6 dB. This is necessary because the power imbalance is not imposed by adjusting HPA gain of the transmitter tower, but by lowering the digital domain signal power.

21.3.1.6 OFDM Modulator

The OFDM modulator is the same as for the SISO profile, i.e., IFFT operation plus guard interval insertion. For the MIMO rate-2 profile, it is duplicated to process the information of each transmit layer.

21.3.2 Receiver

Receiver processes inverse transmitter operations to demodulate and decode corrupted MIMO aerial signal. In next subsections, we describe a possible implementation of a DVB-NGH rate-2 receiver taking into account some memory constraints imposed by the specification.

21.3.2.1 OFDM Decoder

The received signals are processed by the OFDM decoder, which basically removes the guard interval and performs an FFT. In this context, it also outputs CSI samples after channel estimation, which are used by the MIMO demapper to provide reliable information about the transmitted bits. Here, the CSI is formed by the four transmit-to-receive spatial paths.

21.3.2.2 Channel Estimation

The known pilot signals sent multiplexed with the data in a comb-type pilot arrangement are used to estimate the channel at the data positions. To reduce complexity, the two dimensional (2-D) estimation problem can be simplified into two one-dimensional (2 × 1-D) estimation problems with small performance loss [6]. It is expected that receivers will perform temporal interpolation first, followed by the frequency interpolation [7]. For MIXO transmissions, as explained in Chapter 19, the phase for second antenna/transmitter is inverted for alternate pilot carriers. This allows the sum and difference of the two channel responses to be estimated by the first and second antennas, respectively. In this situation, the maximum delay spread that can be supported for a given pilot pattern is halved compared to SIXO transmissions.

21.3.2.3 Frequency De-Interleaver

At the receiver, the frequency de-interleaving is performed for the each data stream as well as for the CSI components. The interleaving is performed on an OFDM symbol basis; therefore, for all the components (i.e., data streams and CSI components), at least one OFDM symbol needs to be stored before de-interleaving process.

21.3.2.4 Time De-Interleaver

Similar to the frequency de-interleaving case, the received data streams and the CSI components are time de-interleaved. The most significant memory requirements arise from the time de-interleaving memory. Here, the receiver has to store six complex values (one complex value for each received signal plus one complex value for each CSI component) compared to the two complex values that need to be stored with SISO transmissions (one complex value for the received signal and one complex value for the CSI component). Same memory constraints apply to the MIMO profile as for the SISO base profile. Therefore, total available memory has to be distributed among the two MIMO branches. The memory required for each branch can be computed as for the SISO base profile (see Chapter 11).

21.3.2.5 MIMO Demodulation

The receiver has to demodulate the received signal to provide soft information to the channel decoder. This task is performed by the MIMO demapper that outputs log-likelihood ratios (LLRs) about the transmitted code bits. The optimal maximum a posteriori (MAP) demapper minimizes the bit error probability. However, the complexity for the computation of each of the LLRs is $O\left(M^{N_T}\right)$, where M is the cardinality of the signal constellation (e.g., $M = 4$ for QPSK) and N_T is the number of transmit antennas. That is, complexity increases exponentially with the number of transmit antennas. To reduce the complexity of the MAP demapper, the max-log approximation replaces the logarithm of a sum of exponentials by a minimum distance calculation. While the performance impact of this approximation is usually small, the complexity still scales exponentially with the number of transmit antennas.

Nonlinear techniques like sphere decoding (SD) further reduce the complexity by finding the most likely transmitted symbol from a subset of the original ML search. SD is based on QR factorization to transform the detection problem into a tree search detection problem. SD algorithms can be mainly categorized into two groups if the goal is to target performance of fixed complexity. The former achieves ML solution with complexity dependent on the channel realization [8]. The latter, while providing fixed complexity, does not always achieve ML solution [9].

Linear demodulators like ZF and MMSE are other kind of efficient receivers with very low computational complexity. They apply a linear equalizer to suppress the stream interference to provide an estimate of the transmitted symbols; therefore, the joint MIMO detection problem is transformed into independent SISO detection problems. The complexity of such procedure comprises the complexity for the symbol estimation $O\left(N_R \cdot N_T^2\right)$ operations plus the complexity to detect all code bits $O(N_R \cdot N_T \cdot M)$, where N_R is the number of receive antennas. As we observe, the complexity for the detection of the independent SISO layers is linear with the number of transmit antennas; hence, a significant complexity reduction can be achieved compared to optimal demapping.

21.3.2.6 Iterative Decoding

In a MIMO BICM system, performance can be further improved by employing iterative decoding [10]. Here, the MIMO demapper and the channel decoder exchange extrinsic information in an iterative fashion as shown in Figure 21.9. We note that iterative decoding affects only the receiver side, and therefore no modification is required in standards and transmitters. However, iterative decoding significantly increases the receiver complexity, making it less suited for mobile devices. The complexity increases linearly with the number of outer iterations (i.e., exchange of information between channel decoder and demapper) due to the repetition of

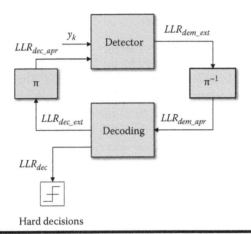

Figure 21.9 Iterative exchange of extrinsic information between MIMO demapper and channel decoder.

MIMO demapping and channel decoder operations. As for non-iterative structures, optimal demapping provides the best performance with iterative detection. However, suboptimal demappers based on linear filtering are still able to exploit the benefits of iterative decoding while keeping reasonable computational complexity [11].

21.4 Performance Simulation Results

In this section, we provide simulation results to illustrate the performance of eSM-PH for DVB-NGH. First, we compare the performance of eSM-PH against other MIMO rate-2 candidates (MIMO SM and MIMO eSM). Next, we study the gain obtained by eSM-PH in comparison with the rate-1 MIMO schemes adopted in DVB-NGH. In addition, results with iterative decoding are provided for two types of MIMO demodulators, i.e., max-log demapper and MMSE demodulator with reduced complexity. Finally, we show the performance loss due to deliberated transmitted power imbalance without and with iterative decoding. We use a 16-QAM symbol constellation on each transmit antenna for eSM-PH and 256-QAM for the rest, i.e., 8 code bits are transmitted per channel use. The schemes are evaluated in the outdoor NGH MIMO channel model [12] with a user velocity of 60 km/h, and the simulations include LDPC codes with a word length size of 16,200 bits.

21.4.1 eSM-PH Performance against Other MIMO Rate-2 Candidates

Figure 21.10 presents frame error rate (FER) vs. CNR in two corner cases of the outdoor NGH MIMO channel model for code rates 7/15 and 2/3. In this case, we present selected quality of service (QoS) of FER 1%. During the DVB-NGH

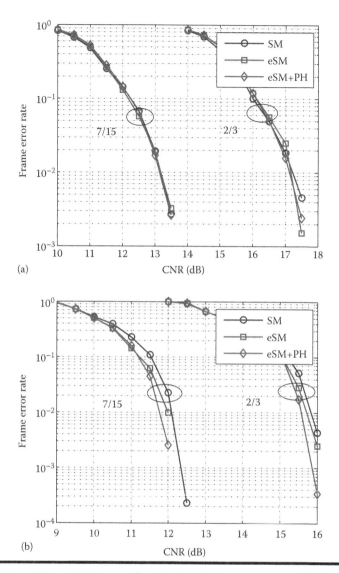

Figure 21.10 FER vs. CNR curve for MIMO NGH corner cases. High correlation case (a) and high *K* case (b).

standardization process, the MIMO rate-2 candidates did not show significant differences under the standard NGH MIMO channel model. For this reason, a set of corner cases were defined to highlight possible differences between the schemes. Corner cases are based on the NGH MIMO channel model but with degenerated characteristics of *K* factor, cross-polarization discrimination (XPD), correlation between transmit and receive antennas, and power imbalance. While the *K* factor describes the power ratio between LOS and non-LOS components, the XPD factor describes the energy coupling between cross-polarized paths. Specifically, the initial overall K factor of 1.0 derived into a low K factor (i.e., overall K factor of 0.0) and into a high K factor (overall K factor of 5.0). The initial XPD value set to 4.0 derived into a low XPD factor of 2 and into a high XPD factor of 8. The initial correlation matrix

$$\mathbf{R}_p = \sigma_{11}^2(\tau_p) \begin{pmatrix} 1.000 & 0.060 & 0.060 & 0.050 \\ 0.060 & 0.250 & 0.030 & 0.050 \\ 0.060 & 0.030 & 0.250 & 0.06 \\ 0.050 & 0.050 & 0.060 & 1.000 \end{pmatrix}, \qquad (21.7)$$

was substituted by a high correlation matrix

$$\mathbf{R}_p = \sigma_{11}^2(\tau_p) \begin{pmatrix} 1.000 & 0.312 & 0.312 & 0.500 \\ 0.312 & 0.250 & 0.125 & 0.312 \\ 0.312 & 0.125 & 0.250 & 0.312 \\ 0.500 & 0.312 & 0.312 & 1.000 \end{pmatrix}. \qquad (21.8)$$

Finally, to study the unintentional power imbalance among transmit antennas, the initial equally powered transmit channel model was replaced by a channel model with 3 dB of power imbalance.

In Figure 21.10, we illustrate as an example the performance of three candidates in the cases of high correlation case (Figure 21.10a) and high K case (Figure 21.10b). For the high correlation case, the tree schemes perform very similarly, despite the slightly better performance of eSM-PH. For the high K case, eSM-PH performs better than the other two schemes. Here, eSM-PH gains 0.24 dB (eSM) and 0.33 dB (SM) for code rate 7/15. On the other hand, the gain of eSM-PH decreases to 0.14 dB (eSM) and 0.26 dB (SM) for code rate 2/3.

21.4.2 Performance of Rate-1 and Rate-2 MIMO in Outdoor NGH MIMO Channel Model

Figure 21.11 presents the system capacity for an FER 1%. The analyzed schemes are SISO, SIMO with two receive antennas, and eSFN, MIMO Alamouti, and eSM-PH with two transmit and two receive antennas. The capacity results include the effect of pilot overhead with the following values used during the NGH standardization process. While a pilot density of 1/12 is assumed for SISO, SIMO, and eSFN, a pilot density of 1/6 is assumed for MIMO Alamouti and eSM-PH. Here, the results highlight the significant gains achieved by the different DVB-NGH MIMO schemes over SISO. Compared with SISO and with 15 dB of average CNR, SIMO provides a 44.7% of

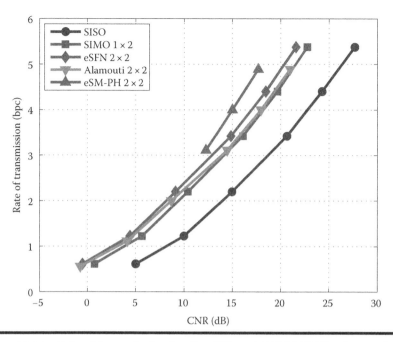

Figure 21.11   Rate of transmission for the different NGH MIMO schemes in the NGH outdoor MIMO channel with 60 km/h speed.

capacity increase (equivalently 4.5 dB of CNR gain at 3.20 bpcu), eSFN provides 57% of capacity increase (5.8 dB of CNR gain at 3.45 bpcu), and eSM-PH provides 81.5% of capacity increase (7.8 dB of CNR gain at 4 bpcu). The performance of MIMO Alamouti lies between SIMO and eSFN due to the effect of increased pilot overhead.

21.4.3 Iterative Decoding Performance

We present eSM-PH performance results with iterative decoding (cf. FER curves labeled with "ID"). Figure 21.12 shows the FER vs. CNR for code rates 1/3, 8/15, and 11/15. For the max-log demapper, the iterative decoding gain increases with increasing code rate, i.e., 1 dB (1/3), 1.1 dB (8/15), and 1.8 dB (11/15). Furthermore, we present results with MMSE demodulator that is able to exploit the benefits of iterative decoding while keeping computational complexity low [11]. In this case, MMSE demodulator gains around 0.7 dB (1/3), 0.9 dB (8/15), and 1.1 dB (11/15) by iterative decoding. However, we observe performance degradation of MMSE demapper compared with the max-log demodulator with the increasing rate. The max-log demapper outperforms the MMSE demapper by about −0.15 dB (0.2 dB), 0.4 dB (0.5 dB), and 1.2 dB (1.9 dB) for (non-)iterative decoding. We note that for all simulated code rates, the MMSE receiver with iterative decoding does not perform worse than the non-iterative max-log receiver.

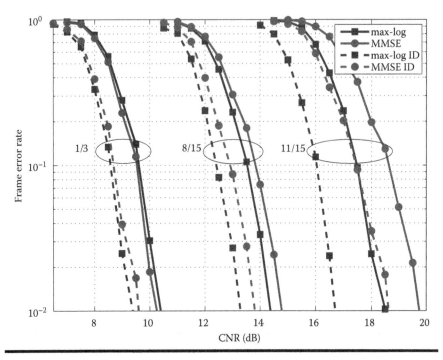

Figure 21.12 **FER vs. CNR of iterative decoding with 8 bpcu and code rates 1/3, 8/15, and 11/15 in the NGH outdoor MIMO channel with 60 km/h speed.**

21.4.4 Performance Results with Transmitted Power Imbalance

Finally, in Figure 21.13, we present the performance degradation in the system capacity due to deliberated power imbalance with (Figure 21.13a) and without (Figure 21.13b) iterative decoding. Here, we use 6 bpcu (QPSK–16-QAM), 8 bpcu (16-QAM–16-QAM), and 10 bpcu (16-QAM–64-QAM), with deliberated power imbalances of 0, 3, and 6 dB. In this case, we use a frequency fast fading model with Rice distribution and spatial correlation between antennas. The selected QoS is FER before BCH 10^{-1}. With 3 dB of power imbalance, the performance loss is 0.3 and 0.2 dB for 8 and 10 bpcu, whereas for 6 bpcu, the performance loss is negligible. With 6 dB of power imbalance, the performance degradation increases to 0.3 dB (6 bpcu), 0.7 dB (8 bpcu), and 0.5 dB (10 bpcu).

We observe a similar situation with iterative decoding as it is shown at the bottom part of Figure 21.13. We note in this case an overall performance improvement of 1 dB in CNR. With 3 dB of power imbalance, the performance loss is 0.2 dB for 8 bpcu while almost no loss is observed for 6 and 10 bpcu. In addition, with 6 dB power imbalance, the performance loss is 0.23 dB (6 bpcu), 0.7 dB (8 bpcu), and 0.6 dB (10 bpcu).

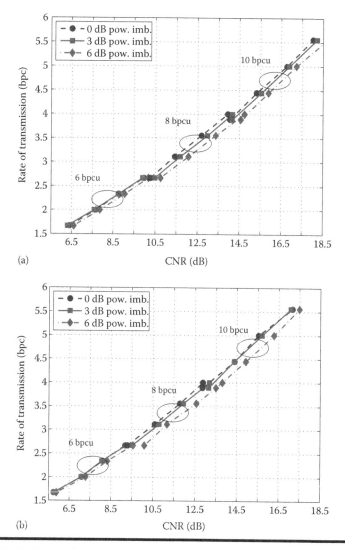

Figure 21.13 Rate of transmission for eSM-PH with deliberated transmitted power imbalances of 0 dB, 3 dB, and 6 dB. Non-iterative decoding (a) and iterative decoding (b).

21.5 Conclusions

eSM-PH MIMO rate-2 code increases the rate of transmission through SM while exploiting the spatial diversity with reasonable decoding computational complexity. Simulation results illustrated an improved performance against other MIMO rate-2 codes, significant gains compared with other MIMO rate-1 codes in the high CNR range, and tailored transmission in the case of deliberated transmitted power imbalance. Furthermore, the detection of eSM-PH can be improved by iterative decoding. In brief, eSM-PH is a promising MIMO scheme for the provision of high data rate services to handheld services in mobile scenarios. However, it requires the implementation of various antennas at both sides of the transmission link. Moreover, additional aspects need to be considered on the deployment of dual polar transmissions in DVB-NGH networks.

References

1. D. Vargas, D. Gozálvez, D. Gómez-Barquero, and N. Cardona, MIMO for DVB-NGH, the next generation mobile TV broadcasting, to appear in *IEEE Communications Magazine*.
2. J. Paulraj, D. A. Gore, R. U. Nabar, and H. Bölcskei, An overview of MIMO communications—A key to gigabit wireless, *Proceedings of the IEEE*, 92(2), 198–218, 2004.
3. J.-C. Belfiore, G. Rekaya, and E. Viterbo, The golden code: A 2×2 full-rate space-time code with nonvanishing determinants, *IEEE Trans. on Information Theory*, 51, 1432–1436, 2005.
4. R. Kobeissi, S. Sezginer, and F. Buda, Downlink performance analysis of full-rate STCs in 2×2 MIMO WiMAX systems, *Proceedings of the IEEE VTC Spring*, Barcelona, Spain, 2009.
5. S. Moon et al., Enhanced spatial multiplexing for rate-2 MIMO of DVB-NGH system, *Proceedings of the IEEE ITC*, Jounieh, Lebanon, 2012.
6. P. Hoeher, S. Kaiser, and P. Robertson, Two-dimensional pilot-symbol-aided channel estimation by Wiener filtering, *Proceedings of the IEEE ICASSP*, Munich, Germany, 1997.
7. ETSI TS 102 831 v1.2.1, Implementation guidelines for a second generation digital terrestrial television broadcasting system (DVB-T2), April 2012.
8. C. Studer, Iterative MIMO decoding: Algorithms and VLSI implementation aspects, PhD. dissertation, ETH Zurich, Switzerland, June 2009.
9. L. G. Barbero, T. Ratnarajah, and C. Cowan, A low-complexity soft-MIMO detector based on the fixed-complexity sphere decoder, *Proceedings of the IEEE ICASSP*, Las Vegas, NV, 2008.
10. B. M. Hochwald and S. T. Brink, Achieving near-capacity on a multiple-antenna channel, *IEEE Transactions on Information Theory*, 51(3), 389–399, March 2003.
11. D. Vargas, A. Winkelbauer, G. Matz, D. Gómez-Barquero, and N. Cardona, Low complexity iterative MIMO receivers for DVB-NGH using soft MMSE demapping and quantized log-likelihood ratios, *Proceedings of the COST IC1004*, Barcelona, Spain, February 2012.
12. P. Moss, T. Y. Poon, and J. Boyer, A simple model of the UHF crosspolar terrestrial channel for DVB-NGH, BBC White Paper, 2011.

NEXT GENERATION HANDHELD DVB TECHNOLOGY

Hybrid Terrestrial– Satellite Profile

IV

Chapter 22

An Overview of the Hybrid Terrestrial–Satellite Profile of DVB-NGH

Vittoria Mignone, David Gómez-Barquero, Damien Castelain, and Marco Breiling

Contents

22.1 Introduction

In order to avoid market fragmentation, as happened with the first-generation mobile broadcasting DVB (Digital Video Broadcasting) standards DVB-H (Handhelds) [1] and DVB-SH (Satellite to Handhelds) [2], the new generation of mobile broadcasting DVB technology DVB-NGH (Next Generation Handheld) targets both sheer terrestrial networks and hybrid terrestrial–satellite networks [3]. DVB-NGH complements the terrestrial coverage (generally in the ultrahigh frequency [UHF] band) with an optional satellite component in the L- or S-frequency bands. In addition to such multifrequency network (MFN) configurations, a single-frequency network (SFN) configuration with both networks operating at the same frequency in the L- or S-bands is also possible.

A hybrid terrestrial–satellite network is probably the most cost-effective network topology for mobile broadcasting. As a general principle, broadcasting is economically advantageous with respect to unicasting, because the infrastructure cost is amortized over the user base. As the user base grows, the cost per subscriber falls. Hence, satellites are ideal to cover very wide areas. As an example, one geostationary satellite is capable of providing coverage to all Western Europe. The approach followed by DVB-NGH is to consider the satellite component as a complement of the terrestrial network, providing nationwide coverage and complementing the terrestrial coverage in low-populated rural areas where the terrestrial network deployment can be too expensive. It should be noted that this approach is completely different from the strategy followed in DVB-SH, which was satellite-centric, with the terrestrial component considered as complementary. However, a major drawback of DVB-SH is that it is necessary to deploy a dedicated terrestrial network in the S-band in urban areas where reception from the satellite is not possible.

The satellite component of DVB-NGH is optional. The hybrid profile includes several technical solutions specifically developed to deal with the characteristic problems of mobile satellite reception and to receive the signals coming from the two networks with a single tuner and to combine them seamlessly. It has been designed with the goal of keeping the maximal commonality with the terrestrial component to ease its implementation at the receiver side. The main elements introduced are as follows:

■ Extended convolutional inter-frame time interleaving (TI; up to approx. 10 s) with fast zapping support using a uniform-late convolutional interleaving profile
■ Single carrier—orthogonal frequency division multiplexing (SC-OFDM) for the satellite component in MFN hybrid networks in order to reduce the peak-to-average power ratio (PAPR) of the transmitted signal
■ An additional preamble OFDM symbol at the beginning of the NGH satellite frames to increase the capacity of the physical layer signaling

Compared to a sheer terrestrial receiver, a satellite receiver requires at least an additional external time de-interleaving (TDI) memory to account for the long TI requirements at the physical layer, SC-OFDM demodulation, and a tuner covering the satellite frequency bands (L and S) (see Figure 22.1).

Due to the extremely long propagation distances and the power limitations at the satellite, mobile satellite links require very robust transmission modes capable of supporting very low signal-to-noise ratios (SNRs). The DVB-NGH satellite profile restricts the modulations to only QPSK and 16QAM. The land mobile satellite (LMS) channel is also characterized by long signal outages due to the blockage of the Line-of-Sight (LoS) with the satellite caused by tunnels, buildings, trees, etc.

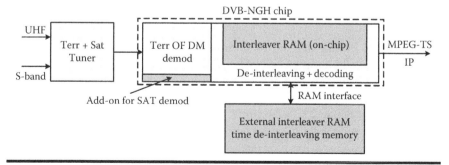

Figure 22.1 **Block diagram of a DVB-NGH receiver supporting both the base sheer terrestrial profile and the hybrid satellite–terrestrial profile.**

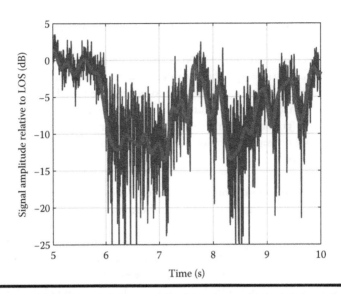

Figure 22.2   Illustration of the LMS suburban channel. Elevation angle 40°, frequency 2.2 GHz, velocity 60 km/h. (From Pérez-Fontán, F. et al., *IEEE Trans. Vehicular Technol.*, 50(6), 1549 © 2001 IEEE.)

The signal received from the satellite is impaired by shadowing effects in the vicinity of the receiver, especially in low elevation angle areas, leading to long fading events and deep attenuations lasting up to several seconds (see Figure 22.2 [4]).

Signal loss events in the order of some seconds can be compensated only by a long time interleaver (e.g., duration in the order of 10 s). TI is a key feature in any mobile broadcasting system to cope with impulse noise and benefit from time diversity. However, it increases the end-to-end latency and the channel change (zapping) time, which is a crucial quality of service parameter, especially for TV use. Generally, it is considered that channel change (zapping) times around 1 s are satisfactory, whereas more than 2 s are felt as annoying [5]. Therefore, fast zapping techniques are a must in mobile TV satellite transmissions in order to keep the zapping time in acceptable values with long TI.

DVB-NGH has adopted for the sheer terrestrial profiles a convolutional interleaver (CI) with a uniform profile for inter-frame TI, keeping the DVB-T2 block interleaver (BI) only for intra-frame interleaving. The hybrid profile of DVB-NGH makes use of an external TDI memory as a complement of the on-chip TDI memory, as shown in Figure 22.1. The size of the on-chip TDI memory is the same as that of the basic terrestrial profile, and it is used only for intra-frame BI, whereas the external TDI memory is used for inter-frame CI. Furthermore, fast zapping support with long TI at the physical layer is enabled using a uniform-late profile of the CI, like in DVB-SH [6]. The uniform-late profile introduces a trade-off between overall performance in mobile channels and performance after zapping.

Another important characteristic in satellite communications is the reduction of the PAPR of the transmitted signal in order to maximize the efficiency of the high-power amplifier on board the satellite. Single carrier modulations such as time division multiplexing (TDM) have been the reference schemes for satellite transmission, e.g., in DVB-S (Satellite) and DVB-S2 (Satellite Second Generation), because of their reduced power fluctuations that allow driving the amplifiers with small input back-offs (IBOs), optimizing the amplifier power efficiency. However, OFDM is the modulation used for digital terrestrial broadcasting. OFDM shows very good performance in terrestrial channels due to the splitting of the available bandwidth into thousands of sub-carriers. For the 6–8 MHz bandwidth typically employed in the UHF band for digital terrestrial TV, the terrestrial propagation channel is frequency selective, but each OFDM sub-carrier experiences a flat channel. Furthermore, its robustness against multipath, which allows constructively combining signals from several transmitters to the total wanted signal, enables SFNs. The criticality of OFDM for satellite transmissions is that it has a high power fluctuation. Due to the independence between the phases of the different sub-carriers, they may combine constructively creating high peaks in the transmitted signal. Hence, the use of OFDM in satellite transmissions requires forcing the high-power amplifiers to work far from the saturation point, for which the transmitted power is maximized, in order to avoid nonlinear distortions. This leads to a reduction of the available link budget. DVB-T2 incorporates two PAPR reduction mechanisms known as tone reservation (TR) and Active Constellation Extension [7]. However, even using these techniques, OFDM requires too large back-offs due to the non-linearities typical of high-power satellite amplifiers.

Within this context, DVB-NGH maintains OFDM as one option for the hybrid profile. The main advantage of using OFDM for the satellite component is that it allows the satellite and terrestrial components to share the same frequency channel forming an SFN. But DVB-NGH also adopts SC-OFDM for the satellite component of hybrid MFN networks. The reason is that it provides an important PAPR gain while preserving many commonalities with OFDM. Hence, the implementation cost of including SC-OFDM in the chips is markedly lower than TDM modulation used in the MFN configuration of DVB-SH.

DVB-NGH has also specified the scheduling of the terrestrial and satellite transmissions in MFN such that parallel reception of both signals is possible for terminals with a single demodulator. This has been made possible thanks to the new logical frame structure of DVB-NGH. For more details on the logical frame structure, see Chapter 14. Parallel reception is particularly useful in the mush area, where the satellite and the terrestrial signals are similarly strong. If one component undergoes deep fading, the receiver can use the other signal to ensure a continuous service. Later on, the situation might be the converse, but still one signal can be received. When a receiver moves from a terrestrially covered area, like an urban area, to satellite coverage, like in the countryside, this parallel reception allows a soft handover over a long time span with seamless transition from terrestrial to satellite coverage.

If either the terrestrial or the satellite signal is clearly dominating all the time, and the other signal is too weak, the receiver needs to receive only the dominating signal.

In DVB-NGH, the physical layer signaling is transmitted like in DVB-T2 in preamble symbols at the beginning of each frame, known as P1 and P2 symbols. The preamble carries a limited amount of signaling data in a very robust way. The frames begin with one P1 symbol, which carries some basic transmission parameters, such as the frame type. Then 1 to 16 P2 symbols follow that carry the rest of the Layer 1 (L1) signaling with the information needed for extracting and decoding the data physical layer pipes (PLPs) from the frames. The P1 symbol provides seven signaling bits, and they have been fully used with DVB-T2, its mobile profile T2-Lite (see Chapter 6), and the single-input single-output (SISO) and multiple-input single-output (MISO) terrestrial profiles of DVB-NGH. For the multiple-input multiple-output (MIMO) terrestrial profile and the hybrid profile of DVB-NGH, an additional preamble P1 (aP1) symbol has been introduced in order to increase the signaling capacity of the P1 symbol. The P1 symbol signals the presence of the aP1 symbol. The aP1 symbol is only transmitted when needed for terrestrial MIMO networks and hybrid terrestrial–satellite networks.

This chapter provides an overview of the hybrid terrestrial–satellite profile of DVB-NGH. The rest of the chapter is structured as follows: Section 22.2 describes the different hybrid network topologies supported in DVB-NGH, SFN and MFN, and describes the signal combination mechanisms. Section 22.3 is devoted to the physical layer signaling of the DVB-NGH satellite component. In Section 22.4, the long TI and fast zapping features are detailed, and selected results are presented illustrating the trade-off between continuous performance and performance after zapping. Section 22.5 introduces the principle of the SC-OFDM modulation, describes its implementation in DVB-NGH, and presents some illustrative results comparing its performance with OFDM. Section 22.6 presents a comparison between the hybrid profile of DVB-NGH and DVB-SH. Finally, the chapter is concluded with Section 22.7.

DVB-NGH also specifies a hybrid terrestrial–satellite MIMO profile with multiple antennas at the transmitters and the receivers. Different cases are possible for this profile, either SFN or MFN. Interested readers on the hybrid MIMO profile are referred to Chapter 24.

22.2 DVB-NGH Hybrid Network Architectures

22.2.1 Single-Frequency Networks and Multifrequency Networks

Generally speaking, hybrid broadcast networks can be classified in two types depending on whether both the satellite and terrestrial network components use the same frequency channel (i.e., SFN) or not (i.e., MFN) (see Figure 22.3).

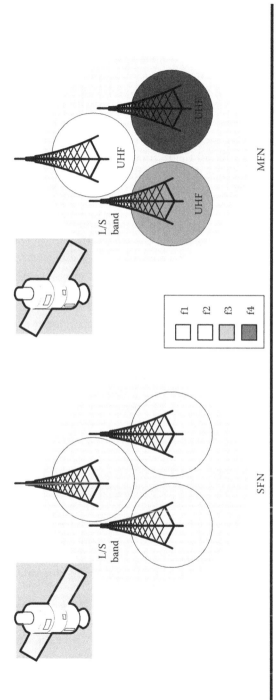

Figure 22.3 Hybrid terrestrial–satellite DVB-NGH SFN and MFN topologies.

In DVB-NGH, the frequency bands possible for the satellite component are the L-band (1452–1492 MHz) and the S-band (2170–2200, 2500–2520, and 2520–2670 MHz), whereas the terrestrial network can operate in addition to the L- and S-bands at the frequency bands VHF and UHF.

In an SFN, all transmitters transmit the same OFDM signal at the very same frequency, and the combination of the signals from several transmitters is done in the air by constructive interference. Transmitters shall be synchronized in both time and frequency, typically using a GPS reference signal. SFN networks generally allow for a high spectral efficiency, but force the signal transmitted by all transmitters to be exactly the same in all the territory. Hence, in a hybrid SFN network, the terrestrial and satellite signals have to be identical. The main benefit of the hybrid SFN network topology in DVB-NGH is that it can be operated in a single frequency channel.

On the other hand, in an MFN network, the different network components transmit at different frequencies. As there is no interference between the signals over the air, each network component may transmit a different signal optimized to the specific propagation conditions. For example, SC-OFDM with low-order modulations can be used for the satellite component optimizing the satellite power usage, and OFDM with higher-order modulations and coding rates can be adopted for the terrestrial component, thanks to the less stringent power requirements and the much lower power loss during the propagation over the relatively short distances, improving the terrestrial spectral efficiency. On the other hand, in an SFN, a compromise has to be sought for the chosen transmission parameters, and most often the most robust parameter from either terrestrial or satellite transmission has to be chosen (e.g., large guard interval [GI] because of terrestrial multipath propagation plus low coding rate because of satellite propagation). As a consequence, it may be possible that an MFN achieves an overall higher spectral efficiency than an SFN.

MFNs also enable the insertion of content specific to each of the two components. Different possible cases can be considered: in the general case, a mixture of global and local content can be transmitted by the satellite and terrestrial networks. Another possible application of this network flexibility is relevant to Scalable Video Coding (SVC) [8]. The SVC base layer could be transmitted as global content, both in the satellite and in the terrestrial network, whereas the SVC enhancement layer would be transmitted only in the terrestrial network. In this case, the two SVC layers should be synchronized.

Nevertheless, the main advantage of the hybrid MFN topology is that there is no need to deploy a dedicated terrestrial network because it is possible to reuse existing DVB-T2 infrastructure to start providing DVB-NGH services in-band a DVB-T2 multiplex. This significantly reduces the investments in network infrastructure required to start providing DVB-NGH services. Thanks to the Future Extension Frames (FEFs) of DVB-T2, it is possible to efficiently share the capacity of one frequency channel in a time division manner between DVB-T2 and

DVB-NGH, with each technology having specific time slots. The satellite component is an independent component, with the possibility to transmit both content common to the terrestrial component and content specific to the satellite. It should be pointed out that the possibility of using FEFs in a DVB-T2 network for DVB-NGH is not valid for the hybrid SFN mode because the satellite component is not present in DVB-T2, and the satellite signal must be continuous (this would result in no signal from the satellite during T2 frames).

22.2.2 Hybrid SFN Deployment

22.2.2.1 System Configuration

In a hybrid SFN, the two network components must transmit the same OFDM waveform. Another constraint posed by the SFN is that all terrestrial SFNs within the satellite footprint have to share exactly the same content (i.e., no local terrestrial or satellite-specific content is possible). The transmission mode cannot be optimized taking into account the particularities of each propagation channel, and the worst case has always to be taken into account. For example, current mobile satellite link budgets require very low spectral efficiencies (around 1 bps/Hz or even lower), which is in most cases too low for the terrestrial component. The terrestrial signal experiences a fast-fading time-dispersive (frequency-selective) channel and requires sufficiently large GI and dense pilot patterns (PPs; in both time and frequency domains), while the satellite propagation channel is generally assumed to be flat and requires no or very small GI and very few pilots. Furthermore, since the OFDM waveform is characterized by high PAPR, the satellite back-off has to be increased to reduce the signal distortion. The joint fulfillment of requirements from the two components results in a reduction of the system capacity and of the available received signal power by the satellite, i.e., in a penalization of the satellite link budget. This has to be traded off against the use of a single frequency channel.

22.2.2.2 Signal Combining

As the signals are transmitted on the same frequency channel, the demodulator sees only one signal that corresponds to the sum of the two transmitted signals. The combining between the two components is thus performed "over the air." Receivers have to demodulate a single signal, and thus only a single demodulation chain is required. Receivers may successfully demodulate the signal when receiving only the terrestrial or the satellite signal, or both.

22.2.2.3 Synchronization Issues

In an SFN, the transmitted signal has to be synchronized in time and frequency at all transmitters such that all received signals add up in a constructive manner. An absolute time reference like a GPS pulse per second (PPS) must be given to all

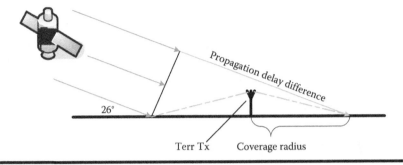

Figure 22.4   Signal propagation delay in a hybrid terrestrial–satellite network.

transmitters. Synchronization of a hybrid SFN network imposes more constraints due to the position and the movement of the satellite [9]. As the satellite covers a large area, it will be associated with multiple terrestrial transmitters. Due the position of the satellite, the satellite propagation delay may differ, and each terrestrial transmitter must be individually synchronized with the satellite. But the propagation delay difference for the different locations in the terrestrial cell can be high (see Figure 22.4). Signal synchronization mechanisms allow synchronizing all terrestrial transmitters with the satellite signal, but propagation delay differences in the different reception locations can be compensated only by means of the OFDM GI [9]. Note that part of this problem can be solved by using sector antennas, where each is synchronized individually.

Moreover, the position of the satellite may vary in time and the transmission time of the satellite signal will also vary. Thus, the synchronization of all terrestrial transmitters must be corrected with the movement of the satellite.

22.2.3 Hybrid MFN Deployment

As stated earlier, MFN networks allow maximum flexibility of configuration. As generally it is foreseen that DVB-NGH will first be launched terrestrially in UHF, possibly in the FEF of DVB-T2 signals, the hybrid satellite component could be added in S- or L-band to complement the terrestrial signal and complete the coverage of the territory, without impacting the terrestrial network configuration. MFN also offers maximum flexibility for the allocation of the services, allowing repeating a selection of the terrestrial services in the satellite frequency, and possibly adding new services, specific for the satellite.

22.2.3.1 Signal Combining

Traditionally, in a hybrid MFN (e.g., in the case of DVB-SH), the receivers require two tuners and two demodulators for parallel reception of both signals in order to combine them. Otherwise, with a single demodulator, it may not be possible to perform a seamless handover when changing from the reception of one signal to the other. DVB-NGH has specified the scheduling of the transmissions of the satellite and terrestrial network components such that the signals are transmitted in different time slots. The new logical frame structure allows the parallel reception and combining of both signals with a single agile tuner, which covers both frequency bands, and a single demodulator.* Parallel means, in this context, that one time slot (e.g., from a terrestrial transmitter) is received after the other (e.g., from the satellite), and the combining is done only after the demodulation. This temporary parallel reception corresponds to a soft handover between sheer terrestrial and sheer satellite reception. This means that there is no hard switching (hard handover) between terrestrial and satellite signals. The handover is not noticeable for the user (i.e., seamless handover).

In DVB-NGH, the combining of terrestrial and satellite signals is carried out not only after the demodulation (as in DVB-SH), but also after FEC decoding, i.e., after the Low Density Parity Check (LDPC) and the BCH decoding. Packet selection is the employed method in order to choose between terrestrial and satellite components, where the packets are either MPEG-TS packets or IP packets.

22.3 Physical Layer Signaling for the Satellite Component of DVB-NGH

The physical layer signaling in DVB-NGH provides to the receivers, on one hand, fast signal detection and scanning, and, on the other hand, all the information needed to access the Layer-2 (L2) signaling and the services themselves. The physical layer signaling is transmitted in the first OFDM symbols of each NGH frame, known as preamble symbols. It is structured in two main parts, the P1 signaling carried in the P1 symbol and the additional P1 symbol aP1, and the L1 signaling carried in the P2 symbols.

22.3.1 P1 Signaling

The P1 symbol is the first OFDM symbol of each frame. It is intended for fast identification of T2-like signals, and it also enables the receiver to detect the beginning of the frame, allowing for both time and frequency synchronization. The P1 symbol carries seven signaling bits that define some essential transmission parameters, such as the preamble format (frame type). In the hybrid profiles of DVB-NGH, the P1 symbol signals the presence of the aP1 symbol just after the P1 symbol, and it signals whether the hybrid SISO or hybrid MIXO profiles are used. The aP1

---

* This has been made possible thanks to the introduction of the concept of the logical channel (LC) and of the LC group. All the logical NGH channel members of an LNC group can be received with a single tuner. For more details on the logical frame structure, see Chapter 14.

symbol provides seven signaling bits, and with this information, the receiver is able to receive the preamble P2 symbols and access the L1 signaling.

The aP1 has the same structure and hence the same properties as the P1 symbol (e.g., robust signal discovery against false detection and resilience to continuous interference). The detection of the preamble P1 and aP1 symbols is very robust, and they can be correctly received even at negative SNR under mobility conditions. Furthermore, the signal detection/acquisition is improved when both P1 and aP1 symbols are used for detection. The aP1 is scrambled with a different pseudorandom sequence, uses a different frequency offset value for the prefix and postfix, and employs a different set of carriers.

For more information on the P1 and aP1 symbols in DVB-NGH, see Chapter 11.

22.3.2 L1 Signaling

The P2 symbols contain all static, configurable, and dynamic L1 signaling. The L1 signaling is divided into two parts: L1-pre and L1-post. The L1-pre signaling signals basic transmission parameters enabling the decoding of the L1-post. The L1-post carries the information needed to identify and decode the data PLPs within the frame. The L1-pre and L1-post signaling fields are independently coded and modulated. The L1-pre is protected with the most robust MODCOD possible: BPSK with coding rate 1/5, while the robustness of the L1-post is a transmission configuration parameter. The transmission mode of the L1-post represents a trade-off between robustness and signaling overhead. Obviously, the L1 signaling needs to be more robust than the data, because it carries information that allows the receivers decoding the data PLPs. It is generally recommended that the physical layer signaling is 3 dB more robust than the data. DVB-NGH has significantly enhanced the transmission robustness of the L1-post, compared to DVB-T2, which allows using very robust MODCODs for the data (i.e., QPSK 1/5).

22.3.2.1 L1 Transmission

DVB-NGH adopts an L1 signaling structure similar to that of DVB-T2, but introduces three new mechanisms in order to improve the L1 robustness: 4k LDPC codes, additional parity (AP), and incremental redundancy (IR). The reduced size of 4k LDPC codes is more suitable for signaling, because it reduces the amount of shortening and puncturing significantly, which degrades the LDPC decoding performance. The AP technique consists of transmitting punctured LDPC parity bits of the L1-post on the previous frame, improving the time diversity and reducing the effective coding rate without affecting the zapping time. The technique IR allows generating AP bits to further increase the L1 transmission robustness. IR extends the 4k LDPC codeword with a coding rate 1/2 into an 8k LDPC codeword, yielding an effective coding rate of 1/4. The resulting codeword has two parts: the first part corresponds to the basic 4k LDPC codeword and the second part corresponds to the additional IR parity bits. In addition to AP and IR, it is possible to use L1-repetition, as in DVB-T2, which consists of transmitting the L1-dynamic signaling information on the current and previous frame.

For more information on the new L1 transmission mechanisms in DVB-NGH, see Chapter 11.

22.3.2.2 Illustrative Performance Evaluation Results

Table 22.1 shows the simulation parameters for the suburban LMS channel model [4]. The QoS metric is the ESR5(20), which represents the percentage of intervals of 20 s with at most 1 erroneous second. The ESR5(20) criterion is usually employed for the LMS channel because it takes into account the time correlation of the channel.

Figure 22.5 shows the robustness of the data path for QPSK 1/5 (i.e., the most robust) and the L1-post signaling with different configuration of the AP technique with L1-repetition and advanced L1 decoding. This decoding technique consists on

Table 22.1   Simulation Parameters for the Hybrid Terrestrial–Satellite Profile of DVB-NGH

Parameter		Value		Parameter	Value
OFDM	FFT size	2048		LMS Channel Model	Perez-Fontan three-state statistical model
	BW	5 MHz	Channel	Center frequency	2.2 GHz
	GI	1/4		Elevation	40°
	PP	PP1		Receiver speed	60 km/h
	FEF + T2 frame	300 ms		Receiver environment	SU (suburban)
Data	MODCOD	QPSK 1/5, QPSK 1/3	L1	Nr. PLPs	8 PLPs
	Time interleaver	100 ms, Type 1 PLP		QoS criteria	ESR5(20)

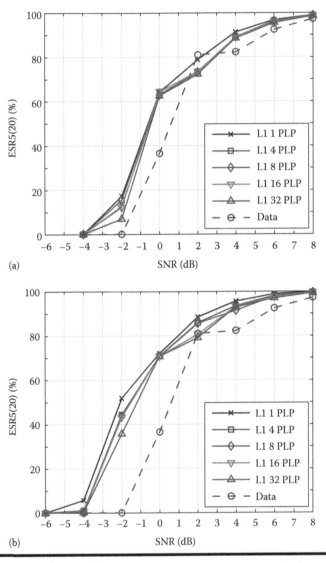

(a)

(b)

Figure 22.5 Robustness of the L1 signaling with repetition and advanced decoding in the satellite LMS suburban channel for different configurations of AP. AP0, AP1, AP2, and AP3 imply 0%, 35%, 70%, and 105% parity bits more. (a) AP0. (b) AP1. (c) AP2. (d) AP3.

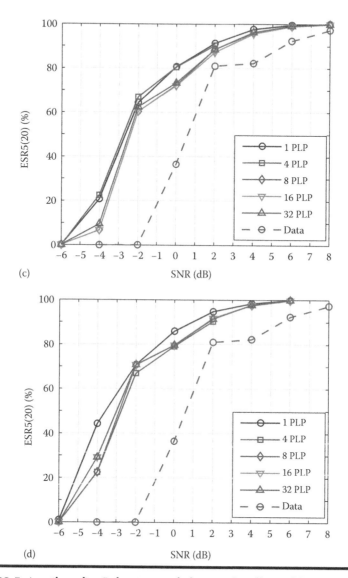

Figure 22.5 (continued)   Robustness of the L1 signaling with repetition and advanced decoding in the satellite LMS suburban channel for different configurations of AP. AP0, AP1, AP2, and AP3 imply 0%, 35%, 70%, and 105% parity bits more. (a) AP0. (b) AP1. (c) AP2. (d) AP3.

reusing the information that has been decoded successfully in previous frames and that is also transmitted in the current frame when decoding the LDPC codewords. In the figure, we can see that the available L1 signaling robustness tools are sufficient to meet the requirements of the satellite data path.

22.4 Forward Error Correction with Long Time Interleaving and Fast Zapping Support

22.4.1 Introduction

DVB-NGH adopted for the sheer terrestrial profiles a CI with a uniform profile for inter-frame TI, keeping the DVB-T2 BI only for intra-frame interleaving. This way, longer TI durations can be achieved for a given TI memory (approximately twice), and lower zapping times and higher service bit rates are supported for the same TI duration. At the transmitter side, the CI operates on units of blocks, each of which is interleaved by a BI. This allows combining the BI and CI into one single interleaver at the receiver, such that each interleaver does not need dedicated memory.

DVB-NGH has also adopted a technique known as adaptive cell quantization. This technique doubles the effective size of the TDI memory* for the more robust modulations, QPSK and 16QAM, because they tolerate a higher quantization noise. This optimization of the TDI memory doubles the maximum TI duration for a given service data rate, or the maximum service data rate for a given TI duration, for both intra-frame and inter-frame interleaving. For more details on the DVB-NGH time interleaver, see Chapter 12.

Despite the improvements of the time interleaver in DVB-NGH, the TDI size specified in the standard for the terrestrial (base and MIMO) profiles is not sufficient to sustain the long interleaving durations required for mobile satellite transmission. Furthermore, the base profile does not incorporate any mechanism to support fast zapping. The hybrid profile of DVB-NGH makes use of an external TDI memory as a complement of the on-chip TDI memory, as shown in Figure 22.1. The on-chip TDI memory is only used for intra-frame BI, and it has the same size of the TDI memory for the base profile (2^{19} cells for QPSK and 16QAM), whereas the external TDI memory is used for inter-frame CI, and it has a size of 2^{22} cells for QPSK and 16QAM. On the other hand, DVB-NGH allows fast zapping support with long TI at the physical layer using a uniform-late profile of the CI like in DVB-SH (see Figure 22.6).

---

* The standard specifies the size of the TDI memory in the receivers in terms of cells (QAM constellation symbols). The size for QPSK and 16QAM is 2^{19} cells in the base profile (recall that the satellite profile includes only these two constellations). For 64QAM and 256QAM, the size is 2^{18} cells.

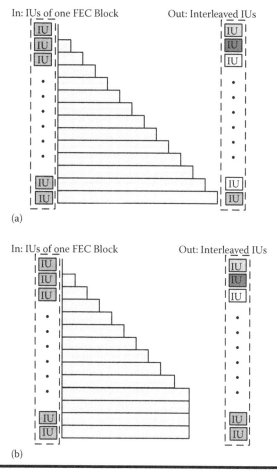

Figure 22.6   Convolutional interleaving profiles: uniform (a) and uniform-late (b).

Table 22.2 shows the maximum service data rate possible for several MODCODs with 10 s TI for the uniform and uniform-late profiles of the CI. The size of the late part of the UL profile is 50% of the codeword, maximum size allowed in DVB-NGH.

22.4.2 Convolutional Interleaving with a Uniform-Late Profile

The CI with a single FEC can provide fast zapping with long TI using a uniform-late profile [6]. The uniform-late profile has the advantage that full protection is progressively and smoothly achieved with time. With this profile, a large portion of the information from the codewords is transmitted in one or multiple

Table 22.2 Maximum Service Data Rate Allowed in the SAT Profile for 10 s Time Interleaving (Approximate Values; The Exact Values Depend on the Logical Frame Duration)

MODCOD	CI Uniform (kbps)	CI Uniform-Late 50% (kbps)
QPSK 1/5	336	671
QPSK 1/3	559	1118
16QAM 1/5	671	1342
16QAM 1/3	1118	2237

late frames with a duration of typically less than 1 s, such that the receiver can start decoding the service immediately after the reception of the late frames (so-called "fast access decoding," because decoding is started before a complete codeword has been received). The robustness after zapping is given by the proportion of code bits transmitted in the late part. The larger the size of the late part, the better the performance after zapping. Table 22.3 shows the correspondence between the amount of information that is carried in the late part, and the equivalent coding rate for fast access decoding. Typical MODCODs for the satellite profile generally employ very robust coding rates like 1/3 or 1/5, in order to enable the reception at low CNR values. The value corresponding to 30% (i.e., 70% of the code bits have not been received) with coding rate 1/3 is not listed in the table, as this would result in fewer code bits received than the number of information bits represented by a codeword (i.e., coding rate < 1), which rules out a successful decoding.

The drawback of the uniform-late profile is that the overall performance in mobile channels is reduced because it results in a nonuniform interleaving of information over time. The size of the late part represents a trade-off between overall performance in mobile channels and performance after zapping. In DVB-SH, it is recommended to employ for the late part an equivalent coding rate 4/5 [6], such that users in good reception conditions (e.g., in LoS with the satellite) can benefit of fast zapping without seriously degrading the

Table 22.3 Equivalent Coding Rate in Fast Access Decoding for the Uniform-Late (UL) CI Profile

Size Late Part	50%	45%	40%	30%
Overall coding rate 1/3	2/3	20/27 (≈11/15)	5/6	—
Overall coding rate 1/5	2/5	4/9	1/2	2/3

continuous performance. In DVB-NGH, the size of the late part is variable, the maximum size being 50%, but it should be taken into account that unlike turbo-codes, LDPC exhibits a larger performance degradation due to puncturing (erasures).

After zapping, the missing parts of the codewords can be considered as erasures of log-likelihood ratios at the input of the LDPC decoder. The information available for LDPC decoding corresponds to the late part of the profile, whereas the erasures correspond to the bits that are transmitted in the uniform part. In the case of DVB-NGH, these erasures are distributed in a pseudorandom manner by the bit and the cell interleaver (see Chapter 11). Since the puncturing pattern is not optimized to the code structure, some performance loss can be expected in fast access decoding when compared to the equivalent non-punctured LDPC code.

22.4.3 Illustrative Performance Evaluation Results

This section presents illustrative results about the performance of long TI in DVB-NGH. Results have been obtained by means of physical layer simulations in the TU6 (Typical Urban 6-path) terrestrial mobile channel and the LMS SU channel [4]. Table 22.4 summarizes the simulation parameters.

22.4.3.1 Performance after Zapping

In Figure 22.7, we show the performance of fast access decoding in the TU6 channel after the reception of only the late frame for the configurations listed in Table 22.3 [10]. We have also represented the curves corresponding to late decoding with coding rates 2/5, 7/15 ($\approx$1/2), 2/3, and 11/15 for the sake of comparison. In the figure, we can see that the performance of fast access decoding diminishes when a lower percentage of information is carried in the late part (i.e., when a more uniform interleaving is performed). Similarly, the performance loss of fast access decoding compared to the equivalent non-punctured code increases with more uniform profiles of interleaving. For coding rate 1/3, the loss with the UL 50% and UL 45% profiles is around 1 and 2 dB respectively compared to the equivalent coding rates 2/3 and 11/15. For coding rate 1/5, the loss with the UL 50%, UL 40%, and UL 30% profiles is around 1, 2, and 3 dB respectively compared to the equivalent coding rates 2/5, 7/15, and 2/3. We can see that the performance degradation in fast access decoding is smaller with lower coding rates, as these can recover from a higher number of erasures in the codewords. For example, going from the UL 50% to the UL 40% profile requires an extra 8 dB to achieve fast access decoding with coding rate 1/3, but only 4 dB with coding rate 1/5.

Table 22.4 Simulation Parameters for Long Time Interleaving

Bandwidth	8 MHz (TU6)	f_{RF}	600 MHz (TU6)
	5 MHz (LMS)		2.2 GHz (LMS)
FFT size	8K (TU6)	FEC codeword	16,200
	2K (LMS)		
Antenna conf.	SISO	Velocity	60 km/h
GI	1/4	Coding rate	1/3, 1/5
Frame duration	200 ms	Sub-slicing	Maximum
Cycle time	1 s	TI	10 s
Constellation	QPSK	Simulation time	1 h
Rotated	Disabled	Service rate	250 kbps
Ch. estimation	Ideal		

Note that the shown performances for fast access decoding apply only in the very first frame immediately after zapping. As more code bits are received for codewords transmitted in later frames, the performance improves during a transitional phase and reaches that of conventional (i.e., late) decoding, as can be seen in the next section.

22.4.3.2 Transitory Performance

Figure 22.8 shows the performance over time with the uniform and uniform-late 50% CI profiles with a TI duration of 10 s. A time-sliced transmission has been assumed, with a cycle time of 1 s. The channel model is again TU6 with 33 Hz Doppler. In the figure, it can be observed that uniform CI profile is not capable of providing fast zapping, whereas with the uniform-late profile, users in good reception conditions can display the service after receiving the first burst of information. For the considered channel model, after 10 s, the performance of both profiles is identical. This is not the case for the LMS channel, as it will be shown in the next section. We can also note that the lower the coding rate, the better the zapping performance for both uniform and uniform-late profiles.

22.4.3.3 Continuous Performance

Figure 22.9 shows simulation results in the LMS SU channel of the late decoding performance for the previously considered configurations with 50% of the code bits

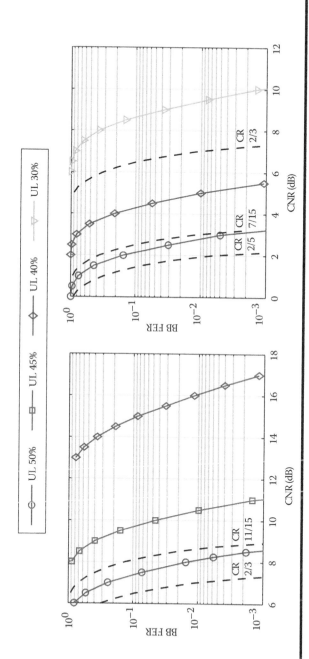

Figure 22.7 Performance after zapping (fast access decoding) with different UL profiles. TU6 channel with 33 Hz Doppler. Left: QPSK 1/3, right: QPSK 1/5.

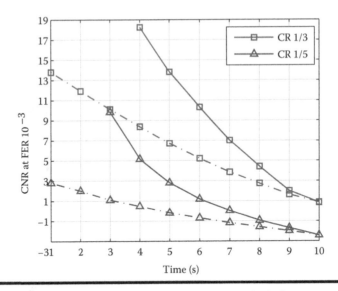

Figure 22.8 **Performance during the transitory period from fast access decoding to decoding of the complete codeword. Continuous lines correspond to uniform and dashed lines correspond to UL profile with 50% of information in the late part. TU6 channel with 33 Hz Doppler.**

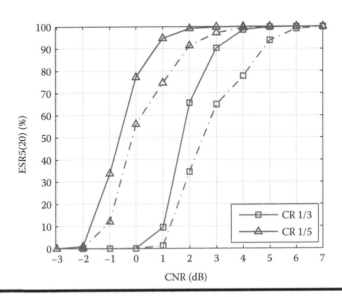

Figure 22.9 **Continuous performance (late decoding) in the LMS SU channel, when 50% of the code bits reside in the late part.**

in the late part. In the figure, we can see that the performance of the CI uniform-late profile is reduced about 1.2 and 1.8 dB compared to the uniform profile for coding rates 1/5 and 1/3, respectively.

22.5 Single Carrier OFDM

22.5.1 SC-OFDM Background

Like most of the wireless systems currently in use, the DVB systems targeting terrestrial transmissions (T, T2, and H) rely on the OFDM modulation. This quasi ubiquity of the OFDM modulation results from several factors: a built-in ability to combat multipath channel degradations, an intrinsic flexibility to map pilot and user/services data in the time and frequency domains, and a reasonable implementation complexity to achieve high spectral efficiencies. Not only for all these reasons but also to maximize the commonalities between T2 and NGH receiver chips, the DVB-NGH standard specifies the OFDM modulation for its terrestrial component. To allow for nationwide coverage and avoid the issues met with the mitigated deployment of the DVB-H and SH systems, the DVB-NGH standard also provisions an optional satellite component. It must be pointed out that there is no sheer satellite mode in the DVB-NGH standard; the satellite component comes in support to the terrestrial transmission, leading to the so-called terrestrial/satellite hybrid profile. It is for instance possible to deploy an SFN network where both the terrestrial and satellite links operate in OFDM in the L- and S-bands according to system parameters specifically defined for the satellite component. Indeed, the satellite component brings its own constraints coming from a specific propagation channel and the technical characteristics of the transponders used in the satellite to forward the signal toward the ground. Information must typically be transmitted with a robust modulation and coding scheme compliant with low SNRs and a long TI to avoid shadowing impairments. If an SFN implementation is not desirable, the DVB-NGH standard also specifies an MFN mode where both components operate in OFDM in two different bands: in the V/UHF band according to terrestrial parameters for the ground link and in the L- or S-band and according to satellite parameters for the space link.

However, the OFDM modulation is not perfectly suited for satellite transmissions. As discussed in Chapter 23, the OFDM modulation shows significant fluctuations of the instantaneous power that prevent from operating the satellite power amplifier (PA) close to the saturation point, i.e., with high power efficiency. But the power consumption and thus the durability of the satellite are directly related to the power efficiency of the on-board amplifiers. For that reason, satellite systems often rely on waveforms with low power fluctuations (generally measured in terms of PAPR). This is the case for the DVB-SH standard that specifies a single carrier (also called TDM) modulation for its sheer satellite component. Interestingly, the DVB-NGH standard introduces a new waveform for satellite transmissions, the so-called SC-OFDM modulation. This new scheme was adopted thanks to its ability to achieve a low PAPR and a high spectral efficiency while showing a lot of implementation commonalities with pure OFDM. The SC-OFDM modulation is actually specified in a second MFN mode where the terrestrial link operates in OFDM. The SC-OFDM modulation is described in detail in Chapter 23. For the consistency of the present chapter, this section provides a brief description of the SC-OFDM scheme.

22.5.2 SC-OFDM Implementation in DVB-NGH

22.5.2.1 DFT Spread Block

The SC-OFDM terminology has actually been devised for the specific case of the DVB-NGH standard. The method is derived from the single-carrier frequency division multiple access (SC-FDMA) used for the uplink transmissions in 3GPP/LTE cellular networks. According to the SC-FDMA technique, each user is allocated with a sub-band of the entire bandwidth with a transmission occurring in OFDM. But unlike in pure OFDM, the symbols to be transmitted are precoded by means of a discrete Fourier transform (DFT) prior modulation. The precoding DFT is computed on the modulated sub-carriers while the modulation IDFT is computed on the whole multiplex with all unused sub-carriers set to zero. The resulting waveform is thus an oversampled version of the original symbols, which insures a low PAPR of the transmitted signal. It is thus possible to operate the PA in the terminals closer to saturation, i.e., with higher power efficiency while benefiting from the zero roll-off spectral efficiency of OFDM. With the ability to significantly reduce the PAPR of OFDM, the SC-FDMA technique clearly appears as a relevant candidate for power-constrained situations. This is particularly true in the context of the DVB-NGH system: it enables to benefit altogether from the low power envelope fluctuations of TDM, the robustness of OFDM in terms of mobility, and the commonalities with OFDM for the implementation of the receiving chips. The SC-FDMA terminology has been replaced by SC-OFDM for NGH as there is no multiple access, but a single signal is transmitted toward all users.

As shown in Figure 22.10, the SC-OFDM transmitter and receiver closely resemble their OFDM counterparts. They mainly differ with the introduction of the DFT and IDFT precoding functions described as spreading and de-spreading in Figure 22.10. The precoding can indeed be viewed as a way of spreading each symbol over the entire spectrum. This leads to the alternative appellation of SC-OFDM as the DFT-spread OFDM modulation. It can be further shown that

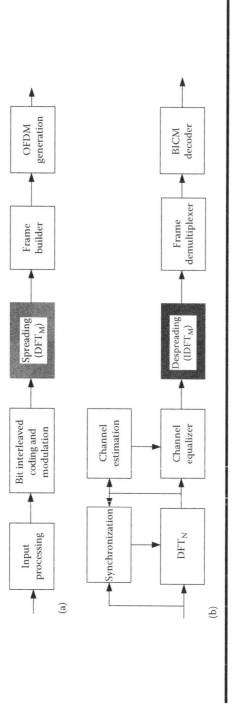

Figure 22.10 Block diagram of the SC-OFDM transmitter (a) and receiver (b).

the overall complexity of the SC-OFDM receiver is not significantly higher than the OFDM one (see Chapter 23).

22.5.2.2 Pilot Pattern

What makes the OFDM modulation particularly suitable for time-varying frequency-selective channels is the ability to insert pilots on whatever cell position in the time and frequency domains. The DVB-T2 and NGH standards specify PPs with reference symbols inserted periodically in time and frequency to allow for a robust channel estimation while not degrading too much the spectral efficiency. In the case of the SC-OFDM modulation, the reduction of the power fluctuation is obtained at the expense of a loss in the OFDM flexibility. Mixing together pilots and spread data would indeed break the good PAPR of the SC-OFDM signal. It shall be mentioned that the 3GPP/LTE specifies a full SC-OFDM pilot inserted every seven symbol periods. Taking into account the Doppler constraints of the DVB-NGH standard, this would lead to inserting a full pilot every six symbol periods, thus doubling the number of pilots with respect to the most robust NGH pilot schemes (PP1 for example). The issue is solved in DVB-NGH through the definition of a new pilot pattern (PP9) that carries an equal amount of user data and reference pilots (see Figure 22.11).

Pilots and data are interleaved in the frequency domain every two sub-carriers in this special OFDM symbol. Data are generated like for a full SC-OFDM symbol but over only $M/2$ cells. The result of the $M/2$-DFT is mapped to the data sub-carriers as shown in Figure 22.12. The sequence of pilot cells is generated from a Zadoff–Chu (ZC) sequence according to the scheme used in the 3GPP/LTE standard. ZC sequences are constant-amplitude zero autocorrelation (CAZAC) sequences both in the time and in the frequency. If $L = M/2$ denotes the ZC sequence length, then

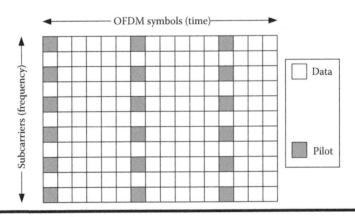

Figure 22.11   NGH pilot structure for SC-OFDM.

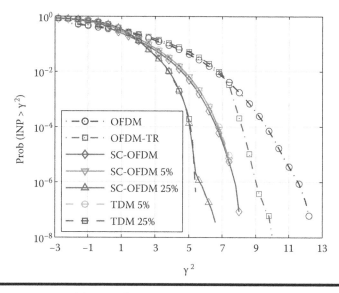

Figure 22.12   CCDF of the INP for SC-OFDM, OFDM, OFDM-TR, and TDM.

the complex value at each position k of each p-root ZC sequence with integer shift l is (the size L of the Zadoff sequence is even here):

$$x_k = e^{-2j\frac{\pi}{L}\left(pk^2 + lk + 0.5k\right)}. \tag{22.1}$$

As only one sequence is needed, $p = 1$ and $l = 0$ have been selected for the PP9 pattern. The sequence is directly generated in the frequency domain and is mapped to the pilot sub-carriers as shown in Figure 22.11.

Recall that computing the N-IDFT of a sequence with zeros regularly inserted one every two samples generates a signal that is twice the repetition of the signal that would be obtained by computing the (N/2)-IDFT on the useful samples. In other words, the OFDM modulation applied onto the PP9 symbols is the sum of two periodic signals, a time-domain ZC sequence and the SC-OFDM signal corresponding to the data symbols. Over each period, the signal is actually the oversampled version of the original signal due to the null sub-carriers on the band edges. Both the ZC sequence and the SC-OFDM symbol (assuming a QPSK modulation) would have a constant envelope if there was no oversampling. The oversampling operation actually introduces variations in the signal amplitude with peaks occurring between each constant envelope samples. The peaks in the sum of the ZC and SC-OFDM signals occur at the same positions. To reduce the amplitude of the peaks, the ZC sequence is time shifted by half a sampling period (+0.5 k term in Equation 22.1).

22.5.3 *Illustrative Performance Results*

The performances of the SC-OFDM modulation are discussed with some details in Chapter 23. This section simply reproduces two results that assess the interest of the SC-OFDM for satellite transmissions. The first result compares the power fluctuations of various waveforms measured in terms of the complementary cumulative density function (CCDF) of the instantaneous normalized power (INP) defined as:

$$\text{CCDF(INP)} = \Pr\left\{ \overbrace{\frac{|v(n)|^2}{\dfrac{1}{N_S}\displaystyle\sum_{k=0}^{N_S-1}|v(k)|^2}}^{INP(v)} > \gamma^2 \right\}, \tag{22.2}$$

where

$v(k)$ is the sampled version of the transmitted signal $v(t)$

N_S is the number of samples considered to evaluate the INP

The CCDF of the INP asymptotically behaves like the CCDF of the PAPR, but it provides a better resolution for the low values of power. Figure 22.12 actually gives the CCDF of the INP for the following modulations: SC-OFDM, OFDM, TDM with different roll-offs, and OFDM-TR, where the latter is a PAPR reduction technique for OFDM specified in DVB-T2 called TR algorithm.

These curves confirm the good PAPR properties of the SC-OFDM that outperforms both the OFDM and OFDM-TR modulation even at low IBOs as we will further discuss later. TDM waveforms exhibit better performance than SC-OFDM but only when the roll-off exceeds 0%, i.e., not for the same spectral occupancy. It occurs that the (SC-)OFDM modulation with rectangular window is spectrally equivalent to TDM with a roll-off of 0%. An evolution of the SC-OFDM, the Extended and Weighted SC-OFDM modulation was devised to allow for controlling the frequency bandwidth just like for TDM. Without going into details on its actual principle, Figure 22.12 shows that the extended SC-OFDM behaves like TDM for the same value of roll-off except for very large power values. But if the application of zero roll-off is readily achieved for SC-OFDM, it implies severe filtering issues for TDM waveforms. With an increasing pressure on the spectrum, it was decided in NGH to keep the SC-OFDM in its original form, i.e., with a roll-off of 0% (smallest bandwidth). The main result of this study is that the SC-OFDM modulation is strictly similar to TDM modulations in terms of PAPR for the same spectral occupancy.

The second result presented here deals with the robustness of the OFDM-based waveforms with respect to the nonlinear degradations of PAs. It is common to illustrate these degradations in terms of total degradation (TD) that measures the overall SNR loss in comparison to a perfect linear amplifier to achieve a given level of bit error rate (BER). As previously mentioned, the power efficiency of an amplifier is optimal when operating at the saturation point. In order to avoid the clipping effect, it is common to operate the PA with some margin from the saturation point, the so-called IBO that constitutes a loss with respect to a perfect amplifier (see Section 23.2). More precisely, the loss is given by the output back-off (OBO) that measures the margin with respect to the output saturation value of the PA. To exploit the full output power, it is not possible to apply small IBOs. The transmitted signal then suffers from clipping that reduces the robustness against noise. It is common to evaluate this degradation as a loss in SNR for a given reference value of target BER in comparison to an ideal amplifier. The TD is commonly expressed as the sum of two terms:

$$TD = OBO + \left[SNR_{PA} - SNR_{linear}\right]_{BER_{ref}} = OBO + \Delta SNR. \quad (22.3)$$

When the back-off is high, there is virtually no in-band distortion and thus no distortion-related BER loss ($\Delta SNR \approx 0 \ll OBO$). When working at low OBO, in-band distortions increase, ΔSNR loss is important, and becomes the predominant term in the TD. There is an optimal working point I_{opt}, which ensures a compromise between OBO and ΔSNR and yields a minimum TD. Figure 22.13 compares the TD and OBO with respect to the IBO for a BER value of 10^{-5} for the OFDM, SC-OFDM, and OFDM-TR modulations and for a configuration with QPSK and

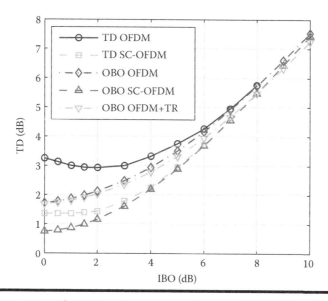

Figure 22.13 TD and OBO performance—Linearized TWTA PA, AWGN channel. OFDM: QPSK 4/9, LDPC 16k, *N* 2048, *GI* 1/32. SC-OFDM: QPSK 4/9, LDPC 16k, *N* 512, *M* 432, *GI* 1/32.

the LDPC code of rate 4/9. It appears from these curves that the optimum IBO for the OFDM is 2 dB leading to a TD of 2.95 dB while the optimum IBO for the SC-OFDM is 1 dB leading to a TD of 1.35 dB. This shows that the SC-OFDM enables operating closer to the saturation point than the OFDM, thus improving the power efficiency of the amplifier. The fact that the TD is smaller by 1.6 dB in favor of SC-OFDM means an increased quality of service (higher robustness), a reduced power consumption, a higher spectral efficiency, or an increased satellite coverage. The main result of this study is that the SC-OFDM modulation leads to an overall better performance than OFDM due to the nonlinear degradations of PAs. It is thus well suited to implement satellite transmissions, especially when PAs are operated at low IBOs.

22.6 Comparison DVB-NGH vs. DVB-SH

The first DVB system for hybrid satellite and terrestrial broadcasting of multimedia services to mobile handheld receivers, namely, the DVB-SH standard, was published in 2007 [2]. DVB-SH was defined as a transmission system for frequencies below 3 GHz suitable for satellite services to handheld devices. The propagation phenomena that characterize the mobile satellite channel drove the choice to the hybrid architecture, the terrestrial networks being the only feasible solution in an urban scenario (in particular for indoor coverage), where satellite propagation is compromised by the presence of obstacles and multipath fading, whereas in rural areas, the lower signal attenuation and the low population density allow for the satellite coverage, with a reduction of the infrastructure costs.

The DVB-SH system relies on a cooperative hybrid satellite/terrestrial infrastructure, where the signals are broadcast to the mobile terminals on two paths (see Figure 22.14):

- A direct path from the broadcast station to the terminals via the satellite.
- An indirect path from the broadcast station to the terminals via terrestrial repeaters that form the Complementary Ground Component (CGC) to the satellite. The CGC can be fed through satellite and/or terrestrial distribution networks.

DVB-SH was designed to assure service continuity between the Space Component and CGC coverage and to work in very different reception conditions related to portable and mobile terminals: indoor/outdoor, urban/suburban/rural, static/mobile conditions. DVB-SH terminals are able to switch the reception to the stronger component (satellite or terrestrial transmitter) or even to combine them when simultaneously available.

In wide sense derived from the DVB-T [11] and DVB-H [1] system specifications, respectively designed for digital television terrestrial broadcasting toward fixed and

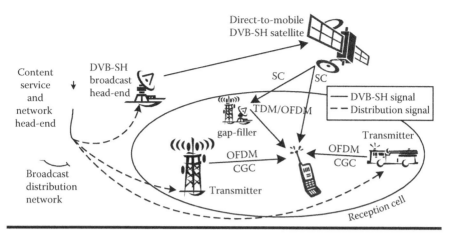

Figure 22.14   DVB-SH system architecture.

mobile terminals, and DVB-S2 [12] for digital satellite broadcasting toward fixed terminals, DVB-SH provides two different transmission modes:

- SH-A, an OFDM mode, for both the satellite and terrestrial paths. This configuration allows the use of the same frequency block for the two paths, for a high spectrum efficiency, and simpler receiver design with single signal reception. Furthermore, at the expenses of an overall tighter synchronization of the terrestrial and satellite networks, the combination of the two signals at the receiver can strengthen the reception in an SFN configuration. As a counterpart, it is not possible to optimize the waveform for the satellite path.

- SH-B, a TDM mode used on the direct satellite path only, together with an OFDM terrestrial path, on a different frequency block. This mode supports code diversity recombination between satellite TDM and terrestrial OFDM modes, to increase the robustness of the transmission in some marginal areas (mainly suburban). Furthermore, it allows the independent optimization of the satellite and terrestrial parameters, so that the on-board satellite amplifiers are able to work closer to the saturation point thanks to the lower PAPR. As a counterpart, it requires a more complex (dual) receiver design.

In order to guarantee a high quality of service to the user also in severe mobile conditions, such as deep fading events, the DVB-SH standard specifies two complementary FEC schemes capable of providing long TI with fast zapping support: one at the physical layer using a turbo-code with a CI and another one at the link layer known as Multi Protocol Encapsulation inter-burst FEC (MPE-iFEC) [6]. According to the type of interleaving employed, receivers are classified into two different classes:

■ Class 1 receivers are based on a short physical layer protection (up to 200 ms) to counteract short fading events, and MPE-iFEC to cope with long fading events.

■ Class 2 receivers are based on a long physical layer protection (up to 30 s) that may be complemented by a link layer MPE-iFEC protection, as in class 1 receivers, for even longer protection.

Class 2 receivers provide the best performance, but they require very large memory sizes (e.g., 256 Mb [6]), making it suitable mainly for car radio applications. Class 1 receivers transfer part of both memory and protection from the physical to the link layer with a suboptimum performance but more appropriate system design for the mobile phone market. The memory size envisaged for this receiver class is in the order of 4 Mb [6]. MPE-iFEC requires significantly less memory because it performs erasure decoding instead of soft decoding. In mobile conditions, it achieves a performance close to single FEC because in the LMS channel, the soft information is unusable when channel attenuation is very deep. However, it does not improve the reception in static channels, and it actually degrades the performance compared to having all the protection at the physical layer.

DVB-NGH was born with a different philosophy: while DVB-SH is a hybrid system where the terrestrial network complements the satellite, DVB-NGH is a terrestrial system with an optional satellite component to help completing the terrestrial coverage. Hence, despite the fact that both systems target hybrid satellite–terrestrial broadcasting to handhelds, different technologies have been adopted for the two standards. The DVB-NGH satellite component maintains maximum commonality with the terrestrial component and introduces technologies optimized for the satellite channel, only when justified by significant improvement of the performance.

Table 22.5 lists the main technologies adopted for the two standards, where it is evident that several differences have been introduced for the satellite link of DVB-NGH in order to improve the overall performance with respect to DVB-SH.

The TDM waveform employed in DVB-SH has been replaced by SC-OFDM in DVB-NGH. SC-OFDM (0% roll-off) is somehow suboptimum to TDM (20% roll-off) in the presence of the PA's nonlinearities (the suboptimality can be roughly quantified in the order of 0.5 dB), but the commonalities to OFDM justify its adoption, for a reduced implementation cost at the receiver side. For a detailed description of SC-OFDM, see Chapter 23.

The FEC adopted by DVB-NGH is the same LDPC plus BCH structure of DVB-T2 with a FEC codeword size of 16,200 bits. The performance is similar to the turbo-code adopted in DVB-SH (less that 1 dB to the Shannon limit). The turbo-code of DVB-SH allows code combining between the terrestrial and satellite signals in hybrid MFN networks, which has a better performance than packet selection used in DVB-NGH. The reason is that the different coding rates

Table 22.5 Comparison DVB-SH vs. Satellite Profile of DVB-NGH

	DVB-SH	*DVB-NGH*
Physical layer FEC	Turbo-code	LDPC + BCH
FEC codeword size	12,282 bits	16k (16,200 bits)
Modulation	OFDM and TDM	OFDM and SC-OFDM
Constellations	OFDM: QPSK and 16QAM	OFDM and SC-OFDM: QPSK and 16QAM (rotated and nonrotated)
	TDM: QPSK, 8PSK, and 16APSK (nonrotated)	
PAPR reduction	None	TR and Active Constellation Expansion (for OFDM)
Memory requirements	Receiver class 1: 417792 IUs (128 code bits each)	2^{22} cells (QPSK or 16QAM symbols) (apply to all PLPs of a service)
	Receiver class 2: 6528 IUs (apply to the whole multiplex)	
Long TI with fast zapping	PHY: CI uniform, early-late and uniform-late	PHY: CI uniform and uniform-late
	LL: MPE-iFEC	
Time slicing	DVB-H-like (delta-T) time slicing and DVB-SH-specific time slicing	DVB-T2-like time slicing and time slicing using the logical channel scheduling
MIXO	No	MISO/OFDM: eSFN, Alamouti
		MIMO/OFDM: eSM + PH
		MIMO/SC-OFDM: SM
Multiple quality of service (robustness, latency)	Link layer with MPE-iFEC	PLPs

(continued)

Table 22.5 (continued)   Comparison DVB-SH vs. Satellite Profile of DVB-NGH

	DVB-SH	*DVB-NGH*
SVC and graceful degradation	No	Physical layer with multiple PLPs
Hybrid signal combining	SFN: OFDM	SFN: OFDM
	MFN: Code combining	MFN: packet selection
Frequency bands	S-band	L- and S-band
Role of terrestrial component	Complementary to satellite CGC	Main component; DVB-NGH can be transmitted inside a FEF of a DVB-T2 signal
IP encapsulation	MPE	GSE
Packet header compression	None	Both TS and IP (ROHC)

are achieved from the same mother code with the lowest coding rate. Another advantage of turbo-codes compared to LDPCs is that they suffer less from performance degradation against erasures. However, DVB-NGH employs rotated constellations that have the potential to improve the performance of long TI by ensuring a more uniform distribution of information in the presence of deep fades [10]. This is especially important for the uniform-late profiles of the convolutional TI, for which the loss of one frame results in a large percentage of information being erased.

Another important advantage of DVB-NGH is the possibility of having different services and even service components with different transmission configurations within the multiplex, thanks to the PLPs. The TDI memory in DVB-NGH is also applied to all PLPs of a service (including all its associated components), rather than the whole multiplex as in DVB-SH. This reduces the size of the required silicon.

DVB-NGH has considerably improved the bandwidth utilization efficiency by designing low overhead protocol stacks. In DVB-NGH, the encapsulation overhead for IP delivery is reduced up to 70% thanks to the use of Generic Stream Encapsulation (GSE) as link layer encapsulation protocol instead of MPE over MPEG-2 Transport Stream (TS). Furthermore, DVB-NGH has specified two packet header compression mechanisms for both the TS and the IP profile. The new TS packet header compression method adopted allows reducing the TS packet overhead from 4 bytes to only 1 byte, providing a 1.1% system capacity increase.

The IP packet overhead can be reduced with Robust Header Compression (ROHC) to approximately 1% of the transmitted data, yielding a capacity increase between 2.5% and 3.5%. For more details on the overhead reduction methods of DVB-NGH, see Chapter 17.

DVB-NGH is also the first mobile broadcast system to specify the use of MIMO antenna configurations in order to overcome the performance limitations of conventional systems with one transmit and one receive antenna without any additional bandwidth or increased transmit power. For details on the hybrid MIMO profile of DVB-NGH, see Chapter 24.

Although DVB-NGH technically outperforms DVB-SH in terms of capacity and coverage, the main advantage of DVB-NGH is that there is no need to deploy a dedicated terrestrial network, because it is possible to reuse existing DVB-T2 infrastructure to start providing DVB-NGH services in-band a DVB-T2 multiplex. Thanks to the FEFs of DVB-T2, it is possible to efficiently share the capacity of one frequency channel in a time division manner between DVB-T2 and DVB-NGH, with each technology having specific time slots.

22.7 Conclusions

The hybrid profile of DVB-NGH has been defined as a good compromise between optimized performance on the LMS channel and commonality to the base terrestrial profile, for minimizing the required design and implementation effort and fostering the introduction in the market of satellite capable demodulators. The satellite component of DVB-NGH allows improving the coverage to rural areas, where the installation of terrestrial networks could be uneconomical. Technologies specified in the standard allow operators to properly configure the system for operation in SFN or MFN, including an external TDI memory and convolutional inter-frame TI with a uniform-late profile in order to support long TI (e.g., around 10 s) with fast zapping, and single carrier OFDM for hybrid MFN networks in order to reduce the PAPR of the transmitted signal. SC-OFDM provides an approximate gain in the order of 2.5 dB in terms of reduced PAPR for high-power satellite amplifiers, which can be directly translated into an increase in the coverage provided by the satellite, achieving a gain of about 1.5 dB in the link budget at low IBOs.

References

1. ETSI EN 302 304 v1.1.1, Digital video broadcasting (DVB); Transmission system for handheld terminals (DVB-H), November 2004.
2. ETSI EN 302 583 v1.1.2, Digital video broadcasting (DVB); Framing structure, channel coding and modulation for satellite services to handheld devices (SH) below 3G, February 2010.

3. DVB Commercial Module sub-group on Next Generation Handheld, Commercial requirements for DVB-NGH, CM-1062R2, June 2009.
4. F. Pérez-Fontán et al., Statistical modelling of the LMS channel, *IEEE Transactions on Vehicular Technology*, 50(6), 1549–1567, 2001.
5. H. Fuchs and N. Färber, Optimizing channel change time in IPTV applications, *Proc. IEEE BMSB*, Las Vegas, NV, 2008.
6. ETSI TS 102 584 v1.1.2, Digital video broadcasting (DVB); DVB-SH implementation guidelines, January 2011.
7. EN 302 755 v1.3.1, *Frame Structure Channel Coding and Modulation for a Second Generation Digital Terrestrial Television Broadcasting System (DVB-T2)*, November 2011.
8. H. Schwarz, D. Marpe, and T. Wiegand, Overview of the scalable video coding extension of the H.264/AVC standard, *IEEE Transactions on Circuits and Systems for Video Technology*, 17(9), 1103–1120, September 2007.
9. S. A. Wilkus et al., Field measurements of a hybrid DVB-SH single frequency network with an inclined satellite orbit, *IEEE Transactions on Broadcasting*, 56(4), 523–531, December 2010.
10. D. Gozálvez, *Combined Time, Frequency and Space Diversity in Multimedia Mobile Broadcasting Systems*, PhD dissertation, Universitat Politèctica de València, Valencia, Spain, June 2012.
11. ETSI EN 300 744 v.1.6.1, Digital video broadcasting (DVB); Framing structure, channel coding and modulation for digital terrestrial television (DVB-T), January 2009.
12. ETSI EN 302 307 v1.2.1, Digital video broadcasting (DVB); Second generation framing structure, channel coding and modulation systems for broadcasting, interactive services, news gathering and other broadband satellite applications, August 2009.

Chapter 23

Single-Carrier OFDM for the Satellite Component of DVB-NGH

Arnaud Bouttier, Damien Castelain,
Cristina Ciochină, and Fumihiro Hasegawa

Contents

23.1 Introduction

One of the main constraints when deploying a broadcasting network is to guarantee the same quality of service to all the subscribers whatever their location in the served area. This constraint is even more stringent when the network is meant to broadcast content to handheld and mobile terminals generally positioned a few feet above the ground, not necessarily in line of sight and possibly inside buildings. For nationwide networks with not densely populated areas, the satellite appears as a technically and economically relevant alternative to the deployment of many towers and gap fillers. To maximize the chances of the DVB-NGH system being widely adopted, the NGH commercial requirements [1] make the provision of an optional satellite component to improve the coverage of the core terrestrial network. The purpose was obviously to avoid the issues met with the DVB-H and DVB-SH systems where the DVB-H has no satellite component and the DVB-SH is a satellite-centric system with an optional terrestrial component. Two kinds of DVB-NGH networks are actually envisaged: terrestrial-only networks described in the so-called core or base profile (see Chapter 7) and mixed terrestrial/satellite networks specified in the hybrid profile (see Chapter 20). A distinctive feature of the DVB-NGH hybrid profile is the specification of a new transmission scheme: the SC-OFDM modulation, a multi-carrier waveform that is particularly well suited for satellite transmissions.

Satellite systems are indeed due to cope with a specific propagation channel but also with the technological constraints of the satellites themselves. On top of those lies the power amplifier (PA) that is due to bring the signal to be transmitted at a level compatible with the receiver sensibility over large areas. In order to guarantee the durability of the satellite, it is critical to keep low the power consumption of the system and thus to optimize the amplifier power efficiency, i.e., to drive the amplifier close to saturation. Single-carrier (SC) modulations also described as time division multiplexing (TDM) modulations have long been the reference scheme for satellite transmissions, e.g., such as in the DVB-S2 system developed in 2003 [2], for their suitability to achieve low power fluctuations compatible with good power efficiency. However, the orthogonal frequency division multiplexing (OFDM) modulation is taking over SC schemes thanks to a better flexibility and a comparatively lower complexity when it comes to compensating for high channel degradations. OFDM is now used in many systems such as 3GPP/LTE cellular networks, IEEE 802.11 WiFi solutions, IEEE 802.16 WiMax systems, etc. Dealing with broadcast, the DVB-SH (Satellite to Handhelds) standard developed in 2007 [3] specifies two modes of operation, one OFDM mode allowing the deployment of single-frequency networks (SFNs) on both the terrestrial and satellite links and a second mode with TDM on the satellite link and OFDM on the terrestrial link for multiple-frequency networks (MFNs).

Despite its obvious advantages, the OFDM modulation is not perfectly well suited for satellite transmissions. Indeed, the more subcarriers are added together, the more the signal behaves like a Gaussian noise with large power fluctuations. The PA shall be operated either in a linear mode far from saturation and thus with a poor power efficiency or in a clipping mode at the cost of significant performance degradations due to saturation. This weakness of OFDM is well known, and several means have been studied to reduce its power fluctuations such as the tone reservation (TR) or active constellation extension (ACE) techniques specified in the DVB-T2 standard [4]. However, as it will be shown in the sequel, the gain of these techniques appears to be marginal in the context of satellite transmissions. More recently, the 3GPP selected the single carrier–frequency division multiple access (SC-FDMA) technique for the uplink of LTE networks in the purpose of reducing the power consumption of handheld devices. The SC-FDMA scheme is actually a derivation of the orthogonal frequency division multiple access (OFDMA), where each user is allocated with a given subset of contiguous OFDM subcarriers. But unlike in OFDMA, the symbols to be transmitted are precoded in the frequency domain by means of a discrete Fourier transform (DFT) prior to OFDM modulation. It is thus possible to benefit from the good PAPR properties of the original symbols to save power in the amplification stage.

It is on the basis of these obvious similarities with the satellite context that the SC-FDMA was proposed as a candidate waveform for the NGH hybrid profile. A broadcast signal is inherently transmitted toward all receivers. The SC-OFDM terminology was selected to identify the specific case where the whole bandwidth is allocated to all users. As shown in this chapter, the SC-OFDM modulation is clearly suitable for satellite transmissions. It was adopted in the DVB-NGH standard thanks to its ability to preserve a lot of commonalities with pure OFDM while benefiting from the low power fluctuations of SC signals. The purpose of this chapter is to introduce the principle of the SC-OFDM modulation and to characterize its properties and performance in the particular case of the NGH hybrid profile.

The rest of the chapter is structured as follows. Section 23.2 deals with the characterization of the PA with respect to satellite transmissions. Section 23.3 briefly describes and compares the merits of the different waveforms used for satellite transmissions. The next section briefly describes the specifications of the SC-OFDM modes in the DVB-NGH standard. The chapter ends with a section dedicated to the performances of the SC-OFDM modulation.

23.2 Power Amplifier Characterization

As mentioned earlier, a PA amplifies the amplitude of the incoming signal so as to bring its power level to a value compatible with the transmission requirements. A perfect PA provides a linear gain whatever the level of the incoming signal. Practically, a PA can only amplify signal up to a maximum value called the saturation amplitude $v_{IN(OUT),Sat}$. A still ideal but more realistic model is the ideal clipper: all peaks above a certain saturation level are clipped, and all the others remain unchanged. It is common to characterize the PA by its AM/AM characteristic, giving the output amplitude $|v_{OUT}|$ as a function of the input amplitude $|v_{IN}|$ (see Figure 23.1).

The impact of the fluctuations in the signal amplitude appears clearly in Figure 23.1 (curve with large dots). If the amplitude of the incoming signal falls in the saturation region, then the output signal will be clipped leading to degradations both in-band with the introduction of noise and also out-of-band

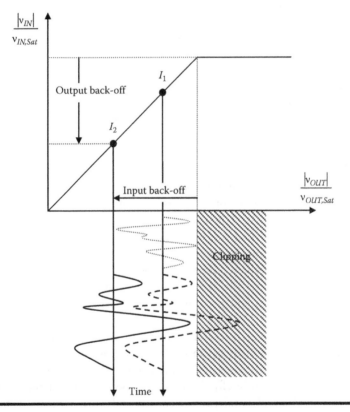

Figure 23.1 Backing-off signals with different dynamic ranges.

with unwanted emissions immediately outside the assigned channel bandwidth (shoulders).

In order to avoid any degradation on the transmitted signal, it is required to operate the PA solely in its linear region. Typically, the average level of the incoming signal is set well below the saturation value so as to leave enough room for higher amplitudes to be amplified without being saturated. The signal is said to be backed off from the saturation point. It is common to define the input back-off (IBO) and output back-off (OBO) with respect to the saturation values as respectively:

$$IBO\big|_{dB} = -10\log_{10}\frac{P_{IN,Avg}}{P_{IN,Sat}}, \tag{23.1}$$

$$OBO\big|_{dB} = -10\log_{10}\frac{P_{OUT,Avg}}{P_{OUT,Sat}}, \tag{23.2}$$

where

$P_{IN(OUT),Sat} = |v_{IN(OUT),Sat}|^2$ represents the input (output) saturation power

$P_{IN(OUT),Avg}$ is the mean power of the signal $v(t)$ at the PA's input (output)

In order to avoid significant degradations, an OFDM signal requires an IBO of 10–15 dBs while an SC signal can afford an IBO of only a few dBs. Using signals with large power fluctuations not only requires high-end PAs with a large linear region but also leads to poor power efficiency. The power efficiency of a PA is measured as the ratio of the radiated output power with respect to the power consumed for operating the PA. The power efficiency, ξ, can be defined as follows:

$$\xi = \frac{P_{OUT,Avg}}{P_{OUT,Avg} + P_{DC}}, \tag{23.3}$$

where P_{DC} is the power consumed by the PA on top of the radiated power (e.g., polarization current). It is clear from this definition that the efficiency of the PA is higher when the average output power is close to the saturation point. For a given transmission, reducing the IBO without saturation makes it possible to either increase the power efficiency (curve with small dots in Figure 23.1) or rely on a PA with reduced constraints in terms of linearity. This is the kind of advantage that led to the selection of the SC-FDMA modulation for the uplink transmission in the 3GPP/LTE cellular system.

The aforementioned analysis clearly shows the importance of the power fluctuations to differentiate the modulations in terms of robustness to PA degradations.

One of the most popular ways of giving a measure of a signal's dynamic range is the peak-to-average power ratio (PAPR). It is used to quantify the envelope excursions of a signal $v(t)$ over a time interval τ [5]:

$$PAPR\left(v(t)\big|_{t\in\tau}\right) = \frac{\max_{t\in\tau}|v(t)|^2}{E\left\{|v(t)|^2_{\,t\in\tau}\right\}}\Bigg|_{dB}. \tag{23.4}$$

The PAPR thus represents the ratio of the maximum instantaneous peak power to the average power of the signal over the observation period τ and is usually expressed in dB. Note that the PAPR provides a partial insight of the signal envelope fluctuations. Other parameters such as in instantaneous normalized power (INP) are considered in this chapter.

23.3 Waveforms for Satellite Mobile Transmissions

23.3.1 TDM

TDM waveforms used in satellite transmissions are actually another name for SC modulations. An SC signal can be described as follows:

$$\bar{y}(t) = \sum_{k=0}^{M-1} \mathrm{Re}\left\{x_k g(t - kT)e^{2j\pi f_0 t}\right\}, \tag{23.5}$$

where
- f_0 is the carrier frequency
- T is the symbol period
- $g(t)$ is the signal waveform with finite energy
- $\{x_k\}$ is a sequence of M discrete symbols carrying the information bits (typically a PSK or QAM alphabet)

Through an appropriate selection of the modulation symbol alphabet and the waveform, it is possible to keep the signal envelope fluctuations low. On the other hand, TDM schemes suffer from strong limitations when it comes to cope with severely time dispersive channels. It can be shown that under some common assumptions, the optimal decoder in the AWGN channel, i.e., the one that minimizes the probability of error, can be simplified as shown in Figure 23.2a. The received signal goes through the matched filter associated to the transmit waveform and is then sampled every T periods at the times that maximize the signal-to-noise ratio (SNR; opening of the eye diagram). Finally, the decision is taken on a symbol-by-symbol basis using a threshold detector. In practice, the demodulator generally delivers

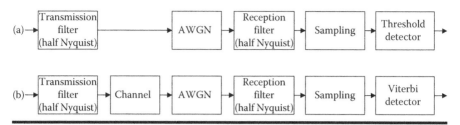

Figure 23.2 **(a) Optimal SC receiver for AWGN channel with Nyquist filtering. (b) Suboptimal SC receiver for selective channel with fixed receive filter.**

soft metrics that improve the performance of the channel decoder. As a matter of fact, the symbol-by-symbol detection can be achieved for any waveform as long as the inter-symbol interference (ISI) introduced by filtering is canceled out at and only at the sampling times. This condition, known as the first Nyquist criterion, is used to reduce the spectral occupancy of the TDM signals with, e.g., the well-known raised-cosine filter. The Nyquist criterion applies on the combined impulse response; the raised-cosine filter shall thus be equally spread between the transmitter and the receiver, preserving the optimality of the matched filter (square root of raised-cosine filter denoted as half-Nyquist in Figure 23.2). The classical square root raised cosine half Nyquist filters are characterized by a parameter, the roll-off, which defines the ratio between the occupied bandwidth and minimum bandwidth required by the signal equal to the modulation rate. For example, a roll-off of 20% means that the occupied bandwidth is 20% larger than the sampling frequency. The smaller the roll-off, the smaller the used bandwidth is. However, decreasing the roll-off has some drawbacks. One of them is that filtering creates peaks in the transmitted signal, i.e., degrades the PAPR, and that these peaks become more pronounced when the roll-off decreases.

If the channel occurs to be dispersive in time, the Nyquist criterion no more holds and the channel also introduces ISI (memory). The detection can no more be applied at the symbol level but in terms of sequences with finite length M, typically using the Viterbi algorithm (maximum likelihood receiver). Figure 23.2b depicts the typical architecture of the Viterbi receiver with a fixed receiving filter that is not matched to the overall transmitting waveform (including the channel), but is selected to whiten the additive noise in the digital domain. Though the Viterbi algorithm reduces the number of sequences to be tested down to $ML-1$ (being L is the length of the discrete-time response of the channel), the complexity of the receiver can become cumbersome when the channel response spreads over several symbol periods (it is generally not the case for satellite transmissions). This situation is typically met for wideband transmissions when the frequency coherence is much smaller than the system bandwidth. One alternative has been devised to cope with those situations: the inverse filtering approach and its dual counterpart, the frequency domain equalization (FDE).

The suboptimal inverse filtering approach basically consists in applying a filter to compensate for the channel frequency selectivity. The problem then reduces to selecting criteria for the definition of the inverse filter. The two most commonly used criteria for the definition of the inverse filter are the zero forcing (ZF, equivalent to the first Nyquist criterion) and minimum mean square error (MMSE) criteria. In both cases, the inverse filter can be divided in two pieces: the matched filter sampled every symbol period so as to maximize the SNR (exhaustive summary) and a digital filter that operates on the down-sampled equivalent channel. Just as for the Viterbi demodulator, the receiving filter is generally kept constant independently of the channel impulse response, leading to the receiver architecture shown in Figure 23.3. The digital inverse filter can be direct, recursive, linear, or even nonlinear (decision feedback equalization).

Figure 23.4 depicts the typical architecture of the TDM modulator and demodulator with more details. After bit interleaved coding and modulation, pilots are inserted and then filtered out. The filtering is generally processed in the digital domain: the signal is oversampled (i.e., zeros are inserted) and then digitally filtered. This implies that after digital-to-analog conversion, the analog filtering (in baseband, IF or RF) will be simplified thanks to this digital filtering.

At the TDM receiver, after analog-to-digital conversion, at a frequency generally higher than the modulation rate, the signal is filtered and then equalized, operations controlled by the synchronization and channel estimation

Figure 23.3   Inverse filtering with a fixed receiving filter and its discrete equivalent model.

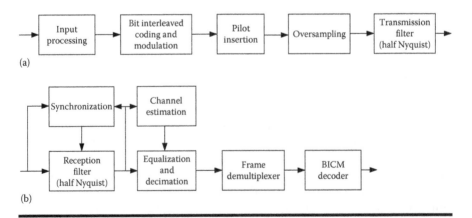

Figure 23.4   TDM transmitter (a) and TDM receiver (b).

modules. After these operations, the signal is down-sampled at the modulation rate and then frame de-multiplexed to be finally decoded.

Whatever the equalizer architecture, when the system bandwidth well exceeds the channel coherence bandwidth (high frequency selectivity), the complexity of the filter may become high with many taps, stability issues, sensitivity to quantification noise, etc. A common alternative is to rely on the FDE. Basically, the FDE demodulator computes the Fourier transform of the signal so as to compensate the effect of the channel in the frequency domain by inverting the channel frequency response in some way. Then, the receiver reconstructs the time domain signal by inverse Fourier transform. FDE simply exploits the fact that the linear convolution in the time domain is transformed into a multiplication in the frequency domain. It would be easier to apply this technique in the discrete time domain, i.e., applying a DFT on the sampled version of the signal. But, in DFT, frequency domain multiplication is equivalent to a circular time domain convolution. As shown in Figure 23.5, a guard interval (GI) populated as a cyclic prefix (CP) with a length longer than the channel response is added in front of each symbol to convert linear convolution into a circular convolution. The general architecture of the FDE demodulator is displayed in Figure 23.6. The size of the DFT shall be large enough to sample the frequency response of the channel well below the coherence bandwidth under the constraint of a quasi-stationary channel response over the observation duration.

The DFT size grows linearly with the length of the channel response. However, the complexity of the DFT per received sample grows log-normally with the DFT size, and therefore the complexity of FDE. The complexity of FDE is thus lower than that of the equivalent time domain equalizer for broadband channel. The SC-DFE solution thus enables maintaining the advantages of SC schemes in terms of low PAPR with a complexity slightly independent of the channel degradations. Nonetheless, the FDE solution also suffers from some drawbacks such as a suboptimal equalization with limited performances on severely frequency selective

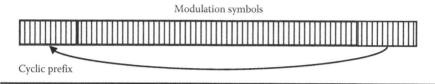

Modulation symbols

Cyclic prefix

Figure 23.5 Addition of a CP to transform the linear convolution into circular convolution for DFT use.

GI insertion (+CP) → Channel → AWGN → GI removal (−CP) → DFT$_N$ → Equalizer → IDFT$_N$ →

Figure 23.6 Principle of the FDE demodulation.

channels, the requirement to demodulate the whole bandwidth whatever the actual bit rate for a given user/service, and the intrinsic limitation of SC modulations to mix data and pilots in the frequency domain (TDM only). The FDE technique consists in projecting the received signal from the time domain to the frequency domain. An alternative would be to assume that the symbols to be transmitted are generated in the frequency domain. Applying an inverse DFT (IDFT) on these samples would project them in the time domain for transmission on the real channel. This is in fact the very principle of the multi-carrier schemes and more specifically the OFDM modulation.

23.3.2 OFDM

The OFDM modulation actually originates from the need to find an alternative to SC schemes for highly selective channels. Troubles occur with SC schemes when the channel delay spread significantly exceeds the symbol duration. The underlying principle of multi-carrier modulations is precisely to modulate in parallel several subcarriers at a much smaller rate, i.e., with an OFDM symbol time much longer than the channel dispersion. Each subcarrier is ultimately affected by a simple fading coefficient that can be compensated for using a single-tap equalizer. To avoid inter-carrier interference (ICI), all the subcarriers shall preferably be transmitted over separated frequency bands, but this would in return lead to a very poor spectral efficiency. The beauty of the OFDM modulation is to enable the transmission on overlapping bandwidths using subcarriers $1/T$ Hertz apart. This is the principle of the OFDM modulation that is defined as:

$$y(t) = \sum_{n=0}^{N-1} \sum_{k=0}^{M-1} x_k^{(n)} \Pi_{[0,T]}(t - kT) e^{2j\pi \frac{nt}{N}}, \qquad (23.6)$$

where $\Pi(t)$ is the rectangular window, equal to one in $[0, T]$, zero elsewhere. It can be shown that the detection can be achieved independently on a symbol basis in both the time (OFDM symbol level) and frequency (subcarrier level) domains. The so-called subcarrier orthogonality brings another advantage: even with a finite-length waveform such as the rectangular waveform, the OFDM modulation achieves a good spectral efficiency, actually similar to the one of an SC modulation with a zero roll-off waveform. The success of the OFDM modulation also results from the possibility to perform the modulation and demodulation in the digital domain by means of a DFT, i.e., with a low implementation complexity.

When the channel is dispersive in time, ISI occurs between each symbol on all the subcarriers. In addition to ISI, this also breaks the orthogonality between the subcarriers, thus introducing ICI. A solution would be to increase the symbol duration to make the channel delay spread negligible. In addition to requiring a large number of subcarriers to achieve high throughputs, this solution is not compatible with time-varying channels where the channel Doppler spread actually defines an upper bound on the symbol period. The most commonly used solution consists in inserting in front of each symbol a GI longer than the typical channel delay spread. The signal is thus transmitted over a longer symbol period with length $T_t = T + T_{GI}$, where T_{GI} is the duration of the GI. The GI serves to mask to the demodulator the transient period between each symbol period. To avoid ISI, the receiver shall synchronize on the received signal so as to apply the matched filtering on the useful part of the symbol (after completion of the GI). In order to make the extension transparent, it is common to populate the GI as a CP that transforms the linear convolution with the channel impulse response into a circular convolution. Thanks to this property, the one-tap equalization originally applicable in the continuous-time domain is also valid in the digital domain. Figure 23.7 depicts the general architecture of respectively an OFDM modulator and demodulator.

Thanks to a proper weighting of the soft metrics sent to the channel decoder, it can be shown that the simple one-tap equalizer is optimal, (ML) in OFDM case, which is not the case for the FDE of SC signals [6]. Along with its intrinsic robustness to channel degradation, one key advantage of the OFDM modulation is to allow for allocating data and pilots in both the time and frequency domains. This is the task of the frame builder in the OFDM transmitter. It is in particular very simple to insert pilots for channel estimation directly in the frequency domain, thus allowing its estimation in the very same frequency domain where equalization is applied. On the other hand, OFDM suffers from a major drawback: it shows important power fluctuations, i.e., a large PAPR. Indeed, each sample is built as the weighted sum of i.i.d. symbols, thus leading to a Gaussian-like signal. It was shown that the instantaneous PAPR (IPAPR) over one symbol duration (see Equation 23.4) behaves as a random variable with a distribution given for large number of subcarriers N by [7]:

$$\Pr\left(\Gamma \leq \gamma\right) \approx 1 - (1 - e^{-\gamma})^N. \tag{23.7}$$

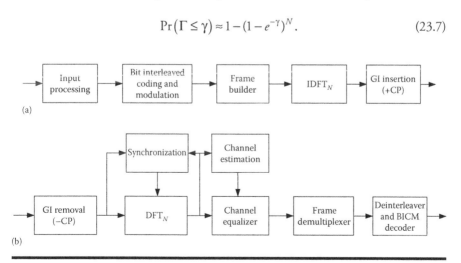

Figure 23.7   OFDM transmitter (a) and OFDM receiver (b).

This shows that the PAPR increases linearly with the number of subcarriers. The maximal value of the IPAPR is:

$$IPAPR_{max} = A_{max}^2 N, \tag{23.8}$$

where A_{max} is the upper bound of the constellation alphabet in modulus. The PAPR tends thus to increase for large constellation such as QAM alphabets. A lot of work has been devoted in the definition of techniques for reducing the PAPR of OFDM signals without altering its performance due to excessive clipping. All these techniques actually show either rather mild performance or a high complexity especially for low PA IBOs. The SC-OFDM is an alternative that intrinsically reduces the PAPR of OFDM.

23.3.3 SC-OFDM

As shown in the previous sections, SC modulations combine low power fluctuations with reasonable receiver complexity using FDE. But SC-FDE schemes do not provide much flexibility to insert pilot within data in the frequency domain and present some loss in spectral efficiency due to the roll-off. Moreover, it is generally needed to double the sampling frequency at the receiver for practical implementation of the demodulation.

The OFDM can actually be interpreted as a derivation of the SC-FDE approach where it becomes possible to allocate data in both the time and frequency domains and where the roll-off can be set to zero. However, this improved flexibility comes at the expense of a high PAPR. The SC-OFDM modulation actually combines the better of the two worlds, the low power fluctuations of the SC modulation and the flexibility and high spectral efficiency of the OFDM modulation. As shown in Figure 23.8, this is obtained by generating the frequency domain samples of the OFDM modulation by means of a DFT applied onto preferably constant amplitude samples. It is thus possible to recover the low PAPR properties of the original symbols in the time domain after OFDM modulation. In the case of the SC-FDE technique, the translation in the frequency domain at the receiver is used to perform the equalization. Similarly for the SC-OFDM, the translation in the frequency domain at the transmitter can be used to perform the subcarrier mapping of user data and reference pilot just like in the OFDM modulation. Moreover, null subcarriers can (must) be inserted at the edges of the frequency multiplex in order to perform a zero roll-off oversampling.

Figure 23.9 illustrates the relationships between the SC-OFDM, OFDM, and SC-FDE techniques. As shown in this figure, the first DFT at the transmitter can be actually be computed over M samples where $M \leq N$. The result of the M-point DFT is then mapped in the central part of the OFDM multiplex padding the

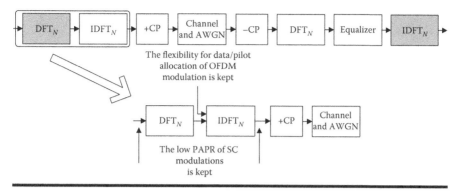

Figure 23.8   From OFDM to SC-OFDM.

unallocated subcarriers with zeros. The resulting signal is simply the zero roll-off oversampled version of the original *M*-sample signal, thus leading to the preservation of the original low PAPR. As explained in the introductory section, the name SC-OFDM was derived from the SC-FDMA terminology used in the 3GPP/LTE system where this technique is used in combination with FDMA in uplink. In the present case, all users can demodulate the signal that is transmitted over the whole bandwidth. Many other approaches are possible: it is possible for instance to alternatively insert $m - 1$ null subcarriers between each sample obtained at the output of the *M*-point DFT. In this case, the resulting signal is periodic as *m* replica of the original signal, thus keeping its low PAPR property. Several other alternatives have been devised to define multiple access schemes in cellular networks (IFDMA, MC-CDMA, SS-MC-MA, etc.). The computation of the first DFT can also be interpreted as a spreading on the symbols to be transmitted. For that reason, SC-OFDM is also known as the DFT-spread OFDM modulation.

The SC-OFDM modulation can be implemented either in the time domain (IFDMA) or in the frequency domain (DFT-spread OFDM). The frequency domain implementation is generally preferred, especially at the receiver. The frequency domain implementation of a SC-OFDM transceiver is described in Figure 23.10. On the transmitter side, it mainly consists in adding a spreading stage, i.e., a DFT of size *M*, to the OFDM transmitter. As already expressed, the SC-OFDM signal can be interpreted as an interpolated version of the signal prior to spreading, the interpolation ratio being equal to N/M. The interpolation filter being performed by DFT/IDFT is about as sharp as possible. Therefore, the SC-OFDM signal is roughly equivalent to a TDM signal with a zero roll-off.

Figure 23.11 depicts the general architecture of the SC-OFDM receiver in the frequency domain. It clearly appears that the SC-OFDM receiver shows a lot of commonalities with the OFDM receiver. There is only one new function, the despreading DFT, and the other modules are only slightly modified with a reasonable

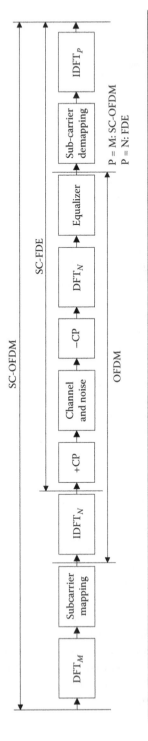

Figure 23.9 **Relationships between the OFDM, SC-FDE, and SC-OFDM modulations.**

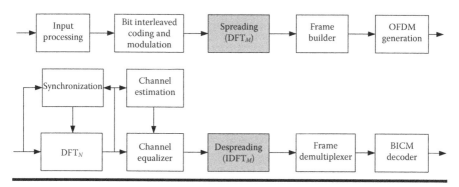

Figure 23.10   SC-OFDM transceiver.

increase in complexity. As an illustration, Table 23.1 compares a possible implementation of the different key functionalities in a multi-carrier receiver for both the OFDM and SC-OFDM modulations. It can be noticed that the functions are very similar except for the occurrence of a summation on the individual OFDM terms when dealing with SC-OFDM.

23.4 SC-OFDM Modulation in DVB-NGH

23.4.1 System Parameters

The DVB-NGH standard specifies a mandatory sheer terrestrial profile (also called core or base profile) and an optional terrestrial/satellite hybrid profile. The hybrid profile is composed of a main component coming from the terrestrial network and an additional component coming from a satellite. The SC-OFDM modulation has been selected with OFDM as the two reference waveforms for the hybrid profile. Besides defining the transmitted waveforms, the hybrid profile defines moreover the mechanisms to receive two waveforms simultaneously and combine their outputs into a single stream. The hybrid waveform can be transparent to the receiver, when an identical signal is transmitted by the terrestrial and satellite transmitters (SFN mode), or two different waveforms (MFN mode) can be used. The following hybrid modes are identified (see Chapter 22):

■ *SFN in OFDM*: In this case, the OFDM satellite parameter set is applicable to both the terrestrial and satellite components.
■ *MFN in OFDM*: In this case, the terrestrial component is built according to the core profile while the satellite component is set according to the OFDM hybrid profile.

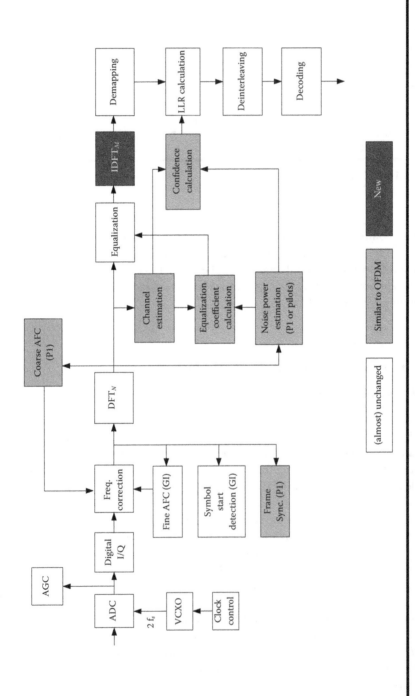

Figure 23.11 SC-OFDM receiver. ADC, analog to digital converter; AFC, automatic frequency control; AGC, automatic gain control; LLR, log likelihood ratio; VCXO, voltage controlled crystal oscillator.

Table 23.1   Complexity of SC-OFDM versus OFDM

	SC-OFDM	OFDM
Demodulation	$r_k'' = DFT\{r_k'\}$	
Equalization (MMSE)	$\hat{s}_k = \dfrac{H_k^*}{\|H_k\|^2 + \sigma^2} r_k''$	
De-spreading	$\hat{x}_n' = IDFT\{\hat{s}_k\}$	$\hat{x}_n' = \hat{s}_n$
Normalization	$\hat{x}_n = \dfrac{\hat{x}_n'}{\tilde{\alpha}}, \quad \tilde{\alpha} = \dfrac{1}{M}\displaystyle\sum_{k=0}^{M-1}\dfrac{\|H_k\|^2}{\|H_k\|^2 + \sigma^2}$	$\hat{x}_n = \dfrac{\hat{x}_n'}{\tilde{\alpha}_n}, \quad \tilde{\alpha}_n = \dfrac{\|H_n\|^2}{\|H_n\|^2 + \sigma^2}$
LLR weighting	$\Gamma_n^j = \dfrac{4\displaystyle\sum_{k=0}^{M-1}\|H_k\|^2}{\sigma^2} LLR_n^j$	$\Gamma_n^j = \dfrac{4\|H_k\|^2}{\sigma^2} LLR_n^j$

- *SFN in SC-OFDM*: This mode concerns only the case of terrestrial gap fillers that amplify the signal from the satellite (in the same frequency). The SC-OFDM satellite component setting is applicable to both the terrestrial and satellite components.
- *MFN in SC-OFDM on the satellite component and OFDM on the terrestrial component*: The terrestrial component is set according to the core profile while the satellite component is set according to the SC-OFDM mode of the hybrid profile.

The satellite component of the hybrid profile is defined for two bandwidths, 2.5 and 5 MHz for a transmission in the L- and S-bands. Table 23.2 describes the main system parameters defined for the SC-OFDM satellite mode.

Table 23.2   Main System Parameters for SC-OFDM Transmissions in DVB-NGH

Bandwidth	2.5 MHz	5 MHz
Sampling freq.	20/7 MHz	40/7 MHz
FFT size (N)	512, 1024	512, 1024, and 2048
GI	1/16 and 1/32 (w.r.t. to N)	
Constellation	QPSK and 16QAM	

23.4.2 Pilot Pattern

One strong advantage of OFDM is the possibility to use pilots scattered in the time and frequency domains, a key feature when it comes to estimate the channel at high speeds, i.e., for large Doppler values. As an illustration, Figure 23.12 depicts the PP1 pilot scheme of the DVB-T2 standard. Note that DVB-T2 specifies eight different pilot patterns, where PP1 already used in DVB-T offers a good compromise between Doppler and long echo robustness (see Table 23.3).

The main issue is that a pilot pattern such as the one in Figure 23.12 cannot be multiplexed with SC-OFDM data without completely degrading the PAPR structure of the transmitted signal. In such a case, the complementary cumulative distribution function (CCDF) of the transmitted signal amplitude is actually similar to the one of an OFDM signal. To solve this issue, the 3GPP/LTE body selected for

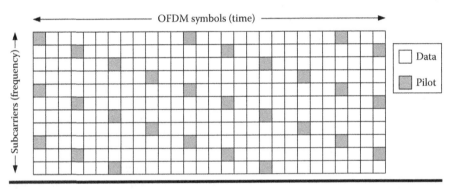

Figure 23.12   DVB-T2 PP1 pilot pattern.

Table 23.3   Parameters Defining the Scattered Pilot Patterns

Pilot Pattern	Separation of Pilot Bearing Carriers (D_x)	Number of Symbols Forming One Scattered Pilot Sequence (D_y)
PP1	3	4
PP2	6	2
PP3	6	4
PP4	12	2
PP5	12	4
PP6	24	2
PP7	24	4
PP9	2	6

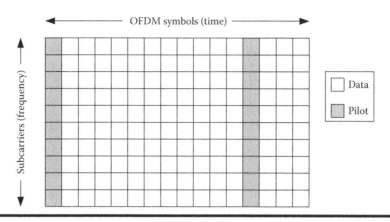

Figure 23.13   LTE-like pilot structure for SC-OFDM.

the uplink a pilot pattern similar to the one depicted in Figure 23.13 (see 3GPP/ LTE TS 36.211 [8]).

According to the 3GPP/LTE approach, a complete SC-OFDM symbol (i.e., all the subcarriers), referenced as a pilot symbol, is constituted solely by pilots and is dedicated to channel estimation and synchronization purposes. Such a symbol is regularly inserted in the frame (twice in each 1 ms subframe made of 14 SC-OFDM symbols). In this case, a classical way of performing channel estimation is represented in Figure 23.14. The channel is estimated at the pilot positions (cells in the DVB terminology), then the noise level is reduced by means of a frequency domain smoothing such as a Wiener filter, and finally the channel is estimated at all positions by performing a time interpolation between two pilot symbols. The 3GPP/LTE scheme shows its limits when it comes to implement time interpolation. In order to maximize the user data throughput, one must limit the number of pilot symbols inserted among data cells. For example, in DVB-T2, the pilot insertion rate per subcarrier depicted in Figure 23.12 is 1/12. If one wants to keep the same throughput with the LTE approach, it implies that a full pilot symbol shall be inserted every 12 OFDM symbols. This limits the capability of the system to follow the channel variations of the signal, for example, due to velocity. According to the

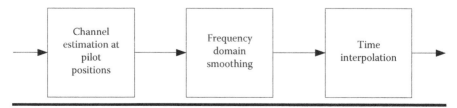

Figure 23.14   Typical channel estimation for multi-carrier modulations (OFDM and SC-OFDM).

Nyquist theorem, if the OFDM rate is $1/T$, a bound on the maximum acceptable Doppler frequency is equal to:

$$f_{max} = \frac{1}{2D_Y T} = \frac{1}{24T},$$ (23.9)

where D_Y (=12) is the number of symbols forming one scattered pilot sequence. The bound is actually even lower when considering practical algorithms to perform time interpolation with a reasonable complexity. It must be understood that this bound is due to the pilot structure itself (Figure 23.13) and not to the related estimation method. With the DVB-T2 scattered pilots shown in Figure 23.12, the bound falls down to $(8T)^{-1}$. The other OFDM pilot patterns specified in the DVB-T2 standard are compared in Table 23.3.

The SC-OFDM mode of the NGH hybrid profile specifies a new pilot pattern (defined as PP9) that divides the bound of Equation 23.9 by two while preserving the low PAPR structure of SC-OFDM. Typically, Zadoff–Chu (ZC) sequences [9] are used as pilot patterns, due to their low PAPR and their good orthogonality and correlation properties. ZC sequences are constant amplitude zero autocorrelation sequences, in both the time and frequency domains. Roughly, the generation of the pilot and data structure can be summarized in three steps:

■ In the last SC-OFDM symbol of each NGH data section made of six symbols (see Figure 23.16), half of the subcarriers are reserved for DFT-spread data, while the other half of subcarriers convey pilots (see Figure 23.15). Data and pilots are frequency multiplexed.

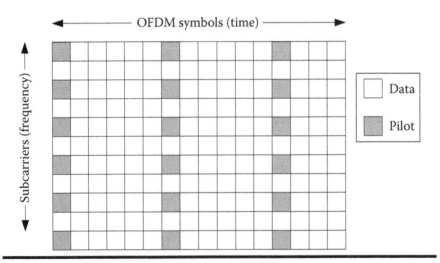

Figure 23.15 NGH pilot structure for SC-OFDM.

▪ For the pilot subcarriers, a constant amplitude sequence such that its DFT counterpart (in the time domain) has also fixed amplitudes is used. If L denotes the ZC sequence length ($L = M/2$), then the complex value at each position k of each p-root ZC sequence with integer shift l is (the size L of the ZC sequence is even here):

$$x_k = e^{-2j\frac{\pi}{L}\left(p\frac{k^2}{2}+lk\right)}. \qquad (23.10)$$

As only one sequence is needed for broadcasting, the index p is set to one ($p = 1$) and the shift l to zero ($l = 0$).

▪ This Zadoff–Chu pilot sequence is modified by adding a half period shift:

$$x_k = e^{-2j\frac{\pi}{L}\left(\frac{k^2}{2}+0.5k\right)}. \qquad (23.11)$$

Constant amplitude should be understood as evaluated on a non-oversampled ZC sequence. After oversampling (OFDM modulation and of digital-to-analog conversion), the ZC sequence is not constant amplitude but has very low envelope fluctuations, lower than the fluctuations of a typical TDM QPSK sequence.

The introduction of the fractional time shift is justified as follows: the global signal, in frequency or time dimensions, corresponds to the sum of two signals, data, and pilots. As the interpolation is a linear process, the global interpolated signal corresponds also to the sum of two signals: the interpolated data signals and the interpolated pilot signals. For the data in the time domain, if no interpolation is performed and if an x-PSK constellation is used, a constant amplitude is obtained every sample period, and the maximum peaks after interpolation are placed just in-between these time instants. The same phenomenon occurs with the classical Zadoff–Chu sequence. Therefore, in this case, we add two signals, the peaks of which are placed at the same positions. By time shifting the Zadoff–Chu sequence of half a sampling period, the peaks of each signal are now interleaved. This implies that the peaks of the global signal, sum of both, are reduced.

If a nonconstant constellation is used, e.g., 16QAM, the amplitude is no longer constant every sample period. However, the peaks will still be placed at the same positions, i.e., just in-between these time instants, and the peaks of the global signal will still be reduced by modifying the original ZC pilot sequence.

Table 23.3 lists the different pilot patterns: PP1 to PP7 for OFDM waveform and PP9 for the SC-OFDM waveform. When comparing the SC-OFDM pattern PP9 to the terrestrial one, it can be noticed that the large D_Y value decrease the Doppler performance. However, it can be noted that

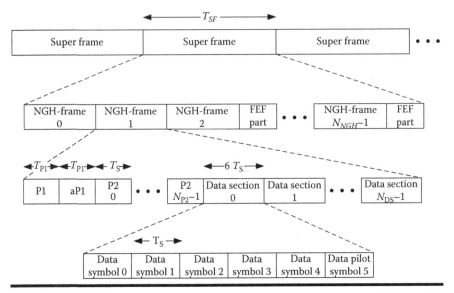

Figure 23.16 DVB-NGH frame structure of the hybrid terrestrial–satellite profile.

- The lower D_X values improve the smoothing filter performance and then decreases the noise on the channel estimates
- The DFT size is generally lower for the satellite component, which will improve Doppler performance
- As the satellite channel is single path, the use of an automatic frequency control (AFC) improves Doppler performance

In accordance with this new pilot scheme, the DVB-NGH specifies a specific frame structure for the hybrid profile (see Figure 23.16). At the top level, the frame structure consists of super-frames, which are divided into NGH frames that contain data sections made of six consecutive SC-OFDM symbols. As shown in Figure 23.16, the last symbol of each data section is a PP9 pilot pattern as described earlier. Note that this structure is typical of the SC-OFDM satellite link. The frame structure for the core profile remains similar to the DVB-T2 structure.

23.4.3 MIMO

With an increasing pressure on both the frequencies (analog switch off and digital dividend) and the bandwidth requirements, it is critical to take the most of the available spectrum. Just like most of the recent wireless systems, the DVB-NGH system relies on multi-antenna schemes to either improve the robustness (spatial diversity) or to increase the capacity (spatial multiplexing) of the broadcasting transmissions (see Chapter 19). The DVB-NGH standard specifies two spatial diversity schemes, a modified Alamouti code and an enhanced SFN solution (the so-called rate 1 schemes are described in Chapter 20). These two schemes are actually part of the core profile along with SISO for sheer terrestrial transmissions. Spatial multiplexing schemes called rate 2 schemes in DVB-NGH are specified in an optional MIMO profile dedicated to the core profile. As described in Chapter 21, the DVB-NGH specifies a 2×2 enhanced spatial multiplexing (eSM) precoding with an optional phase hopping (PH).

In parallel of these schemes dedicated to sheer terrestrial transmissions, the DVB-NGH standard also includes the so-called hybrid MIMO profile that has been devised specifically to facilitate the use of MIMO on the terrestrial and/or satellite elements within a hybrid transmission scenario. This profile encompasses two branches described in the following sections.

23.4.3.1 Hybrid MIMO MFN

The hybrid MIMO MFN describes the case where the satellite and terrestrial parts of the transmission are on different carrier frequencies and do not necessarily share any common frame or symbol timing at the physical layer. At least one of the transmission elements (i.e., terrestrial or satellite) must be made using multiple antennas; otherwise the use case lies within the hybrid profile, not the hybrid MIMO profile. Dealing with the single MFN mode implying SC-OFDM (terrestrial component set according to the core profile and a satellite component in SC-OFDM mode), three MIMO configurations can be considered:

■ The terrestrial component operates in MIMO according to any scheme from the MIMO profile while the SC-OFDM transmission occurs in SISO
■ The terrestrial component operates in SISO according to the base profile while the SC-OFDM transmission occurs in MIMO using a basic SM scheme (no precoding or PH)
■ Both links operate in MIMO, where the terrestrial component operates in MIMO according to any scheme from the MIMO profile, while the satellite link operates in SM mode

The MIMO SM of the hybrid MIMO profile processing consists in transmitting cell pairs (f_{2i}, f_{2i+1}) on the same SC-OFDM symbol and carrier from Tx-1 and Tx-2, respectively:

$$\begin{pmatrix} g_{2i} \\ g_{2i+1} \end{pmatrix} = \begin{pmatrix} f_{2i} \\ f_{2i+1} \end{pmatrix}, \quad i = 0, 1, \ldots, N_{cells}/2 - 1, \tag{23.12}$$

where
i is the index of the cell pair within the FEC block
N_{cells} is the number of cells per FEC block

The insertion of pilots is modified in order to allow for the channel estimation on each transmitter–receiver path. The frequency domain implementation of SC-OFDM makes the SM decoding very simple.

23.4.3.2 Hybrid MIMO SFN

The hybrid MIMO SFN describes the case where the satellite and terrestrial parts of the transmission use the same carrier frequency and radiate synchronized signals intended to create an effective SFN. SC-OFDM is not an option for the hybrid MIMO SFN profile where a synchronized effective SFN transmission can exist only in OFDM mode (terrestrial emitters other than gap fillers implement only OFDM).

23.5 Performance of the SC-OFDM Modulation

The knowledge gathered so far on the SC-OFDM technique has been obtained in the context of broadband cellular networks (SC-FDMA for LTE/UL). It was thus required to check the suitability of the SC-OFDM modulation for satellite broadcasting. In that purpose, the performances of the SC-OFDM modulation have been thoroughly evaluated by means of computer simulations. Some of the most significant results are presented in this section.

23.5.1 Peak-to-Average Power Ratio

Classically, the power fluctuations are measured using the CCDF of the PAPR, defined as [5]:

$$\text{CCDF(PAPR)} = \text{Pr}\left\{\text{PAPR} > \gamma^2\right\}. \qquad (23.13)$$

The parameter γ^2 is a threshold, expressed in dB, and the CCDF value indicates the probability that the PAPR surpasses this threshold. Should we consider a signal (normalized to unitary mean power, for simplicity) passing through an ideal clipper PA, γ^2 has a direct physical interpretation: it can be assimilated to the IBO. Indeed, for a signal working at γ^2 dB of IBO to go into saturation and suffer clipping, it would be needed and necessary that its PAPR be higher than γ^2. The probability in Equation 23.13 is also called clipping probability. Figure 23.17 displays the result of the evaluation of the CCDF of the PAPR with an oversampling factor equal to 4. Figure 23.17 clearly shows the advantage of the SC-OFDM modulation over the OFDM with a gain of 2.8 dB in terms of admissible IBO for a clipping probability of 10^{-3}.

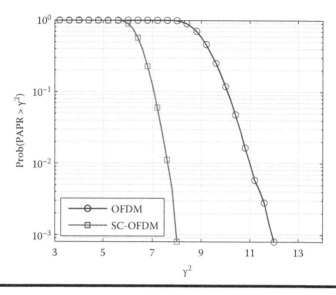

Figure 23.17 CCDF of PAPR for SC-OFDM versus OFDM. QPSK, $N=512$, $M = 432$.

23.5.2 *Instantaneous Normalized Power*

While the CCDF of PAPR is a very popular notion, it has one important drawback. A certain clipping probability ensures that at least one peak per block has an important amplitude and is susceptible to suffer clipping or severe distortion, but gives no information on how many samples in that block are distorted [5]. Yet in practical scenarios, it is of great interest to know how many samples have a certain level and are thus susceptible to be distorted, as all of these samples cause degradation [10]. Indeed, severely clipping one single peak in a large block has a negligible effect on the MER or spectrum shape, while distortion (even mild) of a large number of samples might have unacceptable consequences. From this point of view, it is important to consider a more refined analysis taking into account all the signal samples. This can be done by means of considering the distribution of the INP [5]:

$$
\text{CCDF(INP)} = \Pr\left\{ \frac{\overbrace{|v(n)|^2}^{INP(v)}}{\dfrac{1}{N_S} \displaystyle\sum_{k=0}^{N_S-1} |v(k)|^2} > \gamma^2 \right\},
\qquad (23.14)
$$

where N_S is the number of samples considered to evaluate the INP. Note that the INP is defined in the discrete time domain. One must pay attention that the behavior of a signal in the continuous and discrete time domains may not be the same. Continuous and discrete performances will be equivalent only when oversampling the discrete signal at a sufficient level. The CCDF of INP indicates the probability that the INP at a sample level exceeds a certain threshold γ^2. If we look at the range of important values of γ^2 for the CCDF of PAPR, the probability that one sample in a block exceeds such a level is very weak, and should a sample exceed this level, it is highly likely to be the only one in that block: the CCDF of INP and the CCDF of PAPR asymptotically behave the same. But in the range of lower values of γ^2, the CCDF of INP has a better resolution and shows effects that CCDF of PAPR tends to mask.

Figure 23.18 displays the result of the evaluation of the CCDF of the INP with the same parameters as in the previous section for the OFDM and SC-OFDM modulations. The figure also gives the CCDF of the INP for the TDM waveform with different roll-offs and the PAPR reduction technique for OFDM specified in DVB-T2, the so-called TR algorithm. These curves confirm the good PAPR properties of the SC-OFDM that outperforms both the OFDM and OFDM-TR modulations even at low IBOs. TDM waveforms exhibit better performance than SC-OFDM but only for roll-offs above 0%, i.e., not for the same spectral occupancy. It occurs that the (SC-)OFDM modulation with rectangular window is spectrally equivalent to TDM with a roll-off of 0%. An evolution of the SC-OFDM (Extended and Weighted SC-OFDM) modulation was devised to allow for controlling the frequency bandwidth just like for TDM. Without going into details on its actual principle, Figure 23.18 shows that the

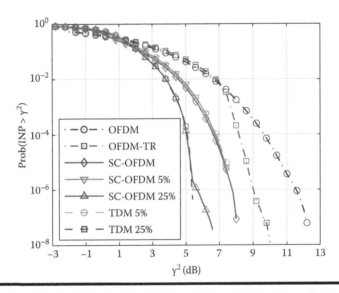

Figure 23.18 CCDF of INP for SC-OFDM, OFDM, OFDM-TR, and TDM. QPSK, $N = 512$, $M = 432$.

extended SC-OFDM behaves like TDM for the same value of roll-off except for very large power values. But if the application of zero roll-off is readily achieved for SC-OFDM, it implies severe filtering issues for TDM waveforms. With an increasing pressure on the spectrum, it was decided to keep the SC-OFDM in its original form, i.e., with a roll-off of 0% (smallest bandwidth). *The main result of this study is that the SC-OFDM modulation is strictly similar to TDM modulations in terms of PAPR for the same spectral occupancy.*

The TR algorithm brings an improvement to OFDM for INP probability less than 4×10^{-3}, corresponding to a signal power greater than 7 dB. For a terrestrial system and linear amplifiers, this brings an improvement if large IBOs are considered, i.e., very low out-of-band emission. If operating a PA with large IBO is not a big issue for terrestrial transmission, it is not at all the case for satellite transmissions where limiting the power consumption is critical. Satellite PAs are often driven with IBOs of a few dBs thus leading to a high saturation of the signal. This is compensated by the use of robust modulation and coding schemes. From Figure 23.18, it is clear that for a satellite system and a foreseen IBO of about 1 dB, TR does not bring any advantage.

23.5.3 Satellite Pilot Pattern (PP9)

As explained in Section 23.4.2, the DVB-NGH standard specifies a pilot pattern specifically dedicated to the SC-OFDM transmission mode. Figure 23.19 plots the CCDF of INP of the hybrid data and pilot SC-OFDM symbols. Simulation

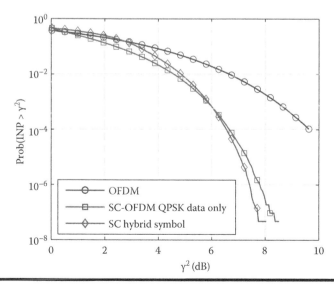

Figure 23.19   CCDF of INP of OFDM, SC-OFDM, and SC-OFDM hybrid pilot. QPSK, $N = 512$, $M = 432$.

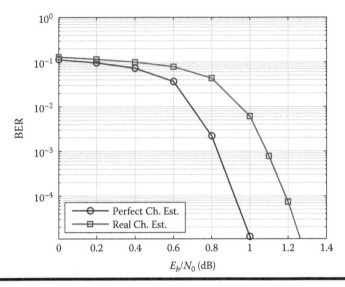

Figure 23.20   Performance of the PP9 pilot scheme with real channel estimation. QPSK 4/9, LDPC 16k, *N* = 512, *GI* = 1/32, AWGN.

parameters considered an FFT size of 0.5k ($N = 512$) and QPSK signal mapping. An oversampling factor of 4 was considered. Results in Figure 23.19 confirm the overall low envelope fluctuations of the hybrid data and pilot SC-OFDM symbol. At a clipping probability per sample of 2×10^{-2}, SC-OFDM outperforms OFDM by 2 dB, and the hybrid symbol has close performance to SC-OFDM, with a slight degradation of 0.4 dB. Since the hybrid symbol appears in a frame once every 6 SC-OFDM symbols, this slight degradation has overall no impact on the performance. At higher clipping probabilities, the performance difference between SC-OFDM and OFDM is even higher, and the hybrid symbol has lower envelope variations than the SC-OFDM data symbol. To illustrate the performance of the PP9 pilot scheme, Figure 23.20 compares the performance of SC-OFDM in an AWGN channel with perfect and real channel estimation. Real channel estimation is carried out according to the scheme introduced in Figure 23.14 with Wiener filtering. The real channel estimation introduces a loss of 1.2 dB, i.e., typical of what is measured in the case of pure OFDM.

23.5.4 Modulation Error Ratio

The CCDF of the PAPR and INP provides a direct insight on the power fluctuations of the signal. As mentioned earlier, satellites often operate with very low IBOs. The PAPR reduction brought by the SC-OFDM modulation with respect to OFDM is thus not sufficient to transmit without saturation. It is thus of great interest to quantify and compare the impact of the saturation on those modulations. The saturation of the incoming signal by the PA introduces within the signal bandwidth interference that can be assimilated to an additional noise. The impact of the PA nonlinearity on the modulation itself (independently from channel coding) is commonly evaluated using the modulation error ratio (MER). The MER basically measures the level of noise introduced by the degradation onto the demodulated constellation samples. The MER is defined in dBs as follows:

$$\text{MER} = 10 \log_{10} \left(\frac{P_{error}}{P_{signal}} \right), \tag{23.15}$$

where

P_{error} is the RMS power of the error vector
P_{signal} is the RMS power of ideal transmitted signal

Note that the MER is closely related to the error vector magnitude (EVM) [7]. The MER was measured for two PAs, an ideal clipper and the linearized TWTA amplifier as used in DVB-S2 [2]. Figure 23.21 displays the result of the evaluation of the MER respectively for the ideal clipping and linearized TWTA PAs with the following parameters: QPSK, $N = 512$ (modulation), $M = 432$ (spreading), and oversampling factor equal to 4.

The MER evaluation confirms that the SC-OFDM is significantly more robust than the OFDM to saturation with a gain of ~3 dBs. The MER evaluation also shows that the TR method does not bring a significant improvement in a satellite environment, i.e., with a linearized amplifier not so linear, whatever the IBO. Even with an ideal amplifier and for small IBO (less than 5 dB), TR does not bring any improvement over OFDM and remains less robust than SC-OFDM, whatever the amplifier.

23.5.5 Bit Error Rate and Total Degradation

When dealing with coded modulations, it is common to evaluate performance in terms of bit error rate (BER) with respect to the SNR. In the present case, the objective is to evaluate the losses due to the PA nonlinear effects in comparison with the ideal case (linear PA without saturation). As previously explained, a PA should ideally be operated close to its saturation point. Assuming a perfect amplifier, the application of an IBO means that the SNR shall be IBO dBs larger to reach the reference level of performance. In practice, it is obviously not possible to increase the SNR—the application of a given IBO actually results in a reduction of the transmitted power and thus of the coverage. Moreover, the AM/AM response of real PAs is not perfectly linear especially close to the saturation region and the actual loss in SNR is given by the OBO associated to the IBO. The OBO thus represents the first cause of in-band degradations when dealing with nonlinear PAs. The second cause

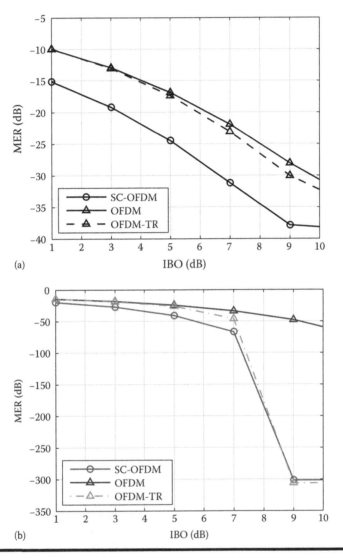

(a)

(b)

Figure 23.21   MER for SC-OFDM, OFDM, and OFDM-TR for a linearized TWTA PA (a) and ideal clipping PA (b).

of degradation is the in-band noise introduced by the saturation applied onto the incoming signal. It is common to evaluate this degradation as the loss in SNR for a given reference value of target BER between the nonlinear amplifier and an ideal linear amplifier for a given OBO [5]:

$$TD = OBO + \left[SNR_{PA} - SNR_{linear}\right]_{BER_{ref}} = OBO + \Delta SNR. \qquad (23.16)$$

When the back-off is high, there is virtually no in-band distortion and thus no distortion-related BER loss ($\Delta SNR \ll OBO$). When working at low OBO, in-band distortions increase, and ΔSNR loss is important and becomes the predominant term in the total degradation (TD). There is an optimal working point I_{opt}, which ensures a compromise between OBO and ΔSNR and yields a minimum TD. Figure 23.22 depicts the BER performance of the OFDM and SC-OFDM modulations assuming an AWGN channel and the linearized TWTA amplifier model.

The BER is given for a set of IBOs as a function of a reference E_b/N_0, i.e., the one that will be measured with a linear PA with the same gain operated at the saturation point of the actual PA or equivalently with no amplifier considering that the PA model is normalized in power with a unitary gain for IBO = 0 dB. This representation directly enables measuring the combined effect of the OBO and interference noise. The TD is simply given by the shift between the ideal reference curve and the measured curve. The TD for a BER value of 10^{-5} has been extracted from those evaluations as shown in Figure 23.23. In order to distinguish the impact of the OBO against the one of the saturation noise, Figure 23.23 also depicts the OBO as a function of the IBO.

It appears from these curves that the optimum IBO for the OFDM is 2 dB leading to a TD of 2.95 dB, while the optimum IBO for the SC-OFDM is 1 dB leading to a TD of 1.35 dB. This shows that the SC-OFDM enables operating closer to the saturation point than the OFDM, thus improving the power efficiency of the amplifier for a reduced consumption or an increased coverage. In addition, this comes with a TD smaller of 1.5 dB in favor of SC-OFDM, which means the ability to increase the quality of service.

The results have been given so far under the assumption of a QPSK modulation alphabet. Figure 23.24 compares the TD measured in the case of the QPSK and 16QAM. Due to its higher PAPR, 16QAM requires an IBO greater than 1 dB with a TD of 2.1 dB, logically but not significantly greater than what can be obtained in QPSK.

23.5.6 Doppler Performance

As mentioned in the DVB commercial requirements, DVB-NGH networks are expected to support moving terminals for speeds up to 350 km/h. The requirement

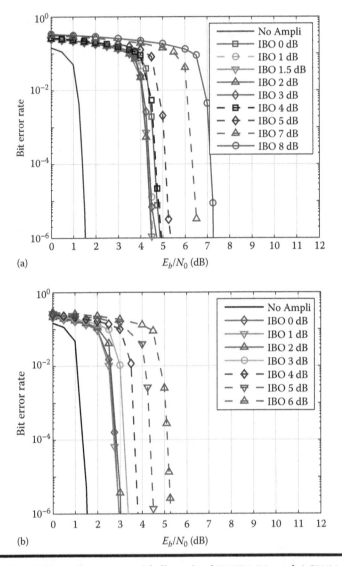

(a)

(b)

**Figure 23.22   BER performance with linearized TWTA PA and AGWN channel.
(a) OFDM, QPSK 4/9, LDPC 16k, $N_{FFT}=2048$, $GI=1/32$. (b) SC-OFDM, QPSK 4/9,
LDPC 16k, $N_{FFT}=512$, $M=432$, $GI=1/32$.**

applies to the satellite segment and thus to the SC-OFDM modulation. To illustrate
the robustness of the SC-OFDM modulation against Doppler degradation, its BER
performances have been evaluated by simulation in the case of a Rice fading ($K=5$)
under the following assumptions: $F_0=2.2$ GHz, $BW=5$ MHz, $N=512$, $GI=1/32$,
QPSK, LDPC 4/9, and a time interleaving of 100 ms. The pilot pattern inserted

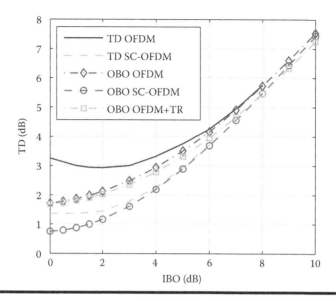

Figure 23.23 **TD and OBO performance—linearized TWTA PA, AWGN channel. OFDM: QPSK 4/9, LDPC 16k, $N = 2048$, $GI = 1/32$. SC-OFDM: QPSK 4/9, LDPC 16k, $N_{FFT} = 512$, $M = 432$, $GI = 1/32$.**

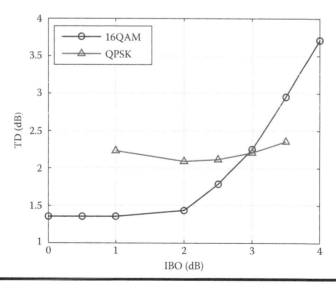

Figure 23.24 **TD and OBO performance—linearized TWTA PA, AWGN channel—QPSK 4/9, LDPC 16k, $N_{FFT} = 512$, $M = 432$, $GI = 1/32$.**

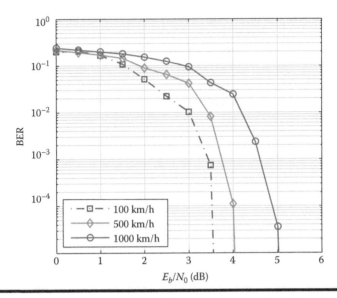

Figure 23.25 Doppler performance of the SC-OFDM modulation, Rice channel ($K=5$) $F_0 = 2.2$ GHz, BW $= 5$ MHz, $N_{FFT} = 512$, $M = 432$, $GI = 1/32$, QPSK 4/9, time interleaver depth 100 ms.

for channel estimation is the PP9 scheme, the only one supported for SC-OFDM satellite transmissions. Figure 23.25 depicts the result of the BER evaluation for three different speeds: 100, 500, and 1000 km/h. Figure 23.25 confirms the robustness of the SC-OFDM modulation to Doppler degradations with a degradation of ~2 dB for a speed of 100 km/h and 3.5 dB for 1000 km/h. Those results are provided without taking into account the gain of 30% typically obtained when considering an elevation angle of 40° for the satellite. SC-OFDM actually behaves closely to OFDM with respect to Doppler in the satellite channel that does not show a lot of frequency selectivity. The support of high speeds here is due to the use of a 512 subcarriers multiplex and the occurrence of the specular component that enables to perform an efficient AFC.

23.5.7 MIMO Performance

As mentioned in Section 23.4.3, a 2×2 spatial multiplexing scheme can be used to perform a MIMO SC-OFDM transmission from the satellite in the hybrid MFN context. The common assumption for rate 2 MIMO in DVB-NGH is the enhanced SM (eSM) scheme where a rotation precoding matrix is applied onto the reference SDM scheme. This scheme was not selected for SC-OFDM transmissions as it breaks the PAPR properties of the modulation. This is shown in Figure 23.26 that

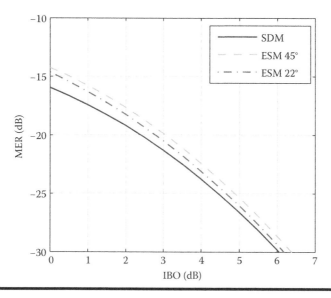

Figure 23.26 MER for eSM and SDM schemes—linearized TWTA PA, QPSK 4/9, $N_{FFT}=512$, $M=432$.

depicts the MER of the transmitted SC-OFDM signal after passing through the linearized TWTA PA model. The simulations have been performed for a 5 MHz bandwidth, an FFT size of 512 points in QPSK, and two angles for the eSM scheme, 22° and 45°. The MER curves show that for the IBO of interest (1 dB), an eSM with an angle of 22° (resp. 45°) brings an MER degradation of about 1.1 dB (resp. 1.6 dB).

The actual performance of the SM scheme on the SC-OFDM has been evaluated by simulation with a modified version of the BBC channel model [11]. The BBC model is a simple hybrid 4×2 SFN model. It has been modified to obtain a 2×2 sheer satellite model with the following parameters:

■ Counter-rotating circular polarization
■ Single-tap Ricean variable $K=5$
■ Satellite cross-polar discrimination: $\alpha=2$

Figure 23.27 depicts the BER performance assuming a speed of 60 km/h, perfect channel estimation, and a simple MMSE MIMO decoding. The SM scheme brings a degradation of only ~0.5 dB with respect to the ideal two SISO parallel transmissions. In comparison to 16QAM SIMO transmission, the SM scheme provides about 1.9 dB of improvement for the same spectral efficiency. Those results are actually close to the ones obtained in OFDM, when no PA is used.

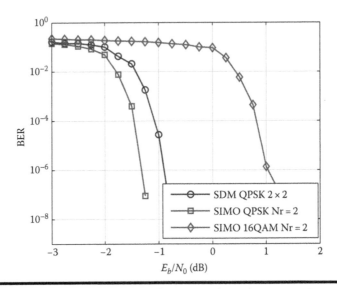

Figure 23.27 **BER performance for the SM scheme in SC-OFDM. BBC 2 × 2 channel model ($K=5$, $\alpha=0.2$, speed 60 km/h), $F_0 = 2.2$ GHz, BW = 5 MHz, $N_{FFT} = 512$, $M = 432$, $GI = 1/32$, QPSK 4/9, time interleaver depth 1 s.**

23.6 Conclusions

This chapter was dedicated to the new waveform selected to implement the transmissions on the satellite link of the DVB-NGH standard, namely, SC-OFDM modulation. This introduction to the SC-OFDM waveform was conducted according to three perspectives. The SC-OFDM is first compared to the two other modulations commonly used for satellite transmissions, the TDM and OFDM waveforms. It is shown that the SC-OFDM modulation enables keeping the low power fluctuations of TDM signals while benefiting from the flexibility and high spectral efficiency of the OFDM modulation. The chapter then focuses on the intrinsic robustness of the SC-OFDM modulation to the PA nonlinear degradations with the ability to operate with a reduced OBO in comparison to OFDM and a TD improved by 1.5 dB. It is thus possible to improve the power efficiency of the PA while improving the coverage. It must be pointed out that this result still holds when considering the PAPR reduction solutions such as the TR approach used in DVB-T2. This kind of solution actually performs well for large IBOs but not for the small IBOs (a few dBs) commonly used in satellite transmissions. Finally, the chapter shows that the SC-OFDM behaves similarly to OFDM when it comes to mitigate the degradation due to the channel and mobility, either in SISO or MIMO. These results clearly demonstrate the advantages of the SC-OFDM modulation in the context of satellite transmissions.

References

1. DVB Commercial Module, www.dvb.org, SB1856, CM-NGH015R1, CM-1062R2 v1.01, Commercial requirements for DVB-NGH, June 2009.
2. ETSI EN 302 307 v1.2.1, Digital video broadcasting; Second generation framing structure, channel coding and modulation systems for broadcasting, interactive services, news gathering and other broadband satellite applications (DVB-S2), August 2009.
3. ETSI EN 302 583 v1.1.2, Digital video broadcasting (DVB); Framing structure, channel coding and modulation for satellite services to handheld devices (SH) below 3 GHz, February 2010.
4. ETSI EN 302 755 v1.3.1, Digital video broadcasting (DVB); Frame structure channel coding and modulation for a second generation digital terrestrial television broadcasting system (DVB-T2), October 2011.
5. C. Ciochină, Physical layer design for the uplink of mobile cellular radiocommunication systems, PhD thesis, Université de Paris-Sud, Orsay, France, 2009. http://www.fr.mitsubishielectric-rce.eu/images/fck_upload/Ciochina_PhD_pdf.pdf
6. B. Le Floch, R. Halbert-Lassalle, and D. Castelain, Digital sound broadcasting to mobile receiver, *IEEE Transactions on Consumer Electronics*, 35(3), 493–503, August 1989.
7. D. Wulich, N. Dinur, and A. Gilinowiecki, Level clipped high order OFDM, *IEEE Transactions on Communications*, 48(6), 928–930, June 2000.
8. 3GPP TS 36.211 v10.4.0, 3rd generation partnership project; Technical specification group radio access network; Evolved Universal Terrestrial Radio Access (E-UTRA); Physical channels and modulation, December 2011.
9. D. C. Chu, Polyphase codes with good periodic correlation properties, *IEEE Transactions on Information Theory*, 18(4), 531–532, July 1972.
10. C. Ciochină, F. Buda, and H. Sari, An analysis of OFDM peak power reduction techniques for WiMAX systems, *Proceedings of the IEEE International Conference on Communications (ICC)*, Istanbul, Turkey, 2006.
11. P. Moss, T. Yeen Poon, and J. Boyer, A simple model of the UHF cross-polar terrestrial channel for DVB-NGH, BBC White Paper WHP205, September 2011.

NEXT GENERATION HANDHELD DVB TECHNOLOGY

Hybrid Terrestrial–Satellite MIMO Profile

V

Chapter 24

Hybrid Satellite–Terrestrial MIMO for Mobile Digital Broadcasting

Tero Jokela, Pantelis-Daniel Arapoglou, Camilla Hollanti, M.R. Bhavani Shankar, and Visa Tapio

Contents

24.1 Introduction

Traditionally, broadcasting services are being offered by either satellite or terrestrial systems. Providing national or regional wide area coverage by satellite is cost efficient, but achieving connectivity to mobile terminals located in densely built environments is not possible due to shadowing of the signal. Terrestrial networks, on the other hand, allow coverage in this environment, but building nationwide terrestrial networks is too expensive. Therefore, hybrid satellite–terrestrial networks are considered currently, complementing the features of both network types.

Further, utilization of multiple antenna multiple-input and multiple-output (MIMO) schemes to improve the communication link performance is exploited in modern communication standards, such as IEEE 802.11n, Long-Term Evolution (LTE)-Advanced, and WiMax. Recently, and building upon the introduction of multiple transmitter antennas in Digital Video Broadcasting–Terrestrial Second Generation (DVB-T2), MIMO for mobile broadcasting has been specified in the DVB-NGH (Next Generation Handheld) standard. With MIMO, the spectral efficiency and diversity of the transmission system can be enhanced. Having higher spectral efficiency means that more information can be transmitted through a given bandwidth and, thus, the broadcasting system can accommodate more services. Increased diversity translates into higher robustness against channel impairments which is important for a mobile (satellite or terrestrial) system.

In this chapter, we consider utilization of hybrid satellite–terrestrial MIMO networks for mobile broadcasting. We unfold the analysis largely based on the DVB-NGH and DVB-SH (Satellite Handheld) specifications, which constitute relevant hybrid mobile broadcasting standards.

The rest of the chapter is structured as follows: Section 24.2 describes hybrid MIMO network architectures based on DVB-NGH and DVB-SH. Section 24.3 discusses MIMO codes for hybrid networks and their properties, taking into account the particularities of combining the terrestrial with satellite transmission. MIMO schemes selected for DVB-NGH are presented together with promising codes that were studied during the DVB-NGH standardization and DVB-SH studies. Section 24.4 presents a performance comparison of the codes designed for DVB-NGH. In parallel to DVB-NGH, there have been efforts to come up with a MIMO extension to the baseline DVB-SH. Various MIMO schemes have been evaluated in the MIMO hardware demonstrator project funded by the European Space Agency, and selected results are presented in Section 24.4. Finally, the chapter is concluded with Section 24.5.

24.2 Hybrid MIMO Network Architectures

The Radiocommunications sector of the International Telecommunications Union (ITU-R) divides mobile satellite services (MSS) using both satellite and terrestrial components into two categories, namely, hybrid and integrated systems. In *integrated* MSS systems, the ground component is complementary and operates as part of the MSS system. This means that the satellite resource and management system also controls the ground component. Further, both satellite and ground components use the same portion of the allocated spectrum. In *hybrid* satellite and terrestrial systems, the subsystems are interconnected but operate independently of each other, i.e., the satellite and terrestrial components have separate network management systems, and they do not necessarily operate over the same spectrum [1]. In terms of spectrum allocation, the first option for accommodating hybrid and integrated systems is the S-band. In Europe, 30 MHz of S-band spectrum (1980–2010 MHz for the uplink and 2170–2200 MHz for the downlink) is allocated to MSS, whereas the corresponding MSS spectrum allocation in the United States is 20 MHz (2000–2020 MHz for the uplink and 2180–2200 MHz for the downlink). Hybrid networks can be configured in two modes: In single-frequency network (SFN), the satellite and terrestrial parts of the transmission utilize the same carrier frequency. In multifrequency network (MFN), the satellite and terrestrial transmitters operate on different carrier frequencies. They do not necessarily share any common frame or symbol timing at the physical layer, but the data payload may be common.

Three types of propagation environments can be coarsely identified for integrated and hybrid systems:

1. In urban areas, the most likely reception path is through the terrestrial base stations, namely, the complementary ground component (CGC). These terrestrial repeaters operate in SFN mode, and the rich scattering environment of the urban environment provides the basis for exploiting diversity and MIMO techniques. Here, the satellite link is available only occasionally. In case the satellite system is configured in SFN mode, the sporadic satellite path access provides an additional contribution when available. Indoor coverage is guaranteed by the dedicated terrestrial repeaters. On the other hand, if the network is configured in MFN mode, the satellite reception path in the receiver may be deactivated to save power, since only a very limited amount of satellite link coverage is available.

2. In suburban environments, the reception scenario is a mix of both a satellite and a terrestrial link. The moderate scattering environment is providing the basis for the MIMO diversity. Both satellite and terrestrial links contribute to the received signal.

3. In rural reception conditions, it can be assumed that mainly the satellite is providing the overall coverage and that the terrestrial repeater links are available only occasionally. Hence, the main reception is provided by the satellite link.

The reception environment of a hybrid network architecture similar to the one described earlier gives rise to a challenging environment for designing appropriate MIMO coding schemes. Only for a small portion of the coverage area, the reception will be truly hybrid. Even when both satellite and terrestrial components are present, the reception between the two components will be power imbalanced. On top of the power imbalance in the direct line-of-sight (LOS) component, there are fundamental differences between satellite component (SC) and CGC in terms of fading channel characteristics [2]. The land mobile satellite (LMS) channel is extremely power-limited with long erasures and necessitates receiving some portion of the LOS component. Furthermore, in contrast to the terrestrial fading channel that is typically frequency selective, the LMS channel is generally modeled as flat. In addition, given that the LMS channel geometry is completely different with respect to the terrestrial one and no scatterers are present near the transmitter, the shape of the Doppler spectrum deviates from that generated assuming a uniform scatterer distribution.

For introducing MIMO into hybrid network architectures, even existing geometries may be exploited by deploying a MIMO technique distributed between the satellite and the terrestrial repeater; this geometry corresponds to a distributed 2×1 MISO dimension. Nevertheless, for grasping the true gains of MIMO technology, a hybrid network should take advantage of more signaling dimensions. This chapter is dealing with dual polarization hybrid MIMO configurations, which is considered advantageous given the size constraints on the satellite, the terrestrial repeater, or the receiver terminal; this geometry then corresponds to a distributed 4×2 MIMO channel in case of SFN or two parallel 2×2 MIMO channels in case of MFN. The choice of dual polarization scheme employed in the satellite, the terrestrial repeater, and the user terminal (UT) can be either dual linear (LP) or dual circular (CP). In practice, terrestrial dual polarization MIMO systems opt for the linear pair due to the legacy from single polarization systems. On the other hand, for MSS systems operating at L- or S-band, the use of CP instead of LP is preferred to counteract the effects of Faraday rotation. In Reference 3, it has been shown that the choice of polarization for the satellite signal can be made independently according to an LP to CP baseband converter at the UT. Interestingly, this combining does not incur a 3 dB polarization loss as for the single polarization case.

24.2.1 Existing Hybrid Paradigm: DVB-SH

One of the first comprehensive standardization efforts in the area of hybrid satellite–terrestrial communications is the DVB-SH standard [4–6] aiming at providing ubiquitous Internet Protocol (IP)-based multimedia services to a variety of mobile and fixed terminals having compact antennas with very limited directivity at frequencies below 3 GHz. The system coverage is obtained by combining the satellite and terrestrial signals.

In Figure 24.1, a typical DVB-SH system is shown as it is presented in Reference 5. It is based on a hybrid architecture combining the SC and the CGC. The ground component consists of terrestrial repeaters fed by a broadcast distribution network. In DVB-SH, three kinds of terrestrial repeaters are envisaged (see Figure 24.1):

■ TR(a) are broadcast infrastructure transmitters that complement reception in areas where satellite reception is difficult, for example, in urban areas.
■ TR(b) are personal gap fillers of limited coverage providing local retransmission, on-frequency (SFN), and/or with frequency conversion (MFN). Typical application for gap fillers is to provide indoor coverage.
■ TR(c) are mobile broadcast infrastructure transmitters creating a "moving complementary infrastructure" on board moving platforms (cars, trains, bus).

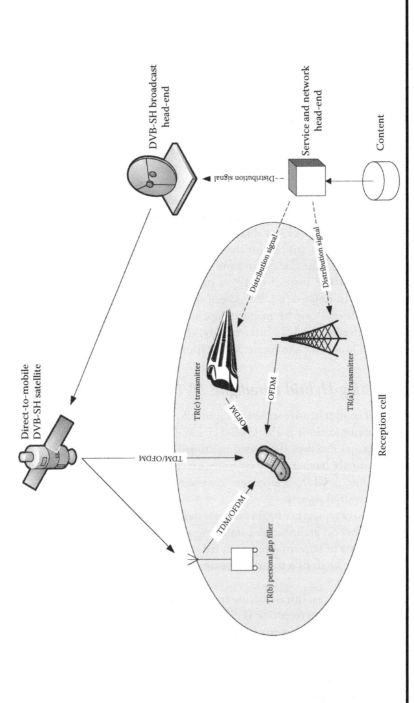

Figure 24.1 Overall DVB-SH transmission system reference architecture.

Concerning the DVB-SH physical layer waveform characteristics, the main points are summarized as follows [7]:

- System components: SC can be orthogonal frequency division multiplexing (OFDM) in the SH-A option or time division multiplexing (TDM) in the SH-B option, whereas the CGC is always in OFDM to cope with the terrestrial frequency-selective channel. Thus, SH-A is employed in an SFN configuration, while MFN architecture is used in SH-B.
- Modulation formats: SH-B TDM supports QPSK, 8PSK, and 16APSK; SH-A OFDM supports QPSK and 16-QAM.
- For OFDM, various FFT sizes (2k, 4k, 8k) are available. For TDM, various roll-off values (0.15, 0.25, 0.35) are available.
- Physical layer coding: In all SH-A/B cases, a turbo forward error correcting (FEC) scheme is adopted with a variety of coding rates spanning from 1/5 to 2/3.
- Receiver classes: Two receiver classes are possible. Class 1 has short physical layer time interleaver (200 ms), Class 2 time interleaver size can extend to 10 s. Class 1 is often combined with a link layer SH-specific FEC named iFEC, which can span tens of seconds.

24.2.2 Extension to a Hybrid MIMO Paradigm: DVB-NGH

DVB-NGH specifies four different profiles: Base (see Chapter 7), terrestrial MIMO (see Chapter 19), hybrid terrestrial–satellite (see Chapter 22), and hybrid MIMO. In this chapter, the profile for hybrid terrestrial and satellite MIMO transmissions is considered. As specified for this profile in DVB-NGH, the satellite and terrestrial transmitters can be utilized to form a hybrid MIMO network. The DVB-NGH physical layer specification is based on DVB-T2, and the main physical layer properties of the DVB-NGH system are listed as follows:

- System components: OFDM for the terrestrial part: in hybrid situation, satellite uses OFDM when operating in SFN and SC-OFDM (single-carrier OFDM) can be used for the SC in MFN (see Chapter 23).
- Modulation formats: QPSK, 16-QAM, 64-QAM, and 256-QAM for terrestrial and QPSK and 16-QAM for satellite. Thus, given that the satellite link is power limited, in hybrid SFN the modulation is limited to QPSK and 16-QAM.
- For terrestrial component OFDM, various FFT sizes (1k, 2k, 4k, 8k, 16k) are possible. For SC using OFDM, FFT sizes 1k and 2k are possible. Further, when SC-OFDM is used for satellite also, 0.5k FFT size is available.
- Physical layer coding: concatenated low-density parity check (LDPC) and BCH FEC are adopted with a variety of effective coding rates spanning from 1/5 to 3/4.

■ Physical layer time interleaving utilizing 2^{19} cells (constellation symbols) for QPSK and 16-QAM is available. For higher modulation orders, that is, for 64-QAM and 256-QAM, 2^{18} cells are available. For receivers capable of receiving hybrid transmissions, additional external interleaving memory of size 2^{21} cells can be utilized.

The hybrid MIMO network in DVB-NGH can be formed to operate either as MFN or SFN. They do not necessarily share any common frame or symbol timing at the physical layer, but the data payload may be common. At least either terrestrial or satellite transmission is made using multiple polarizations for the transmissions to qualify as hybrid MIMO; otherwise, the transmission would be covered by the hybrid (SISO) profile. Synchronized signals are transmitted to create an effective SFN. In hybrid SFN, MIMO precoding may exist in conjunction with eSFN (enhanced SFN) preprocessing. It is a transmit diversity scheme and described further in Chapter 20. MIMO coding is applied within or across the satellite and terrestrial transmission elements. When single-input single-output (SISO) and MIMO frames are mixed, MIMO coding is utilized during MIMO frames, while during the SISO frames, eSFN is specified. MIMO schemes can be utilized on a physical layer pipe (PLP) basis, and they are never used for preamble OFDM symbols P1 and aP1. The MIMO codes for different transmission antenna configurations are presented in Section 24.3.

For the reception of rate 1 MIMO codes, dual polarized receive antenna is recommended, but single polarization antenna is sufficient. For rate 2, however, the receiver must be equipped with a cross-polar pair of antennas. Out of the various possible configurations, in this chapter, we evaluate the performance of selected space–time codes assuming dual polarized transmission for both the satellite and the terrestrial components and also assuming that the terminals are equipped with a dual polarized receiver. This leads to a distributed MIMO channel of dimension 4×2 in SFN configuration. Space–time (ST) codes that, due to the broadcasting nature of the system, do not make any use of channel knowledge at the transmitter are therefore preferred.

24.3 MIMO Codes for Hybrid Networks

MIMO codes, also known as space–time codes, describe the transmission format of different data symbols on multiple streams. These codes are designed to address one or both of the following criteria [8]:

■ Multiplexing different data streams to enable high throughput
■ Spreading the symbol information across the multiple streams over multiple channel uses in order to enable appropriate diversities

The interested reader is referred to Reference 8 for further details and references on space–time coding. While several flavors of space–time codes exist, we focus on the block codes in this chapter. This is due to the presence of elegant code constructions with optimal properties and, possibly, low complexity decoding. It should be recalled that receiver complexity is also an important design parameter toward ensuring a low-cost UT.

Two variants of block codes exist:

1. Space–time block coding (STBC) is a block processing scheme where a block of data is transformed into a matrix for transmissions over antennas and time. These blocks of data are decoded independently.
2. On the other hand, MIMO-OFDM systems use space-frequency block coding (SFBC) to exploit frequency selectivity. SFBCs differ from STBCs in the sense that the former transmits code symbols over antennas and OFDM subcarriers while the latter uses antennas and time instants.

While ST coding has been used for terrestrial/satellite only [3,8], their use in hybrid systems (as depicted in Figure 24.2) is relatively new. ST coding techniques for single stream/transmitter are considered in Reference 10 assuming an MFN architecture. ST coding of signals from a satellite using dual polarization and multiple terrestrial repeaters employing a single polarization is considered in the recent work for an SFN architecture [11]. These works devise constructions based on two characteristic formulations designed for wireless communications:

■ *Spatial multiplexing (SM)* introduces no redundancy (and hence low diversity) and exhibits a high throughput. While maximum likelihood (ML) decoder is the optimal receiver, simpler linear receiver based on zero-forcing

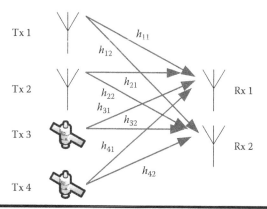

Figure 24.2   **4 × 2 hybrid MIMO antenna configuration.**

or minimum mean square error (MMSE) criterion can be used with a performance loss. The Vertical-Bell Labs Layered Space–Time (V-BLAST) scheme uses such a multiplexing centric code.

■ *Alamouti code.* In contrast to SM, Alamouti code provides the maximum diversity. On the other hand, it employs two channel uses for transmitting two symbols resulting in a throughput of one symbol per channel use (same as a SISO system). The elegance of the code lies in the fact that the ML receiver simplifies to single symbol decoding, thereby making the receiver less complex. Due to these advantages, Alamouti code has been incorporated in the terrestrial standards like IEEE WiMAX [12] and LTE.

While the earlier approaches use available codes as such, a systematic approach would follow the traditional way of obtaining code design criteria and then constructing appropriate STBC. With this aim, the next section deals with the performance criteria used in hybrid MIMO code design.

24.3.1 Hybrid Code Design Criteria

The dual polarized hybrid system can be abstracted as a 4 transmit, 2 receive MIMO system. The hybrid space–time code X is composed of two block codes: a 2×2 code X_S transmitted from the satellite and X_T transmitted from the terrestrial repeater.

24.3.1.1 Design Criteria for SFN

In SFN, the hybrid code takes the form, $X = (X_S \ X_T)_{2 \times 4}$ and its construction follows the guidelines of joint MIMO code design. Both the satellite and the terrestrial site transmit only global information, i.e., signals from both sites contain essentially the same data. For such a joint MIMO transmission, the primary design criterion for an ST code **C** is known to be *full diversity*, meaning for a lattice code that

$$det(X^\dagger X) \neq 0 \tag{24.1}$$

for all $0 \neq X \in$ **C**, where **C** denotes the code in use. Here $X^\dagger$ denotes the conjugate transpose of X and the product $X^\dagger X$ refers either to $X^\dagger X$ or $XX^\dagger$, whichever has smaller size. Note that the bigger matrix will always have zero determinant. Once achieved, the next criterion tells us to *maximize the minimum determinant* of the code:

$$det_{min}(\mathbf{C}) = \inf_{X \neq 0} |det(X^\dagger X)|, X \in \mathbf{C}. \tag{24.2}$$

If further the minimum determinant of the lattice is nonzero, meaning that it can always be bound from below by a strictly positive constant, it is said that the code has a *nonvanishing determinant* (NVD). NVD implies a nonvanishing coding gain.

24.3.1.2 Design Criteria for MFN

For MFN employing multiple radio frequencies, the criteria change. There, we should ideally have an NVD for the product:

$$det_{min}\left(X_S^\dagger X_S\right)det_{min}\left(X_T^\dagger X_T\right) \tag{24.3}$$

instead of the whole matrix $X=\begin{pmatrix}X_S\\X_T\end{pmatrix}$. This can be seen by transforming the parallel channel into an equivalent channel, which corresponds to the space–time matrix

$$X'=\begin{pmatrix}X_S & 0\\0 & X_T\end{pmatrix}. \tag{24.4}$$

It is then obvious that we have to consider the product of the determinants instead of the joint determinant. We denote the NVD described by Equation 24.3 as *parallel NVD*.

24.3.1.3 MFN vs. SFN: Decoding Complexity Perspective

Let us consider the case of four $(2+2)$ transmit antennas and two receive antennas. In SFN, the 4×2 channel matrix is of the form

$$H=\left(H_S \quad H_T\right)_{2\times4}, \tag{24.5}$$

where the index S stands for the satellite-destination path and the index T for the terrestrial-destination path. In MFN, the channel matrix becomes

$$H=\begin{pmatrix}H_S & 0\\0 & H_T\end{pmatrix}_{4\times4} \tag{24.6}$$

due to the fact that the satellite and terrestrial antennas do not cross-couple in frequencies. As a drawback, also the total bandwidth increases.

According to the earlier dense channel matrix structure in SFN, the only complexity reductions are provided by those that can possibly be embedded in the STBC code itself. By this we mean certain orthogonality relations preserved by any channel matrix, similar to the Alamouti code. One option of course is to use suboptimal decoders, but that will have a price in terms of degradation of performance.

In MFN, however, if we consider the R matrix of the QR decomposition of the equivalent channel matrix, we observe a significant complexity reduction regardless of the STBC code structure due to the sparsity of the channel matrix:

$$R_{MFN} = \begin{pmatrix} R_{11} & 0 \\ 0 & R_{22} \end{pmatrix}_{8\times 8}, \tag{24.7}$$

i.e., the upper right quarter becomes all zeros, making the code two-group decodable [13] and (at least) halving the complexity.

ST coding based on lattice codes [14] has been widely considered in the literature toward achieving desired diversity and rates. A similar approach could also be considered for the hybrid MIMO codes. Before discussing novel codes based on lattice structures, we briefly include the crucial terminology on lattice codes in the following section for the ease of reading. The reader is referred to Reference 14 for a thorough introduction to lattice codes.

24.3.2 Lattice Space–Time Codes

Definition 24.1 *For our purposes, a lattice Λ is a discrete abelian subgroup of a complex matrix space:*

$$\Lambda = \left\{ \sum_{i=1}^{s} a_i B_i \mid a_i \in \mathbf{Z} \right\} \subset M_{n\times T}(\mathbf{C}), \tag{24.8}$$

where the elements $B_1, \ldots, B_s \in M_{n\times T}(\mathbf{C})$ are linearly independent, i.e., form a lattice basis over Z, and $s \leq 2nT$ is called the *rank* of the lattice.

If the lattice has a $\mathbf{Z}[j]$-basis, we may also write

$$\Lambda = \left\{ \sum_{i=1}^{s/2} a_i B_i' \mid a_i \in \mathbf{Z}[j], j = \sqrt{-1} \right\} \subset M_{n\times T}(\mathbf{C}). \tag{24.9}$$

In practice, n corresponds to the total number of transmit antennas, and T is the ST block length corresponding to the number of channel uses required to transmit a codeword. The lattice basis matrices are often referred to as *linear dispersion* or *weight* matrices.

Definition 24.2 *A space–time lattice code* C *is a finite subset of a lattice, usually formed by taking linear combinations of the basis elements with coefficients from a certain finite set S. For instance, S can be a PAM alphabet corresponding to a Z-basis, or a QAM alphabet corresponding to a* **Z**[*j*]*-basis. With this notation, more formally we define*

$$\mathsf{C} = \{X \in \Lambda \mid a_i \in S\} \subseteq \Lambda \subseteq M_{n \times T}(\mathbf{C}). \tag{24.10}$$

24.3.2.1 Maximum-Likelihood Decoding and Fast Decodability

Maximum-likelihood decoding amounts to searching the code C for the codeword

$$Z = argmin\{\|Y - HX\|_F^2\}_{X \in \mathsf{C}}, \tag{24.11}$$

closest to the received matrix *Y* with respect to the squared Frobenius norm $\|\cdot\|_F^2$.

Consider C of Q-rank *k*, i.e., each codeword *X* is a linear combination $\sum_{i=1}^{k} B_i g_i$ of generating matrices $B_1, \ldots, B_k$, weighted by coefficients $g_1, \ldots, g_k$, which are PAM information symbols. The matrices $B_1, \ldots, B_k$ therefore define our code. Each $n_r \times T$ matrix HB_i corresponds, via vectorization, to a vector $\mathbf{b}_i \in \mathbf{R}^{2Tn_r}$ obtained by stacking the columns followed by separating the real and imaginary parts of HB_i. We define the (generating) matrix

$$B = (\mathbf{b}_1, \mathbf{b}_2, \ldots, \mathbf{b}_k) \in M_{2Tn_r \times k}(\mathbf{R}), \tag{24.12}$$

so every received codeword can be represented as a real vector *B***g**, with $\mathbf{g} = (g_1, \ldots, g_k)^T$ having coefficients in the real alphabet *S* in use.

Now finding $argmin\{\|Y - HX\|_F^2\}_{X \in \mathsf{C}}$ becomes equivalent to finding $argmin\{\|\mathbf{y} - B\mathbf{g}\|_E^2\}_{\mathbf{g} \in |S|^k}$ with respect to Euclidean norm, where **y** is the vectorization of the received matrix *Y*. The latter search is performed using a real sphere decoder [15], with the complexity of exhaustive search amounting to $|S|^k$, as the coefficients of *g* run over all the values of *S*. The complexity of decoding can, however, be reduced if the code has additional structure [13]. Performing a QR decomposition of *B*, *B* = *QR*, with $Q^\dagger Q = I$, reduces finding $argmin\{\|\mathbf{y} - B\mathbf{g}\|_E^2\}_{\mathbf{g}}$ to minimizing

$$\|\mathbf{y} - QR\mathbf{g}\|_E^2 = \|Q^\dagger\mathbf{y} - R\mathbf{g}\|_E^2, \tag{24.13}$$

where *R* is an upper right triangular matrix. The number and structure of zeros of the matrix *R* may improve the *decoding complexity* (formally defined in Reference 16 to be the minimum number of vectors *g* over which the difference in Equation 24.13 must be computed). When the structure of the code allows for the degree (i.e., the exponent of $|S|$) of decoding complexity to be less than the rank of the code, we say that the code is *fast-decodable*.

More precisely, we have the following definitions from Reference 13.

Definition 24.3 *A space–time code is said to be fast-decodable if its R matrix has the following form:*

$$R = \begin{pmatrix} \Delta & R_1 \\ 0 & R_2 \end{pmatrix}, \tag{24.14}$$

where

  Δ is a diagonal matrix
  R_2 is upper triangular

The authors of Reference 13 give criteria when the zero structure of R coincides with that of M, where M is a matrix capturing information about orthogonality relations of the basis elements of B_i:

$$M_{k,l} = \| B_k^\dagger B_l + B_l^\dagger B_k \|_F. \tag{24.15}$$

In particular, Reference 13 Lemma 2 shows that if M has the structure

$$M = \begin{pmatrix} \Delta & R_1 \\ V_2 & V_3 \end{pmatrix}, \tag{24.16}$$

where Δ is diagonal, then

$$R = \begin{pmatrix} \Delta & R_1 \\ 0 & R_2 \end{pmatrix}. \tag{24.17}$$

We could thus rephrase Definition 24.3 in terms of M.

24.3.2.2 Field Norm and Code Rate

The notion of relative field norm is used in Reference 17 to prove the NVD property of the proposed constructions. Let us briefly recall it to ease the reading. Let E/F be a Galois extension with a Galois group $\mathrm{Gal}(E/F) = \{\sigma_1, \ldots, \sigma_n\}$, where $n = [E{:}F]$.

Definition 24.4 *Let $a \in E$. The relative field norm of $N_{E/F}:E \to F$ is defined as the product of the algebraic conjugates:*

$$N_{E/F}(a) = \sigma_1(a) \cdots \sigma_n(a). \tag{24.18}$$

We may abbreviate $N(a) = N_{E/F}(a)$, if there is no danger of confusion.

Let us denote the ring of algebraic integers of a field F by O_F. According to basic algebra, we have the following.

Proposition 24.1 *If $a \in O_E\backslash\{0\}$, then $N_{E/F} \in O_F\backslash\{0\}$. In particular, if $F = \mathbf{Q}$ or $F = \mathbf{Q}(\sqrt{-m})$ with m a square free integer, then $N_{E/F}(a) \geq 1$ for any $a \in O_E$.*

As there are multiple definitions of rate in the literature, we specify here our definition of the code rate. As before, let n denote the total number of transmit antennas and T the block length of the ST matrix.

Definition 24.5 *If a space–time code matrix $X_{n \times T}$ contains s independent PAM symbols, or equivalently s/2 independent QAM symbols, the ST code rate is*

$$R = \frac{s}{2n} \tag{24.19}$$

complex dimensions (read, independent symbols) per channel use.

24.3.3 MIMO Codes Studied during DVB-NGH Standardization

Various codes both for the sheer terrestrial and for the hybrid MIMO profiles based on the described lattice structure were proposed during the DVB-NGH standardization. The proposed codes can be used in both SFN and MFN. In MFN, there is an obvious loss in frequency utilization compared to the same code to be used in SFN, as the total used bandwidth is increased. On the other hand, one can get significant complexity savings in the decoding process as compared to SFN. For example, single-symbol decodability in MFN is possible for some of the proposed codes.

The structure of the codes allows for different antenna combinations, i.e., one or two antennas (polarizations) at the satellite and one to three antennas (polarizations) at the terrestrial site. The usual assumption arising from the existing technology and devices is two receive antennas (polarizations). Codes were proposed for both the cases when satellite and terrestrial transmit exactly the same (global) content, and when they both transmit some local content in addition to the global content.

24.3.3.1 L_2: A Full-Diversity Rate-1 Quasi-Orthogonal Code

The following rate-1 code is similar to Jafarkhani's quasi-orthogonal code presented in Reference 18, except that the proposed code has full diversity and hence better (actually nonvanishing) coding gain. The L_2 code has the following properties:

- Quasi-orthogonal structure
- Full diversity unlike Jafarkhani's code (i.e., more robust)
- Nonvanishing coding gain
- Rate 1, enabling high-quality reception also in the presence of correlation
- Shadow protection
- Complex sphere decoding with reduced hard decision (hd) complexity $\leq 2|S|^2$ in SFN
- Complex sphere decoding with reduced hd complexity $\leq 4|S|$ in MFN

The encoding matrix is

$$
X_{L_2} = \begin{pmatrix} c_1 & jc_2 & -c_3^* & -c_4^* \\ c_2 & c_1 & jc_4^* & -c_3^* \\ c_3 & jc_4 & c_1^* & c_2^* \\ c_4 & c_3 & -jc_2^* & c_1^* \end{pmatrix} \begin{matrix} -\text{satellite} \\ -\text{satellite/terrestrial} \\ -\text{terrestrial} \\ -\text{terrestrial} \end{matrix} \quad (24.20)
$$

where

   $j = \sqrt{-1}$, c_i are complex integers (e.g., QAM symbols)
   c^* denotes the complex conjugate of c

More details of this code can be found in Reference 19.

In MFN, the R matrix of the QR decomposition of the equivalent channel matrix is diagonal (after swapping the second and third columns of X_{L_2}), i.e., the code becomes single-complex-symbol decodable, and we can decode each QAM symbol independent of each other. This, however, requires the channel to remain (nearly) constant for the duration of one codeword.

Assuming there are two satellite antennas and two terrestrial antennas, the first two rows are transmitted by the satellite and the latter two by terrestrial antennas. The block structure of the code provides robustness in the case of deep shadowing and low rate $R = 1$ is advantageous in the presence of high correlation. One may also use this code with 1 satellite antenna + 3 terrestrial antennas in an obvious way. If transposed, there is no need for encoding on board the satellite, as imposed by the majority of satellite systems being transparent.

24.3.3.2 MUMIDO Code: Good Compromise between Complexity, Rate, and Delay

This code has been published in Reference 20. The MUMIDO code has the following properties.

- Diversity two, rate two
- Shadow protection

Decoding complexity $\leq 2|S|^3$ in SFN, and $\leq 4|S|$ in MFN (the R-matrix is again diagonal). The 4×2 encoding matrix is

$$
X_{MUMIDO} = \begin{pmatrix} c_1 + c_3\zeta & -c_2^* + c_4^* j\zeta \\ c_2 + c_4\zeta & c_1^* - c_3^* j\zeta \\ c_1 - c_3\zeta & -c_2^* - c_4^* j\zeta \\ c_2 - c_4\zeta & c_1^* + c_3^* j\zeta \end{pmatrix}_{4 \times 2}
\begin{matrix} -\text{satellite} \\ -\text{satellite/terrestrial} \\ -\text{terrestrial} \\ -\text{terrestrial} \end{matrix}, \tag{24.21}
$$

where

$j = \sqrt{-1}$, $\zeta = e^{2j\pi/8}$, and c_i are complex information symbols (QAM symbols)
c^* denotes the complex conjugate of c

24.3.3.3 RESM: Restricted Enhanced Spatial Multiplexing

Using the previous notation, the following form of SM with reduced decoding complexity of at most $|S|^2$ was proposed by R. Vehkalahti. The QAM symbols c_1, c_2 are mapped as follows:

$$
(c_1, c_2) \mapsto X_{RESM} = \begin{pmatrix} c_1 + c_2\zeta \\ c_1^* - c_2^* j\zeta \\ c_1 - c_2\zeta \\ c_1^* + c_2^* j\zeta \end{pmatrix}
\begin{matrix} -\text{satellite} \\ -\text{satellite/terrestrial} \\ -\text{terrestrial} \\ -\text{terrestrial} \end{matrix}. \tag{24.22}
$$

24.3.3.4 C_1 Code

This code has been published in Reference 21. Let $\zeta = e^{2j\pi/5}$ and $r = (8/9)^{1/4}$. The encoding matrix is

$$
X_{C_1} = M(x_0, x_1, x_2, x_3) = \begin{pmatrix}
x_0 & -r^2(x_1)^* & -r^3\sigma(x_3) & -r\sigma(x_2)^* \\
r^2 x_1 & (x_0)^* & r\sigma(x_2) & -r^3\sigma(x_3)^* \\
r x_2 & -r^3(x_3)^* & \sigma(x_0) & -r^2\sigma(x_1)^* \\
r^3 x_3 & r(x_2)^* & r^2\sigma(x_1) & \sigma(x_0)^*
\end{pmatrix},
$$
(24.23)

where

$$
x_i = x_i(y_1, y_2, y_3, y_4) = y_1(1 - \zeta) + y_2(\zeta - \zeta^2) + y_3(\zeta^2 - \zeta^3) + y_4(\zeta^3 - \zeta^4),
$$

and

$$
\sigma(x_i) = y_1(1 - \zeta^3) + y_2(\zeta^3 - \zeta) + y_3(\zeta - \zeta^4) + y_4(\zeta^4 - \zeta^2).
$$

The variables y_i in each of the x_i range over a certain PAM set, so that the code encodes overall 16 independent PAM symbols. In other words, a PAM vector $(a_1, \ldots, a_{16})$ is mapped into a 4×4 matrix:

$$
\sum_{i=1}^{16} a_i B_i,
$$
(24.24)

where the basis matrices of B_i of the code are

$$
B_1 = M(x_0(1,0,0,0),0,0,0),
$$
$$
B_2 = M(x_0(0,1,0,0),0,0,0),
$$
$$
\vdots
$$
$$
B_{16} = M(0,0,0,x_3(0,0,0,1)).
$$

This code can be used similar to L_2 but provides higher rate $R = 2$. In SFN, the decoding complexity is $|S|^6$, but in MFN only $|S|^2$. More specifically, the complexity is 12 or 4 real dimensions: recall that this code does not admit $\mathbf{Z}[j]$-basis that would allow us to use 8 QAM symbols, but uses 16 PAM symbols instead for linearly combining $\mathbf{Z}$-basis matrices.

24.3.3.5 Other 4 Tx Codes

If we want to have simple Alamouti encoding at the satellite site while avoiding repetition, we can use the code described in Reference 17:

$$
X = \begin{pmatrix} X_S \\ X_T \end{pmatrix} = \begin{pmatrix} a & -b^* \\ b & a^* \\ \sigma(a) & -b^* \\ b & \sigma(a)^* \end{pmatrix},
\tag{24.25}
$$

where $a \in \mathbf{Z}[\zeta_8]$, $b \in \mathbf{Z}[j]$, and $\sigma:\zeta_8 \mapsto -\zeta_8$ is the generator of the cyclic Galois group of the field extension $\mathbf{Q}(\zeta_8)/\mathbf{Q}(j)$. The code has an "intermediate" rate of $R = 3/2$ QAM symbols per channel use.

24.3.3.6 3 Tx Codes

Next, we describe a code with an intermediate rate $R = 3/2$, while maintaining joint NVD. Let $a \in \mathbf{Q}(\zeta_8)$, and let $\sigma:\zeta_8 \mapsto -\zeta_8$ be the generator of the cyclic Galois group of $\mathbf{Q}(\zeta_8)/\mathbf{Q}(j)$. In Reference 17 it is proposed to use the encoding matrix:

$$
X_{L_3} = \begin{pmatrix} X_S \\ X_T \end{pmatrix} = \begin{pmatrix} a & b \\ \sigma(a) & -b^* \\ b & \sigma(a)^* \end{pmatrix},
\tag{24.26}
$$

where $b \in Z[j]$. This code has a joint NVD in SFN and full diversity (but not parallel NVD) in MFN.

24.3.4 MIMO Codes Selected for DVB-NGH Hybrid Profile

When designing MIMO codes for hybrid scenario, the particular system characteristics analyzed previously should be taken into account. Also, considering the SFN situation, it is important to take into account the amount of pilot symbols required for channel estimation. In the four-transmitter case, if channel responses for all channels are required, the pilot density is four times the density required for the same estimation accuracy to the one transmitter (SISO) case. For example, assuming that the required density of pilots over total bits transmitted for a certain mobility is equal to 1/12, then for four-transmitter MIMO, the density would already be 1/3, eating out a great share

Table 24.1 MIMO Schemes for Different Hybrid Scenarios

#Tx	Terrestrial	Satellite	Schemes
2	Single polarization	Single polarization	Rate 1: eSFN, Alamouti
3	Dual polarization	Single polarization	Rate 1: eSFN, Alamouti + QAM
			Rate 2: VMIMO
3	Single polarization	Dual polarization	Rate 1: eSFN, Alamouti + QAM
			Rate 2: VMIMO
4	Dual polarization	Dual polarization	Rate 1: eSFN, Alamouti + Alamouti
			Rate 2: eSM + PH terr + eSM + PH sat (+eSFN on Sat)

of bandwidth available for the payload data. For this reason, in the schemes that were finally selected for DVB-NGH, the satellite transmitters simply repeat the transmission of the terrestrial transmitters. Thus, assuming that the relative delay of the received signals is within the guard interval of OFDM symbols, coherent reception is possible. Another consideration motivating this choice is the need to guarantee that a satisfactory receiver performance is possible even if one of the streams is lost (a highly probable event in such mobile systems). Furthermore, due to the presence of the SC, system elements that are not usually present in typical MIMO wireless systems, like the long time interleaving and the strong nonlinear behavior of the satellite high power amplifier, have an impact on the design and choice of appropriate MIMO technique. The combinations of different forms of diversities, such as long time interleaving and MIMO, are studied in Reference 22.

For optimal utilization of the hybrid network, MIMO codes for two to four transmitters are considered in DVB-NGH. Both terrestrial and satellite transmitters can utilize one or two polarizations. The selected codes for SFN scenarios are presented in Table 24.1. For MFN, when SC-OFDM is used, plain SM is used instead of eSM.

24.3.4.1 Single Satellite and Single Terrestrial Polarization

When single polarization is available at both satellite and terrestrial sites, Alamouti coding for the signal can be performed. Also eSFN mechanism can be used to form a simple hybrid SFN network.

24.3.4.2 Single Satellite and Dual Terrestrial Polarization

When the terrestrial transmissions are dual polarized and the satellite utilizes only a single polarization, a simple combination of QAM and Alamouti transmission called "Alamouti + QAM" [17] can be utilized:

$$X = \begin{pmatrix} X_S \\ X_T \end{pmatrix} = \begin{pmatrix} a & b \\ a & b \\ -b^* & a^* \end{pmatrix},$$ (24.27)

where *a* and *b* are QAM symbols. In this code, the first row corresponds to satellite transmitter that transmits the same symbols as one terrestrial polarization. Sphere decoding complexity for this code is M, i.e., equal to the constellation size. This code satisfies both the joint and the parallel NVD. Namely, a straightforward calculation results in

$$det(X^{\dagger}X) \geq 2$$ (24.28)

and

$$det(X_S^{\dagger}X_S)det(X_T^{\dagger}X_T) \geq 1.$$ (24.29)

Also eSFN can be used. If rate 2 transmission is required virtual MIMO (VMIMO) can be used. In VMIMO the single polarized transmitter emulates at the transmitter side an optimized 2×1 channel while the dual polarized transmitter emits rate 2 MIMO.

24.3.4.3 Dual Satellite and Single Terrestrial Polarization

When the terrestrial transmissions use single polarization and satellite transmissions are dual polarized, Alamouti + QAM can be utilized. Now only the roles of terrestrial and satellite transmitters are reversed as compared to the code presented for the case of dual terrestrial and single satellite polarization. Also using eSFN and VMIMO is possible.

24.3.4.4 Dual Satellite and Dual Terrestrial Polarization

In this case both terrestrial and satellite transmissions utilize dual polarization. For rate 1, the straightforward solution is to transmit the same Alamouti blocks from both satellite and terrestrial transmitters. Further, eSFN can be used for satellite to enhance the operation. This code is called "Alamouti + Alamouti":

$$
X = \left(\frac{X_S}{X_T} \right) = \begin{pmatrix} a & b^* \\ b & -a^* \\ \hline a & b^* \\ b & -a^* \end{pmatrix}_{4 \times 2}
\tag{24.30}
$$

In addition to rate 1, eSM is specified for rate 2 (further described in Chapter 21) in DVB-NGH:

$$
X = \begin{pmatrix} x_1 \\ x_2 \end{pmatrix} = \sqrt{2} \begin{pmatrix} \sqrt{\beta} & 0 \\ 0 & \sqrt{1-\beta} \end{pmatrix} \begin{pmatrix} \cos\theta & \sin\theta \\ \sin\theta & -\cos\theta \end{pmatrix} \begin{pmatrix} \sqrt{\alpha} & 0 \\ 0 & \sqrt{1-\alpha} \end{pmatrix} \begin{pmatrix} a \\ b \end{pmatrix}
\tag{24.31}
$$

where a and b are QAM symbols. The eSM processing may be applied for 4, 6, 8 bits/cell unit (bpcu). The eSM encoding process shall use defined parameter values for each use according to Table 24.2.

Further, phase hopping is applied:

$$
\begin{pmatrix} g_1 \\ g_2 \end{pmatrix} = \begin{pmatrix} 1 & 0 \\ 0 & e^{j\,\Phi PH(i)} \end{pmatrix} \begin{pmatrix} x_1 \\ x_2 \end{pmatrix}, \quad i = 0, 1, \ldots, N_{cells}/2 - 1
\tag{24.32}
$$

where
N_{cells} is the number of cells required to transmit one LDPC block
$\Phi PH(i) = 2\pi/9$ is the phase change for cell pair i

Table 24.2 eSM Parameters for Hybrid MIMO Profile

n_{bpcu}	Modulation		β	θ (°)	α
4	x_1 (Tx1)	QPSK	0.50	67.50	0.50
	x_2 (Tx2)	QPSK			
6	x_1 (Tx1)	QPSK	0.50	45	0.44
	x_2 (Tx2)	16-QAM			
8	x_1 (Tx1)	16-QAM	0.50	57.76	0.50
	x_2 (Tx2)	16-QAM			

It is initialized to 0 at the beginning of each FEC block and is incremented by $2\pi/9$ for every cell pair. The satellite transmits the same signal except for the eSFN predistortion.

24.3.5 Codes Applicable to MIMO Extension of DVB-SH

The code design for the MIMO extension of DVB-SH focused on the situation where the two transmitters employ identical data. A DVB-SH/A configuration was considered where the two transmitters employ OFDM signaling. Toward this, the following structures were envisaged.

24.3.5.1 Distributed Spatial Multiplexing

This is a simple construction where the same spatially multiplexed streams are transmitted from satellite and terrestrial base stations. The transmissions from the two links are viewed as multipath components, thereby providing diversity. Unlike in the earlier constructions, both X_S and X_T are 2×1 vectors and the code takes the form

$$X = \left(\frac{X_S}{X_T} \right) = \begin{pmatrix} a \\ b \\ a \\ b \end{pmatrix}_{4 \times 1} \tag{24.33}$$

This code entails the joint decoding of two symbols.

24.3.5.2 Distributed Spatial Multiplexing + Block Alamouti

Another construction involves the use of SM between colocated antennas (polarizations) and Alamouti coding over satellite and terrestrial links. The resulting 4×2 code takes the form

$$X = \left(\frac{X_S}{X_T} \right) = \begin{pmatrix} a & b^* \\ c & d^* \\ \hline b & -a^* \\ d & -c^* \end{pmatrix}_{4 \times 2} \tag{24.34}$$

This code is envisaged to provide additional coding benefits over the simple SM. However, unlike the traditional Alamouti code, it is not amenable to single symbol decodability: a joint decoding of four symbols need to be used for extracting optimal performance. This increases the receiver complexity.

24.3.5.3 Space Frequency Codes

The constructions presented earlier employ coding, if any, across polarizations, time, and/or links. Efforts to include the frequency selectivity have not been considered. While the satellite channels are usually frequency flat, there is a relative delay between the reception of the satellite and terrestrial links. For the considered MIMO extension of DVB-SH/A with identical data transmitted over the two links, the relative delay results in a frequency-selective channel for each subcarrier of the OFDM symbol [9]. The availability of frequency selectivity motivates the use of SFBCs as they have been considered in literature to exploit frequency diversity in MIMO-OFDM systems.

One approach towards the construction of SFBCs involves use of available STBCs, with the time dimension replaced by subcarrier indices. Based on this approach, the well-known Golden codes [23] have been considered in Reference 9 to exploit polarization diversity and frequency selectivity. The code takes the form

$$
X = \left(\frac{X_S}{X_T} \right) = \begin{pmatrix} \overset{k_1}{a} & \overset{k_2}{b} \\ c & d \\ \hline a & b \\ c & d \end{pmatrix}_{4 \times 2}
\tag{24.35}
$$

where k_1 and k_2 are the subcarrier indices of the same OFDM symbol and $G = \begin{pmatrix} a & b \\ c & d \end{pmatrix}$ constitutes a Golden code [23] with $a = \left[\alpha(s_1 + s_2\theta) / \sqrt{5} \right]$, $b = \left[\alpha^* (s_1 + s_2\theta^*)/\sqrt{5} \right]$, $c = \left[j\alpha^* (s_3 + s_4\theta^*)/\sqrt{5} \right]$, $d = \left[\alpha(s_3 + s_4\theta)/\right]\sqrt{5}$, $\theta = \left[(1 + \sqrt{5})/2 \right]$, $\theta^* = \left[(1 - \sqrt{5})/2 \right]$, $\alpha = 1 + j(1 - \theta)$, and $\alpha^* = 1 + j(1 - \theta^*)$.

It is shown in Reference 9 that the proposed SFBC achieves full polarization diversity and that subcarrier separation $k_1 - k_2$ along with the relative delay between the links determines the coding gain [8]. Towards achieving higher spectral efficiencies, all the subcarriers of the OFDM symbol need to be used. This entails the use of $N/2$ Golden codes (each for satellite and terrestrial) in SFBC mode if the OFDM symbol contains N subcarriers. The subcarrier indices for each of the $N/2$ Golden codes need to be chosen to maximize the coding gain. An algorithm to obtain these indices without the knowledge of the relative delay has been proposed in Reference 9, and advantages due to coding gain enhancement have been depicted for various channel models in Reference 9. Similar to distributed SM + block Alamouti, joint decoding of four symbols is needed for extracting optimal performance. This clearly increases the receiver complexity.

24.4 Performance Evaluation

A performance comparison of hybrid MIMO schemes in a 4×2 hybrid SFN network environment is now presented. First, simulations for the codes that are selected for DVB-NGH and the most promising codes studied along the course of standardization process are presented. Further, performance simulations in the context of a possible MIMO extension to DVB-SH system are presented.

24.4.1 Performance Simulations for a DVB-NGH-Based Hybrid MIMO System

24.4.1.1 DVB-NGH Simulation System Description

The simulation system for hybrid MIMO codes consists of main DVB-NGH functionality. The SFBC encoder is located right after the frequency interleaving and before pilot insertion and inverse fast Fourier transform (IFFT) modules. The functional block diagram of the DVB-NGH simulator is shown in Figure 24.3.

The BCH code specified in DVB-NGH standard is omitted in the simulations, as its effect on the performance is small in the bit error rate (BER) region studied here.* Since only QPSK and 16-QAM are available to both SC and CGC, in the simulations presented here, these two constellations are considered. Another reason for avoiding higher-order constellations is that the complexity of sphere decoding for the SFBCs increases rapidly when increasing the number of constellation points.

The SFBC encoding is performed in the simulator after the interleaving stages to have the SFBC block as "tightly packed" as possible, meaning mapped to neighboring subcarriers. In DVB-NGH, the encoding is done before the frequency interleaving, but then the frequency interleaver operates on pairs of cells. This tight packing is important for optimal performance of the SFBC codes, as the channel response for all the subcarriers carrying one block should remain unchanged, or in practice, with minimal change. Sometimes there is a pilot carrier between the carriers of one SFBC block slightly increasing the separation.

After the SFBC encoding, the OFDM symbols are generated by transforming the frequency domain signal to the time domain by performing IFFT and adding guard interval (cyclic prefix). These operations are performed for each transmitting antenna. The pilots are not inserted in the simulation system, as ideal channel

---

* Its main purpose is error floor removal at low BER.

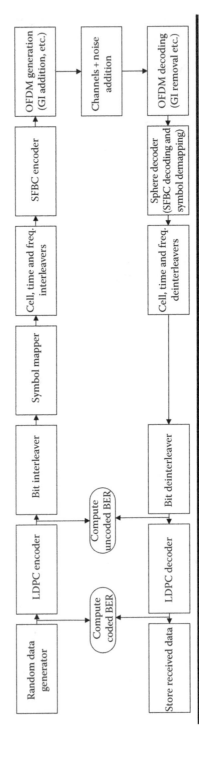

Figure 24.3   Functional block diagram of the used DVB-NGH simulator.

knowledge is assumed. Therefore, the effect of channel estimation algorithm is outside the scope of this chapter.

In the receiver, the SFBC is decoded by soft output sphere decoder with Max-Log log-likelihood ratio (LLR) demapper [24]. Thus, constellation demapping and SFBC decoding are performed jointly. Iterative MIMO-LDPC decoding is not used in the simulations. Belief-propagation-based iterative algorithm is used for soft decision decoding of the LDPC codes. Low rate codes (3/15, 5/15, and 6/15) are included in the DVB-NGH as operation point of a mobile system should be at low SNR, which translates into utilization of strong FEC codes and low-order modulation.

24.4.1.2 Channel Models

In addition to the traditional Rayleigh independently and identically distributed (i.i.d.) channel, a hybrid channel model used also in the standardization process of DVB-NGH is considered. The channel model is fully described in Chapter 19. This channel model comprises terrestrial dual polarized channel model generated based on field measurements with cross-polar 2×2 MIMO [25]. Outdoor channel with eight taps is considered for the terrestrial part. The first tap is a pure LOS component. Correlation between the paths of 2×2 channels is present. The correlation matrix for the outdoor channel is [25]

$$R = \begin{pmatrix} 1.00 & 0.06 & 0.06 & 0.05 \\ 0.06 & 0.25 & 0.03 & 0.05 \\ 0.06 & 0.03 & 0.25 & 0.06 \\ 0.05 & 0.05 & 0.06 & 1.00 \end{pmatrix}$$

The cross-polar discrimination (XPD) factor is 6 dB (4 in linear scale). For the satellite part, the Rice channel for representing the LOS nature of the satellite transmission is used. The observed power imbalance between the satellite and terrestrial signals is an adjustable parameter (λ). The channel block in Figure 24.3 evaluates the channel from each transmit antenna to each receive antenna by assuming a fixed total transmit power per segment.

24.4.1.3 Simulation Results

The performance comparison of 4×2 schemes in SFN configuration is shown here. The SFBCs considered are Alamouti + Alamouti, L_2, PH-eSM, SM 4×2, and MUMIDO. Alamouti + Alamouti, L_2, and SM 4×2 have been studied in a terrestrial 4×2 environment in Reference 26. First, the performance of the codes is presented in commonly used Rayleigh i.i.d. channel. Further, the hybrid channel described earlier with Doppler frequency of 33.3 Hz and power balance $\lambda = 0.2$ is used, meaning that the received satellite signal is five times weaker than the terrestrial. Out of the considered scenarios, this might correspond to the urban reception scenario where terrestrial network is denser. The other system parameters used in the simulations are shown in Table 24.3.

The comparison of the SFBCs in Rayleigh i.i.d. channel without the LDPC decoding using 16-QAM modulation is shown in Figure 24.4. For this comparison, hard decisions are made at the output of the sphere decoder in order to calculate BER.

It is observed that the performance of PH-eSM and SM 4×2 are rather similar as could be expected in this rich fading channel. The SM 4×2 code transmitting

Table 24.3 DVB-NGH System Parameters Used in the Simulations

Bandwidth	8 MHz	LDPC length	16k
FFT size	2k	Code rates	3/15, 5/15, 6/15, 7/15, 9/15
Guard interval	1/4	Channel estimation	Ideal
Modulation	QPSK, 16-QAM	Time interleaver	2^{18} cells

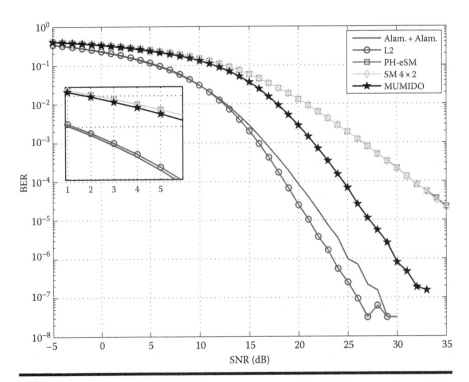

Figure 24.4   SFBC performance comparison in Rayleigh i.i.d. channel, 16-QAM.

the same two symbols from both terrestrial and satellite simultaneously is taken as a reference rate 2 code. Also it is observed that the performance of the MUMIDO code is better than the performance of the other two rate 2 codes. On the other hand, MUMIDO does not fulfill the criterion of transmitting similar signals from both terrestrial and satellite arms. Thus, the amount of required pilots for channel estimation is increased as compared to the other rate 2 alternatives. Moreover, it is observed that L_2 outperforms the Alamouti + Alamouti transmission. Again, when using the L_2, satellite and terrestrial stations transmit different signals, thus affecting the amount of required pilots. When combined with LDPC, the operating point of the system corresponds to a rather low SNR, where the difference of, for example, Alamouti + Alamouti and L_2 is small and actually favors Alamouti + Alamouti as is seen in Figure 24.4.

Similar comparison for the hybrid channel using QPSK modulation is presented in Figure 24.5. For the hybrid channel, fairly similar observations as for the Rayleigh i.i.d. can be made. The performance of Alamouti + Alamouti and L_2 is rather similar in hybrid channel. MUMIDO begins to outperform PH-eSM at around 7 dB. It should be noted that Alamouti + Alamouti and L_2 represent rate 1 codes offering half the spectral efficiency of the other codes that are of rate 2.

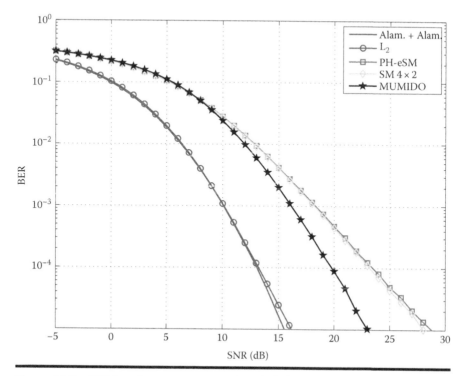

Figure 24.5 SFBC performance comparison in hybrid channel, QPSK.

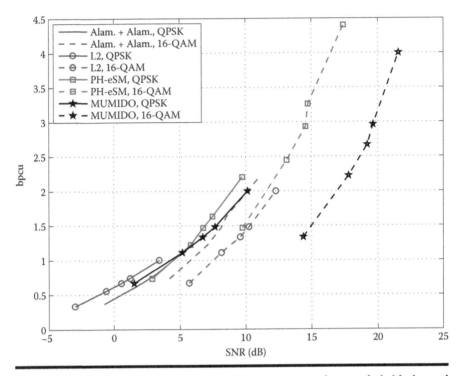

Figure 24.6   Performance comparison of 4 × 2 MIMO schemes, hybrid channel model ($\lambda = 0.2$).

Next, let us compare the performance of the SFBCs combined with LDPC FEC in the hybrid channel. An illustrative mechanism for this comparison is by spectral efficiency in terms of bpcu vs. SNR curves. Such curves are presented in Figure 24.6, where the spectral efficiency is calculated assuming a pilot density of 1/12 for Alamouti + Alamouti and PH-eSM, while for L_2 and MUMIDO, the density is assumed to be doubled to 1/6, as the channel from the satellite needs to be estimated separately.

To obtain the capacity curves, LDPC code rates 3/15, 5/15, 6/15, 7/15, and 9/15 for all studied SFBCs are simulated and the SNR for reaching BER 10^{-4} is searched. Thus, each point in the curves corresponds to SNR required to reach the BER criterion. This BER criterion was selected as a compromise between simulation time and accuracy.

It is observed that the L_2 with QPSK outperforms Alamouti + Alamouti and also MUMIDO outperforms PH-eSM with QPSK and low code rate. With 16-QAM, on the other hand, the situation is reversed; that is, Alamouti + Alamouti outperforms L_2 and PH-eSM outperforms MUMIDO. In the standardization effort, many channel situations were considered (e.g., different correlation between the terrestrial paths, different values for λ, etc., and some described in Chapter 21) in addition to the configuration presented here. The codes performing best on average over various scenarios were finally selected.

From a MIMO code design point of view, this kind of simulation studies are useful, as the operation point of the whole system is affected by other error control mechanisms present in the system. For example, the use of strong LDPC codes in combination with time interleaving drives the operation point of the system to be at low SNR. For example, in Figure 24.4, the L_2 code outperforms the Alamouti + Alamouti at high SNR, but at low SNR (where the system is operational due to LDPC coding), the Alamouti + Alamouti shows better performance.

24.4.2 Performance Simulations for a DVB-SH-Based Hybrid System

The transmitter supporting MIMO extension to DVB-SH implementing the typical OFDM physical layer elements is depicted in Figure 24.7. Note that each stream represents a polarization in the dual polarization setup and the per-stream processing (except for MIMO encoding) mimics the DVB-SH/A chain. The transmitter architecture is same for both the terrestrial and satellite transmissions.

Packetization function includes mode and stream adaptation followed by a turbo encoder. Time interleavers specified in Reference 5 are used. Additional operations of bit demultiplexing and symbol interleaving precede modulation and MIMO encoding. Further, the pilot sequences are modified in comparison to Reference 5 to distinguish between the two streams. These modified pilots are then inserted, and the standard operations of IFFT and cyclic prefix insertion are performed.

The receiver implements the core physical layer functionalities of Reference 5 on each polarization and is depicted in Figure 24.8. These are similar to the transmitter and hence will not be described. Subsequent to baseband processing and synchronization, MIMO channel estimation and decoding are implemented as shown in Figure 24.8.

The MIMO receiver demodulates the received streams and generates the LLRs for the turbo decoder on each stream. Linear, reduced-complexity ML (based on k-best algorithm) and full-complexity ML MIMO receivers are used in this study.

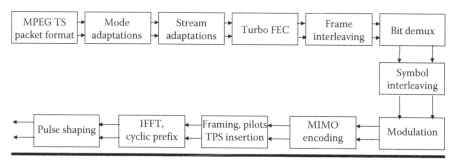

Figure 24.7 **Block diagram of DVB-SH transmitter adapted to support MIMO extension.**

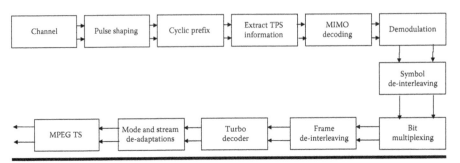

Figure 24.8 Block diagram of DVB-SH receiver adapted to support MIMO extension.

24.4.2.1 Channel Models

In terms of channel modeling, a more sophisticated formulation as compared to DVB-NGH, a dual polarization MIMO LMS channel model is employed.

For the terrestrial dual polarization MIMO channel, the Winner II model is employed, whereas for the satellite dual polarization MIMO channel, the model described in Reference 27 and based on a first-order Markov chain is employed. More details on the channel parameters used in these simulations are presented in Tables 24.4 and 24.5.

24.4.2.2 Simulation Results

Distributed 4×2 MIMO SM and the distributed 4×2 MIMO SM + block Alamouti discussed in Section 24.3.5 were considered for simulations under the MIMO hardware (HW) project [28]. Simulation results on these hybrid MIMO codes are presented in Figure 24.9. Both cases have been evaluated assuming an optimal detector, and in addition, the performance of the distributed 4×2 MIMO SM with block Alamouti coding has also been evaluated with nonoptimal receiver (list sphere detector, LSD, with list size 16).

From Figure 24.9, it can be observed that the more sophisticated coding in the distributed 4×2 MIMO SM + block Alamouti provides only marginal gain with respect to the distributed 4×2 MIMO SM. These results suggest that the strong FEC coding together with the long interleaving over this particular 4×2 hybrid MIMO setting renders the addition of the Alamouti coding over the pure SM not attractive. More simulation results for MIMO extension of the DVB-SH systems can be found from Reference 28.

The results in Figure 24.9 are also presented in terms of the error second ratio ESR5(20), which is seconds in error over an observation period. ESR5 corresponds to 1 s with error over a 20 s observation period, hence 5% of the time.

Table 24.4 MIMO Channel Characteristics Employed for the Terrestrial Component (WINNER II Model for Suburban Area)

Parameter	Value Used	Parameter	Value Used
Carrier frequency	S-band (2.2 GHz)	Antenna patterns	Ideal isotropic pattern with unit gain for copolarization
Vehicle speed	60 km/h	Rx-antenna XPD	15 dB
Propagation environment	Suburban macrocell	Number of clusters	LOS: 15
			NLOS: 14
Large-scale parameters update interval	3 m	Number of rays per cluster	20
Tx/Rx-antenna array	Two colocated elements with linear (H/V) elements		

Table 24.5 MIMO LMS Channel Characteristics

Parameter	Value Used	Parameter	Value Used
Reference	[27]	Large-scale correlation	~0.9
Simulation time	3600 s	Antenna XPD	15 dB
Carrier frequency	S-band (2.2 GHz)	Environment XPD	5.5 dB
Vehicle speed in km/h	60 km/h	Very large-scale propagation model	2-State first-order Markov model
Polarization	Right hand/left hand CP	Small-scale correlation	0.5
Propagation environment	Tree-shadowed	Loo distribution triplet	As in Tables III and IV of Reference 2
Elevation angle	40	Doppler spectrum	Low-pass filter

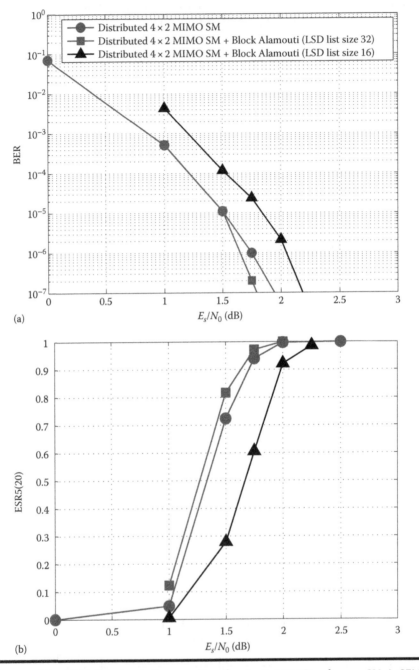

(a)

(b)

Figure 24.9 DVB-SH simulation results of a 4×2 MIMO scheme, SH-A SFN, QPSK 1/3, uniform long, WIM2 + LMS-ITS 60 km/h. (a) Bit error rate (BER) performance comparison. (b) Erroneous seconds ratio (ESR) performance comparison.

24.5 Conclusions

In this chapter, hybrid satellite–terrestrial MIMO for digital broadcasting was considered. To gain insight into the hybrid network architectures, two hybrid paradigms, namely, DVB-SH and DVB-NGH, were described. Further, design criteria for the MIMO codes for hybrid networks were studied, and basic concepts regarding their design were presented. The codes selected for the DVB-NGH standard together with the ones that were studied during the standardization but were not included in the standard were detailed. In addition to DVB-NGH, codes investigated for the MIMO extension of DVB-SH system were described. Finally, simulations of the MIMO code performance in hybrid environment were presented for both DVB-NGH and DVB-SH.

References

1. S. Kota, G. Giambene, and S. Kim, Satellite component of NGN, integrated and hybrid networks, *International Journal of Satellite Communications and Networking*, 29(3), 191–208, May 2011.
2. R. Prieto-Cerdeira, F. Perez-Fontan, P. Burzigotti, A. Bolea-Alamanac, and I. Sanchez-Lago, Versatile two-state land mobile satellite channel model with first application to DVB-SH analysis, *International Journal of Satellite Communications and Networking*, 28(56), 291–315, Sept. 2010.
3. P.-D. Arapoglou, P. Burzigotti, M. Bertinelli, A. Bolea-Alamanac, and R. De Gaudenzi, To MIMO or not to MIMO in mobile satellite broadcasting systems, *IEEE Transactions on Wireless Communications*, 10(9), 2807–2811, Sept. 2011.
4. ETSI EN 302 307 v1.1.1, Digital video broadcasting (DVB): Second generation framing structure, channel coding and modulation for broadcasting, interactive services, news gathering and other broadband satellite applications, June 2004.
5. ETSI TS 102 584, Guidelines for implementation for satellite services to handheld devices (SH) below 3 GHz, June 2010.
6. P. Kelley, Overview of the DVB-SH specifications, *International Journal of Satellite Communications and Networking*, 27(4/5), 198–214, May 2009.
7. A. Bolea Alamanac et al., Performance validation of the DVB-SH standard for satellite/terrestrial hybrid mobile broadcasting networks, *IEEE Transactions on Broadcasting*, 57(4), 802–825, Dec. 2011.
8. B. Vucetic and J. Yuan, *Space-Time Coding*, Wiley, New York, 2003.
9. M. R. Bhavani Shankar, P.-D. Arapoglou, and B. Ottersten, Space-frequency coding for dual polarized hybrid mobile satellite systems, *IEEE Transactions on Wireless Communications*, 11(8), 2806–2814, 2012.
10. H. W. Kim, K. Kang, and D.-S. Ami, Distributed space-time coded transmission for mobile satellite communication using ancillary terrestrial component, *Proceedings of the IEEE International Conference on Communications (ICC)*, Glasgow, U.K., June 2007.
11. A. I. Perez-Neira et al., MIMO channel modeling and transmission techniques for multi-satellite and hybrid satellite-terrestrial mobile networks, *Elsevier Physical Communication*, 4(2), 127–139, June 2011.

12. IEEE 802.16-2009, IEEE standard for local and metropolitan area networks, part 16: Air interface for broadband wireless access systems, 2009.

13. G. R. Jithamithra and B. S. Rajan, Minimizing the complexity of fast sphere decoding of STBCs, *Proceedings of the IEEE International Symposium on Information Theory*, St. Petersburg, Russia, Aug. 2011.

14. F. Oggier and E. Viterbo, Algebraic number theory and code design for Rayleigh fading channels, *Communications and Information Theory*, 1(3), 333–416, 2004.

15. E. Viterbo and J. Boutros, A universal lattice code decoder for fading channel, *IEEE Transactions on Information Theory*, 45(7), 1639–1642, July 1999.

16. E. Biglieri, Y. Hong, and E. Viterbo, On fast-decodable space-time block codes, *IEEE Transactions on Information Theory*, 55(2), 524–530, Feb. 2009.

17. C. Hollanti, R. Vehkalahti, and Y. Nasser, Algebraic hybrid satellite-terrestrial space-time codes for digital broadcasting in SFN, *Proceedings of the IEEE Workshop on Signal Processing Systems (SIPS)*, Beirut, Lebanon, 2011.

18. H. Jafarkhani, A quasi-orthogonal space-time block code, *IEEE Transactions on Communications*, 49(1), 1–4, Jan. 2001.

19. C. Hollanti, J. Lahtonen, and H.-F. Lu, Maximal orders in the design of dense space-time lattice codes, *IEEE Transactions on Information Theory*, 54(10), 4493–4510, Oct. 2008.

20. R. Vehkalahti and C. Hollanti, Reducing complexity with less than minimum delay space-time lattice codes, *Proceedings of the IEEE Information Theory Workshop (ITW)*, Paraty, Brazil, 2011.

21. R. Vehkalahti, C. Hollanti, and F. Oggier, Fast-decodable asymmetric space-time codes from division algebras, *IEEE Transactions on Information Theory*, 58(4), 2362–2385, Apr. 2012.

22. D. Gozálvez, Combined time, frequency and space diversity in multimedia mobile broadcasting systems, PhD dissertation, Universitat Politècnica de València, Valencia, Spain, 2012.

23. J.-C Belfiore, G. Rekaya, and E. Viterbo, The golden code: A 2×2 full-rate space-time code with non-vanishing determinants, *IEEE Transactions on Information Theory*, 51(4), 1432–1436, Apr. 2005.

24. C. Studer and H. Bölcskei, Soft input soft output single tree search sphere decoding, *IEEE Transactions on Information Theory*, 56(10), 4827–4842, Oct. 2010.

25. P. Moss, T. Yeen Poon, and J. Boyer, A simple model of the UHF cross-polar terrestrial channel for DVB-NGH, BBC Research White Paper WHP205, Sept. 2011.

26. T. Jokela, C. Hollanti, J. Lahtonen, R. Vehkalahti, and J. Paavola, Performance evaluation of 4×2 MIMO schemes for mobile broadcasting, *Proceedings of the IEEE International Symposium on Broadband Multimedia Systems and Broadcasting (BMSB)*, Erlangen, Germany, 2011.

27. K. P. Liolis, J. Gomez-Vilardebo, E. Casini, and A. Perez-Neira, Statistical modeling of dual-polarized MIMO land mobile satellite channels, *IEEE Transactions on Communications*, 58(11), 3077–3083, Nov. 2010.

28. M. R. Bhavani Shankar et al., MIMO extension to DVB-SH, *Proceedings of the International Communication Satellite Systems Conference*, Ottawa, Ontario, Canada, 2012.

Index